Yanovitskij

Light Scattering in Inhomogeneous Atmospheres

Springer
Berlin
Heidelberg
New York
Barcelona
Budapest
Hong Kong
London
Milan
Paris
Santa Clara
Singapore
Tokyo

Edgard G. Yanovitskij

Light Scattering in Inhomogeneous Atmospheres

Translated by Sergij Ginsheimer
and Oleg Yanovitskij

With 37 Figures and 54 Tables

Springer

Professor Edgard G. Yanovitskij

Laboratory of Radiative Transfer Theory
The Main Astronomical Observatory
Ukrainian National Academy of Sciences, Golosiiv
252650, Kyiv-22, Ukraine

Translators:

Sergij Ginsheimer
Mayakovsky Str. 4b, Apt. 4
253217, Kyiv, Ukraine

Oleg Yanovitskij
Volgograds'ka Str. 15, Apt. 59
252141, Kyiv, Ukraine

Library of Congress Cataloging-in-Publication Data

Yanovitskij, Edgard G., 1937-
Light scattering in inhomogeneous atmospheres / Edgard G. Yanovitskij.
p. cm.
Includes bibliographical references and index.
ISBN 3-540-61362-5 (hardcover : alk. paper)
1. Planets--Atmospheres. 2. Light--Scattering. 3. Radiative transfer. I. Title.
QB603.A85Y36 1997
551.5'65--dc20 96-24479
CIP

ISBN 3-540-61362-5 Springer-Verlag Berlin Heidelberg New York

Typesetting: Camera-ready by author
Cover design: *design & production* GmbH, Heidelberg

SPIN 10481397 55/3144 – 5 4 3 2 1 0 – Printed on acid-free paper

Preface

This book is aimed at studying the scattering of monochromatic radiation in plane inhomogeneous media. We are dealing with the media whose optical properties depend on a single spatial coordinate, namely of a depth. The most widely known books on radiation transfer, for instance

1. S. Chandrasekhar, *Radiative Transfer*, Oxford, Clarendon Press, 1950, (RT),
2. V.V. Sobolev, *Light Scattering in Planetary Atmospheres*, New York, Pergamon Press, 1975, (LSPA),
3. H.C. van de Hulst, *Multiple Light Scattering. Tables, Formulas and Applications.* Vol. 1,2, New York, Academic Press, 1980, (MLS),

treat mainly the homogeneous atmospheres. However, as known, the actual atmospheres of stars and planets, basins of water, and other artificial and natural media are not homogeneous. This book deals with the model of vertically inhomogeneous atmosphere, which is closer to reality than the homogeneous models.

This book is close to the aforementioned monographs in its scope of problems and style. Therefore, I guess that a preliminary knowledge of the contents of these books, particularly of the book by Sobolev, would facilitate the readers' task substantially. On the other hand, all concepts, problems, and equations used in this book are considered in full in Chap. 1. So, it will be possible for those readers who do not possess the above knowledge to understand this book.

A general idea about the content of the book can be gained from both the Introduction and the Table of Contents. It is worth noticing here that references are not usually provided throughout the book; they can be found in bibliographical comments and additions at the end of each part of the book. The unique exceptions are the closing chapter and the aforementioned monographs. These monographs are briefly referred to as RT, LSPA, and MLS, and these references are provided whenever it becomes necessary.

This book is of a certain interest for astrophysicists and geophysicists who study the optics of the atmosphere and the ocean, as well as for other scientists who work with solving the transfer equation. It may also be useful for students specializing in these areas.

The book is primarily intended for theoreticians. However, scientists working with experiments interpreting the experimental data may also find it as rather helpful for two reasons at least. The first is that the book can be used as a reference book to find the necessary formulas. Second, there are tables of functions and parameters in the Appendix, which are used to find the solutions to a number of problems of radiation transfer in both homogeneous and the simplest inhomogeneous models of atmospheres. Tables are given for a rather large variety of optical media. Furthermore, the closing chapter includes a number of examples of how to apply the theory developed in this book to the specific problems of the optics of atmospheres of stars and planets.

Although, as an employee of the Main Astronomical Observatory of the Ukrainian Academy of Sciences in Kiev, I am a follower of academician Sobolev, who heads Leningrad's school in theoretical astrophysics for more than forty years. The impact of Sobolev and his scientific school has always been very important for me. Academician Sobolev has looked through the manuscript of this book and made a number of important comments that have helped me to improve the presentation of the material. It was also very important that I discussed my results many times at seminars at the Laboratory for Theoretical Astrophysics of the Astronomical Observatory of Leningrad University. Always frank and open, these discussions considerably promoted my research and helped me to achieve the scientific results, many of which constitute a considerable part of this monograph. I am particularly grateful to V.V. Ivanov and D.I. Nagirner. Numerous discussions with V.V. Ivanov allowed me to elaborate upon a new approach to the problems of transfer theory, which, in turn, influenced this book. M.I. Mishchenko's assistance was very helpful in the preliminary editing of the text of the book.

I am pleased to express my gratitude to all these persons.

* * *

It has been over six years since the above lines were written. There have been a lot of changes in the world, in the souls of people and states, since then. For instance, I am proud of having become a citizen of my own national state – Ukraine.

However, each good thing must be paid for. The price, for me, was that my monograph was not published in its original form. It was supposed to be published by the Moscow publishing house "Nauka" in the second quarter of 1992. Those plans were never realized. For this reason I took steps in 1994 which led to the publication of an English translation of the original monograph by one of the European publishing houses. Next I had to withdraw the monograph from the "Nauka" publishing house in 1995. My decision was encouraged by the polite consent of the Springer-Verlag publishing house in the person of Prof. Dr. W. Beiglböck. Springer-Verlag gave me the possibility of publishing this book. It also gave the financial support for the translation into English and for typesetting this monograph.

Compared with the first draft, this monograph has, naturally, undergone some minor changes. The new bibliography extends until 1994. I have also added sections 1.11, 4.10, 6.10, and 14.4.1. However, in many other respects, the monograph remains almost unchanged.

As long ago as the end of 1989 Dr. D.I. Nagirner, having attentively read the monograph, made some emendations and significant notes, which were all taken into account. Prof. V.V. Ivanov, Dr. T. Kostiuk, and Dr. M.I. Mishchenko helped me in getting some bibliographical materials. We were also given considerable assistance by Dr. E.S. Kryachko, Dr. A.C. Balke, and Dr. M.I. Mishchenko in the language editing of some chapters of the book. Finally, I would like to emphasize the decisive role of my son Oleg. Not only was he an active organizer of the translation of my monograph into English, but he has also taken upon himself troublesome secretarial duties connected with publishing. I am is sincerely grateful to every aforementioned person.

Kiev, March 1996 Edgard Yanovitskij

P.S. I would like to express my sincere appreciation to Prof. H.C. van de Hulst who familiarized himself with this book during its development. He has also suggested some important remarks and corrections.

Contents

Basic Notation

A	– surface albedo, Sect. 5.3.
$A(\mu_0)$	– plane albedo of an atmosphere, Sect. 5.3.
$A(\mu_{0,0})$	– plane albedo for the atmosphere with optical thickness τ_0.
A_s $A_s(\tau_0)$	– spherical albedo of an atmosphere, Sect. 1.9.
A_g	– geometrical albedo of an atmosphere, Sect. 1.9.
$B_\infty(\tau, \mu)$	– source function in the problem for two-sided infinity, Sect. 1.7.
$B(\tau, \mu)$	– source function in the Milne problem, Sect. 1.6.
$B(\tau, \mu, \varphi)$	– source function in the problem with a parallel external flux, Sect. 1.5.
$B^m(\tau, \mu, \mu_0)$ $B^m(\tau, \mu, \mu_0; \tau_0)$	– mth azimuthal harmonics of source function, Sect. 1.5.1.
$B(\tau, \mu, \mu_0; \tau_0)$	– zeroth azimuthal harmonics of source function.
C	– asymptotic constant, Sect. 3.11.3.
c_j	– see (2.15).
$E^\downarrow(\tau, \mu_0)$ $E^\uparrow(\tau, \mu_0)$	– illumination from above and from below of the horizontal area in the atmosphere at optical depth τ, Sect. 1.8.
g	– parameter of the Henyey-Greenstein phase function, see (1.15).
$H_\tau(\mu_0)$	– projection of flux vector onto τ axis, Sect. 1.8.
$I_\infty(\tau, \mu)$	– radiation intensity in the problem for two-sided infinity, Sect. 1.7.
$\tilde{I}_{ij}(\tau, \mu)$	– the same quantity for two adjacent half-spaces, Sect. 7.1.
$I(\tau, \mu)$	– radiation intensity in the Milne problem, Sect. 1.6.
$I(\tau, \mu, \mu_0)$	– radiation intensity in the parallel flux problem, Sect. 1.5.
$I^m(\tau, \mu, \mu_0)$ $I^m(\tau, \mu, \mu_0; \tau_0)$	– mth azimuthal harmonics of radiation intensity, Sect. 1.5.1.
$I(\tau, \mu, \mu_0)$ $I(\tau, \mu, \mu_0; \tau_0)$	– zeroth azimuthal harmonics of radiation intensity.
$i(\mu) = I_\infty(0, \mu)$	– Sects. 2.2, 2.3.
$i_{ij}(\mu) = \tilde{I}_{ij}(0, \mu)$	– Sect. 7.1.
k	– diffusion exponent, Sect. 2.2
k_0	– Sect. 2.5.

M	– asymptotic constant (Sect. 2.4) or M integral, see (11.37).
N	– asymptotic constant, Sect. 3.11.3. For inhomogeneous atmosphere – see (11.49).
$\tilde{N}_{\tau_0} = \tilde{N}(\tau_0)$	– the same quantity for inverted atmosphere – see (12.39).
$P_n(\mu)$	– nth order Legendre polynomial.
$P_n^m(\mu)$	– associated Legendre function.
$r(\tau_0)$	– reflection factor for optically thick atmosphere, see (4.46) and (4.53).
$t(\tau_0)$	– transmission factor for optically thick atmosphere, see (4.46) and (4.54).
$t^*(\tau_0)$	– modified transmission factor, see (12.51).
$U(\mu_0, \tau_0)$	– absorption function, Sect. 4.8.1.
$U_s(\tau_0)$	– see (4.82).
$u(\mu)$	– escape function, Sect. 1.6.
$\tilde{u}_{\tau_0}(\mu)$	– escape function for inverted atmosphere, Sect. 11.7.
$u_0(\mu, \tau_0)$	– see (4.73).
$V(\mu_0, \tau_0)$	– relative illumination of lower atmosphere boundary, Sect. 1.9.
$V_s(\tau_0)$	– see (4.82).
$v_0(\mu)$	– see (3.98).
$v_0(\mu, \tau_0)$	– see (4.78).
$w_0(\mu, \tau_0)$	– see (4.74).
$x_n(\tau)$	– coefficient of the expansion of phase function in a series of Legendre polynomials, see (1.53).
x_{nj}	– the same coefficient for layer j in multilayer atmosphere, Sect. 6.7.
$x_i^m(\tau)$	– see (1.57).
$\bar{x}_1(\tau)$	– see (1.117).
$y_n(\tau)$	– coefficient of the expansion of $I_\infty(\tau, \mu)$ in a series of Legendre polynomials, Sect. 12.1.
z	– geometrical depth in atmosphere, Sect. 1.4.
α	– truncation parameter of inhomogeneous atmosphere, Sect. 9.2.
$\left.\begin{matrix}\alpha(\boldsymbol{r})\\ \alpha(z)\end{matrix}\right\}$	– volume extinction coefficient, Sect. 1.2.
γ	– angle of scattering, Sect. 1.2; see also (1.43).
γ_0	– asymptotic constant, see (3.62).
ε_0	– asymptotic constant, see (3.84).
ζ	– asymptotic constant, see (4.54e).
θ	– angle between the direction of light propagation and internal normal to the upper boundary of the atmosphere, Sect. 1.5.
θ_0	– angle of incidence of light on the upper boundary of atmosphere in the problem with parallel external flux, Sect. 1.5.
$\theta(\mu)$	– unit step function, see (1.46).

$\left.\begin{matrix}\kappa(\boldsymbol{r})\\ \kappa(z)\end{matrix}\right\}$	– volume absorption coefficient, Sect. 1.2.
λ	– single scattering albedo, Sect. 1.2.
$\tilde{\lambda}$	– wavelength, Sect. 1.2.
$\rho(\mu, \mu_0, \varphi)$	– reflection coefficient, see (1.95) and (1.97)
$\left.\begin{matrix}\mu = \cos\theta\\ \mu_0 = \cos\theta_0\end{matrix}\right\}$	– Sect. 1.5.
ν	– light frequency.
$\left.\begin{matrix}\rho^m(\mu, \mu_0)\\ \rho^m(\mu, \mu_0; \tau_0)\end{matrix}\right\}$	– mth azimuthal harmonics of reflection coefficient, Sect. 1.9.
$\left.\begin{matrix}\rho(\mu, \mu_0)\\ \rho(\mu, \mu_0; \tau_0)\end{matrix}\right\}$	– zeroth azimuthal harmonics of reflection coefficient, Sect. 1.9.
$\left.\begin{matrix}\sigma(\boldsymbol{r})\\ \sigma(z)\end{matrix}\right\}$	– volume scattering coefficient, Sect. 1.2.
$\sigma(\mu, \mu_0, \varphi)$	– transmission coefficient, see (1.96) and (1.98).
$\sigma^m(\mu, \mu_0; \tau_0)$	– mth azimuthal harmonics of transmission coefficient, Sect. 1.9.
$\sigma(\mu, \mu_0; \tau_0)$	– zeroth azimuthal harmonics of transmission coefficient, Sect. 1.9.
τ	– optical depth, Sect. 1.4.
τ_0	– optical thickness of atmosphere, Sect. 1.4.
τ^e	– extrapolated length, Sect. 4.8.1.
τ_0^e	– see (4.56).
τ_j	– optical thickness of the layer j in multilayer atmosphere, Sect. 6.1.
φ	– azimuth of the direction of radiation propagation, Sect. 1.5.
φ_0	– azimuth of radiation incident on atmosphere, Sect. 1.5.
$\varphi_i^m(\mu, \tau_0)$	– generalized Ambartsumian φ function, Sect. 9.1.
$\varphi(\mu)$	– Ambartsumian function, Sect. 7.3.
$\chi(\gamma)$	– phase function, Sect. 1.2.
$\chi_{\mathrm{R}}(\gamma)$	– Rayleigh phase function, see (1.9).
$\chi_{\mathrm{H}}(\gamma)$	– Henyey-Greenstein phase function, see (1.15).
$\psi_i^m(\mu, \tau_0)$	– generalized Ambartsumian ψ function, Sect. 9.1.
$\boldsymbol{\omega}$	– unit vector of the direction of radiation propagation, Sect. 1.1.

Introduction

We embark with a brief excursion into the history of the formation and development of transfer theory.

The work of Chvolson [1] was devoted to the scattering of light in opal glasses. He pioneered the transfer equation in the integral form, which was the starting point for further development of the theory by Schuster, Schwarzschild, Milne, and Eddington. This was quite natural because the transfer of radiation in external regions of most cosmic objects determines the nature of their spectra, which are used to analyze the physical conditions there.

The newly derived equations have required us to develop new methods of solution. In particular, the study of the Milne equation led to the Wiener–Hopf method for solving the integral equations of this type. Hopf's monograph [2] summarizes the achievements made in the analysis of the transfer equation during the 1920s to the beginning of 1930s.

A new stage in the development of the theory was commenced in the beginning of the 1940s by Ambartsumian's proposing a new approach [3] based on the invariance principles. Further development of this fundamental idea was given in the works by Chandrasekhar [4] and others.

A large variety of problems in transfer theory were considered by Sobolev [5] (transfer of radiation in frequencies of lines, non-stationary radiation field, etc.). Later on Sobolev generalized a number of results obtained in [5] for isotropic scattering (namely, linear equations for transmission and reflection coefficients, the structure of the kernel resolvent for the integral transfer equation, asymptotic separation of angular variables in optically thick layers) to non-isotropically scattering plane medium [6], which is a typical model of a planetary atmosphere.

The two-volume monograph by van de Hulst [7] must be considered as the fundamental reference book on the theory of monochromatic light scattering. It contains a large number of tables and figures and, therefore, is very useful for those readers who study applied aspects of transfer theory.

A further development of Ambartsumian's invariance principles was achieved in Bellman's method of invariant imbedding [8], which was later used to build up the algorithms for the solution of many problems in diverse areas of science and technology [9, 10].

Many mathematical methods and their applications to the problems of the transfer of monochromatic radiation can be found in books by Kourganoff

[11] and Busbridge [12]. The monograph by Ivanov [13] gives an elaborate treatment of the problems related to transfer theory in the frequencies of spectral lines.

Until now, we have dealt with the solution of the scalar transfer equation, i.e., the equation that makes no allowances for the polarization of radiation. The vector version of the transfer equation is provided in the monograph by Chandrasekhar [4], where one can find the solution for the angular Rayleigh matrix (see also [5], Chap. 5 and [14], Chap. 5). The detailed description of the state of the problem can be found in the review by Hansen and Travis [15] and also in the theses by de Rooij [16], de Haan [17], and Wauben [18], which were completed under the supervision of Professor Hovenier. The problems of the scattering and propagation of radiation in anisotropic media were treated in the monograph by Dolginov, Gnedin and Silantiev [19]; the analysis of polarization characteristics of the radiation of various celestial objects can also be found there.

So far, we have discussed the works dealing with the problems of the transfer theory in its astrophysical aspect. However, one should remember that the development of the above areas was substantially promoted by that of similar problems in the theory of neutron diffusion (see [20]–[23]). Transfer theory also describes the propagation of X-rays and gamma radiation in a substance (see, for instance, [22, 24]) and the process of complex heat transfer [25]–[27]. Transfer theory also plays an important role in geophysics under the study of light scattering in the atmosphere or oceans [28]–[37]. And finally, it should be stressed that the theory of transfer has kinship with the kinetic equation. A typical equation of the linear transfer theory is, in fact, a particular case of the Boltzman equation [38].

The transfer equation, like all other important equations of mathematical physics, is always under attention in mathematics. Achievements in this area are expounded, for instance, in the monograph by Vladimirov [39] and books by Maslennikov [40, 41], Ershov and Shikhov [42, 43], and Germogenova [44].

The method of solving the problems of transfer theory pioneered by Case is worthy of special attention. Case introduced singular eigenfunctions of the linear transfer equation in plane media, which considerably extended the scope of problems that can be solved exactly. The detailed description of this method is given in the monograph by Case and Zweifel [45].

Finally, one should note the following. The application of modern methods of transfer theory to particular problems is impossible without using computers and developing the corresponding numerical algorithms. These are particularly analyzed in the fundamental monograph by Marchuk and Lebedev [46]. Modern numerical and semi-analytical methods applicable to solving one- and many-dimensional transfer equations are presented in the monograph by Sushkevich, Strelkov, and Ioltukhovskii [47].

The above discussion gives a clue about a wide range of areas of modern physics (note that we have listed just a few of them) where transfer theory can be applicable. This complexity of its problems shows the importance of solving any particular problem: once it is solved, the problem and its solution can be utilized in many areas. As we have already noted, transfer theory is of the special significance for astrophysics, especially for the study of physical properties of planetary atmospheres.

This book includes three parts dedicated to homogeneous, multilayer, and inhomogeneous atmospheres. Let me say a few words about the reasons for my choice.

Fast progress in exploring outer space makes it very important to study the physical conditions existing on planets. The celestial bodies within the solar system are the first objects to study and, therefore, after man and his devices left the Earth's atmosphere, the amount and accuracy of the information on these bodies rose dramatically. So, on one hand, the exploration of outer space poses a number of problems and, on the other hand, it provides us with valuable information to resolve those problems. Also, the construction of new powerful telescopes and the use of the latest developments of radioelectronics resulted in an increase in quality and quantity of terrestrial observations of planetary surfaces.

The interpretation of the data of observation requires further development of the theory, first and foremost the theory of radiation energy transfer. Notwithstanding all notable achievements, there are still areas where work is just beginning now, or where considerable progress has just been reached by exploiting new approaches to the problem in question. The problem of radiative transfer in a plane *inhomogeneous* atmosphere falls into the first category. Such a model of the atmosphere looks rather close to the reality when compared with standard models. The problem of evaluating the *internal* radiation fields in homogeneous atmospheres and multilayer atmospheres falls into the second category. This problem acquired special significance after the optical probing of the Venusian atmosphere. One may think that we are not far from the time when the same optical probing will be done in atmospheres of giant planets. Interpretation of these important and, no doubt, expensive results will demand the latest and most accurate methods of transfer theory.

Proceeding the content of the book, I should give readers more details about the aforementioned areas.

All important results obtained by the beginning of the 1970s in the theory of anisotropic scattering in plane homogeneous atmospheres were described by Sobolev in LSPA. There was only one section about inhomogeneous atmospheres in the book that corresponds to the status of theory at that time. This section deals with the coefficients of transmission and reflection by a plane layer. Chapter 9 of the present book gives the elaborate treatment of this problem for the case of anisotropic scattering. The theory of asymptotic separation of angular variables was developed for anisotropically scattering atmospheres at that time, and concluding his monograph, Sobolev stressed the importance of generalizing theory to inhomogeneous cases. This generalization is the main concern of the third part of this book. It is based on the generalized invariance principle (see Sect. 1.12) reformulated for inhomogeneous atmospheres. This principle allows us also to obtain the other results of the theory of anisotropic light scattering in plane homogeneous atmospheres being generalized to inhomogeneous cases.

This seems to be the right place to make some comments about the problem of evaluating the internal fields of radiation in homogeneous atmosphere for anisotropic scattering. This problem and the closely related ones constitute a major content of the first part of this book. Like many others, it can hardly be solved without use of modern computers. Therefore, it makes sense to develop the theoretical approaches which may provide the efficient methods for numerical calculations. The generalized invariance principle was formulated in the middle of the 1970s. With regard to the radiation fields in plane homogeneous atmospheres, it provided a way to construct a very simple and efficient algorithm for the numerical calculation of radiation fields in homogeneous semi-infinite atmospheres (see Sect. 3.10). The use of this principle allows us to formulate explicitly the invariance relation for homogeneous layer of finite thickness (see Sect. 4.1). This results in a set of integral equations used to determine the radiation intensity (see Sect. 4.7) and, what is more important, in the rather simple and effective method for calculating the internal radiation fields. The latter one is used within the well-known van de Hulst doubling method (see Sects. 4.5, 4.6 of the present book).

There are in fact two additional reasons to include a rather large section on transfer theory in a homogeneous media into a book devoted to inhomogeneous atmospheres. One is the didactic. Investigating a homogeneous atmosphere, the reader may handle the invariance relations much used here. I express a hope that handling these relations facilitates the reader in perceiving the similarity between the theories of scattering in homogeneous and inhomogeneous atmospheres.

The second reason is that the complete solution of the problem of light scattering in a plane homogeneous layer permits us to construct a rigorous algorithm to calculate the radiation field in a multilayer atmosphere as the simplest model of inhomogeneous media. The latter is just what the second part of the book is concerned with.

Nearly conservative scattering (i.e., the situation when the coefficient of true absorption is much lower than the scattering coefficient) is of importance for practical application of the theory. It is essential to emphasize that in this case a number of quantities determining the radiation field in an optically thick layer may be represented correctly by simple rigorous asymptotical formulas (a part of them being expressed explicitly). This is especially important for solving the inverse problems of the optics of atmosphere. Some examples of these formulas are present in the LSPA [Sects. 2.5, 2.6, 3.11 (homogeneous atmosphere) and Chap. 12 (inhomogeneous atmosphere)]. Simple asymptotic formulas are derived for the azimuth-averaged radiation intensity, flux, spherical albedo, etc.

Considerable attention is paid in the first part of the book to the description of the numerical methods used to calculate the radiation fields in homogeneous and multilayer cloudy atmospheres. Algorithms and programs based on these methods have been exploited for many years in the Main Astronomical Observatory of the National Academy of Sciences of Ukraine, where I work, as well as in the other Institutes of the CIS where astronomy is studied. It is the main reason why I include a number tables of functions and parameters into the present book. These tables may be helpful in interpreting the observational data. Note that I use two kinds of numbering for tables in this book. Tables in the text are numbered using arabic numbers, whereas those in the Appendix have Roman numbers.

The third part of the book hardly concerns numerical methods for the solution of the problem at all. The reason for this is that it is almost impossible to construct a universal numerical method without defining the way the optical properties depend on depth. The main aim of this part is to develop the analytical theory of radiative transfer in inhomogeneous media that relies on imposing fewer restrictions on the behavior of the most important functions determining the optical properties on various depths. The rigorous relationships derived there appear to be very useful for building up numerical methods for particular problems of transfer theory in inhomogeneous medium. In some cases, they make it possible to predict the general structure of the solution and control the accuracy at each stage of the calculations via some rigorous integral relationships, which is very important even for such a universal numerical tool as the Monte-Carlo method.

The closing chapter of the book concerns the description of applications of theoretical results to a number of practical problems in astrophysics and the optics of the atmosphere. In particular, I give the rigorous asymptotic theory of the formation of a weak absorption line in multiple scattering in an inhomogeneous atmosphere of large optical thickness. Also, some areas of science and technology are discussed, where the theory developed may be applied.

I shall make one more remark in conclusion. Recently, Minin has published a book [14] entirely devoted to multiple light scattering in planetary atmospheres. There, the main attention is on the consistent expounding of the theoretical results of the Leningrad school of radiative transfer theory. The monograph by Minin may be treated as a useful additional source of information on various areas of light scattering in planetary atmospheres. It can also be regarded as an introduction to the areas that are not touched upon in the present book (scattering of polarized radiation, nonstationary fields of radiation, heat transfer).

1. Basic Concepts, Equations and Problems

This introductory chapter discusses the scattering of radiation in a planetary atmosphere. Its results refer equally to the whole complex of physical problems described by the similar equations. Most of the equations and problems are formulated here for the case of an inhomogeneous stratified atmosphere whose optical properties depend on one spatial variable. Thus, throughout the book we will constantly refer to the formulas contained in this chapter.

1.1 Intensity of Radiation

In the theory of multiple scattering of light, radiation is treated as a gas of photons consisting of individual quanta. Photons are viewed as particles that move with the velocity of light c. Accordingly, each photon is characterized by the following three parameters:

1. the ***energy*** $\mathcal{E}$ or the frequency ν, related by $\mathcal{E} = h\nu$ (h being Planck's constant);
2. ***radius vector*** $\boldsymbol{r}$, the three components of which specify the location of a photon in space relative to some frame of reference;
3. the ***two angular variables*** that give the direction of motion of the photon. This direction we will characterize by the unit vector $\boldsymbol{\omega}$.

The set of six variables – frequency, three spatial coordinates and two angular variables (e.g. angles θ and φ in a spherical coordinate system) – define a six-dimensional phase space. Each photon is represented by a point in this space.

The transfer theory is a statistical theory in which radiation is considered to be a statistical ensemble of photons which are described by the parameters enumerated above. A complete description of this ensemble is given by the corresponding ***distribution function*** $f(\nu, \boldsymbol{r}, \boldsymbol{\omega}, t)$. The value $f(\nu, \boldsymbol{r}, \boldsymbol{\omega}, t)\mathrm{d}\nu\, \mathrm{d}V\, \mathrm{d}\omega$ is the number of photons with frequencies from ν to $\nu + \Delta\nu$ located at a time t in a volume $\mathrm{d}V$ near a point $\boldsymbol{r}$ and moving within a solid angle $\mathrm{d}\omega$ around direction $\boldsymbol{\omega}$. If the phase density f does not depend on time t, the radiation field is said to be ***steady***. We will only consider such fields.

We turn now to another characteristic commonly used to describe the radiation field. Let us consider an elementary area $\mathrm{d}S$ the position of which is defined by the vector $\boldsymbol{r}$. Let there be photons with frequencies from ν to $\nu + \mathrm{d}\nu$, moving through this area in directions lying within solid angle $\mathrm{d}\omega$ about a vector $\boldsymbol{\omega}$ normal to $\mathrm{d}S$. In this case the energy $d\mathcal{E}_\nu$ transferred by these photons through the area $\mathrm{d}S$ during a time interval from t to $t + \mathrm{d}t$ is, by definition, equal to

$$\mathrm{d}\mathcal{E}_\nu = I_\nu(\boldsymbol{r}, \boldsymbol{\omega})\mathrm{d}\nu\,\mathrm{d}S\,\mathrm{d}\omega\,\mathrm{d}t, \tag{1.1}$$

where the quantity $I_\nu(\boldsymbol{r}, \boldsymbol{\omega})$ is called the ***intensity of radiation***. It has the dimension of $\mathrm{erg\,cm^{-2}s^{-1}Hz^{-1}ster^{-1}}$.

Since the volume $\mathrm{d}V = \mathrm{d}Sc\,\mathrm{d}t$, and the energy of one photon is $h\nu$, the same value $\mathrm{d}\mathcal{E}_\nu$ can be defined in terms of the distribution function f in the following way:

$$\mathrm{d}\mathcal{E}_\nu = h\nu f(\nu, \boldsymbol{r}, \boldsymbol{\omega})\mathrm{d}\nu\,\mathrm{d}Sc\,\mathrm{d}t\,\mathrm{d}\omega. \tag{1.2}$$

Comparing (1.1) and (1.2), we have:

$$I_\nu(\boldsymbol{r}, \boldsymbol{\omega}) = ch\nu f(\nu, \boldsymbol{r}, \boldsymbol{\omega}). \tag{1.3}$$

Thus, both the intensity of radiation I_ν and the phase density f completely describe the radiation field. We have to note that the intensity of radiation or the phase density will completely characterize the radiation field only if the polarization effects need not be taken into account. Otherwise, instead of I_ν the ***Stokes vector***, $\boldsymbol{I}_\nu$ has to be considered. However, in this book we will deal only with the ***scalar*** radiative theory which is concerned solely with the quantity I_ν, and it may be said that a number of results presented here can be quite easily (at least in a formal way) extended to the vector case.

As is clear from (1.3), for a given frequency ν the intensity of radiation I_ν may be considered, up to a constant factor, as the probability density of the orientation distribution of the photons. Hence it follows from here that an important property of the intensity is its independence of the distance between the source and observer, provided there are no energy sources and sinks on the line of sight between them. This enables the intensity I_ν to be measured directly for *extended* (as we have to know $\mathrm{d}\omega$) sources of radiation such as planets, the Sun, nebulae, etc.

Other characteristics of the radiation field, for example illumination, flux and so on, can be expressed in terms of I_ν. Their description will be given below.

1.2 Interaction of Radiation with Matter

We will assume that the interaction of radiation and matter does not give rise to a change in the photon frequency, i.e., we will study ***monochromatic light***

scattering. This assumption is definitely not valid for frequencies in spectral lines. However, this assumption realistically describes the problem of light scattering by particles with sizes comparable to the wavelength of radiation, the neutron transfer theory in the constant cross section approximation, etc.

Furthermore, we will assume that the physical properties of the diffuse medium depend only on spatial coordinates, whereas the process of interaction of the radiation with an elementary volume of the medium does not depend on the direction of illumination. In other words, we suppose that the elementary volume has the property of ***isotropy***. For example, the elementary volume containing a spatially disordered set of homogeneous spherical particles of different sizes will be isotropic. One can also imagine that the volume contains a sufficiently large set of arbitrarily shaped particles oriented in such a way that this ensemble of particles will be isotropic.

In the theory of monochromatic light scattering the optical properties of an elementary volume of the medium are assumed to be given. They are described by three quantities:

1. ***Volume extinction coefficient*** $\alpha(\boldsymbol{r})$ which is defined in the following way. In the vicinity of the point $\boldsymbol{r}$ we consider a plane layer of geometrical thickness $\mathrm{d}s$ containing this point (the orientation of the layer does not matter: a result of isotropy). Consider radiation incident normally on the boundary of this layer near the point $\boldsymbol{r}$. As a result of interaction with the matter, a fraction of the energy incident on the layer equal to $\alpha(\boldsymbol{r})\mathrm{d}s$ will be removed. This interaction can be of two kinds. A certain part of the energy undergoes ***true absorption***, i.e., it is converted to other forms of energy (for example, heat). We will denote this part by $\kappa(\boldsymbol{r})\mathrm{d}s$, where the quantity $\kappa(\boldsymbol{r})$ is called the ***volume coefficient of true absorption*** or just the ***volume absorption coefficient***. Another part of the radiation of the incident beam changes its direction of propagation upon the interaction with matter, i.e., it is scattered. We will denote this part by $\sigma(\boldsymbol{r})\mathrm{d}s$ and the quantity $\sigma(\boldsymbol{r})$ is called the ***volume scattering coefficient***. It is obvious that ***volume extinction coefficient*** $\alpha(\boldsymbol{r})$ equals
$$\alpha(\boldsymbol{r}) = \sigma(\boldsymbol{r}) + \kappa(\boldsymbol{r}). \tag{1.4}$$
Note that the quantity $\alpha(\boldsymbol{r})$ is sometimes called the absorption coefficient. The dimension of the quantities α, σ and κ is cm^{-1}.
2. The dimensionless quantity
$$\lambda(\boldsymbol{r}) = \frac{\sigma(\boldsymbol{r})}{\alpha(\boldsymbol{r})} = \frac{\sigma(\boldsymbol{r})}{\sigma(\boldsymbol{r}) + \kappa(\boldsymbol{r})} \tag{1.5}$$
is the probability that a photon survives the act of scattering and is called the ***single scattering albedo*** (it is also often denoted by a, ω or ω_0). When $\lambda(\boldsymbol{r}) = 1$, we refer to the case of ***conservative scattering*** or ***pure scattering***.

3. The ***phase function*** $\chi(\gamma, \boldsymbol{r})$. The quantity $\chi(\gamma, \boldsymbol{r})\mathrm{d}\omega/4\pi$ is the probability that radiation is scattered by the elementary volume into a solid angle $\mathrm{d}\omega$ about a direction forming an angle γ with the direction of the incident radiation. Thus,

$$\int_{(4\pi)} \chi(\gamma, \boldsymbol{r}) \frac{\mathrm{d}\omega}{4\pi} = 1. \tag{1.6}$$

The axial symmetry of the phase function is due to the assumption that the elementary volume of medium is isotropic. Otherwise, the quantity χ (often also designated as x, P etc.) would also depend on the direction of the incident radiation. Taking into account that $\mathrm{d}\omega = 2\pi \sin\gamma \mathrm{d}\gamma$, (1.6) can be rewritten as follows

$$\frac{1}{2} \int_0^\pi \chi(\gamma, \boldsymbol{r}) \sin\gamma \mathrm{d}\gamma = 1. \tag{1.7}$$

The quantities α, λ and χ completely determine the optical properties of the elementary volume of the medium. These quantities depend on the frequency of the radiation. Hereafter we will not indicate this dependence, but it will always be implied.

Now we will discuss the methods for calculating the quantities α, λ, and χ. If a planetary atmosphere consists only of molecules, one can assume that $\kappa = 0$ in continuum and hence $\alpha = \sigma$. In this case the Rayleigh-Cabannes theory says that at a height h above the planetary surface

$$\sigma(h) = 24\pi^3 \left(\frac{r_\mathrm{s} \bar{M}}{\tilde{\lambda}^2 N_\mathrm{A}} \right)^2 N(h) \frac{6(1+\Delta)}{6 - 7\Delta}, \tag{1.8}$$

where $N(h)$ is the number of molecules per unit volume, $\tilde{\lambda}$ is the wavelength of light, $\bar{M}$ is the mean molecular weight of the gases comprising the atmosphere, r_s is the specific refraction, N_A is the Avogadro number, and Δ is the molecular depolarization factor. As we see, $\sigma(h) \sim \lambda^{-4}$. The values of $\bar{M}$, r_s, and Δ for some gases are given in Table 1.1.

The phase function in this case has the form

$$\chi_\mathrm{R}(\gamma) = \frac{3}{4} \left(1 + \cos^2\gamma\right) \tag{1.9}$$

and is usually called the Rayleigh phase function (the subscript R indicates this fact).

If an atmosphere also contains aerosols (liquid droplets, dust, etc.), the calculation of the above-mentioned values is noticeably more complicated. If the particles can be considered uniform spheres with a given complex refractive index and radius, the quantities α, λ and $\chi(\gamma)$ can be calculated using

Table 1.1

Gas	Chem. formula	$\bar{M}$, mol	$r_s, \mathrm{cm}^3\mathrm{g}^{-1}$	$\Delta \times 10^2$
Air	—	28.97	0.151	4.0
Carbon dioxide	CO_2	44.01	0.152	7.2
Oxygen	O_2	32.00	0.126	6.4
Nitrogen	N_2	28.1	0.159	3.6
Hydrogen	H_2	2.016	1.03	0.9
Gelium	He	4.003	0.127	0.03
Methane	CH_4	16.04	0.413	2×10^{-3}
Ammonia	NH_3	17.03	0.325	1.0
Argon	Ar	39.95	0.106	5×10^{-4}
Neon	Ne	20.18	0.05	0.01

the Mie theory. Light scattering by nonspherical particles can be computed using modern computational methods. We do not intend to describe these methods here. The reader is referred to the corresponding publications (see the list of references in the bibliographic notes to Part I of book).

The scattering volume element of a real medium can contain particles of various sizes (so-called polydisperse systems of particles). Therefore, to calculate light scattering by an elementary volume, we have to know the particle size distribution function.

It is obvious that distribution functions obtained from observations are inconvenient to use in calculations. Moreover, these functions are usually unknown for astrophysical objects. For this reason, when studying polydisperse systems, we can use analytical expressions as more or less correct approximations of real distributions.

We will give two examples of the most often used distributions.

1. ***Gamma distribution***

$$f_g(r) = \frac{r^{(1-3b)/b} e^{-r/ab}}{(ab)^{(1-2b)/b} \Gamma(1/b - 2)}, \tag{1.10}$$

where

$$a = r_{\mathrm{eff}} = \frac{\int\limits_0^\infty r^3 f_g(r)\,\mathrm{d}r}{\int\limits_0^\infty r^2 f_g(r)\,\mathrm{d}r}, \tag{1.11}$$

$$b = v_{\mathrm{eff}} = \frac{\int\limits_0^\infty (r - r_{\mathrm{eff}})^2 f_g(r) r^2\,\mathrm{d}r}{r_{\mathrm{eff}}^2 \int\limits_0^\infty r^2 f_g(r)\,\mathrm{d}r}. \tag{1.12}$$

Here $\Gamma(x)$ is the gamma function and $f(r)\mathrm{d}r$ represents a fraction of the particles with radii in the interval from r to $r + \mathrm{d}r$ so that the normalizing condition

$$\int_0^\infty f(r)\,\mathrm{d}r = 1 \tag{1.13}$$

is satisfied. As we see, the distribution function $f_{\mathrm{g}}(r)$ is characterized by two parameters: effective radius r_{eff} and effective variance v_{eff}.

2. ***Log normal distribution*** is given by

$$f_{\ln}(r) = \frac{1}{\sqrt{2\pi}\sigma r} \exp\left(-\frac{\ln^2(r/r_0)}{2\sigma^2}\right), \tag{1.14}$$

where σ^2 is the dispersion of the logarithm of particle radii, r_0 is the geometric mean radius. This particle size distribution function also has two parameters.

Thus, the phase function is defined by four parameters (real and imaginary parts of complex refractive index, mean radius and dispersion) even in the simplest case of the polydisperse Mie problem, where the light is scattered by an elementary volume which contains a system of uniform spherical particles with given particle size distribution. Theoretical study of the extent to which the diffuse radiation field is affected by the shape of the phase function is complicated by the large number of parameters. In order to approximate the phase functions of any degree of elongation however, they can sometimes be represented by simple one-parameter approximation formulas. The so-called Henyey-Greenstein phase function

$$\chi_{\mathrm{H}}(\gamma) = \frac{1 - g^2}{(1 + g^2 - 2g\cos\gamma)^{3/2}}, \tag{1.15}$$

appears to be the most convenient one. Here $|g| < 1$ is a parameter. The closer g is to 1, the larger is the elongation of phase function in the forward direction $\gamma = 0$. This phase function has four important properties, namely:

1. It approximates real phase functions observed in the Earth atmosphere reasonably well.
2. It is one-parameter.
3. It may be easily expanded in Legendre polynomials $P_n(\cos\gamma)$:

$$\chi_{\mathrm{H}}(\gamma) = \sum_{n=0}^{\infty} (2n + 1) g^n P_n(\cos\gamma), \tag{1.16}$$

which is essential to the theory of multiple light scattering.

4. The simple recurrence relation

$$(2m + 1)\chi_{\mathrm{H}}^{m+2}(\mu, \mu_0) = \frac{4a(m + 1)}{b}\chi_{\mathrm{H}}^{m+1}(\mu, \mu_0) - (2m + 3)\chi_{\mathrm{H}}^{m}(\mu, \mu_0), \tag{1.16a}$$

$m = 0, 1, 2, \dots$
takes place for azimuthal components $\chi_{\mathrm{H}}^{m}(\mu, \mu_0)$ (see (1.51) below). Here

$$a = 1 + g^2 - 2g\mu\mu_0, \qquad b = 2g\sqrt{(1-\mu^2)(1-\mu_0^2)}.$$

For $m = 0, 1$

$$\chi_{\mathrm{H}}^{0}(\mu, \mu_0) \equiv \chi_{\mathrm{H}}(\mu, \mu_0) = \frac{1-g^2}{\pi} \int_0^{\pi} \frac{\mathrm{d}\varphi}{(a \pm b\cos\varphi)^{3/2}}$$
$$= \frac{2(1-g^2)\sqrt{a+b}}{\pi(a^2-b^2)} E\left(\sqrt{\frac{2b}{a+b}}\right),$$

$$\chi_{\mathrm{H}}^{1}(\mu, \mu_0) = \frac{1-g^2}{\pi} \int_0^{\pi} \frac{\cos\varphi \,\mathrm{d}\varphi}{(a \pm b\cos\varphi)^{3/2}}$$
$$= \frac{2(1-g^2)}{\pi(\pm b)\sqrt{a+b}} \left[K\left(\sqrt{\frac{2b}{a+b}}\right) - \frac{a}{a-b} E\left(\sqrt{\frac{2b}{a+b}}\right)\right],$$

where $b \geq 0$ and $K(x)$ and $E(x)$ are the full elliptic integrals of the first and second kind, respectively.

In order to achieve a better approximation of the real phase function, it is possible to use a combination of two expressions of type (1.15) with different values of the parameter g:

$$\chi(\gamma) = c\chi_{\mathrm{H}}(\gamma, g_1) + (1-c)\chi_{\mathrm{H}}(\gamma, g_2), \tag{1.17}$$

where, as a rule, $g_1 > 0$, $g_2 \leq 0$, and the parameter c is within the range $0 \leq c \leq 1$. In this case the phase function of (1.17) has three parameters. Most of van de Hulst's calculations in MLS deal with the phase function of (1.15).

1.3 Radiative Transfer Equation

We will assume that the scattering law for an elementary volume of the medium [i.e. the quantities $\alpha(\boldsymbol{r})$, $\lambda(\boldsymbol{r})$ and $\chi(\gamma, \boldsymbol{r})$] is known and the energy sources in it are specified. In this case it is possible to find the diffuse radiation field in the medium, i.e. the intensity of radiation $I(\boldsymbol{r}, \boldsymbol{\omega})$. We will now obtain the equation defining this quantity.

First of all, along with the extinction coefficient $\alpha(\boldsymbol{r})$, we will introduce the ***emission coefficient*** $\varepsilon(\boldsymbol{r}, \boldsymbol{\omega})$. The quantity $\varepsilon(\boldsymbol{r}, \boldsymbol{\omega})\,\mathrm{d}V\,\mathrm{d}\omega\,\mathrm{d}\nu\,\mathrm{d}t$ is, by definition, the energy emitted by the elementary volume $\mathrm{d}V(\boldsymbol{r})$ within a solid angle $\mathrm{d}\omega$ around the direction $\boldsymbol{\omega}$ in a frequency interval from ν to $\nu+\mathrm{d}\nu$ during the time period $\mathrm{d}t$. It has the dimension of $\mathrm{erg\,cm^{-3}s^{-1}Hz^{-1}ster^{-1}}$.

The quantity $\varepsilon(\boldsymbol{r},\boldsymbol{\omega})$ can be conceived of as consisting of two parts. The first part $[\varepsilon_1(\boldsymbol{r},\boldsymbol{\omega})]$ results from reemission of the energy falling on the volume $\mathrm{d}V$ from all directions. The second part, which is denoted $\varepsilon_0(\boldsymbol{r},\boldsymbol{\omega})$, is determined by the internal energy sources (primary, or true sources) of the elementary volume. In other words,

$$\varepsilon(\boldsymbol{r},\boldsymbol{\omega}) = \varepsilon_0(\boldsymbol{r},\boldsymbol{\omega}) + \varepsilon_1(\boldsymbol{r},\boldsymbol{\omega}). \tag{1.18}$$

Let us find the quantity $\varepsilon_1(\boldsymbol{r},\boldsymbol{\omega})$. We assume that the volume $\mathrm{d}V$ is illuminated by radiation with intensity $I(\boldsymbol{r},\boldsymbol{\omega}')$. Then, the energy absorbed by this volume per unit frequency interval per unit time from the direction $\boldsymbol{\omega}'$ is equal to

$$\alpha(\boldsymbol{r})I(\boldsymbol{r},\boldsymbol{\omega}')\,\mathrm{d}\omega'\mathrm{d}V.$$

Part of this energy is totally absorbed (true absorption). Another part is scattered in all directions. Specifically, part of the energy that is equal to $\lambda(\boldsymbol{r})\chi(\boldsymbol{\omega}\boldsymbol{\omega}',\boldsymbol{r})\mathrm{d}\omega/4\pi$ where $\boldsymbol{\omega}\boldsymbol{\omega}' = \cos\gamma'$ and γ' is the angle between the directions of the incident and scattered radiation is scattered in the direction $\boldsymbol{\omega}$.[1] As a result, we obtain the following expression for the energy reemitted from the direction $\boldsymbol{\omega}'$ in the direction $\boldsymbol{\omega}$

$$\alpha(\boldsymbol{r})I(\boldsymbol{r},\boldsymbol{\omega}')\,\mathrm{d}\omega'\mathrm{d}V\,\lambda(\boldsymbol{r})\chi(\boldsymbol{\omega}\boldsymbol{\omega}',\boldsymbol{r})\frac{\mathrm{d}\omega}{4\pi}.$$

Since the volume $\mathrm{d}V$ is illuminated from all directions, in order to find $\varepsilon_1(\boldsymbol{r},\boldsymbol{\omega})$ this expression must be integrated over all directions $\boldsymbol{\omega}'$. Thus, instead of (1.18) we obtain

$$\varepsilon(\boldsymbol{r},\boldsymbol{\omega}) = \lambda(\boldsymbol{r})\alpha(\boldsymbol{r})\int\limits_{(4\pi)} I(\boldsymbol{r},\boldsymbol{\omega}')\chi(\boldsymbol{\omega}\boldsymbol{\omega}',\boldsymbol{r})\frac{\mathrm{d}\omega'}{4\pi} + \varepsilon_0(\boldsymbol{r},\boldsymbol{\omega}). \tag{1.19}$$

Equation (1.19) expresses the quantity $\varepsilon(\boldsymbol{r},\boldsymbol{\omega})$ in terms of the intensity of radiation $I(\boldsymbol{r},\boldsymbol{\omega})$. As was stated before, the quantities $\alpha(\boldsymbol{r})$, $\lambda(\boldsymbol{r})$ and $\chi(\gamma,\boldsymbol{r})$, which define optical properties of the medium, are assumed to be known.

We will now derive the equation which determines changes in the intensity of radiation as it passes through a medium. Let $I(\boldsymbol{r},\boldsymbol{\omega})$ be the intensity of radiation at a point $\boldsymbol{r}$ with the direction $\boldsymbol{\omega}$. Then, $I(\boldsymbol{r},\boldsymbol{\omega})\mathrm{d}S\,\mathrm{d}\nu\,\mathrm{d}\omega\mathrm{d}t$ is, by definition, the energy passing during the time $\mathrm{d}t$ through the area $\mathrm{d}S$ which is perpendicular to the direction $\boldsymbol{\omega}$ in the frequency interval from ν to $\nu+\mathrm{d}\nu$ within the solid angle $\mathrm{d}\omega$. Take the point $\boldsymbol{r}'$ at a distance $\mathrm{d}s$ from the point $\boldsymbol{r}$ in the direction $\boldsymbol{\omega}$ so that $\boldsymbol{r}' = \boldsymbol{r}+\boldsymbol{\omega}\mathrm{d}s$ (Fig. 1.1). As a result, the quantity

[1] In this book we will denote the arguments of the phase function in various ways: $\chi(\gamma,\boldsymbol{r})$, $\chi(\boldsymbol{\omega}\boldsymbol{\omega}',\boldsymbol{r})$, $\chi(\cos\gamma,\boldsymbol{r})$, $\chi(\gamma,\tau)$, $\chi(\cos\gamma,\tau)$, etc. The reader should, however, bear in mind that in all cases the same function is meant. The first argument denotes the direction of scattering and the second one denotes the place where scattering takes place. For a homogeneous medium the second argument is absent.

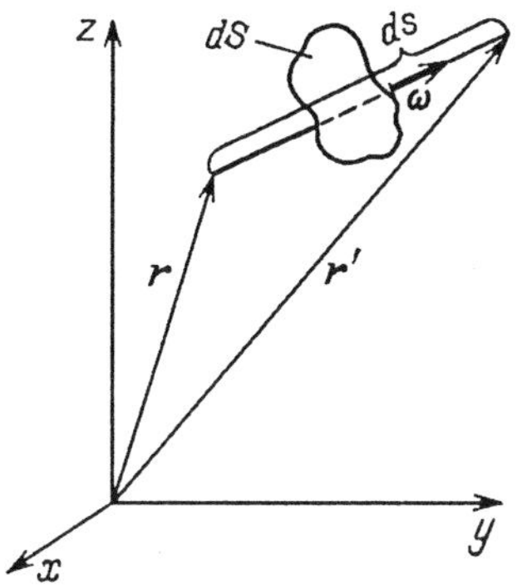

Fig. 1.1. On the derivation of the radiative transfer equation

$$[I(\boldsymbol{r}',\boldsymbol{\omega})-I(\boldsymbol{r},\boldsymbol{\omega})]\,\mathrm{d}S\,\mathrm{d}\nu\,\mathrm{d}t\,\mathrm{d}\omega=[I(\boldsymbol{r}+\boldsymbol{\omega}\,\mathrm{d}s,\boldsymbol{\omega})-I(\boldsymbol{r},\boldsymbol{\omega})]\,\mathrm{d}S\,\mathrm{d}\nu\,\mathrm{d}t\,\mathrm{d}\omega \tag{1.20}$$

will characterize the change in intensity at the distance $\mathrm{d}s$ in the direction $\boldsymbol{\omega}$.

This change results from two factors. Part of the energy will be absorbed at the distance $\mathrm{d}s$. Obviously, it is equal to

$$-\,\alpha(\boldsymbol{r})\,\mathrm{d}s\,I(\boldsymbol{r},\boldsymbol{\omega})\,\mathrm{d}S\,\mathrm{d}\nu\,\mathrm{d}\omega\,\mathrm{d}t. \tag{1.21}$$

At the same time in the elementary volume $\mathrm{d}V=\mathrm{d}s\,\mathrm{d}S$, energy equal to

$$\varepsilon(\boldsymbol{r},\boldsymbol{\omega})\,\mathrm{d}s\,\mathrm{d}S\,\mathrm{d}\nu\,\mathrm{d}\omega\,\mathrm{d}t \tag{1.22}$$

will be emitted during the time $\mathrm{d}t$. Thus, from (1.20–22) we have the following relation for the energy balance

$$\begin{aligned}&[I(\boldsymbol{r}+\boldsymbol{\omega}\,\mathrm{d}s,\boldsymbol{\omega})-I(\boldsymbol{r},\boldsymbol{\omega})]\,\mathrm{d}S\,\mathrm{d}\nu\,\mathrm{d}t\,\mathrm{d}\omega\\&=-\alpha(\boldsymbol{r})\,\mathrm{d}s\,I(\boldsymbol{r},\boldsymbol{\omega})\,\mathrm{d}S\,\mathrm{d}\nu\,\mathrm{d}\omega\,\mathrm{d}t+\varepsilon(\boldsymbol{r},\boldsymbol{\omega})\,\mathrm{d}s\,\mathrm{d}S\,\mathrm{d}\nu\,\mathrm{d}\omega\,\mathrm{d}t.\end{aligned}$$

Thus, it follows that

$$I(\boldsymbol{r}+\boldsymbol{\omega}\,\mathrm{d}s,\boldsymbol{\omega})-I(\boldsymbol{r},\boldsymbol{\omega})=-[\alpha(\boldsymbol{r})I(\boldsymbol{r},\boldsymbol{\omega})-\varepsilon(\boldsymbol{r},\boldsymbol{\omega})]\,\mathrm{d}s. \tag{1.23}$$

Expanding $I(\boldsymbol{r}+\boldsymbol{\omega}\mathrm{d}s,\boldsymbol{\omega})$ according to the Taylor formula, we obtain

$$\cos\theta_1\frac{\partial I(\boldsymbol{r},\boldsymbol{\omega})}{\partial x}+\cos\theta_2\frac{\partial I(\boldsymbol{r},\boldsymbol{\omega})}{\partial y}+\cos\theta_3\frac{\partial I(\boldsymbol{r},\boldsymbol{\omega})}{\partial z}=-\alpha(\boldsymbol{r})I(\boldsymbol{r},\boldsymbol{\omega})+\varepsilon(\boldsymbol{r},\boldsymbol{\omega}), \tag{1.24}$$

where $\cos\theta_i$ are direction cosines of the vector $\boldsymbol{\omega}$:

$$\boldsymbol{\omega}=\boldsymbol{i}\cos\theta_1+\boldsymbol{j}\cos\theta_2+\boldsymbol{k}\cos\theta_3. \tag{1.25}$$

Equation (1.24) may be written in the more compact form

$$\boldsymbol{\omega}\nabla I(\boldsymbol{r},\boldsymbol{\omega})+\alpha(\boldsymbol{r})I(\boldsymbol{r},\boldsymbol{\omega})=\varepsilon(\boldsymbol{r},\boldsymbol{\omega}). \tag{1.26}$$

Taking into account that the emission coefficient ε is expressed in terms of the intensity by (1.19), the final equation defining the intensity of radiation as a function of the optical properties of the medium may be written as follows:

$$\boldsymbol{\omega}\nabla I(\boldsymbol{r},\boldsymbol{\omega}) + \alpha(\boldsymbol{r})I(\boldsymbol{r},\boldsymbol{\omega}) = \lambda(\boldsymbol{r})\alpha(\boldsymbol{r}) \int\limits_{(4\pi)} I(\boldsymbol{r},\boldsymbol{\omega}')\chi(\boldsymbol{\omega}\boldsymbol{\omega}',\boldsymbol{r})\frac{\mathrm{d}\omega'}{4\pi} + \varepsilon_0(\boldsymbol{r},\boldsymbol{\omega}). \tag{1.27}$$

This is the sought-for ***equation of radiative transfer*** in an inhomogeneous medium of arbitrary shape that emits and scatters radiant energy.

The radiative transfer equation must now be supplemented with appropriate boundary conditions. Let a system occupy a volume bounded by a convex surface. We will assume this surface to be nonreflecting. In this case the boundary condition has the form

$$I(\boldsymbol{r}_0,\boldsymbol{\omega}) = I_0(\boldsymbol{r}_0,\boldsymbol{\omega}) \qquad (\text{at } \boldsymbol{\omega}\boldsymbol{n} < 0), \tag{1.28}$$

where $\boldsymbol{r}_0$ is a radius-vector of an arbitrary point on the boundary surface, $\boldsymbol{n}$ is the unit vector normal to the boundary in the outward direction at the point $\boldsymbol{r}_0$, $I_0(\boldsymbol{r}_0,\boldsymbol{\omega})$ is the intensity of radiation at the point $\boldsymbol{r}_0$ in the direction $\boldsymbol{\omega}$ radiated by external sources. Condition (1.28) must be changed if the boundary reflects part of the incident radiation (e.g. water surface) or if the boundary is not convex.

1.4 Radiative Transfer Equation in a Stratified Medium

Generally the analytical investigation of (1.27) is too complicated a problem. Therefore we will consider here only a particular case of (1.27) where radiative transfer takes place in a *stratified medium.* By a stratified medium we mean a plane-parallel atmosphere whose optical properties depend only on one coordinate z which is perpendicular to the layers (see Fig. 1.2). In many cases this geometrical model of the medium gives a good approximation for real physical systems both in nature (stellar and planetary atmospheres, ocean) and in laboratories.

We denote the angle between the positive direction of the z axis and the direction of the radiation by θ; φ is the asimuth. Let one of the boundaries of the layer be at the level $z = 0$ and another boundary be at the level $z = z_0$ (see Fig. 1.2), so that the geometrical thickness of the layer is equal to z_0. Instead of the geometrical depth z, we introduce the ***optical depth***

$$\tau = \int\limits_0^z \alpha(z')\,\mathrm{d}z'. \tag{1.29}$$

At $z = z_o$ this quantity is called the ***optical thickness*** of the layer and is denoted by τ_0. As a result, the transfer equation (1.27) may be rewritten in the form

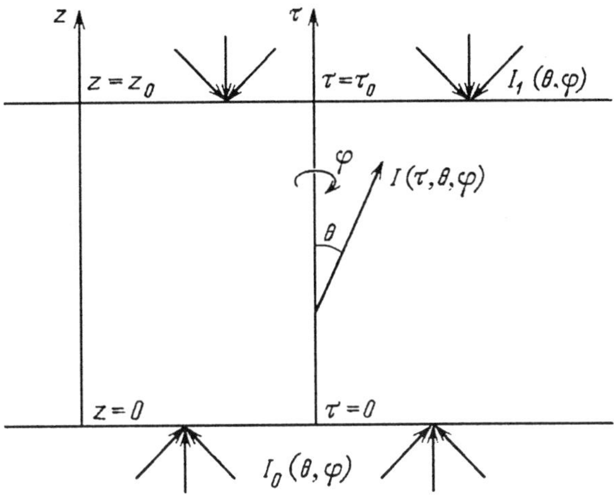

Fig. 1.2. Radiative transfer in a stratified atmosphere

$$\cos\theta \frac{\mathrm{d}I(\tau,\theta,\varphi)}{\mathrm{d}\tau} + I(\tau,\theta,\varphi)$$
$$= \frac{\lambda(\tau)}{4\pi} \int\limits_0^{2\pi} \mathrm{d}\varphi' \int\limits_0^{\pi} I(\tau,\theta',\varphi')\chi(\cos\gamma',\tau)\sin\theta'\mathrm{d}\theta' + \frac{\varepsilon_0(\tau,\theta,\varphi)}{\alpha(\tau)}, \quad (1.30)$$

where

$$\cos\gamma' = \cos\theta\cos\theta' + \sin\theta\sin\theta'\cos(\varphi - \varphi'). \quad (1.31)$$

As to the boundary condition (1.28), it is replaced in this case by two boundary conditions:

$$I(0,\theta,\varphi) = I_0(\theta,\varphi), \qquad \theta \in \left[0, \frac{\pi}{2}\right], \quad (1.32)$$

$$I(\tau_0,\theta,\varphi) = I_1(\theta,\varphi), \qquad \theta \in \left[\frac{\pi}{2}, \pi\right]. \quad (1.33)$$

Here I_0 and I_1 are given intensities of the radiation incident on the boundaries of the layer. It should be emphasized that these quantities are assumed to be independent of the horizontal coordinates x and y. The problem of solving (1.30) at zero boundary conditions and the arbitrarily prescribed emission coefficient $\varepsilon_0(\tau,\theta,\varphi)$ which stems from internal sources of radiation will be called the ***general problem***. If internal sources are absent and the boundary conditions are given by expressions (1.32) and (1.33), solving the radiative transfer equation reduces to the so-called ***parallel external flux problem***. The ***Milne problem*** and the ***problem for two-sided infinity*** are closely related to this.

Let us consider these problems in order of decreasing complexity and importance. As to the general problem, it will be discussed in detail in Sections 1.11, 3.12 and 4.10.

1.5 The Parallel External Flux Problem

For clarity, we will limit ourselves to a planetary atmosphere. We begin by assuming that the atmosphere thickness is much smaller than the radius of the planet. This will allow us to consider a medium consisting of plane-parallel layers instead of a spherical atmosphere. Furthermore, the optical properties of the atmosphere α, λ and $\chi(\gamma)$ will be assumed to depend only on one spatial variable z. Thus, we assume that the atmosphere is horizontally homogeneous.

Let us also assume that the atmosphere is illuminated only by parallel solar radiation (i.e., the Sun will be considered as an infinitely distant point source of radiation) and that internal energy sources are absent ($\varepsilon_0 \equiv 0$). At this time we will not consider light reflected by the surface of the planet.

This problem will be called the ***parallel external flux problem***. There are different names for this problem. For example, it is called the basic problem in LSPA by Sobolev. It is also sometimes called the diffuse reflection problem. As will become clear later, a number of more complicated problems may be reduced to this problem.

We denote the angle between the outward normal to the upper boundary of the atmosphere and the direction toward the Sun (Sun zenith angle, see Fig. 1.3) by θ_0. Let πF be the illumination of the surface perpendicular to the incident radiation at the top of the atmosphere, and let φ_0 be the azimuth of the incident radiation. It is obvious that the intensity of the direct solar radiation at the optical depth τ in the atmosphere is defined by the expression

$$\pi F \delta(\cos\theta - \cos\theta_0)\delta(\varphi - \varphi_0)\mathrm{e}^{-\tau/\cos\theta_0}, \tag{1.34}$$

where δ is Dirac delta-function. Let us assume that in (1.30) (where we have already set $\varepsilon_0 = 0$)

$$I(\tau,\theta,\varphi) = I_1(\tau,\theta,\varphi) + \pi F \delta(\cos\theta - \cos\theta_0)\delta(\varphi - \varphi_0)\mathrm{e}^{-\tau/\cos\theta_0}. \tag{1.35}$$

The quantity I_1 here is the intensity of the *diffuse* radiation which is determined only by those photons which experienced at least one act of scattering. In this case (1.30) may be rewritten as follows:

$$\cos\theta \frac{\mathrm{d}I_1(\tau,\theta,\varphi)}{\mathrm{d}\tau} + I_1(\tau,\theta,\varphi) = B(\tau,\theta,\varphi), \tag{1.36}$$

$$\begin{aligned} B(\tau,\theta,\varphi) = {} & \frac{\lambda(\tau)}{4\pi}\int_0^{2\pi}\mathrm{d}\varphi'\int_0^{\pi} I_1(\tau,\theta',\varphi')\chi(\cos\gamma',\tau)\sin\theta'\,\mathrm{d}\theta' \\ & + \frac{\lambda(\tau)F}{4\pi}\chi(\cos\gamma,\tau)\mathrm{e}^{-\tau/\cos\theta_0}, \end{aligned} \tag{1.37}$$

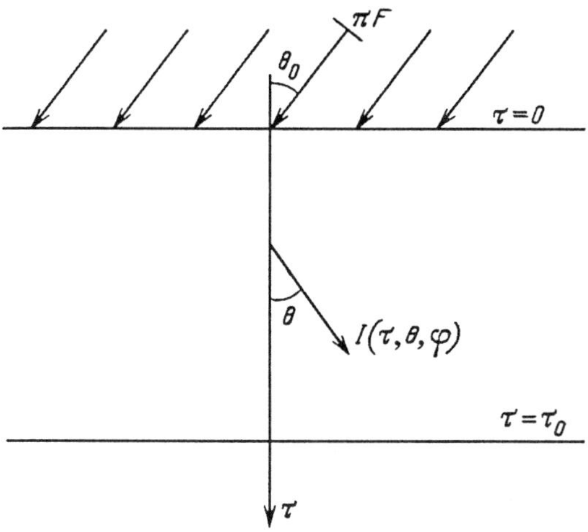

Fig. 1.3. The parallel external flux problem

where

$$\cos\gamma = \cos\theta\cos\theta_0 + \sin\theta\sin\theta_0\cos(\varphi - \varphi_0) \tag{1.38}$$

and

$$B = \frac{\epsilon_1}{\alpha}. \tag{1.39}$$

This quantity B is generally known as the ***source function***; it has the same dimension as the radiation intensity. Its physical meaning is easily understood from the fact that $B\,\mathrm{d}\tau = \varepsilon_1\,\mathrm{d}z$, i.e., it is the amount of radiant energy which is radiated by an atmosphere layer of thickness $\mathrm{d}z$ through a unit area in a unit frequency interval during a unit time within a unit solid angle around the direction defined by the angles θ and φ.

Hereafter, we will omit the subscript "1" in the expression for intensity, i.e., by the quantity $I(\tau,\theta,\varphi)$ we will mean (unless specially noted) the intensity of the diffuse radiation. Furthermore we set $F = 1$, i.e., the illumination of the top boundary of atmosphere is assumed to be equal $\pi\cos\theta_0$. This should not lead to confusion if we remember that, when using the final expressions for the radiation intensity, *they must be multiplied by a given factor* F. Finally, we will use the designations

$$\mu = \cos\theta, \qquad \mu_0 = \cos\theta_0, \tag{1.40}$$

where $\mu \in [-1,+1]$, $\mu_0 \in [0,1]$. As a result, (1.36–38) take the form:

$$\mu\frac{\mathrm{d}I(\tau,\theta,\varphi)}{\mathrm{d}\tau} + I(\tau,\theta,\varphi) = B(\tau,\theta,\varphi), \tag{1.41}$$

$$B(\tau,\mu,\varphi) = \frac{\lambda(\tau)}{4\pi}\int_0^{2\pi}\mathrm{d}\varphi'\int_{-1}^{+1} I(\tau,\mu',\varphi')\chi(\gamma',\tau)\,\mathrm{d}\mu' + \frac{\lambda(\tau)}{4\pi}\chi(\gamma,\tau)\mathrm{e}^{-\frac{\tau}{\mu_0}} \tag{1.42}$$

and

$$\cos\gamma = \mu\mu_0 + \sqrt{(1-\mu^2)(1-\mu_0^2)}\cos(\varphi-\varphi_0). \tag{1.43}$$

In what follows, instead of the designations $I(\tau,\mu,\varphi)$ (or correspondingly $B(\tau,\mu,\varphi)$), the extended form of these quantities will be used, namely, $I(\tau,\mu,\mu_0,\varphi-\varphi_0)$ and $B(\tau,\mu,\mu_0,\varphi-\varphi_0)$. By doing so, we emphasize their dependence on the *parameters* μ_0 and φ_0.

It is necessary to add boundary conditions to the equations thus obtained. Since the layer is not illuminated by the *diffuse* radiation either from below of from above, the above-mentioned conditions have the form ($\mu \in [0,1]$):

$$I(0,\mu,\varphi) = 0; \qquad I(\tau_0,-\mu,\varphi) = 0. \tag{1.44}$$

Thus, solving the parallel external flux problem is reduced to finding the radiation intensity with the aid of (1.41) and (1.42) under the boundary conditions (1.44). If the optical thickness of the layer is infinitely large ($\tau_0 = \infty$), such an atmosphere is called ***semi-infinite***. For this atmosphere, instead of the second boundary condition of (1.44) we require that the solution of the transfer equation at $\tau \to \infty$ be *limited.*

If the quantities $\lambda(\tau)$ and $\chi(\gamma,\tau)$ do not depend on the optical depth, the atmosphere will be called ***homogeneous***. Otherwise, we have an ***inhomogeneous*** medium (or, more specifically, an inhomogeneous stratified atmosphere).

The solution of the parallel external flux problem also immediately gives the solution of a more common problem, namely, determining the radiation field in a plane layer, illuminated at both boundaries by radiation of a given intensity which does not depend on the horizontal coordinate. To be more specific, we assume that only one of the boundaries ($\tau = 0$) of the layer is illuminated by radiation with intensity $I_0(\mu,\varphi)$ ($\mu \in [0,1]$). In this case we denote the total intensity (i.e., the intensity of direct + diffuse radiation) by $I^*(\tau,\mu,\varphi)$ ($\mu \in [-1,+1]$). It is apparent that the quantity I^* consists of two components: the intensity of *direct* radiation which is attenuated in the layer of optical thickness τ, and *diffuse* radiation which results from the irradiation of the boundary $\tau = 0$ by light with the intensity $I_0(\mu,\varphi)$. In other words:

$$I^*(\tau,\mu,\varphi) = I_0(\mu,\varphi)e^{-\frac{\tau}{\mu}}\theta(\mu) + \frac{1}{\pi}\int_0^{2\pi} d\varphi' \int_{-1}^{+1} I_0(\mu',\varphi')I(\tau,\mu,\mu',\varphi-\varphi')\,d\mu', \tag{1.45}$$

where $\theta(\mu)$ is the unit step function

$$\theta(\mu) = \begin{cases} 1 & \mu \geq 0, \\ 0 & \mu < 0. \end{cases} \tag{1.46}$$

The second term on the right hand side of (1.45) results from considering the intensity $I_0(\mu,\varphi)$ incident from the outside as consisting of the superposition of the parallel beams distributed over angles θ and φ in accordance with the distribution function $I_0(\mu, \varphi)$. In particular, if this function is given by (1.34) without the exponential factor then instead of (1.45), we obtain the already known expression (1.35).

The simple relationship (1.45) plays an important role in the transfer theory. We will repeatedly use this relation throughout the book.

1.5.1 Azimuthal Harmonics of the Radiation Intensity

We have seen above that the radiation intensity depends on one spatial variable τ, two angular variables μ and φ and a number of parameters: μ_0, φ_0, τ_0, including quantities which characterize $\lambda(\tau)$ and the phase function. When solving a transfer equation, it is clearly desirable to reduce the number of these variables. In particular, it turns out that the radiation intensity and the source function in most cases can be represented by Fourier expansions in terms of azimuthal variables. Namely,

$$I(\tau, \mu, \mu_0, \varphi - \varphi_0) = I^0(\tau, \mu, \mu_0) + 2 \sum_{m=1}^{\infty} I^m(\tau, \mu, \mu_0) \cos m\,(\varphi - \varphi_0), \quad (1.47)$$

$$B(\tau, \mu, \mu_0, \varphi - \varphi_0) = B^0(\tau, \mu, \mu_0) + 2 \sum_{m=1}^{\infty} B^m(\tau, \mu, \mu_0) \cos m\,(\varphi - \varphi_0). \quad (1.48)$$

As a result, we obtain separate equations for each of the azimuthal harmonics I^m and B^m,

$$\mu \frac{\mathrm{d}I^m(\tau, \mu, \mu_0)}{\mathrm{d}\tau} + I^m(\tau, \mu, \mu_0) = B^m(\tau, \mu, \mu_0), \quad (1.49)$$

$$B^m(\tau, \mu, \mu_0) = \frac{\lambda(\tau)}{2} \int_{-1}^{+1} I^m(\tau, \mu', \mu_0) \chi^m(\mu, \mu'; \tau)\, \mathrm{d}\mu' + \frac{\lambda(\tau)}{4} \chi^m(\mu, \mu_0; \tau) \mathrm{e}^{-\frac{\tau}{\mu_0}}, \quad (1.50)$$

where

$$\chi^m(\mu, \mu_0; \tau) = \frac{1}{2\pi} \int_0^{2\pi} \chi(\cos\gamma, \tau) \cos m\varphi \, \mathrm{d}\varphi. \quad (1.51)$$

The boundary conditions of (1.44) must be rewritten as follows (for $\mu \in [0, 1]$):

$$I^m(0, \mu, \mu_0) = 0; \qquad I^m(\tau_0, -\mu, \mu_0) = 0. \quad (1.52)$$

In (1.50–52) we will omit the superscript "0", i.e., we will write I instead of I^0, B instead of B^0.

To make sure that representations (1.47–50) are possible, we suppose that the phase function is expanded in Legendre polynomials $P_n(\cos\gamma)$, i.e.

$$\chi(\cos\gamma, \tau) = \sum_{n=0}^{\infty} x_n(\tau) P_n(\cos\gamma). \quad (1.53)$$

This assumption is usually used in radiative transfer theory. However, if $\cos\gamma$ is given by (1.43), the following addition theorem for spherical functions is true:

$$P_n(\cos\gamma) = P_n(\mu)P_n(\mu_0) + 2\sum_{m=1}^{n} \frac{(n-m)!}{(n+m)!} P_n^m(\mu)P_n^m(\mu_0)\cos m\,(\varphi-\varphi_0), \tag{1.54}$$

where $P_n^m(\mu)$ are associated Legendre functions. By substituting (1.54) into (1.53) and interchanging the order of the summation, we have

$$\chi(\cos\gamma,\tau) = \chi(\mu,\mu_0;\tau) + 2\sum_{m=1}^{n} \chi^m(\mu,\mu_0;\tau)\cos m\,(\varphi-\varphi_0), \tag{1.55}$$

where

$$\chi^m(\mu,\mu_0;\tau) = \sum_{i=m}^{\infty} x_i^m(\tau)P_i^m(\mu)P_i^m(\mu_0), \tag{1.56}$$

$$x_i^m(\tau) = x_i(\tau)\frac{(i-m)!}{(i+m)!}. \tag{1.57}$$

The zeroth azimuthal harmonic here has no superscript, i.e., we denoted $\chi^0(\mu,\mu_0;\tau) = \chi(\mu,\mu;\tau)$.

We note that in the case of a homogeneous atmosphere there is no need to use the expansion (1.56) for the Henyey-Greenstein phase function (1.15) (or the superposition (1.17)). To calculate the azimuthal harmonics $\chi_n^m(\mu,\mu_0)$ one can use a simple recursion relation (1.16a), which substantially speeds up calculation especially in the case of strongly anisotropic scattering (i.e., when $(1-g) \ll 1$).

Taking into account that

$$\frac{1}{2\pi}\int_0^{2\pi} \cos m\,(\varphi'-\varphi_0)\cos n\,(\varphi-\varphi')\,\mathrm{d}\varphi' = \begin{cases} 0, & m \neq n, \\ \frac{1}{2}\cos m(\varphi-\varphi_0) & m = n \neq 0, \\ 1 & m = n = 0, \end{cases} \tag{1.57a}$$

substitution of (1.47),(1.48), and (1.55) into (1.41) and (1.42) gives (1.49) and (1.50). The upper infinite limit in the sum (1.53) is generally replaced by some number N, i.e., it is assumed that N components of the sum approximate an actual phase function adequately. As a result, instead of (1.41) and (1.42) one must solve $N+1$ subsystems of (1.49) and (1.50).

1.5.2 Integral Equation for the Source Function

Equation (1.41) along with (1.42) is an ***integro-differential equation***. It determines the radiation intensity in a medium. However, from (1.41) and

(1.42) it is possible to obtain one Fredholm ***integral equation*** of the second kind for the source function $B(\tau, \mu, \varphi)$. In order to do this we must solve (1.41) for $I(\tau, \mu, \varphi)$ and substitute the solution into (1.42).

From (1.41) with the boundary conditions (1.44), we have

$$I(\tau, \mu, \varphi) = \int_0^\tau B(\tau', \mu, \varphi) e^{-\frac{\tau-\tau'}{\mu}} \frac{d\tau'}{\mu}, \tag{1.58}$$

$$I(\tau, -\mu, \varphi) = \int_\tau^{\tau_0} B(\tau', -\mu, \varphi) e^{-\frac{\tau'-\tau}{\mu}} \frac{d\tau'}{\mu}, \quad (\mu \in [0, 1]). \tag{1.59}$$

Expressions (1.58) and (1.59) have a very simple physical meaning. Let us examine, for instance, the expression (1.58) for the intensity of downward radiation. The quantity $B(\tau', \mu, \varphi)\frac{d\tau'}{\mu}$ is the intensity of radiation which comes from the layer $d\tau'$ in the direction $\boldsymbol{\omega} = (\mu, \varphi)$. When this radiation arrives at the level τ, it will be attenuated in the atmosphere by a factor $e^{-\frac{\tau-\tau'}{\mu}}$. Hence, the intensity at the level τ is equal to $B(\tau', \mu, \varphi)e^{-\frac{\tau-\tau'}{\mu}}\frac{d\tau'}{\mu}$. Since all the elementary layers $d\tau'$ which are located in the interval from $\tau' = 0$ to $\tau' = \tau$ contribute to the intensity of the downward radiation at the level τ, it is necessary to integrate the above-mentioned expression over τ' within these limits to find the quantity $I(\tau, \mu, \varphi)$

Substitution of (1.58) and (1.59) into (1.42) gives ($\mu \in [-1, +1]$),

$$B(\tau, \mu, \varphi) = \frac{\lambda(\tau)}{4\pi} \int_0^{2\pi} d\varphi' \left[\int_0^1 \chi(\gamma', \tau)\, d\mu' \int_0^\tau B(\tau', \mu', \varphi') e^{-\frac{\tau-\tau'}{\mu'}} \frac{d\tau'}{\mu'} \right.$$
$$\left. - \int_{-1}^0 \chi(\gamma', \tau)\, d\mu' \int_\tau^{\tau_0} B(\tau', \mu', \varphi') e^{-\frac{\tau-\tau'}{\mu'}} \frac{d\tau'}{\mu'} \right] + \frac{\lambda(\tau)}{4} \chi(\gamma, \tau) e^{-\frac{\tau}{\mu_0}}. \tag{1.60}$$

After determining the function B from (1.60), the radiation intensity can be found from (1.58) and (1.59). Alternatively, having the radiation intensity $I(\tau, \mu, \varphi)$, one can find the source function with the aid of the expression (1.42). Hence, the problems of determining the radiation intensity and the source function are equivalent.

Similarly, from (1.49) and (1.50) under the conditions (1.52) we have:

$$I^m(\tau, \mu, \mu_0) = \int_0^\tau B^m(\tau', \mu, \mu_0) e^{-\frac{\tau-\tau'}{\mu}} \frac{d\tau'}{\mu}, \tag{1.61}$$

$$I^m(\tau, -\mu, \mu_0) = \int_\tau^{\tau_0} B^m(\tau', -\mu, \mu_0) e^{-\frac{\tau'-\tau}{\mu}} \frac{d\tau'}{\mu}, \quad (\mu \in [0, 1]). \tag{1.62}$$

By substituting expressions (1.61) and (1.62) into (1.50), we obtain the integral equation for the m-th azimuthal harmonic of the source function ($\mu \in [-1, +1]$):

$$B^m(\tau, \mu, \mu_0) = \frac{\lambda(\tau)}{2} \int_0^1 \frac{d\mu'}{\mu'} \left[\int_0^\tau \chi^m(\mu, \mu'; \tau)\, B^m(\tau', \mu', \mu_0) e^{-\frac{\tau - \tau'}{\mu'}}\, d\tau' \right.$$

$$\left. + \int_\tau^{\tau_0} \chi^m(\mu, -\mu'; \tau) B^m(\tau', -\mu, \mu_0) e^{-\frac{\tau' - \tau}{\mu'}}\, d\tau' \right] + \frac{\lambda(\tau)}{4} \chi^m(\mu, \mu_0; \tau) e^{-\frac{\tau}{\mu_0}}. \tag{1.63}$$

Solving this equation enables us to calculate the corresponding azimuthal harmonic of the radiation intensity with the aid of (1.61) and (1.62).

As we see, each azimuthal harmonic of the source function $B^m(\tau, \mu, \mu_0)$ depends on two variables τ and μ and a parameter μ_0. However, by substituting (1.56) into (1.63), we can see that

$$B^m(\tau, \mu, \mu_0) = \sum_{i=m}^{N} x_i^m(\tau) P_i^m(\mu) B_i^m(\tau, \mu_0), \tag{1.64}$$

where the function $B_i^m(\tau, \mu_0)$ is determined by the following system of equations:

$$B_i^m(\tau, \mu_0) = \frac{\lambda(\tau)}{2} \sum_{j=m}^{N} \left[\int_0^\tau x_j^m(\tau') B_j^m(\tau', \mu_0) E_{ij}^m(\tau - \tau')\, d\tau' \right.$$

$$\left. + (-1)^{i+j} \int_\tau^{\tau_0} x_j^m(\tau') B_j^m(\tau', \mu_0) E_{ij}^m(\tau' - \tau)\, d\tau' \right] + \frac{\lambda(\tau)}{4} P_i^m(\mu_0) e^{-\frac{\tau}{\mu_0}}, \tag{1.65}$$

where $i = m, m+1, \ldots, N$, and

$$E_{ij}^m(\tau) = \int_0^1 P_i^m(\mu) P_j^m(\mu) e^{-\frac{\tau}{\mu}} \frac{d\mu}{\mu}. \tag{1.66}$$

Thus, we represent the function $B^m(\tau, \mu, \mu_0)$, which depends on two variables, in terms of $(N - m + 1)$ functions $B_i^m(\tau, \mu_0)$. These functions depend only on single variable τ (μ_0 is a parameter).

For complete solution of the problem, i.e., in order to find the function $B(\tau, \mu, \varphi)$, the system of equations (1.65) has to be solved for all m. For any given N, the total number of functions $B_i^m(\tau, \mu_0)$ to be found is $\frac{1}{2}(N+1)(N+2)$. When N is large, solving the problem by finding the functions B_i^m from equations (1.65) becomes difficult in practice. A simpler method for determining $B^m(\tau, \mu, \mu_0)$ in a homogeneous atmosphere is presented in LSPA, Chap. 5.

1.6 The Milne Problem

Let us suppose that we have a semi-infinite stratified atmosphere (Fig. 1.4). Imagine that the sources of radiation of infinite power are present in this atmosphere at an infinitely large depth. The radiation intensity in such a medium does not depend on azimuth[2], i.e., the problem has an *axial symmetry* and the radiative transfer equation can be written as follows:

$$\mu\frac{\mathrm{d}I(\tau,\mu)}{\mathrm{d}\tau}+I(\tau,\mu)=B(\tau,\mu), \tag{1.67}$$

$$B(\tau,\mu)=\frac{\lambda(\tau)}{2}\int\limits_{-1}^{+1}I(\tau,\mu')\chi(\mu,\mu';\tau)\,\mathrm{d}\mu', \tag{1.68}$$

where

$$\chi(\mu,\mu_0;\tau)=\frac{1}{2\pi}\int\limits_{0}^{2\pi}\chi(\gamma,\tau)\,\mathrm{d}\varphi \tag{1.69}$$

is the phase function averaged over the azimuth.

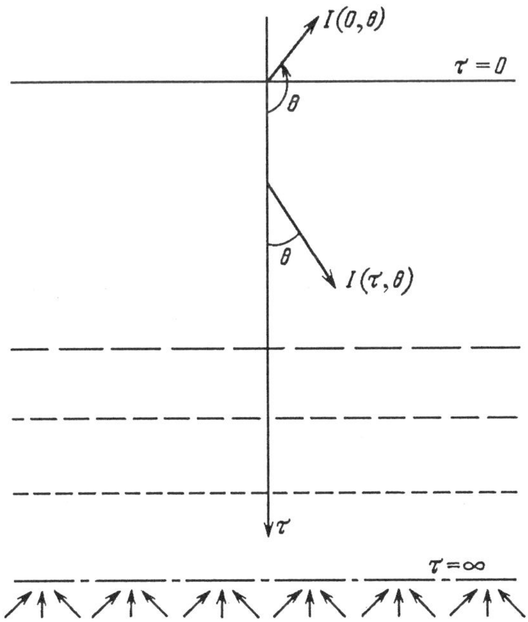

Fig. 1.4. The Milne problem

[2] Generally speaking, one can formally consider the Milne problem for certain numbers of azimuthal harmonics of the intensity. However, this question will not be treated here. For details, see [48].

Finding the nontrivial solution to (1.67–68) under the boundary condition

$$I(0,\mu) = 0 \qquad \text{at } \mu \in (0,1] \tag{1.70}$$

represents the ***Milne problem***. Since this equation is homogeneous, we must specify the normalizing condition for its solution. This will be done somewhat later. For the moment we draw the reader's attention to the following important fact. The Milne problem does not have a solution for any arbitrary assumption about the functions $\lambda(\tau)$ and $\chi(\gamma,\tau)$. For example, let us assume that in the layer of finite optical depth τ_0, $\lambda(\tau) \neq 0$ ($\tau \in [0,\tau_0]$) and $\lambda(\tau) \equiv 0$ at $\tau > \tau_0$. The Milne problem, in this case, is meaningless.

Now we consider another example, more instructive from our standpoint. Let us suppose that in an atmosphere with a geometrical thickness z_0 (respective optical thickness is equal to τ_0) for all $\tau \geq 0$

$$\lambda(\tau) = \mathrm{e}^{-m\tau}, \tag{1.70a}$$

where $m \geq 0$. In such a case we have

$$\int_0^{\tau_0} \lambda(\tau)\,\mathrm{d}\tau = \frac{1}{m}\left(1 - \mathrm{e}^{-m\tau_0}\right),$$

with $\tau_0 = \tau_0^{\mathrm{s}} + \tau_0^{\mathrm{a}}$, where, in accordance with (1.4), τ_0^{s} and τ_0^{a} denote the scattering and absorbing components of the optical thickness, respectively. On the other hand, taking into account (1.5) and (1.29), we find

$$\int_0^{\tau_0} \lambda(\tau)\,\mathrm{d}\tau = \int_0^{z_0} \sigma(z)\,\mathrm{d}z = \tau_0^{\mathrm{s}}.$$

Thus, for an arbitrary $m > 0$ one can write

$$\tau_0^{\mathrm{s}} = \frac{1}{m}\left(1 - \mathrm{e}^{-m\tau_0}\right) \leq \frac{1}{m}.$$

Consequently, in this case the value of τ_0^{s} is finite for *any geometrical thickness* of the atmosphere $z_0 > 0$ (including $z_0 = \infty$).

By this we mean that if $\lambda(\tau)$ is determined by (1.70a), the Milne problem has meaning only when $m \to 0$ (in fact, at $m \ll 1$). Therefore, in what follows, when we discuss the Milne problem for a medium whose optical properties depend on τ, we always assume that *functions $\lambda(\tau)$ and $\chi(\gamma,\tau)$ are such that they enable the existence of a nontrivial solution for this problem.*

For example, let us consider a semi-infinite atmosphere with a layer of arbitrary (but finite) optical thickness τ_0 with functions $\lambda(\tau)$ and $\chi(\gamma,\tau)$ ($\tau \in [0,\tau_0]$)dependent arbitrarily on τ, and at $\tau > \tau_0$ the medium becomes homogeneous with $\lambda \neq 0$. For such a half-space the solution to the Milne problem to all appearances exist. In fact it is possible to model a great variety of real cases using this model.

From (1.67) and (1.70) we have (at $\mu \in [0,1]$):

$$I(\tau,\mu) = \int\limits_0^\tau B(\tau',\mu)\mathrm{e}^{-\frac{\tau-\tau'}{\mu}}\frac{\mathrm{d}\tau'}{\mu}, \tag{1.71}$$

$$I(\tau,-\mu) = \int\limits_\tau^\infty B(\tau',-\mu)\mathrm{e}^{-\frac{\tau'-\tau}{\mu}}\frac{\mathrm{d}\tau'}{\mu}. \tag{1.72}$$

By substituting Eqs (1.71) and (1.72) into (1.68), we arrive at the equation for the source function in the Milne problem:

$$B(\tau,\mu) = \frac{\lambda(\tau)}{2}\int_0^1 \chi(\mu,\mu';\tau)\,\mathrm{d}\mu' \int_0^\tau B(\tau',\mu')\mathrm{e}^{-\frac{\tau-\tau'}{\mu'}}\frac{\mathrm{d}\tau'}{\mu'}$$

$$+\frac{\lambda(\tau)}{2}\int_0^1 \chi(\mu,-\mu';\tau)\,\mathrm{d}\mu' \int_\tau^\infty B(\tau',-\mu')\mathrm{e}^{-\frac{\tau-\tau'}{\mu'}}\frac{\mathrm{d}\tau'}{\mu'}, \quad (\mu \in [-1,+1]). \tag{1.73}$$

In what follows, when we consider this problem, we will often deal with the intensity of radiation that emerges from the atmosphere. We denote this quantity by $u(\mu)$. Following van de Hulst we call it the ***escape function***. It is obvious that

$$u(\mu) = \int\limits_0^\infty B(\tau,-\mu)\mathrm{e}^{-\frac{\tau}{\mu}}\frac{\mathrm{d}\tau}{\mu}, \qquad (\mu \in [0,1]). \tag{1.74}$$

1.7 The Problem for Two-Sided Infinity

Let us imagine that we add to the semi-infinite layer as considered in the Milne problem an other *identical* one. In such an infinite atmosphere we reverse the direction of counting optical depths but keep the origin of the coordinate system at the same place (see Fig. 1.5 and Fig. 1.4). As a result, optical properties of the medium are symmetrical with respect to the reference point, i.e.,

$$\lambda(-\tau) = \lambda(\tau); \qquad \chi(\gamma,\tau) = \chi(\gamma,-\tau), \tag{1.75}$$

and sources of unlimited power are situated at minus infinity. The problem of determining the intensity $I_\infty(\tau,\mu)$ in such a medium is called the ***problem for two-sided infinity***. In this case, (1.67) and (1.68) would remain as before:

$$\mu\frac{\mathrm{d}I_\infty(\tau,\mu)}{\mathrm{d}\tau} + I_\infty(\tau,\mu) = B_\infty(\tau,\mu), \tag{1.76}$$

$$B_\infty(\tau,\mu) = \frac{\lambda(\tau)}{2}\int\limits_{-1}^{+1} I_\infty(\tau,\mu')\chi(\mu,\mu';\tau)\,\mathrm{d}\mu'. \tag{1.77}$$

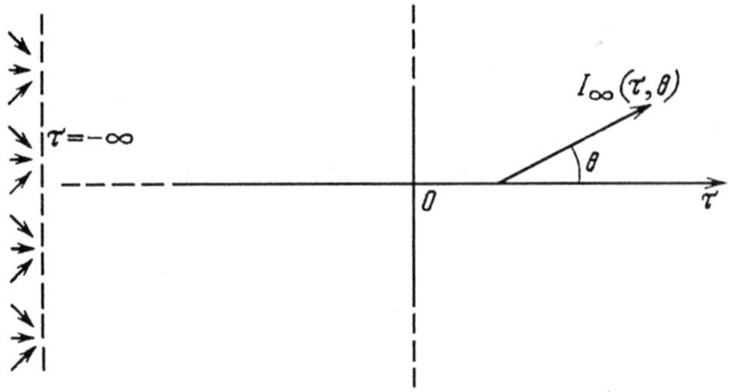

Fig. 1.5. The problem for two-sided infinity

When solving (1.76), the quantity $I_\infty(\tau,\mu)$ is required to be limited at $\tau \to \infty$, and the normalizing condition

$$\frac{\lambda(0)}{2}\int_{-1}^{+1} I_\infty(0,\mu)\,\mathrm{d}\mu = 1. \tag{1.77a}$$

must be satisfied. As follows from (1.76) and (1.77) the integral equation for $B_\infty(\tau,\mu)$ has the form:

$$B_\infty(\tau,\mu) = \frac{\lambda(\tau)}{2}\int_0^1 \chi(\mu,\mu';\tau)\,\mathrm{d}\mu' \int_{-\infty}^{\tau} B_\infty(\tau',\mu')\mathrm{e}^{-\frac{\tau-\tau'}{\mu'}}\frac{\mathrm{d}\tau'}{\mu'}$$
$$+\frac{\lambda(\tau)}{2}\int_0^1 \chi(\mu,-\mu';\tau)\,\mathrm{d}\mu' \int_{\tau}^{\infty} B_\infty(\tau',-\mu')\mathrm{e}^{-\frac{\tau'-\tau}{\mu'}}\frac{\mathrm{d}\tau'}{\mu'}. \tag{1.78}$$

All three problems formulated above (see Sects. 1.5 – 1.7) are closely related. In some instances the solution to our main problem – the parallel external flux problem – can be expressed in terms of solutions to more simple problems, namely, the Milne problem and the problem for two-sided infinity. This will be discussed in detail in subsequent chapters.

1.8 Radiation Flux

An important quantity which characterizes the radiant energy flowing through a unit area per unit time is the ***radiation flux vector***:

$$\boldsymbol{H}(\boldsymbol{r}) = \int_{(4\pi)} \boldsymbol{\omega}' I(\boldsymbol{r},\boldsymbol{\omega}')\,\mathrm{d}\omega'. \tag{1.79}$$

Vectors $\boldsymbol{r}$ and $\boldsymbol{\omega}'$ define the position and orientation of the unit area. If we denote the angle between the vectors $\boldsymbol{\omega}'$ and $\boldsymbol{\omega}$ by β, so that $\cos\beta = \boldsymbol{\omega}\boldsymbol{\omega}'$, the projection of the flux vector $\boldsymbol{H}$ on the direction $\boldsymbol{\omega}$ is

$$H(\boldsymbol{r}) = \int\limits_{(4\pi)} I(\boldsymbol{r}, \boldsymbol{\omega}) \cos\beta \, \mathrm{d}\omega' . \tag{1.80}$$

This quantity is called the ***radiation flux*** in this direction. It has the dimension of $\mathrm{erg \cdot cm^{-2} \cdot s^{-1} \cdot Hz^{-1}}$.

The radiation flux through an arbitrarily oriented area is easily expressed in terms of the fluxes in the directions of coordinate axes x, y, z, whose direction cosines are equal to $\cos\theta_1$, $\cos\theta_2$ and $\cos\theta_3$, respectively. Since vectors $\boldsymbol{\omega}$ and $\boldsymbol{\omega}'$ can be represented as

$$\boldsymbol{\omega} = \boldsymbol{i}\cos\theta_1 + \boldsymbol{j}\cos\theta_2 + \boldsymbol{k}\cos\theta_3,$$

$$\boldsymbol{\omega}' = \boldsymbol{i}\cos\theta_1' + \boldsymbol{j}\cos\theta_2' + \boldsymbol{k}\cos\theta_3',$$

where $\boldsymbol{i}, \boldsymbol{j}, \boldsymbol{k}$ are unit vectors of the coordinate axes, we can write

$$\cos\beta = \boldsymbol{\omega}\boldsymbol{\omega}' = \cos\theta_1\cos\theta_1' + \cos\theta_2\cos\theta_2' + \cos\theta_3\cos\theta_3' . \tag{1.81}$$

As a result, from (1.80) and (1.81) we find

$$H(\boldsymbol{r}) = H_x(\boldsymbol{r})\cos\theta_1 + H_y(\boldsymbol{r})\cos\theta_2 + H_z(\boldsymbol{r})\cos\theta_3, \tag{1.82}$$

where H_x, H_y and H_z are fluxes in the directions of their respective axes, i.e.,

$$H_{x,y,z} = \int\limits_{(4\pi)} I(\boldsymbol{r}, \boldsymbol{\omega}') \cos\theta_{1,2,3}' \, \mathrm{d}\omega' . \tag{1.83}$$

Let us apply the equations obtained above to the parallel external flux problem. We direct the z-axis to be parallel to the τ-axis, and the x-axis is placed in the plane of incidence of the solar radiation (the azimuth for this plane is $\varphi = \varphi_0$). Then, using the spherical coordinate system (θ', φ'), we have

$$\cos\theta_1' = \sin\theta' \cos(\varphi' - \varphi_0), \quad \cos\theta_2' = \sin\theta' \sin(\varphi' - \varphi_0), \quad \cos\theta_3' = \cos\theta' = \mu' . \tag{1.84}$$

We substitute expressions (1.35), (1.47), and (1.84) into (1.83). Taking into account (1.57a), for the total fluxes (i.e., fluxes of direct + diffuse radiation) in the direction of the coordinate axes, we obtain

$$\frac{H_x(\tau, \mu_0)}{4\pi} = \frac{1}{2}\int\limits_{-1}^{+1} I^1(\tau, \mu', \mu_0)\sqrt{1 - \mu'^2}\, \mathrm{d}\mu' + \frac{1}{4}\sqrt{1 - \mu_0^2}\mathrm{e}^{-\tau/\mu_0}, \tag{1.85}$$

$$H_y(\tau, \mu_0) = 0, \tag{1.86}$$

$$\frac{H_\tau(\tau, \mu_0)}{4\pi} = \frac{1}{2}\int\limits_{-1}^{+1} I(\tau, \mu', \mu_0)\mu' \, \mathrm{d}\mu' + \frac{1}{4}\mu_0 \mathrm{e}^{-\tau/\mu_0} . \tag{1.87}$$

Thus, in this case the radiative flux vector lies in the vertical plane with the azimuth $\varphi = \varphi_0$. If the medium scatters isotropically, then $I^1 \equiv 0$. Consequently, the direction of the flux vector in this medium is always coincident with the direction of the τ-axis, and its absolute value is determined by (1.87).

The quantities H_x and H_τ are directly expressed in terms of the components $B_i^m(\tau, \mu_0)$ of the source function. To be certain that this is correct, we substitute expansions (1.56) and (1.64) into (1.50). Comparing the result with (1.85) and (1.87), we obtain

$$H_x(\tau, \mu_0) = \frac{4\pi}{\lambda(\tau)} B_1^1(\tau, \mu_0), \tag{1.88}$$

$$H_\tau(\tau, \mu_0) = \frac{4\pi}{\lambda(\tau)} B_1^0(\tau, \mu_0). \tag{1.89}$$

Besides the flux, the concept of ***illumination*** is used in astrophysics, geophysics etc. Illumination of the horizontal area from above and from below is determined by the formulas

$$\frac{E^{\downarrow}(\tau, \mu_0)}{4\pi} = \frac{1}{2}\int_0^1 I(\tau, \mu', \mu_0)\mu' \, d\mu' + \frac{1}{4}\mu_0 e^{-\tau/\mu_0}, \tag{1.90}$$

$$\frac{E^{\uparrow}(\tau, \mu_0)}{4\pi} = \frac{1}{2}\int_0^1 I(\tau, -\mu', \mu_0)\mu' \, d\mu', \tag{1.91}$$

respectively. In atmospheric physics these quantities are sometimes called the downward and upward flux respectively. From comparison of (1.90), (1.91) and (1.87), as well as from physical considerations, it follows that

$$H_\tau(\tau, \mu_0) = E^{\downarrow}(\tau, \mu_0) - E^{\uparrow}(\tau, \mu_0). \tag{1.92}$$

We emphasize once again that the quantities H_τ and $E^{\downarrow}$ are the ***total flux*** and the ***total illumination***. They include both direct (the second summand in the right-hand side of (1.87) and (1.90)) and diffuse radiation.

1.9 Characteristics of Radiation at the Boundaries of an Atmosphere. The Problem of Diffuse Reflection and Diffuse Transmission of Light

For astrophysical and geophysical applications, the intensities of radiation emerging from the atmosphere are of special interest. These quantities are easily expressed in terms of the source function $B(\tau; \mu, \varphi)$. By setting $\tau = 0$ and $\tau = \tau_0$ in (1.59) and (1.58) respectively, we find ($\mu \in [0, 1]$)

$$I(0, -\mu, \mu_0, \varphi) = \int_0^{\tau_0} B(\tau'; -\mu, \mu_0, \varphi) e^{-\frac{\tau'}{\mu}} \frac{d\tau'}{\mu}, \tag{1.93}$$

$$I(\tau_0, \mu, \mu_0, \varphi) = \int_0^{\tau_0} B(\tau'; \mu, \mu_0, \varphi) e^{-\frac{\tau_0 - \tau'}{\mu}} \frac{d\tau'}{\mu}. \tag{1.94}$$

The dependence of quantities I and B on the parameter μ_0 is specifically indicated here. Instead of the radiation intensities which are defined by (1.93) and (1.94), the ***reflection coefficient*** (ρ) and the ***transmission coefficient*** (σ) are more commonly used. They are given by the formulas

$$I(0, -\mu, \mu_0, \varphi) = \rho(\mu, \mu_0, \varphi)\mu_0, \tag{1.95}$$

$$I(\tau_0, \mu, \mu_0, \varphi) = \sigma(\mu, \mu_0, \varphi)\mu_0, \tag{1.96}$$

respectively. From a comparison of (1.93) and (1.94) with (1.95) and (1.96) we immediately have

$$\rho(\mu, \mu_0, \varphi)\mu_0 = \int_0^{\tau_0} B(\tau; -\mu, \mu_0, \varphi) e^{-\frac{\tau}{\mu}} \frac{d\tau}{\mu}, \tag{1.97}$$

$$\sigma(\mu, \mu_0, \varphi)\mu_0 = \int_0^{\tau_0} B(\tau; \mu, \mu_0, \varphi) e^{-\frac{\tau_0 - \tau}{\mu}} \frac{d\tau}{\mu}. \tag{1.98}$$

Determination of the reflection and transmission coefficients for a plane atmospheric layer is called the ***problem of diffuse reflection and transmission***. When considering a semi-infinite atmosphere ($\tau_0 = \infty$), we can only formulate the problem of diffuse reflection implying the problem of finding the reflection coefficient ρ. It is obvious that the problem of diffuse reflection and transmission is the parallel external flux problem for the boundaries of layer, i.e., it is a particular case of the latter one.

Since the radiation intensity is expanded in Fourier series in azimuthal variable (see (1.47)), the reflection and transmission coefficients can be written in the same manner. They have the following form:

$$\rho(\mu, \mu_0, \varphi - \varphi_0) = \rho(\mu, \mu_0) + 2 \sum_{m=1}^{\infty} \rho^m(\mu, \mu_0) \cos m(\varphi - \varphi_0), \tag{1.99}$$

$$\sigma(\mu, \mu_0, \varphi - \varphi_0) = \sigma(\mu, \mu_0) + 2 \sum_{m=1}^{\infty} \sigma^m(\mu, \mu_0) \cos m(\varphi - \varphi_0). \tag{1.100}$$

We denote here $\rho^0(\mu, \mu_0) = \rho(\mu, \mu_0)$, $\sigma^0(\mu, \mu_0) = \sigma(\mu, \mu_0)$.

An important property of the reflection and transmission coefficients are the ***reciprocity relations***. Namely, for a homogeneous layer of the medium, it turns out that

$$\left.\begin{aligned} \rho(\mu, \mu_0, \varphi - \varphi_0) &= \rho(\mu_0, \mu, \varphi_0 - \varphi), \\ \sigma(\mu, \mu_0, \varphi - \varphi_0) &= \sigma(\mu_0, \mu, \varphi_0 - \varphi). \end{aligned}\right\} \tag{1.100a}$$

The first of these relations is also valid for an inhomogeneous plane atmosphere; however, in this case the second one appears to be somewhat more complicated (for more details, see Sect. 1.11).

For the atmospheric boundaries, (1.90) and (1.91) take the form

$$\frac{E^{\downarrow}(\tau_0,\mu_0)}{4\pi\mu_0} = \frac{1}{2}\int_0^1 \sigma(\mu',\mu_0)\mu'\,\mathrm{d}\mu' + \frac{1}{4}\mathrm{e}^{-\tau_0/\mu_0}, \tag{1.101}$$

$$\frac{E^{\uparrow}(0,\mu_0)}{4\pi\mu_0} = \frac{1}{2}\int_0^1 \rho(\mu',\mu_0)\mu'\,\mathrm{d}\mu', \tag{1.102}$$

respectively. The first of these equations determines the illumination of the lower atmospheric boundary from above by direct as well as diffuse radiation, whereas the second one gives the illumination of the top boundary from below by diffuse radiation. The illumination E_0 of the top atmospheric boundary by the direct solar radiation is defined by the quantity $E_0 = \pi\mu_0$ for the normalization chosen (see Sect. 1.5). In this case the ratio

$$\frac{E^{\uparrow}(0,\mu_0)}{E_0} = A(\mu_0) \tag{1.103}$$

is the so-called ***plane albedo*** of the atmosphere. It follows from (1.102) and (1.103) that

$$A(\mu_0) = 2\int_0^1 \rho(\mu',\mu_0)\mu'\,\mathrm{d}\mu'. \tag{1.104}$$

This quantity characterizes the reflectivity of the planetary atmosphere for given conditions of illumination (i.e., when μ_0 is known). Furthermore, we need the ratio $E^{\downarrow}(\tau_0,\mu_0)/E_0 = V(\tau_0,\mu_0)$. It follows from (1.101) that

$$V(\mu_0,\tau_0) = 2\int_0^1 \sigma(\mu',\mu_0)\mu'\,\mathrm{d}\mu' + \mathrm{e}^{-\tau_0/\mu_0}. \tag{1.105}$$

In addition to the plane albedo, the ***spherical albedo*** A_s is used in planetary studies. It represents the ratio of the energy reflected by the planet to the solar energy falling on it. It is easy to show (LSPA, Sect. 1.5) that

$$A_\mathrm{s} = 4\int_0^1 \mu\,\mathrm{d}\mu \int_0^1 \rho(\mu,\mu')\mu'\,\mathrm{d}\mu'. \tag{1.106}$$

The difference $(1-A_\mathrm{s})$ integrated over all wavelengths determines the fraction of the radiant energy which is absorbed by the planetary atmosphere or in a more general case, also by the planetary surface. The importance of knowing the spherical albedo lies, in particular, in the fact that this quantity ultimately determines the global energetics of the atmosphere and, consequently, the dynamical processes in it.

Finally the ***geometrical albedo*** A_g plays an important role in planetary physics. This is considered to be the ratio of the illumination from the planet in full phase (i.e., when the whole illuminated hemisphere of the planet is visible from the Earth) to the illumination produced by a plane, absolutely white surface of the same radius as the planet placed at the same position. The surface is assumed to reflect light according to Lambert's law. It can be shown (see LSPA, Sect. 9.2) that the geometrical albedo is expressed in terms of the reflection coefficient $\rho(\mu, \mu_0, \varphi)$ as follows:

$$A_g = 2 \int_0^1 \rho(\mu, \mu, \pi) \mu^2 \, d\mu. \tag{1.107}$$

For the outer planets (Mars, Jupiter, Saturn and so on) the quantity A_g can be determined *directly* from terrestrial observations.

1.10 The Flux Integral and *K* Integral

A variety of integral relations play an important role in solving the radiative transfer equation. They will be discussed in detail in the subsequent chapters of this book. The so-called ***flux integral*** and K ***integral*** are particular well-known cases of these relations. Considering the importance of these integrals, we will obtain them now, prior to deriving the more general integral relations mentioned above.

We multiply both parts of the transfer equation (1.41) by an elementary solid angle $d\omega$ and integrate over all directions. Taking into account (1.42) and (1.87) and the normalizing condition for the phase function (1.6), we obtain

$$\frac{1}{4\pi} \frac{dH_\tau(\tau, \mu_0)}{d\tau} = -[1 - \lambda(\tau)] J(\tau, \mu_0), \tag{1.108}$$

where

$$J(\tau, \mu_0) = \frac{1}{2\pi} \int_{-1}^{+1} I(\tau, \mu', \mu_0) \, d\mu' + \frac{e^{-\tau/\mu_0}}{4\pi} \tag{1.109}$$

is the ***mean intensity of the direct + diffuse radiation***. If in (1.108) we convert the optical depth τ to the geometrical depth z, using (1.29), and taking into account (1.5), then, instead of (1.108), we have

$$\frac{1}{4\pi} \frac{dH_\tau(\tau, \mu_0)}{dz} / J(\tau, \mu_0) = -\kappa(z). \tag{1.110}$$

This relation represents a remarkable example of a solution in explicit form to one of the ***inverse problems*** of atmospheric optics. Indeed, from the data of optical probing of the atmosphere we can find, in principle, the rate of change of the radiation flux with depth z as well as the mean intensity, to determine immediately the z-dependence of the volume absorption coefficient.

Let us return, however, to (1.108). Let the atmosphere be conservatively scattering at any depth ($\lambda(\tau) \equiv 1$). Then, instead of (1.108), we have

$$\frac{\mathrm{d}H_{0\tau}(\tau, \mu_0)}{\mathrm{d}\tau} = 0, \tag{1.111}$$

while it follows that

$$H_{0\tau}(\tau, \mu_0) = C, \tag{1.112}$$

where C is an arbitrary integration constant, and the subscript "0" denotes that this is the conservative case.

Relation (1.112) represents one of the desired integrals, the ***flux integral.*** It expresses the constancy of the total radiation flux in a purely scattering atmosphere, including an inhomogeneous one.

Next, we multiply both parts of (1.41) by $\mu\,\mathrm{d}\omega$ and integrate over all directions. Taking into account that from (1.56) at $m = 0$

$$\frac{1}{2}\int\limits_{-1}^{+1} \chi(\mu, \mu'; \tau)\mu'\,\mathrm{d}\mu' = \frac{1}{3}x_1(\tau)\mu, \tag{1.113}$$

and using (1.42) and (1.87), we obtain

$$\frac{\mathrm{d}K(\tau, \mu_0)}{\mathrm{d}\tau} = -\left[1 - \frac{\lambda(\tau)x_1(\tau)}{3}\right] H_\tau(\tau, \mu_0). \tag{1.114}$$

We introduce here the notation

$$\frac{K(\tau, \mu_0)}{4\pi} = \frac{1}{2}\int\limits_{-1}^{+1} I(\tau, \mu, \mu_0)\mu^2\,\mathrm{d}\mu + \frac{1}{4}\mu_0^2 \mathrm{e}^{-\tau/\mu_0}. \tag{1.115}$$

We set $\lambda(\tau) \equiv 1$ in (1.114) and substitute the result into (1.112). As a consequence, after integration we have

$$\frac{K_0(\tau, \mu_0)}{4\pi} = -C\tau\left[1 - \frac{\bar{x}_1(\tau)}{3}\right] + C_1, \tag{1.116}$$

where

$$\bar{x}_1(\tau) = \frac{1}{\tau}\int\limits_0^\tau x_1(t)\,\mathrm{d}t, \tag{1.117}$$

and C_1 is one more arbitrary constant.

The relationship (1.116) gives us another important integral, namely the so-called ***K* integral**. The integration constants C and C_1 are usually determined from boundary conditions, i.e., the radiation intensity on the boundaries is considered to be known. For more details, see LSPA, Sect. 1.6 and Sect. 3.11.5 of this book.

1.11 Green Function and Reciprocity Relations

The Green function $G(\tau, \boldsymbol{\omega}; \tau_1, \boldsymbol{\omega}_1)$ of the radiative transfer equation obeys, by definition, the following equation:

$$\mu \frac{\partial G(\tau, \boldsymbol{\omega}; \tau_1, \boldsymbol{\omega}_1)}{\partial \tau} + G(\tau, \boldsymbol{\omega}; \tau_1, \boldsymbol{\omega}_1)$$
$$= \frac{\lambda(\tau)}{4\pi} \int_{(4\pi)} \chi(\boldsymbol{\omega}\boldsymbol{\omega}', \tau) G(\tau, \boldsymbol{\omega}'; \tau_1, \boldsymbol{\omega}_1)\, \mathrm{d}\omega' + 2\pi\delta(\tau - \tau_1)\delta(\boldsymbol{\omega} - \boldsymbol{\omega}') \quad (1.118)$$

with the boundary conditions

$$\left.\begin{aligned} G(0, \boldsymbol{\omega}; \tau_1, \boldsymbol{\omega}_1) &= 0 \qquad \text{at } \mu > 0, \\ G(\tau_0, \boldsymbol{\omega}; \tau_1, \boldsymbol{\omega}_1) &= 0 \qquad \text{at } \mu < 0, \end{aligned}\right\} \quad (1.119)$$

where $\tau, \tau_1 \in (0, \tau_0)$. Consequently, as seen from (1.118), the Green function G for a plane anisotropically scattering inhomogeneous medium is the intensity of radiation which propagates at the depth $\tau \geq 0$ in the direction $\boldsymbol{\omega} = (\mu, \varphi)$. The medium contains a plane unidirectional source placed at the depth τ_1 which radiates in the direction $\boldsymbol{\omega}_1 = (\mu_1, \varphi_1)$.

The Green function satisfies the ***reciprocity relation*** which expresses the invariance of the radiation field when the source and the detector of photons are interchanged. Specifically,

$$G(\tau, \boldsymbol{\omega}; \tau_1, \boldsymbol{\omega}_1) = G(\tau_1, -\boldsymbol{\omega}_1; \tau, -\boldsymbol{\omega}). \quad (1.120)$$

This relation is a consequence of the invariance of the multiple light scattering process described by the transfer equation with respect to the time inversion. It is clear that the probability of realizing one or other photon trajectory does not depend on the direction of motion (forward or backward) of the photon along this trajectory. On this basis, considering the transition of a photon from the point $(\tau_1, \boldsymbol{\omega}_1)$ of the phase space to the point $(\tau, \boldsymbol{\omega})$ along all possible trajectories, we obtain the reciprocity relation. A rigorous formal proof of the reciprocity relation is given, for instance, in [45], Sect. 2.7.

The ***surface Green functions*** $G(\tau, \boldsymbol{\omega}; 0, \boldsymbol{\omega}_1)$ and $G(\tau, \boldsymbol{\omega}; \tau_0, \boldsymbol{\omega}_1)$ are related to the diffuse radiation intensity as follows ($\mu_1 \geq 0$):

$$G(\tau, \boldsymbol{\omega}; 0, \boldsymbol{\omega}_1)\mu_1 = 2I(\tau, \boldsymbol{\omega}, \boldsymbol{\omega}_1) + 2\pi\delta(\mu - \mu_1)\delta(\varphi - \varphi_1)\mathrm{e}^{-\frac{\tau}{\mu_1}}, \quad (1.121)$$

$$G(\tau,-\boldsymbol{\omega};\tau_0,-\boldsymbol{\omega}_1)\mu_1 = 2\tilde{I}(\tau,-\boldsymbol{\omega},\boldsymbol{\omega}_1)+2\pi\delta(\mu-\mu_1)\delta(\varphi-\varphi_1)\mathrm{e}^{-\frac{\tau_0-\tau}{\mu_1}}, \quad (1.122)$$

where $\tilde{I}$ denotes the radiation intensity in the parallel external flux problem when a layer is illuminated on the $\tau = \tau_0$ side, i.e., *from below.* It is obvious that if the medium is *homogeneous*, then $\tilde{I}(\tau_0 - \tau, -\boldsymbol{\omega}, \boldsymbol{\omega}_1) = I(\tau, \boldsymbol{\omega}, \boldsymbol{\omega}_1)$.

By analogy with the definitions of (1.95) and (1.96) we introduce the definitions for the reflection and transmission coefficients with the expressions:

$$\left.\begin{aligned} I(0,-\boldsymbol{\omega},\boldsymbol{\omega}_1) &= \rho(\boldsymbol{\omega},\boldsymbol{\omega}_1)\mu_1, \\ I(\tau_0,\boldsymbol{\omega},\boldsymbol{\omega}_1) &= \sigma(\boldsymbol{\omega},\boldsymbol{\omega}_1)\mu_1, \\ \tilde{I}(\tau_0,\boldsymbol{\omega},\boldsymbol{\omega}_1) &= \tilde{\rho}(\boldsymbol{\omega},\boldsymbol{\omega}_1)\mu_1, \\ \tilde{I}(0,-\boldsymbol{\omega},\boldsymbol{\omega}_1) &= \tilde{\sigma}(\boldsymbol{\omega},\boldsymbol{\omega}_1)\mu_1. \end{aligned}\right\} \quad (1.123)$$

Then, using (1.121–123) and the reciprocity relation (1.120), we arrive at the following reciprocity relations:

$$\left.\begin{aligned} \rho(\boldsymbol{\omega},\boldsymbol{\omega}_1) &= \rho(\boldsymbol{\omega}_1,\boldsymbol{\omega}), \\ \tilde{\rho}(\boldsymbol{\omega},\boldsymbol{\omega}_1) &= \tilde{\rho}(\boldsymbol{\omega}_1,\boldsymbol{\omega}), \\ \sigma(\boldsymbol{\omega},\boldsymbol{\omega}_1) &= \tilde{\sigma}(\boldsymbol{\omega}_1,\boldsymbol{\omega}), \end{aligned}\right\} \quad (1.124)$$

which play a very important role when the problem of diffuse reflection and transmission is being solved. We will use them throughout the book. The reader can find further details in RT, Sects. 13, 31 and 52, as well as in MLS, Chap. 3.

As follows from the definition, the Green function is the perfect tool for finding solutions to the *general problem*, which was formulated at the end of Sect. 1.4. We return to (1.30), and denote

$$\frac{\epsilon_0(t,\theta,\varphi)}{\alpha(t)} \equiv g(t,\boldsymbol{\omega}).$$

The function $g(t,\boldsymbol{\omega}_1)\mathrm{d}t$ is the amount of radiant energy emitted by *internal* sources lying at the depth t in an elementary volume with a unit cross-section and optical thickness $\mathrm{d}t$, this radiation being emitted in the direction $\boldsymbol{\omega}_1$ during the unit time interval. Then the solution of the general problem, i.e., finding the radiation intensity $I_\mathrm{s}(\tau,\boldsymbol{\omega})$ in a medium, is given by the expression

$$I_\mathrm{s}(\tau,\boldsymbol{\omega}) = \frac{1}{2\pi}\int_0^{\tau_0} g(t,\boldsymbol{\omega}')\,\mathrm{d}t \int_{(4\pi)} G(\tau,\boldsymbol{\omega};t,\boldsymbol{\omega}')\,\mathrm{d}\omega'. \quad (1.125)$$

This radiation originates from the internal sources. Thus, the knowledge of the Green function gives the complete solution to the general problem. However, it will be shown later for the case of a homogeneous atmosphere

(Sects. 3.12.5 and 4.10) and a multilayer atmosphere (Sect. 6.10) that by making some general assumptions, finding the function $I_s(\tau, \omega)$ can be reduced to solving the parallel external flux problem, a much simpler problem.

1.12 Invariance Principles

The concept of invariance, that is, the invariability of processes when certain operations are being done, plays a very important role in physics. The statement that something remains unchanged when certain operations are being done is called the ***invariance principle***. The corollaries of such a principle, sometimes called ***invariance relations***, are usually of less wide applicability. The significance of invariance principles can be put down to the fact that often enough they allow us to obtain very important (and, in certain cases, apparently very complicated) relations without applying more routine mathematical methods which would, as a rule, involve a lot of technical problems.

It was V.A. Ambartsumian who introduced invariance principles into transfer theory. When we consider the diffuse reflection of light from a semi-infinite homogeneous atmosphere, ***the Ambartsumian's invariance principle*** expresses the following obvious fact: the addition to such an atmosphere of a layer with an arbitrary thickness and the same optical properties does not result in any change in the intensity of the diffusely reflected radiation. Therefore, we can obtain an integral equation defining the intensity of outgoing radiation in the following way. Assuming the layer to be infinitely thin we calculate all changes caused by the layer and then set them equal to zero (see Sect. 3.5)

To consider a plane layer of finite thickness, Ambartsumian generalized his invariance principle as follows: if a layer of optical thickness $\Delta\tau$ is added to one side of the medium and the same layer is removed from the other side, then there are no changes in the radiation intensity at the atmospheric boundaries.

Now one can introduce the following definition, which we will call the ***generalized invariance principle***.

Let an arbitrary plane atmosphere $\mathcal{A}$ (infinite, semi-infinite or finite) include arbitrarily distributed sources whose power depends on only one spatial coordinate, τ. (The optical properties of the medium are assumed also to depend on only one spatial coordinate, namely τ). Single out two levels in $\mathcal{A}$, τ_1 and τ_2, and consider the atmosphere $\mathcal{B}$ lying between them (see Fig. 1.6a).

It is obvious that, for each $\tau \in [\tau_1, \tau_2]$, the intensity of radiation (or source function) in $\mathcal{A}$ is, as the same time, the intensity of radiation (or source function) in the atmosphere $\mathcal{B}$ the boundaries of which are illuminated by the intensity of total (i.e., direct + diffuse) radiation entering $\mathcal{A}$.

It is important here that the above definition of the general invariance principle is also suitable for an *inhomogeneous atmosphere*. Apparently, this

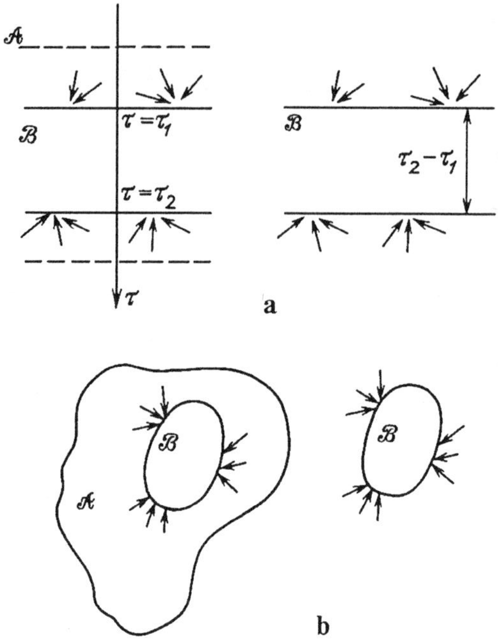

Fig. 1.6. The generalized invariance principle

is the reason for some remarkable similarities between results obtained for homogeneous and inhomogeneous atmospheres.

Note the following important fact. The above definition of the generalized invariance principle is, broadly speaking, applicable to an atmosphere of *arbitrary* geometry. Indeed, if the atmosphere $\mathcal{A}$ has an arbitrary shape and geometry and the convex domain $\mathcal{B}$ is a part of $\mathcal{A}$, then the radiation field in $\mathcal{B}$ is obviously *invariant* when $\mathcal{B}$ is taken out of $\mathcal{A}$ (the translation of $\mathcal{B}$) and the sources at the boundary of $\mathcal{B}$ are simultaneously switched on. It is supposed that the intensity of the sources equal to the radiation intensity in $\mathcal{A}$ at the boundary of the domain $\mathcal{B}$ (see Fig. 1.6b). The point is that the atmosphere $\mathcal{B}$ "knows" about the existence of $\mathcal{A}$ only due to the radiation leaving $\mathcal{A}$ through the boundary of the domain $\mathcal{B}$.

The physical considerations constituting the generalized invariance principle are, of course, rather simple and have been used by many researchers to study various problems in the theory of radiative transfer in inhomogeneous media. Nonetheless, the importance here is that the application of this principle allows us to obtain directly the invariance relations from which the solutions of some problems of the transfer theory follow. The reader will have a number of opportunities to see it since the generalized invariance principle will be used frequently in this book.

Part I

HOMOGENEOUS ATMOSPHERE

2. Radiation Field in an Infinite Atmosphere

We embark this chapter with the simple problem for two-sided infinity, previously formulated in Sect. 1.7. Let us recall its formulation for a homogeneous atmosphere.

We are given the infinite plane medium whose optical properties are characterized by a single scattering albedo λ and a phase function $\chi(\gamma)$. "Priming" sources of radiation of unlimited output are placed at minus infinity (see Fig. 1.5). As follows from (1.76–77a), the solution of this problem reduces to that of transfer equation

$$\mu \frac{\mathrm{d}I_\infty(\tau,\mu)}{\mathrm{d}\tau} + I_\infty(\tau,\mu) = \frac{\lambda}{2}\int_{-1}^{+1} I_\infty(\tau,\mu')\chi(\mu,\mu')\,\mathrm{d}\mu' \tag{2.1}$$

under the normalizing condition

$$\frac{\lambda}{2}\int_{-1}^{+1} I_\infty(0,\mu)\,\mathrm{d}\mu = 1. \tag{2.2}$$

The boundary condition will be discussed below.

This problem is of importance not just due to its simplicity. The key point is that a knowledge of the radiation field in infinite media facilitates essentially a study of the radiation field in deep layers of a semi-infinite medium, and in atmospheres of large optical thickness. All of this will be thoroughly discussed later. Let us now deal with the simplest case.

2.1 Conservatively Scattering Atmosphere

Let us put $\lambda = 1$ in (2.1) and (2.2). Then, taking into account the normalizing condition (see (1.7)) for the phase function

$$\frac{1}{2}\int_{-1}^{+1} \chi(\mu,\mu')\,\mathrm{d}\mu' = 1, \tag{2.3}$$

one readily finds $I_\infty(\tau,\mu) \equiv 1$.

Thus, treating the conservatively scattering infinite atmosphere with *arbitrary* phase function and sources placed at minus infinity, the radiation field does not depend on the spatial coordinate τ, and, moreover, it is *isotropic*. For example when flying through dense clouds[1], particularly in winter when land is covered with snow, one can not sensibly determine the direction of light sources by brightness. This situation can be modelled quite well by the problem for two-sided infinity with conservative scattering.

2.2 General Case

Let us assume $\lambda \in [0,1)$. Since the medium absorbs, we can demand from the physical considerations that

$$I_\infty(\tau,\mu) \to 0 \quad \text{at} \quad \tau \to +\infty. \tag{2.4}$$

Taking the boundary condition into account, we seek the solution of (2.1) in the following form:

$$I_\infty(\tau,\mu) = i(\mu)\mathrm{e}^{-k\tau}, \tag{2.5}$$

where k is positive. Substituting (2.5) into (2.1), we obtain

$$i(\mu)(1-k\mu) = \frac{\lambda}{2}\int\limits_{-1}^{+1} i(\mu')\chi(\mu,\mu')\,\mathrm{d}\mu' \tag{2.6}$$

and, instead of (2.2), we have

$$\frac{\lambda}{2}\int\limits_{-1}^{+1} i(\mu)\,\mathrm{d}\mu = 1. \tag{2.7}$$

The function $i(\mu)$ represents the radiation intensity at the boundary of two half-spaces, found by solving (2.6). Substitution of this solution into (2.7) results in the ***characteristic equation*** for determining k.

For example, under the isotropic scattering, we find from (2.6) and (2.7):

$$i(\mu) = (1-k\mu)^{-1}. \tag{2.8}$$

Substitution of (2.8) into (2.7) gives the characteristic equation

$$\frac{\lambda}{2k}\ln\frac{1+k}{1-k} = 1 \quad \text{or} \quad k = \tanh\frac{k}{\lambda}. \tag{2.9}$$

[1] It is well known that droplets forming clouds do not absorb light in the visible region of the spectrum.

k is commonly called the ***diffusion exponent***. Physically, it determines the relative rate of attenuation of the radiation flux (or intensity) in an infinite medium. In fact, from (1.87) and (2.5), we have the flux $H_\infty(\tau)$:

$$H_\infty(\tau) = \frac{4\pi}{\lambda} c_1 \mathrm{e}^{-k\tau}, \tag{2.10}$$

where

$$c_1 = \frac{\lambda}{2} \int_{-1}^{+1} i(\mu)\mu \, \mathrm{d}\mu. \tag{2.11}$$

Hence, we find that $k = -H'_\infty(\tau)/H_\infty(\tau)$.

From (2.6) and (2.7) it follows that first, if k is the solution of the characteristic equation, then $-k$ obeys this equation as well, and second, since the radiation intensity is positive, $|k| < 1$.

The problem for two-sided infinity is the typical eigenvalue problem. It appears that in the general case of an arbitrary phase function, the characteristic equation has a discrete spectrum of eigenvalues k_j $(j = 1, 2, \ldots, n)$ in the interval $[-1, +1]$. They are symmetrical about zero. The set of eigenfunctions (or ***modes***) $i_j(\mu)$ corresponds to this spectrum. In the case of conservative scattering $(\lambda = 1)$, the characteristic equation always has the root $k = 0$.

In what follows, we will only be interested in the minimal root of the characteristic equation in the interval $[0, 1]$, i.e., the root $k_1 \equiv k$, and the related first eigenfunction $i_1(\mu) \equiv i(\mu)$ (or, as is sometimes said, the ***principal mode*** of the radiation intensity). The reason is that other modes will be attenuated relatively fast in so far as $k_1 < k_2 < \ldots < k_n$. These modes will not usually concern us in our study. Notice that the characteristic equation (2.9) for isotropic scattering has a single root in the interval $[0, 1]$, and, consequently, the eigenfunction (2.8) is unique.

2.3 Characteristic Equation and Method of Its Solution

We expand the function $i(\mu)$ in a series of Legendre polynomials $P_n(\mu)$. Then, by substituting the expansion into (2.6) and using the recursion formula

$$(n+1)P_{n+1}(\mu) + nP_{n-1}(\mu) = (2n+1)P_n(\mu)\mu \tag{2.12}$$

and expression (see (1.56))

$$\chi(\mu, \mu_0) = \sum_{j=0}^{\infty} x_j P_j(\mu) P_j(\mu_0), \tag{2.13}$$

we obtain

$$i(\mu) = \frac{1}{\lambda} \sum_{j=0}^{\infty} (2j+1) c_j P_j(\mu). \tag{2.14}$$

The coefficients

$$c_j = \frac{\lambda}{2}\int\limits_{-1}^{+1} i(\mu)P_j(\mu)\,\mathrm{d}\mu \tag{2.15}$$

obey the recursion relation

$$(j+1)c_{j+1} + jc_{j-1} = \frac{c_j}{k}(2j+1-\lambda x_j), \tag{2.16}$$

with $c_0 = 1$ as follows from (2.7) and (2.15). The quantity c_1 is equal to

$$c_1 = \frac{1-\lambda}{k} \tag{2.17}$$

and determines the radiation flux at the half-space interface, as seen from (2.10).

If we introduce

$$r_j = c_j/c_{j-1} \qquad (j = 1, 2, 3, \ldots), \tag{2.18}$$

then from (2.16), we have

$$(j+1)r_{j+1} = \frac{2j+1-\lambda x_j}{k} - \frac{j}{r_j}, \tag{2.19}$$

or

$$r_j = \frac{jk}{2j+1-\lambda x_j - (j+1)kr_{j+1}}. \tag{2.20}$$

Taking into account that $r_1 = (1-\lambda)/k$, we find from (2.20) the characteristic equation in the form of a continued fraction:

$$1-\lambda = \cfrac{k^2}{3-\lambda x_1 - \cfrac{4k^2}{5-\lambda x_2 - \cfrac{9k^2}{7-\lambda x_3 - \ldots}}}. \tag{2.21}$$

Its solution gives the diffuse exponent k.

Let us assume that the limit $\lim_{j\to\infty} r_j = r$ exists. Then, from (2.19) under the assumption that $x_j/j \to 0$ with $j \to \infty$, we find

$$r^2 - 2\frac{r}{k} + 1 = 0 \tag{2.22}$$

whence

$$r = \frac{1-\sqrt{1-k^2}}{k} = \frac{k}{1+\sqrt{1-k^2}}. \tag{2.23}$$

It is not difficult to obtain the asymptotic formula for the quantity r_n when $n \gg 1$. Using (2.19) and (2.23) and representing $r_n = r(1+a/n+b/n^2+\ldots)$, we readily obtain

$$r_n = r\left[1 - \frac{1}{2n}\left(1 - \frac{\lambda x_{n-1}}{\sqrt{1-k^2}}\right) + \frac{k}{8n^2 r\sqrt{1-k^2}}\left(1 - \frac{\lambda x_{n-1}}{\sqrt{1-k^2}}\right)^2 + O\left(\frac{1}{n^3}\right)\right]. \tag{2.24}$$

The characteristic equation (2.21) is typical to determine k. However, as k tends to one, the convergence of this continued fraction becomes poor. In this case, the following trick may be used to determine k. Let $k^{(1)}$ be the value of k found to the first order approximation via (2.21) for example. Then, setting $j = n - 1 \gg 1$ in (2.20) and using (2.24) for finding r_n, we calculate $r_{n-1}, r_{n-2}, \ldots$ by means of (2.20) ("downward descent") until we find r_2 and $r_1 = (1-\lambda)/k^{(2)}$, where $k^{(2)}$ is the value of k to the second order. Hence, from (2.20), we have

$$\left[k^{(2)}\right]^2 + 2r_2(1-\lambda)k^{(2)} - (1-\lambda)(3-\lambda x_1) = 0,$$

or explicitly,

$$k^{(2)} = \sqrt{r_2^2(1-\lambda)^2 + (1-\lambda)(3-\lambda x_1)} - r_2(1-\lambda).$$

Using this new refined value of k, we repeat all the steps from the beginning until this procedure converges.

Values of k for the Henyey-Greenstein phase function (1.15) and different values of parameter g are given in Table I in the Appendix. Values of k for the Rayleigh phase function (1.9) indicated by R, are presented in the same Table.

When calculating the eigenvalues k_j ($j = 1, 2, \ldots,$), it is important to keep the following in mind. Let us designate the coefficient c_j involved in (2.16) by $c_j(1/k, \lambda)$. Then, it may be shown that the number of roots n of the characteristic equation in the interval $[0,1]$ is equal to the number of roots in the interval $[0, \lambda]$ for the equation

$$\Psi(1, \lambda) = 0, \tag{2.25}$$

where

$$\Psi(\mu, \lambda) = \frac{\lambda}{2}\sum_{l=0}^{\infty} x_l c_l(\mu, \lambda) P_l(\mu) \tag{2.26}$$

is the ***characteristic function***. This important function is discussed in more detail in RT, Sect. 37 and, particularly, in LSPA, Chap. 5.

When the diffusion exponent is obtained, the coefficient c_j is readily calculated with the help of the recursion relation (2.12), then we find the radiation intensity $i(\mu)$ using (2.14). In this way, the solution of the problem is completed. We note that the function $i(\mu)$ can be represented in alternative form:

$$i(\mu) = \frac{\sum_{j=0}^{\infty} x_j c_j P_j(\mu)}{1 - k\mu}. \tag{2.27}$$

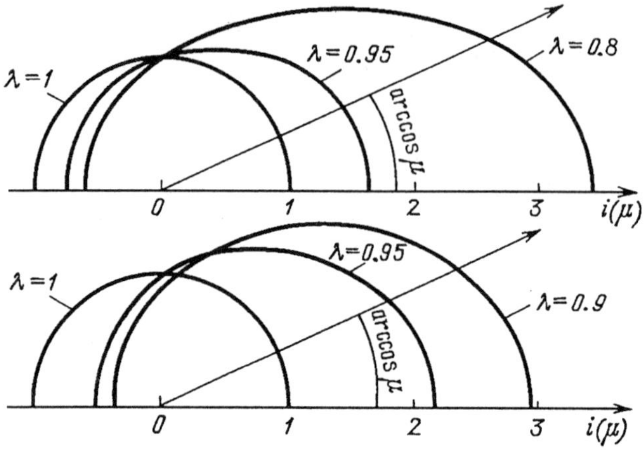

Fig. 2.1. Function $i(\mu)$ at different λ for the Henyey-Greenstein phase function (1.15). Isotropic scattering $g = 0$ (top figure); $g = 0.75$ (bottom figure)

This expression is a direct consequence of (2.6), (2.13), and (2.15). Equation (2.27) is better suited for calculating $i(\mu)$, since $x_j \to 0$ at $j \to \infty$, and consequently, fewer terms are added than if (2.14) is used.

Polar diagrams of the radiation intensity $i(\mu)$ are shown in Fig. 2.1 for two phase functions: $g = 0$ (isotropic scattering, upper figure) and $g = 0.75$, at various values of λ. It is seen that the degree of anisotropy of the radiation field grows as: 1) forward elongation of the phase function increases; 2) absorption in the medium increases (parameter λ decreases).

2.4 Normalization Constant M

As will be elucidated in the next chapter, we often need the quantity

$$M = 2\int_{-1}^{+1} i^2(\mu)\mu \, \mathrm{d}\mu. \tag{2.28}$$

Inasmuch as function $i(\mu)$ may be considered to be known, the above-mentioned constant can be found using direct integration. However, the quantity M can be represented as a quadratic form of the coefficients c_j which are determined by the recurrence relation (2.16). It is possible to obtain at least four different representations for M. Let us derive them.

Instead of one of the functions $i(\mu)$ we substitute in (2.28) the expression (2.14). Taking into account (2.12) and (2.15), we obtain

$$M = \frac{4}{\lambda^2}\sum_{j=0}^{\infty}[(j+1)c_{j+1} + jc_{j-1}]c_j. \tag{2.29}$$

Using (2.16), we find from (2.29) that

$$M = \frac{4}{\lambda^2 k} \sum_{j=0}^{\infty} (2j + 1 - \lambda x_j) c_j^2. \tag{2.30}$$

Let us multiply both parts of (2.16) by c_j and perform the summation over all j. As a result we have,

$$\frac{1}{k} \sum_{j=0}^{\infty} (2j + 1 - \lambda x_j) c_j^2 = \sum_{j=0}^{\infty} (j+1) c_j c_{j+1} + \sum_{j=0}^{\infty} j c_j c_{j-1} = 2 \sum_{j=0}^{\infty} (j+1) c_j c_{j+1}.$$

Thus, considering (2.30), we obtain

$$M = \frac{8}{\lambda^2} \sum_{j=0}^{\infty} (j+1) c_j c_{j+1}. \tag{2.31}$$

The right-hand side of (2.31) may finally be rewritten as

$$\sum_{j=0}^{\infty} (j+1) c_j c_{j+1} = [c_1(c_0 + 2c_2) + c_3(3c_2 + 4c_4) + c_5(5c_4 + 6c_6) + \ldots].$$

Then, using (2.16), we find

$$M = \frac{8}{\lambda^2 k} \sum_{j=0}^{\infty} (4j + 3 - \lambda x_{2j+1}) c_{2j+1}^2. \tag{2.32}$$

Finally, representing the right-hand side of (2.30) as two sums where summation is performed separately over even and odd indices, and using (2.32), we find

$$M = \frac{8}{\lambda^2 k} \sum_{j=0}^{\infty} (4j + 1 - \lambda x_{2j}) c_{2j}^2. \tag{2.33}$$

It can easily be seen that $M \to 0$ if $\lambda \to 1$ and $M \to \infty$ if $\lambda \to 0$. The numerical values of M for the Henyey-Greenstein and Rayleigh phase functions are given in Table II in the Appendix. Formula (2.32) is used to calculate these values. This formula contains the most rapidly convergent series compared to the other representations of M. The first term on the right-hand side of (2.32), namely

$$M \simeq \frac{8(1-\lambda)^2(3 - \lambda x_1)}{\lambda^2 k^3}, \tag{2.34}$$

gives the value $M = 20.2$, even for $\lambda = 0.7$; $g = 0.9$ (in this case $k = 0.4325$ according to Table I), whereas the exact value is $M = 25.03$ (see Table II). Two terms in (2.32) give $M = 24.76$. This formula becomes more accurate for smaller g and $(1 - \lambda)$.

2.5 Radiation Field with Nearly Conservative Scattering

This case will attract our particular attention throughout the book. There are two reasons for this: first, media with $(1-\lambda) \ll 1$, i.e., where the absorption of a photon is almost always followed by its reemission, frequently occur in nature. Second, nearly conservative scattering is the most complicated case in the practice of numerical calculations of radiation fields.

All is greatly simplified in the problem for two-sided infinity. Indeed, using (2.16), (2.17), and (2.21), one can find explicitly an arbitrary number of terms of the expansions of k and c_j in series, in terms of powers of $(1-\lambda)^{1/2}$. In particular, we have

$$k^2 = (1-\lambda)(3-x_1)\left\{1 + \left(\frac{x_1}{3-x_1} - \frac{4}{5-x_2}\right)(1-\lambda) + \frac{36(1-\lambda)^2}{(5-x_2)^2}\left[1 - \frac{3-x_1}{7-x_3} - \frac{5-x_2}{3(3-x_1)}\right] + O[(1-\lambda)^3]\right\} \quad (2.35)$$

or

$$k = [(1-\lambda)(3-x_1)]^{1/2}\left\{1 + \frac{1}{2}\left(\frac{x_1}{3-x_1} - \frac{4}{5-x_2}\right)(1-\lambda) + \frac{18}{(5-x_2)^2}\left[1 - \frac{3-x_1}{7-x_3} - \frac{5-x_2}{3(3-x_1)} - \left(\frac{5-x_2}{12}\right)^2\left(\frac{x_1}{3-x_1} - \frac{4}{5-x_2}\right)^2\right](1-\lambda)^2 + O\left[(1-\lambda)^3\right]\right\}. \quad (2.36)$$

Then, using (2.16), (2.17), and (2.36), after cumbersome but simple manipulations, we obtain

$$c_1 = \left(\frac{1-\lambda}{3-x_1}\right)^{1/2}\left\{1 - \frac{1}{2}\left(\frac{x_1}{3-x_1} - \frac{4}{5-x_2}\right)(1-\lambda) - \frac{18}{(5-x_2)^2}\left[1 - \frac{3-x_1}{7-x_3} - \frac{5-x_2}{3(3-x_1)} - \frac{(5-x_2)^2}{48}\left(\frac{x_1}{3-x_1} - \frac{4}{5-x_2}\right)^2\right](1-\lambda)^2 + O\left[(1-\lambda)^3\right]\right\}, \quad (2.37)$$

$$c_2 = \frac{2(1-\lambda)}{5-x_2}\left\{1 - \frac{(7-x_3)x_2 - 9(3-x_1)}{(5-x_2)(7-x_3)}(1-\lambda) + O\left[(1-\lambda)^2\right]\right\}, \quad (2.38)$$

$$c_3 = \frac{6(3-x_1)^{1/2}}{(5-x_2)(7-x_3)}(1-\lambda)^{3/2} + O\left[(1-\lambda)^{5/2}\right]. \quad (2.39)$$

It is seen that the order of vanishing of the coefficient c_n is $(1-\lambda)^{n/2}$. Then, employing (2.14), we obtain the following explicit expression for the function $i(\mu)$:

$$i(\mu) = 1 + \frac{3k_0}{3-x_1}\mu + \left[1 + \frac{10P_2(\mu)}{5-x_2}\right]\frac{k_0^2}{3-x_1} + 3\left\{\frac{3}{2}(2-x_1)\mu + \frac{2(3-x_1)^2}{5-x_2}\left[\frac{7P_3(\mu)}{7-x_3} + \frac{\mu}{3-x_1}\right]\right\}\frac{k_0^3}{(3-x_1)^3} + O(k_0^4), \tag{2.40}$$

where

$$k_0 = \left[(1-\lambda)(3-x_1)\right]^{1/2}. \tag{2.41}$$

Similarly, with the help of (2.34) and (2.36), we obtain the following asymptote of M,

$$M = \frac{8k_0}{3-x_1}\left\{1 + \left[\frac{3}{2}(2-x_1) + \frac{(11-x_2)(3-x_1)}{5-x_2}\right]\frac{k_0^2}{(3-x_1)^2}\right\} + O(k_0^5). \tag{2.42}$$

Simpler formulas for $i(\mu)$ and M are often used,

$$i(\mu) = 1 + \frac{3k_0}{3-x_1}\mu + O(k_0^2), \tag{2.43}$$

$$M = \frac{8k_0}{3-x_1} + O(k_0^3). \tag{2.44}$$

We further suggest that the problem for two-sided infinity with given λ and $\chi(\gamma)$ is solved. This implies that the function $i(\mu)$ and the diffusion exponent k are known. The values of the function $i(\mu)$ for the Henyey-Greenstein phase function at $g = 0.25; 0.5; 0.75; 0.9$, and various λ are given in Table XXXV in the Appendix.

3. Semi-Infinite Medium

This chapter deals with solving the parallel external flux problem, and the Milne problem. We rely on the generalized invariance principle formulated in Sect. 1.12. We also describe some concrete methods for numerical solution of these problems.

The model of a semi-infinite atmosphere is of considerable practical interest. To zero order, the atmosphere of many planets can be regarded as semi-infinite media. Moreover, a knowledge of the radiation field in a semi-infinite medium substantially facilitates solving the parallel external flux problem for atmospheres of large optical thickness (see the next chapter). We will also consider the case of nearly conservative scattering ($1 - \lambda \ll 1$) which is of great practical importance. For low true absorption, it appears that some characteristics of the radiation field at all optical depths are directly expressed in terms of radiation intensity for conservative scattering.

The present chapter ends with the algorithm for solving the general problem of radiative transfer theory, based on the so-called Q representation of the Green function for the transfer equation.

3.1 Invariance Relation for the Parallel External Flux Problem

Let us consider a plane atmosphere ($\tau_0 = \infty$), semi-infinite and illuminated by parallel rays. All notation is standard and given in Sect. 1.5 (see also Fig. 3.1). Let us put an imaginary new boundary at an arbitrary level τ_1 ($\tau_1 \geq 0$) in the medium. We intend to find the intensity of diffuse radiation at an arbitrary depth $\tau_1 + \tau$ ($\tau \geq 0$).

Let us now remove a layer of optical thickness τ_1 from the atmosphere under study. Then, the generalized invariance principle (see Sect. 1.12) suggests the following: if the upper boundary of this truncated atmosphere is illuminated by the same parallel external flux but attenuated by $\mathrm{e}^{-\tau_1/\mu_0}$, and diffuse radiation of intensity $I(\tau_1, \mu, \mu_0, \varphi - \varphi_0)$, then the radiation field in the truncated atmosphere at depth τ (τ is measured from the new boundary) is equal to that in the original medium at the depth $\tau_1 + \tau$. In other words, allowing for (1.45), one can write the following ***invariance relation*** ($\mu \in [-1, +1]$, $\mu_0 \in [0, 1]$):

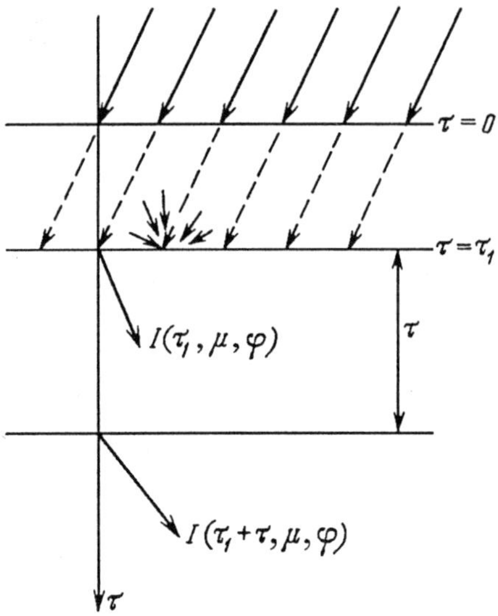

Fig. 3.1. On derivation of the invariance relation 3.1

$$
\begin{aligned}
I(\tau_1+\tau,\mu,\mu_0,\varphi-\varphi_0) &= I(\tau,\mu,\mu_0,\varphi-\varphi_0)\mathrm{e}^{-\tau_1/\mu_0} \\
&+ I(\tau_1,\mu,\mu_0,\varphi-\varphi_0)\mathrm{e}^{-\tau/\mu}\theta(\mu) \\
+\tfrac{1}{\pi}\int_0^{2\pi}\mathrm{d}\varphi'\int_0^1 & I(\tau_1,\mu',\mu_0,\varphi'-\varphi_0)I(\tau,\mu,\mu',\varphi-\varphi')\,\mathrm{d}\mu' . \qquad (3.1)
\end{aligned}
$$

This relation links nonlinearly the intensities at two arbitrary depths with that at the summed depth. $\theta(\mu)$ is a unit step function (see (1.46)).

Let us discuss once more the physical meaning of the terms on the right-hand side of (3.1). The first term arises due to the upper boundary being illuminated by direct radiation attenuated by $\mathrm{e}^{-\tau/\mu_0}$. The second is caused by the fact that this boundary is also illuminated by diffuse radiation of intensity $I(\tau_1,\mu,\mu_0,\varphi-\varphi_0)$. Finally, the third term is in fact the intensity of diffuse radiation that illuminates the boundary τ_1, attenuated by $\mathrm{e}^{-\tau/\mu}$ in passing through the layer of optical thickness τ.

Using the expansion (1.47) of the radiation intensity in terms of azimuth, one derives the following expression for the m-th harmonic of intensity:

$$
\begin{aligned}
I^m(\tau_1+\tau,\mu,\mu_0) &= I^m(\tau,\mu,\mu_0)\mathrm{e}^{-\tau_1/\mu_0} \\
+2\int_0^1 I^m(\tau_1,\mu',\mu_0)I^m(\tau,\mu,\mu')\,\mathrm{d}\mu' &+ I^m(\tau_1,\mu,\mu_0)\mathrm{e}^{-\tau/\mu}\theta(\mu). \qquad (3.2)
\end{aligned}
$$

It is the above expression that is used extensively below.

It is convenient occasionally to use the alternative form of this relation that can be obtained by passing to the source function B. For this, we multiply both sides of (3.2) by $\frac{\lambda}{2}\chi^m(\mu,\mu_1)$ and integrate with respect to μ from -1 to $+1$. Using (1.5) and after some algebra, we get the equation:

$$B^m(\tau_1+\tau,\mu,\mu_0) = B^m(\tau,\mu,\mu_0)\mathrm{e}^{-\tau_1/\mu_0} + 2\int\limits_0^1 I^m(\tau_1,\mu',\mu_0)B^m(\tau,\mu,\mu')\,\mathrm{d}\mu'. \tag{3.3}$$

Its physical meaning is clear. The source function at the depth $\tau+\tau_1$ is composed of two terms: the first one is the source function in the truncated atmosphere, which is determined by the radiation illuminating the upper boundary of the atmosphere (the first term in (3.3)). The second term is caused by the diffuse radiation of intensity $I^m(\tau_1,\mu,\mu_0)$.

3.2 The Milne Problem

In the Milne problem the removal of a layer of an arbitrary optical thickness from the atmosphere surface leaves the relative spatial-angular distribution of radiation in the atmosphere unchanged because this results in an atmosphere analogous to the original one. The only thing that may change is the absolute value of intensity owing to the fact that the "new" boundary approaches "priming" radiation sources. In other words, if $I(\tau,\mu)$ is the intensity in the original atmosphere at depth τ, then after removing a layer of thickness τ_1 from the surface, the intensity at the same depth τ measured from the new boundary is equal to $f(\tau_1)I(\tau,\mu)$, where f is an unknown function. Successive removal of layers of thickness τ_1 and τ_2 yields an atmosphere with the intensity at depth τ being equal to $f(\tau_2)f(\tau_1)I(\tau,\mu)$. On the other hand, the same atmosphere can be obtained by truncating a layer of thickness $\tau_1+\tau_2$, and the intensity at depth τ is equal to $f(\tau_1+\tau_2)I(\tau,\mu)$. Hence, $f(\tau_1+\tau_2) = f(\tau_1)f(\tau_2)$. Therefore $f(\tau) = \mathrm{e}^{k\tau}$, where k is an unknown constant. This constant is equal to the smallest absolute value of the roots of the characteristic equation (2.21), i.e., to the diffusion exponent. This will be proved in Sect. 3.9.

The above considerations can be summarized as follows. In the Milne problem the removal of a layer of optical thickness τ from the atmosphere surface, or, equivalently, its ***truncation*** (the term is employed throughout the book), increases the intensity of radiation by $\mathrm{e}^{k\tau}$ at all depths. Because $k=0$ for a conservatively scattering atmosphere, this truncation does not influence the radiation field.

Let us proceed to deriving the invariance relations. We remove a layer of optical thickness τ_1 from the atmosphere, which makes the radiation intensity higher at all depths, and introduce sources of intensity $I(\tau_1,\mu)$ at the new boundary. The generalized invariance principle and (1.45) then yield,

$$I(\tau_1+\tau,\mu) = I(\tau,\mu)\mathrm{e}^{k\tau_1} + 2\int_0^1 I(\tau_1,\mu')I(\tau,\mu,\mu')\,\mathrm{d}\mu' + I(\tau_1,\mu)\mathrm{e}^{-\tau/\mu}\theta(\mu), \tag{3.4}$$

$\mu \in [-1,+1]$, $\tau, \tau_1 \geq 0$. In terms of the source functions, it is rewritten as follows:

$$B(\tau_1+\tau,\mu) = B(\tau,\mu)\mathrm{e}^{k\tau_1} + 2\int_0^1 I(\tau_1,\mu')B(\tau,\mu,\mu')\,\mathrm{d}\mu'. \tag{3.5}$$

One easily sees that both (3.3) and (3.5) can be directly derived from the transfer equation. However, the physically obvious method of derivation that follows directly from the generalized invariance principle is preferred. Equations (3.2) and (3.4) are the basic ones for further derivations. One should remember that (3.2) and (3.3) are valid for all azimuthal harmonics, while (3.4) and (3.5) hold only for the zeroth harmonic.

3.3 Relationship Between the Milne Problem and the Parallel External Flux Problem

This very important relationship is easily established via the generalized invariance principle. Let us consider again the problem for two-sided infinity (Fig. 1.5). Let us take first the left half-space ($\tau < 0$). The radiation field there is provided by sources placed at minus infinity and by the radiation of intensity $i(-\mu)$ illuminating the boundary. On the other hand, we are aware that this radiation field is determined by (2.5). Thus, we have

$$i(-\mu)\mathrm{e}^{k\tau} = MI(\tau,\mu) + 2\int_0^1 I(\tau,\mu,\mu')i(-\mu')\,\mathrm{d}\mu' + i(-\mu)\mathrm{e}^{-\tau/\mu}\theta(\mu), \tag{3.6}$$

where M is a normalizing factor. Similarly, for the right-side half-space ($\tau > 0$) we have

$$i(\mu)\mathrm{e}^{-k\tau} = 2\int_0^1 I(\tau,\mu,\mu')i(\mu')\,\mathrm{d}\mu' + i(\mu)\mathrm{e}^{-\tau/\mu}\theta(\mu). \tag{3.7}$$

Equations (3.6) and (3.7) are called ***the Ivanov relations***.

Rewritten in terms of the source function, (3.6) and (3.7) take the form

$$(1+k\mu)i(-\mu)\mathrm{e}^{k\tau} = MB(\tau,\mu) + 2\int_0^1 B(\tau,\mu,\mu')i(-\mu')\,\mathrm{d}\mu', \tag{3.8}$$

$$(1-k\mu)i(\mu)\mathrm{e}^{-k\tau} = 2\int_0^1 B(\tau,\mu,\mu')i(\mu')\,\mathrm{d}\mu'. \tag{3.9}$$

The following formulas are called the ***Sobolev–van de Hulst relations***:

$$i(\mu) = Mu(\mu) + 2\int_0^1 \rho(\mu,\mu')i(-\mu')\mu'\,\mathrm{d}\mu', \tag{3.10}$$

$$i(-\mu) = 2\int_0^1 \rho(\mu, \mu')i(\mu')\mu' \, \mathrm{d}\mu'. \tag{3.11}$$

They are derived from (3.6) and (3.7) at $\tau = 0$ ($\mu \in [0, 1]$). The escape function $u(\mu)$ (i.e., the boundary value of the Milne intensity $I(0, -\mu) = u(\mu)$) is normalized as

$$2\int_0^1 u(\mu)i(\mu)\mu \, \mathrm{d}\mu = 1. \tag{3.12}$$

Allowing for (3.12) and using (3.10) and (3.11), one arrives at the conclusion that the constant M is determined by (2.28).

The conservative limit of (3.11) is the condition that the plane albedo of a semi-infinite atmosphere (see 1.104) is equal to unity:

$$A_0(\mu) = 2\int_0^1 \rho_0(\mu, \mu')\mu' \, \mathrm{d}\mu' = 1. \tag{3.13}$$

Subscript "0" is usually used hereafter to denote conservative scattering. Using (2.43) and (2.44), one derives the conservative limit of (3.10) from (3.10), (3.11).

$$u_0(\mu) = \frac{3}{4}\left[\mu + 2\int_0^1 \rho_0(\mu, \mu')\mu'^2 \, \mathrm{d}\mu'\right] = \frac{3}{2}\int_0^1 \rho_0(\mu, \mu')(\mu + \mu')\mu' \, \mathrm{d}\mu'. \tag{3.14}$$

Equations (3.13) and (3.14) are the particular cases of more general relations derivable from (3.6) and (3.7) as $\lambda \to 1$. Namely,

$$I_0(\tau, \mu) = \frac{3}{4}\left[\left(1 - \frac{x_1}{3}\right)\tau - \mu + 2\int_0^1 I_0(\tau, \mu, \mu')\mu' \, \mathrm{d}\mu' + \mu \mathrm{e}^{-\tau/\mu}\theta(\mu)\right], \tag{3.15}$$

$$2\int_0^1 I_0(\tau, \mu, \mu') \, \mathrm{d}\mu' = 1 - \mathrm{e}^{-\tau/\mu}\theta(\mu). \tag{3.16}$$

The asymptotic expressions of (2.43) and (2.44) are to be used to derive (3.15).

Equations (3.6) and (3.15) show that if the intensity in the parallel external flux problem is known, then the Milne intensity can be found. One may also demonstrate that the above problems are equivalent if the scattering is isotropic. Of course, $i(\mu)$ and k are assumed to be known for $\lambda \neq 1$. Thus, in a sense the Milne problem "does not exist" as a separate entity.

Equation (3.7), as well as its particular cases, (3.11) and (3.16), can be regarded as integral relations, which are satisfied by azimuthally averaged radiation intensity in the parallel external flux problem. As we shell see below, these relations are helpful in both evaluating and controlling the accuracy of the above intensity.

3.4 Corollaries

The following corollaries result from the invariance relations, (3.1–5). They are of considerable interest. In fact, they had been found long before the more general relations were derived. The simplest form of (3.1) and (3.4) is that is obtained at $\tau = 0$. Namely, from (3.1) and (3.4), one derives

$$I^m(\tau, -\mu, \mu_0) = \mathrm{e}^{-\tau/\mu_0}\rho^m(\mu, \mu_0)\mu_0 + 2\int_0^1 I^m(\tau, \mu', \mu_0)\rho^m(\mu, \mu')\mu'\,\mathrm{d}\mu', \tag{3.17}$$

$$I(\tau, -\mu) = \mathrm{e}^{k\tau}u(\mu) + 2\int_0^1 I(\tau, \mu')\rho(\mu, \mu')\mu'\,\mathrm{d}\mu'. \tag{3.18}$$

This form of (3.1) and (3.4) is the most widely known one (see, RT, Sect. 29; the latter one is written there for a particular case of conservative scattering). Equations (3.17) and (3.18) are invariance relations in Chandrasekhar's form.

Another important particular form of the invariance relations is derived assuming τ_1 to be an infinitesimal quantity (equal to $\mathrm{d}\tau$). In terms of the source functions, they take the form

$$\frac{\mathrm{d}B^m(\tau, \mu, \mu_0)}{\mathrm{d}\tau} = -\frac{1}{\mu_0}B^m(\tau, \mu, \mu_0) + 2\int_0^1 B^m(\tau, \mu, \mu')B^m(0, \mu', \mu_0)\frac{\mathrm{d}\mu'}{\mu'}, \tag{3.19}$$

$$\frac{\mathrm{d}B(\tau, \mu)}{\mathrm{d}\tau} = kB(\tau, \mu) + 2\int_0^1 B(\tau, \mu, \mu')B(0, \mu')\frac{\mathrm{d}\mu'}{\mu'}. \tag{3.20}$$

Here, as follows from (1.63) and (1.68),

$$B^m(0, \mu, \mu_0) = \frac{\lambda}{2}\mu_0\int_0^1 \chi^m(\mu, -\mu')\rho^m(\mu', \mu_0)\,\mathrm{d}\mu' + \frac{\lambda}{4}\chi^m(\mu, \mu_0), \tag{3.21}$$

$$B(0, \mu) = \frac{\lambda}{2}\int_0^1 \chi(-\mu, \mu')u(\mu)\,\mathrm{d}\mu'. \tag{3.22}$$

The simplest way to derive (3.19) is to differentiate both sides of (3.3) with respect to τ_1 and then to put $\tau_1 = 0$. One should now use the transfer equation, which results in the following expression for $\mu \in (0, 1]$:

$$\left.\frac{\mathrm{d}I^m(\tau, \mu, \mu_0)}{\mathrm{d}\tau}\right|_{\tau=0} = \frac{1}{\mu}B^m(0, \mu, \mu_0). \tag{3.23}$$

Equation (3.20) is obtained from (3.5) similarly.

Equation (3.19) is the invariance relation in the Ambartsumian's form. Both (3.19) and (3.20) are used, for instance, in LSPA (Sects. 2.2, 2.3), where they are derived in another way.

In terms of intensities, the invariance relations (3.19) and (3.20) are of considerable interest. To obtain the first one, it is sufficient to differentiate both sides of (3.2) with respect to τ_1 and to put then $\tau_1 = 0$. This results in

$$\frac{\mathrm{d}I^m(\tau,\mu,\mu_0)}{\mathrm{d}\tau} = -\frac{1}{\mu_0}I^m(\tau,\mu,\mu_0) + 2\int_0^1 I^m(\tau,\mu,\mu')B^m(0,\mu',\mu_0)\frac{\mathrm{d}\mu'}{\mu'} + \frac{1}{\mu}B^m(0,\mu,\mu_0)\mathrm{e}^{-\tau/\mu}\theta(\mu). \tag{3.24}$$

Similarly, from (3.4) one derives

$$\frac{\mathrm{d}I(\tau,\mu)}{\mathrm{d}\tau} = kI(\tau,\mu) + 2\int_0^1 I(\tau,\mu,\mu')B(0,\mu')\frac{\mathrm{d}\mu'}{\mu'} + \frac{1}{\mu}B(0,\mu)\mathrm{e}^{-\tau/\mu}\theta(\mu). \tag{3.25}$$

Eliminating the derivative from (3.24), (3.25) with the help of the transfer equation, we obtain ($\tau \geq 0$, $\mu \in [-1,+1]$):

$$\left(\frac{1}{\mu_0} - \frac{1}{\mu}\right) I^m(\tau,\mu,\mu_0) = -\frac{1}{\mu}B^m(\tau,\mu,\mu_0) + 2\int_0^1 I^m(\tau,\mu,\mu')B^m(0,\mu',\mu_0)\frac{\mathrm{d}\mu'}{\mu'} + \frac{1}{\mu}B^m(0,\mu,\mu_0)\mathrm{e}^{-\tau/\mu}\theta(\mu), \tag{3.26}$$

$$(1+k\mu)I(\tau,\mu) = B(\tau,\mu) - 2\mu\int_0^1 I(\tau,\mu,\mu')B(0,\mu')\frac{\mathrm{d}\mu'}{\mu'} - B(0,\mu)\mathrm{e}^{-\tau/\mu}\theta(\mu). \tag{3.27}$$

Equations (3.26) and (3.27) are of considerable importance for the following reasons. First, they involve neither the integrals nor the derivatives with respect to the spatial variable τ. If $B^m(\tau,\mu,\mu_0)$ and $B(\tau,\mu)$ in the first terms on the right-hand sides are substituted with their expressions

$$B^m(\tau,\mu,\mu_0) = B^m(0,\mu,\mu_0)\mathrm{e}^{-\tau/\mu_0} + 2\int_0^1 I^m(\tau,\mu',\mu_0)B^m(0,\mu,\mu')\,\mathrm{d}\mu', \tag{3.28}$$

$$B(\tau,\mu) = B(0,\mu)\mathrm{e}^{k\tau} + 2\int_0^1 I(\tau,\mu')B(0,\mu,\mu')\,\mathrm{d}\mu', \tag{3.29}$$

which follow from (3.3) and (3.5) at $\tau = 0$, respectively, then we arrive at the integral equations for the intensity at depth τ. They involve only the integrals over angular variables. In other words, if the intensity at the boundary — the reflection coefficient $\rho^m(\mu,\mu_0)$ — is known, the intensity at the arbitrary depth τ can be found by solving the integral equations (3.26) and (3.27) *without determining the radiation field at other depths.* The intensity, and consequently, the source function (see 3.21 and 3.22) at the boundary, can be found by the method described in the next section.

Second, (3.26–29) suggest that to solve these equations, we need only to find the intensity of descending radiation ($\mu \geq 0$), after which, the intensity of ascending radiation can easily be determined via (3.17) and (3.18). This substantially reduces the calculation procedure.

Another and possibly the most efficient method of evaluating the radiation field inside the atmosphere is expounded in Sect. 3.10. It is based on a particular form of the invariance relation (3.2).

Concluding this section, we take a look at another and more special case of the invariance relation for the Milne problem. Putting $\tau = 0$ in (3.27), one finds

$$(1 - k\mu)u(\mu) = B(0, -\mu) + 2\mu \int_0^1 \rho(\mu, \mu')B(0, \mu')\,\mathrm{d}\mu'. \tag{3.30}$$

Substituting $B(0, \mu)$ with its expression (3.22) and using (3.24) with $m = 0$, one derives

$$(1 - k\mu)u(\mu) = 2 \int_0^1 B(0, \mu', \mu)u(\mu')\,\mathrm{d}\mu'. \tag{3.31}$$

This is the invariance relation for the Milne problem in the Ambartsumian's form.

3.5 Ambartsumian's Equation for the Reflection Coefficient and a Method for Its Solution

As a result of putting $\tau = 0$ in (3.26) and use the definition (1.95), one gets ($\mu \in [0, 1]$):

$$(\mu+\mu_0)\rho^m(\mu, \mu_0) = B^m(0, -\mu, \mu_0)+2\mu \int_0^1 B^m(0, \mu, \mu')\rho^m(\mu, \mu')\,\mathrm{d}\mu'. \tag{3.32}$$

Substituting (3.21) into the above equation, one arrives at the Ambartsumian's integral equation for the reflection coefficient

$$\begin{aligned}(\mu + \mu_0)\rho^m(\mu, \mu_0) &= \frac{\lambda}{4}\chi^m(-\mu, \mu_0)\\ &+\frac{\lambda}{2}\mu_0 \int_0^1 \chi^m(\mu, \mu')\rho^m(\mu', \mu_0)\,\mathrm{d}\mu' + \frac{\lambda}{2}\mu \int_0^1 \chi^m(\mu_0, \mu')\rho^m(\mu', \mu)\,\mathrm{d}\mu'\\ &+\lambda\mu\mu_0 \int_0^1 \rho^m(\mu, \mu')\,\mathrm{d}\mu' \int_0^1 \chi^m(\mu', -\mu'')\rho^m(\mu'', \mu_0)\,\mathrm{d}\mu''.\end{aligned} \tag{3.33}$$

Function $\rho^m(\mu, \mu_0)$, which depends on two angular variables, can be expressed in terms of the auxiliary functions of a single angular variable. To do this, one uses the expression (1.56) for $\chi^m(\mu, \mu_0)$. Upon substituting it into (3.21), one gets

$$B^m(0, \mu, \mu_0) = \frac{\lambda}{4} \sum_{i=m}^{\infty} x_i^m P_i^m(\mu)\varphi_i^m(\mu_0), \tag{3.34}$$

where

$$\varphi_i^m(\mu) = P_i^m(\mu) + 2\mu \int_0^1 \rho^m(\mu', \mu)P_i^m(-\mu')\,\mathrm{d}\mu'. \tag{3.35}$$

Inserting (3.34) into (3.32) and using the symmetry of the reflection coefficient, $\rho^m(\mu,\mu_0)=\rho^m(\mu_0,\mu)$, one ultimately finds

$$\rho^m(\mu,\mu_0)=\frac{\lambda}{4}\sum_{i=m}^{\infty}(-1)^{i+m}x_i^m\frac{\varphi_i^m(\mu)\varphi_i^m(\mu_0)}{\mu+\mu_0}. \tag{3.36}$$

The substitution of (3.36) into (3.35) yields a system of nonlinear integral equations that determine $\varphi_i^m(\mu)$.

These functions are usually called ***Ambartsumian's functions***. Assuming the sum of n terms of (1.56) to be an adequate approximation for the phase function, we need to know $n-m+1$ functions $\varphi_i^m(\mu)$ to determine $\rho^m(\mu,\mu_0)$. On the other hand, the number of functions necessary to find the total reflection coefficient $\rho(\mu,\mu_0,\varphi)$ is $\frac{1}{2}(n+1)(n+2)$.

If the phase function is not very extended (i.e. n is not too large), the Sobolev method (see LSPA, Chap. 5) can be used to find $\rho^m(\mu,\mu_0)$. In the opposite case (i.e. when n is large), it is more convenient to solve (3.33) directly. Now we proceed to the description of the numerical method applicable to its solution.

For $m\geq 1$, the solution of (3.33) can easily be found by the method of successive approximations using the free term of (3.33), i.e.,

$$\frac{\lambda}{4}\frac{\chi^m(-\mu,\mu_0)}{\mu+\mu_0}, \tag{3.37}$$

as the initial approximation. Convergence of iterations improves with the increase of m and decrease of the elongation of the phase function.

For $m=0$, the standard scheme of successive approximations works well only if λ is sufficiently far from unity. This method becomes, in fact, inapplicable when $1-\lambda\ll 1$.

The physical reason of this is explained as follows: roughly speaking, at the n-th iteration, the method of successive approximations takes into account only the photons undergoing n scatterings. It is clear that under conservative scattering, for even a very large n there are many photons that undergo more than n scatterings before leaving medium and, thus, contribute to $\rho(\mu,\mu_0)$ (they could not "die" any other way). That is why the problem consists in finding an efficient procedure to accelerate a convergence of iterations for $\lambda\simeq 1$.

The zeroth azimuthal harmonic of the reflection coefficient $\rho(\mu,\mu_0)$ must satisfy not only the Ambartsumian equation (3.33) but also the integral relation (3.11), where function $i(\mu)$ is assumed to be known. The algorithm of the solution of (3.33) is constructed in such a way that each successive approximation satisfies (3.11). By this precaution it is ensured that photon scatterings of *all* multiplicities are taken into account. Indeed, for conservative scattering this procedure guarantees that the plane albedo is strictly equal to unity (see (3.13)). As a result, the current $\rho(\mu,\mu_0)$ satisfies both

(3.33) and (3.11) within the accuracy taken after small number of iterations. For more details, see [49].

This method was used to calculate the reflection coefficient for a semi-infinite atmosphere with the Rayleigh phase function (in tables, R stands for the phase function) and also with the Henyey-Greenstein phase function for $g = 0(0.25)0.75(0.05)0.9$. The particles albedo was taken equal to 1; 0.999; 0.995; 0.99; 0.98; 0.95; 0.9(−0.1)0.7.

The Tables III–XVI of the Appendix contain the results of calculations of the distribution of brightness $\bar{r}(\mu)$ over planetary disk at phase angle $\alpha = 0$ ($\mu = \mu_0$), i.e., the quantity

$$\bar{r}(\mu) = \frac{\rho(\mu, \mu, \pi)}{\rho(1, 1, \pi)}. \tag{3.38}$$

These tables also contain the values of escape function $u(\mu)$ which have been calculated according to (3.10) ($\lambda \neq 1$) and (3.14) ($\lambda = 1$). The reflection coefficient in the center of disk, i.e. $\rho(1, 1, \pi)$, is given in Table XVIII.

Geometric albedo A_{g} and spherical albedo A_{s}, which are also important characteristics of planetary atmosphere, are calculated with the help of (1.107), (1.106) and presented in the Tables XX and XIX.

3.6 Some Integral Relations Involving Escape Functions

These relations themselves are of interest. Furthermore, we will need them in future considerations.

The first relation can be derived as follows: substitute the expression for $B(0, \mu, \mu_0)$, which is (3.21) with m=0, into (3.31). Multiply both sides of the above equality by $2i(-\mu)$ and integrate the product with respect to μ from 0 to 1. Using (3.10) one finds

$$\begin{aligned} 2\int_0^1 u(\mu) i(-\mu)\,\mathrm{d}\mu - kN &= \lambda \int_{-1}^{+1} i(\mu)\,\mathrm{d}\mu \int_0^1 \chi(\mu, -\mu') u(\mu')\,\mathrm{d}\mu' \\ &\quad -\lambda M \int_0^1 u(\mu)\,\mathrm{d}\mu \int_0^1 \chi(\mu, -\mu') u(\mu')\,\mathrm{d}\mu', \end{aligned} \tag{3.39}$$

where

$$N = 2\int_0^1 u(\mu) i(-\mu) \mu\,\mathrm{d}\mu. \tag{3.40}$$

Making use of (2.6) one ultimately derives

$$\lambda M \int_0^1 u(\mu)\,\mathrm{d}\mu \int_0^1 \chi(\mu, -\mu') u(\mu')\,\mathrm{d}\mu' = 2kN, \tag{3.41}$$

instead of (3.39). Formula (3.22) makes it possible to put (3.41) into a simpler form as follows

$$M \int_0^1 u(\mu) B(0, \mu) \,\mathrm{d}\mu = kN.$$

Numerical values of N (this quantity is often encountered later on) for the Rayleigh and Henyey-Greenstein phase functions with various g are provided in Table XII.

Multiply the transfer equation (1.49) (at $m = 0$) by $2u(\mu)\mu$ and integrate the product with respect to μ from 0 to 1. Using the representation (3.28) of the source function one obtains

$$2 \frac{\mathrm{d}}{\mathrm{d}\tau} \int_0^1 I(\tau, \mu', \mu) u(\mu') \mu' \,\mathrm{d}\mu' + 2 \int_0^1 I(\tau, \mu', \mu) u(\mu') \,\mathrm{d}\mu'$$
$$= 2\mathrm{e}^{-\tau/\mu} \int_0^1 B(0, \mu', \mu) u(\mu') \,\mathrm{d}\mu'$$
$$+2 \int_0^1 I(\tau, \mu'', \mu) \,\mathrm{d}\mu'' \cdot 2 \int_0^1 B(0, \mu', \mu'') u(\mu') \,\mathrm{d}\mu'.$$

Introducing (3.31) into the above expression, one finds

$$\frac{\mathrm{d}}{\mathrm{d}\tau} \left[2 \int_0^1 I(\tau, \mu', \mu) u(\mu') \mu' \,\mathrm{d}\mu' \right] + 2k \int_0^1 I(\tau, \mu', \mu) u(\mu') \mu' \,\mathrm{d}\mu'$$
$$= (1 - k\mu) \mathrm{e}^{-\tau/\mu} u(\mu).$$

This formula gives one more of the relations sought, namely

$$2 \int_0^1 I(\tau, \mu', \mu) u(\mu') \mu' \,\mathrm{d}\mu' = \left(\mathrm{e}^{-k\tau} - \mathrm{e}^{-\tau/\mu} \right) u(\mu)\mu. \tag{3.42}$$

This relation involves the intensity of descending radiation. The intensity of ascending radiation satisfies a more complicated equation:

$$2 \int_0^1 I(\tau, -\mu', \mu) u(\mu') \mu' \,\mathrm{d}\mu' = \left[2 \int_0^1 I(\tau, -\mu') \rho(\mu, \mu') \mu' \,\mathrm{d}\mu' - I(\tau, \mu) \right] \mu. \tag{3.42a}$$

This equation follows from (3.52a) at $\tau_1 = 0$.

Similarly, using the transfer equation in the Milne problem (1.67) and (3.29), (3.41), we have

$$M \int_0^1 I(\tau, \mu) u(\mu) \mu \,\mathrm{d}\mu = N \sinh k\tau. \tag{3.43}$$

Let us find the conservative limits ($\lambda = 1$) of (3.41) and (3.43). It follows from (3.40), (3.12), and (2.43) that $N \to 1$ when $\lambda \to 1$. Then, from (2.44) we obtain

$$4 \int_0^1 u_0(\mu) \,\mathrm{d}\mu \int_0^1 \chi(\mu, -\mu') u_0(\mu') \,\mathrm{d}\mu' = 3 - x_1, \tag{3.44}$$

$$8\int_0^1 I_0(\tau,\mu)u_0(\mu)\mu\,\mathrm{d}\mu = (3-x_1)\tau, \tag{3.45}$$

instead of (3.41) and (3.43)

Equations (3.42–45) can, in particular, be used to control the accuracy of the solution of the transfer equation. We will see in the next section that these equations are special cases of more general relations.

3.7 Integrals of the Transfer Equation

Integrals of the transfer equation play a very important role in the theory of scattering of light in plane atmospheres. Some integrals, namely the ***flux integral*** and K ***integral*** (see Sect. 1.10, also RT, Sects. 8, 10; LSPA, Sect. 1.6) have long been known. Now we are to derive a number of integrals of the transfer equation for a semi-infinite atmosphere which are quadratic with respect to intensity. Aforementioned classical integrals are the particular cases of these integrals.

1. Let us consider relations (3.10), (3.11), (3.17), and (3.18). Their common feature is that they all involve the integration of the reflection coefficient, which is known to be symmetrical with respect to μ and μ'. Taking this into account, we multiply (3.17) by $2I^m(\tau_1,\mu,\mu_1)$ and integrate the product with respect to μ from 0 to 1 ($\tau_1 \geq 0$). Using (3.17) twice and performing simple manipulations, we obtain
$$2\int_{-1}^{+1} I^m(\tau,-\mu',\mu_0)I^m(\tau_1,\mu',\mu_1)\mu'\,\mathrm{d}\mu'$$
$$= I^m(\tau_1,-\mu_0,\mu_1)\mathrm{e}^{-\tau/\mu_0}\mu_0 - I^m(\tau,-\mu_1,\mu_0)\mathrm{e}^{-\tau_1/\mu_1}\mu_1. \tag{3.46}$$
This is the one of the integrals sought. Putting either $\tau=0$ or $\tau_1=0$ in (3.46) one derives (3.17). If we simultaneously set $\tau=\tau_1$ and $\mu=\mu_1$, (3.46) becomes an identity. For other integrals of this type, see (4.35) and (4.36).
2. Performing similar manipulations with (3.18) and taking (3.43) into account, one finds
$$M\int_{-1}^{+1} I(\tau,-\mu)I(\tau_1,\mu)\mu\,\mathrm{d}\mu = N\sinh[k(\tau_1-\tau)]. \tag{3.47}$$
This integral involves only the intensity in the Milne problem. At $\tau=0$ (or $\tau_1=0$) (3.47) gives relation (3.43).
3. We set $m=0$ and multiply (3.17) by $2I(\tau_1,\mu)\mu$. Integrating the product with respect to μ from 0 to 1 and using (3.18) and (3.42), we obtain
$$2\int_{-1}^{+1} I(\tau,-\mu',\mu)I(\tau_1,\mu')\mu'\,\mathrm{d}\mu' = I(\tau_1,-\mu)\mathrm{e}^{-\tau/\mu}\mu - u(\mu)\mu\mathrm{e}^{k(\tau_1-\tau)}. \tag{3.48}$$
Putting $\tau_1=0$ in (3.48) one derives (3.42); for $\tau=0$ one obtains (3.18).

Finally, the appropriate combinations of (3.10), (3.11) and (3.17), (3.18) give the following four integrals

$$2\int_{-1}^{+1} I(\tau,\mu',\mu)i(-\mu')\mu'\,\mathrm{d}\mu' = -i(-\mu)\mu\mathrm{e}^{-\tau/\mu}, \tag{3.49}$$

$$2\int_{-1}^{+1} I(\tau,\mu',\mu)i(\mu')\mu'\,\mathrm{d}\mu' = \left[Mu(\mu)\mathrm{e}^{-k\tau} - i(\mu)\mathrm{e}^{-\tau/\mu}\right]\mu, \tag{3.50}$$

$$2\int_{-1}^{+1} I(\tau,-\mu)i(-\mu)\mu\,\mathrm{d}\mu = N\mathrm{e}^{-k\tau}, \tag{3.51}$$

$$2\int_{-1}^{+1} I(\tau,-\mu)i(\mu)\mu\,\mathrm{d}\mu = \mathrm{e}^{k\tau}. \tag{3.52}$$

One can write one more integral, which is provided without proof

$$2\int_{-1}^{+1} I(\tau,-\mu',\mu)I(\tau_1,-\mu')\mu'\,\mathrm{d}\mu' = [I(\tau_1,\mu)\mathrm{e}^{-\tau/\mu}$$

$$-I(\tau+\tau_1,\mu) + 2\int_0^1 I(\tau+\tau_1,-\mu')\rho(\mu,\mu')\mu'\,\mathrm{d}\mu']\mu, \tag{3.52a}$$

The following equations are the conservative limits of (3.49) and (3.52) (or (3.50) and (3.51)):

$$2\int_{-1}^{+1} I_0(\tau,\mu',\mu)\mu'\,\mathrm{d}\mu' + \mu\mathrm{e}^{-\tau/\mu} = 0, \tag{3.53}$$

$$2\int_{-1}^{+1} I_0(\tau,\mu)\mu\,\mathrm{d}\mu = -1, \tag{3.54}$$

These formulas (3.53–54) are the flux integrals for the Milne problem and the parallel external flux problem respectively (see (1.87)). The minus unity in (3.54) means that the flux of radiation is directed outward. Equations (3.49–52) can also be used to derive the expressions for the K integral for nearly conservative scattering. This problem will be considered later (see Sect. 3.11.5).

Also of certain interest is the conservative limit of (3.47), which looks as follows:

$$8\int_{-1}^{+1} I_0(\tau,-\mu)I(\tau_1,\mu)\mu\,\mathrm{d}\mu = (3-x_1)(\tau-\tau_1). \tag{3.55}$$

This means that the optical distance between two levels can, up to constant factor $3-x_1$, be found by measuring the Milne intensity at these levels.

To conclude the section one should note that the integrals of the transfer equation, i.e. (3.46–52), together with relation (3.7) are very useful in solving corresponding problems of the theory of light scattering.

3.8 Separation of Variables. Angular Relaxation of Photons

Many problems of multiple light scattering are difficult to solve because the radiation intensity depends on many variables and parameters. For example, the parallel external flux problem involves one spatial (τ) and two angular (μ and φ) variables; it also involves parameters μ_0 and λ. For anisotropic scattering we have a number of quantities x_n characterizing the phase function. In considering a layer of finite optical thickness, one has to allow for one more parameter, namely optical thickness τ_0 of atmosphere. Thus, even if we were able to calculate the radiation intensity at a "low" price, it would be extremely difficult to analyze the results obtained. Therefore, the cases when the number of variables can be reduced are very important for the theory of radiative transfer. Usually this reduction is due to the separation of variables and parameters.

There are two kinds of the separation of variables, namely, rigorous separation and asymptotic separation. We have already encountered the first kind in Sect. 3.5, when we found that the reflection coefficient, or its m-th harmonic to be more exact, can be represented by a superposition of functions each depending on only one variable (see (3.36)). This separation is based on the expansion of the radiation intensity in terms of azimuthal harmonics which, in turn, is due to the expansion of the phase function in terms of Legendre polynomials.

In Sect. 3.4 we found that positive and negative domains of the variable μ (see (3.26–29) and relevant comments) get separated. This could also be considered as a kind of rigorous separation of angular variables.

The nature of asymptotic separation of variables is different from that of the rigorous one. When a photon undergoes multiple scattering, ***angular relaxation*** takes place, i.e., the photon gradually "forgets" its initial direction of movement, rather like a pedestrian does after having been captured by a chaotic crowd. The angular relaxation is a very fast process and depends on extension of the phase function. Even for very extended phase functions after a small number of scatterings the information about the initial direction of movement is almost completely lost. For example, in the case of Henyey-Greenstein phase function (1.15), the degree of isotropization of radiation after n scatterings is determined by the following expression:

$$\chi_H(\gamma, g^n) = \frac{1 - g^{2n}}{(1 + g^{2n} - 2g^n \cos\gamma)^{3/2}},$$

which follows from (3.108a).

The angular relaxation of photons is a very general phenomenon and can be observed in both homogeneous and inhomogeneous media. Furthermore, even geometrical features have no significant influence on the process of angular relaxation. The only important thing is that a photon should undergo sufficiently many scatterings.

The above-mentioned points sometimes allow us to guess the form of the solution of a problem without solving the problem itself, which may turn out to be very difficult. A knowledge of the form of the solution may substantially facilitate the solution of the problem. Examples of this will be given in the next section as well as in other sections of the book.

3.9 Radiation Field in Deep Atmosphere Layers

Let us consider the parallel external flux problem and try to find out how the intensity $I(\tau, \mu, \mu_0, \varphi - \varphi_0)$ varies with depth τ.

First of all, a photon "forgets" its initial azimuth $\varphi - \varphi_0$ as the depth increases. This implies that at sufficiently large depth, the radiation intensity becomes independent of the azimuth of its propagation. Hence, we need only consider the zeroth harmonic of the radiation intensity, i.e., the function $I(\tau, \mu, \mu_0)$. The other harmonics are negligibly small.

One can speak about the azimuth of the photon propagation if there is a singled-out vertical plane. For the parallel external flux problem this is the plane where the source of radiation (sun) is placed. When τ is sufficiently large, the observer cannot observe the source and, therefore, cannot speak about the azimuth.

Secondly, the photon "forgets" the initial direction of movement μ_0 for sufficiently large depths. As a result, the relative angular structure of the radiation field becomes identical to that of the problem for two-sided infinity, i.e., it is characterized by function $i(\mu)$. In other words, for $\tau \gg 1$ one can write

$$I^{\text{as}}(\tau, \mu, \mu_0) = C(\tau, \mu_0) i(\mu), \tag{3.56}$$

where $C(\tau, \mu_0)$ is a normalizing factor to be found below. Superscript "as" refers to sufficiently large optical depths ($\tau \gg 1$). In order to find C, we substitute (3.56) into the *rigorous* relation (3.50). Taking (2.28) into account, as well as the fact that for $\tau \gg 1$ the second term on the right-hand side of (3.50) should be discarded[1], we ultimately arrive at the following Sobolev asymptotic formula:

$$I^{\text{as}}(\tau, \mu, \mu_0) = i(\mu) u(\mu_0) \mathrm{e}^{-k\tau} \mu_0. \tag{3.57}$$

One can easily check that (3.57) satisfies not only the integral relations (3.50), (3.49), and (3.7) but also the radiative transfer equation.

With regard to the radiation intensity in the Milne problem, the following van de Hulst formula can be easily derived by substituting (3.57) into (3.6):

$$MI^{\text{as}}(\tau, \mu) = i(-\mu)\mathrm{e}^{k\tau} - Ni(\mu)\mathrm{e}^{-k\tau} \tag{3.58}$$

which holds for $\tau \gg 1$. N is determined by (3.40).

---

[1] As $k \in [0, 1]$ and $\mu^{-1} \geq 1$, $\exp(-k\tau) \gg \exp(-\tau/\mu)$ for $\tau \gg 1$.

The structure of (3.58) can be also determined from physical arguments. Indeed, the radiation field at large depths in the Milne problem is affected by the following two factors. The first is that the sources lying at infinitely large depth contribute to the radiation field with an intensity proportional to $i(-\mu)$. The second factor is that the radiation field is also influenced by "negative" sources due to the photon outflow through the boundary $\tau = 0$. The contribution of the "negative" sources is proportional to $i(\mu)$. Finally, one can write

$$I^{\text{as}}(\tau, \mu) = C_1(\tau) i(-\mu) + C_2(\tau) i(\mu). \tag{3.59}$$

Unknown quantities $C_1(\tau)$ and $C_2(\tau)$ can be easily found with the help of integrals (3.51) and (3.52). As a result, we again arrive at the formula (3.58).

Equation (3.57) suggests that if the scattering is conservative ($\lambda = 1$), then the radiation field becomes isotropic

$$I_0^{\text{as}}(\tau, \mu, \mu_0) = u(\mu_0)\mu_0, \tag{3.60}$$

i.e., it becomes independent of the spatial coordinate τ. For the Milne problem, the use of (3.15) yields

$$I_0^{\text{as}}(\tau, \mu) = \frac{3}{4}\left[\left(1 - \frac{x_1}{3}\right)\tau - \mu + \gamma_0\right], \tag{3.61}$$

where

$$\gamma_0 = \int_0^1 u_0(\mu)\mu^2 \, d\mu. \tag{3.62}$$

One sees that the complete isotropization does not occur, although, being very close to the isotropic one, the radiation field grows with τ. The numerical calculations of $\delta_0 = 2\gamma_0$ for both the Rayleigh and Henyey-Greenstein phase functions are presented in the Table XXI for different values of g. Thus, it can be seen that the dependence of δ_0 on the phase function is very weak.

Now is the time to prove that k in (3.4) plays the role of the diffusion exponent. For this, we use the asymptotic formulas (3.57) and (3.58), where k is, as can be easily proved, the above quantity. Substituting (3.57) and (3.58) into (3.4) and assuming $\tau_1, \tau_2 \gg 1$, one arrives at the conclusion that the equality holds only if the quantity k in (3.4) is in fact the diffusion exponent.

Let us now summarize all these results before proceeding to the next section. Sect. 3.5 dealt with how to calculate the reflection coefficient, i.e., the radiation intensity at the boundary of the medium ($\tau = 0$). Formulas (3.57) and (3.58) allow us to evaluate the radiation field in deep layers ($\tau \gg 1$). Hence, the only thing to be done is to develop a way of calculating the intensity at a surface layer of the atmosphere. This is what we will do in the next section. And in closing, we should note that the estimation of the accuracy of (3.57) and (3.58) will be given in Sect. 4.8.4.

3.10 Doubling Formula. Radiation Field in an Atmospheric Surface Layer

The basic invariance relation (3.2) is a theorem of addition, that enables one to find the intensity at a depth $\tau + \tau_1$ by relying on the integration and the known intensities at the depths τ and τ_1. Neither iterations nor solution of equations are necessary for this method.

The ***doubling formula***

$$I^m(2\tau, \mu, \mu_0) = I^m(\tau, \mu, \mu_0) \left[\mathrm{e}^{-\tau/\mu_0} + \theta(\mu)\mathrm{e}^{-\tau/\mu} \right] + 2\int_0^1 I^m(\tau, \mu', \mu_0) I^m(\tau, \mu, \mu') \, \mathrm{d}\mu' \tag{3.63}$$

is an important particular case of the basic invariance relation. This formula follows from (3.2) for $\tau_1 = \tau$. It can be used to calculate radiation fields in semi-infinite homogeneous atmospheres.

The general scheme of the method is very simple. First we find an azimuthal harmonic $\rho^m(\mu, \mu_0)$ of the reflection coefficient (it can, for instance, be done by the method described in Sect. 3.5). Then we choose a sufficiently small initial τ and calculate $I^m(\tau, \mu, \mu_0)$ by the formulas of single or double scattering. For this case, these formulas are relatively simple and allow us to express $I^m(\tau, m, \mu_0)$ in terms of the intensity at the lower boundary of a layer of thickness τ, which is illuminated by direct radiation from above, and by the radiation diffusely reflected from the semi-infinite medium beneath, from below. The subsequent application of the doubling formula (3.63) allows us to find the intensity at depths of 2τ, 4τ, $8\tau, \ldots$, etc. Combining (3.63) and (3.2) we can construct an algorithm providing intensities at equidistant depths with an arbitrary step τ. When the optical depth is large, it is more sensible, of course, to use the asymptotic formula (3.57) to calculate the zeroth harmonic of the intensity.

It is very important that the above relations allow us to check the "integral" accuracy of the calculations with respect to both angular variables, μ and μ_0. For the zeroth azimuthal harmonic of intensity we can use (3.7), (3.42), (3.46), (3.49), and (3.50) for this purpose. Equation (3.26) together with (3.28) and (3.46) can be used for the harmonics of higher order ($m \geq 1$).

It is also important that this method of calculation requires the intensity only of *descending* radiation ($\mu \in [0, 1]$) to be known, since the intensity of ascending radiation can easily be found by means of (3.17). For more details see [50], where the author, Zh. Dlugach, has shown that the method of calculation described above is highly efficient and very stable from a numerical point of view.

Now we are to analyze some of the results described in [50], which will allow us to get a general idea about the way the radiation field varies in the surface layer of a semi-infinite atmosphere.

The calculations were made for the Henyey-Greenstein phase function (1.15) for $g = 0; 0.5; 0.75$ and 0.85. The initial τ was taken as 2^{-14} and the integration was carried out by the Gaussian method with eight (for $g \leq 0.75$) and twelve (at $g = 0.85$) nodes. The harmonics $\rho^m(\mu, \mu_0)$ were calculated as described in Sect. 3.5 with λ being equal to 1 and 0.9.

For extended phase functions, I^m turned out to be almost independent of the particular value of g. Therefore, we present the results only for $g = 0.75$. For several cases we compare the anisotropic scattering with isotropic scattering ($g = 0$) to get a general idea about the peculiarities of the radiation field brought about by the anisotropy of scattering. Slight absorption ($\lambda = 0.9$) does not result in any qualitative changes. Therefore, the figures to be presented mostly relate to the case of conservative scattering. Nonetheless, all the peculiarities caused by absorption will be discussed whenever necessary.

We will consider the zeroth azimuthal harmonic and higher harmonics ($m \geq 1$) separately since their behavior differ substantially.

a. Azimuth-averaged radiation intensity (zeroth azimuthal harmonic). Figures 3.2 and 3.3 show how the logarithm of the ratio of the zeroth harmonic of the intensity of descending radiation $I(\tau, \mu, \mu_0)$ to $I(\tau, \mu_c, \mu_0)$ for conservative scattering depends on μ with $\mu_c = 0.980\ldots$ and various $\tau = 2^n$. The case of $\mu_0 = 0.408\ldots$ is presented in Fig. 3.2 whereas that of $\mu_0 = 0.980\ldots$ is shown in Fig. 3.3. The values of μ_c and μ_0 are the corresponding nodes of the Gaussian quadrature formula. The solid lines correspond to $g = 0.75$ and the broken ones to $g = 0$.

The figures suggest that for small τ the intensity strongly depends on μ, and the calculations show that this dependence grows even stronger with g. This behavior may be ascribed to the form of the phase function, since single scattering plays an important role for small τ. Since for the spherical phase function the ratio $i(\tau, \mu, \mu_0)/I(\tau, \mu_c, \mu_0)$ depends only weakly on μ_0, the profiles are provided only for $\mu_0 = 0.408\ldots$. On the other hand, the comparison between Fig. 3.2 and Fig. 3.3 suggests that there is a substantial dependence on μ_0 if the phase function is extended.

The dependence on μ becomes weaker with the increase of the optical depth since the role of single scattering becomes less important. The figures show that for isotropic scattering with $\tau \simeq 2$, and for $g = 0.75$ with $\tau \simeq 4$, the regime turns out to be close to an asymptotic. Equation (3.57) provides three correct significant digits for $\tau = 8$ with $g = 0$ and for $\tau = 16$ with $g = 0.75$. So, if the phase function becomes more extended, the optical depth rises where the regime approaches an asymptotic one. This corresponds with what should be expected, of course.

The quantities $I(\tau, \mu, \mu_0)/I(\tau, \mu_c, \mu_0)$ for ascending radiation with $\lambda = 1$ are given in Fig. 3.4 ($g = 0$) and Fig. 3.5 ($g = 0.75$). The solid lines correspond to $\mu_0 = 0.408\ldots$, and the broken ones to $\mu_0 = 0.980\ldots$. It can be easily seen that the zeroth harmonic of the ascending radiation depends on μ in a relatively weak manner even for small τ. This can be ascribed to the fact

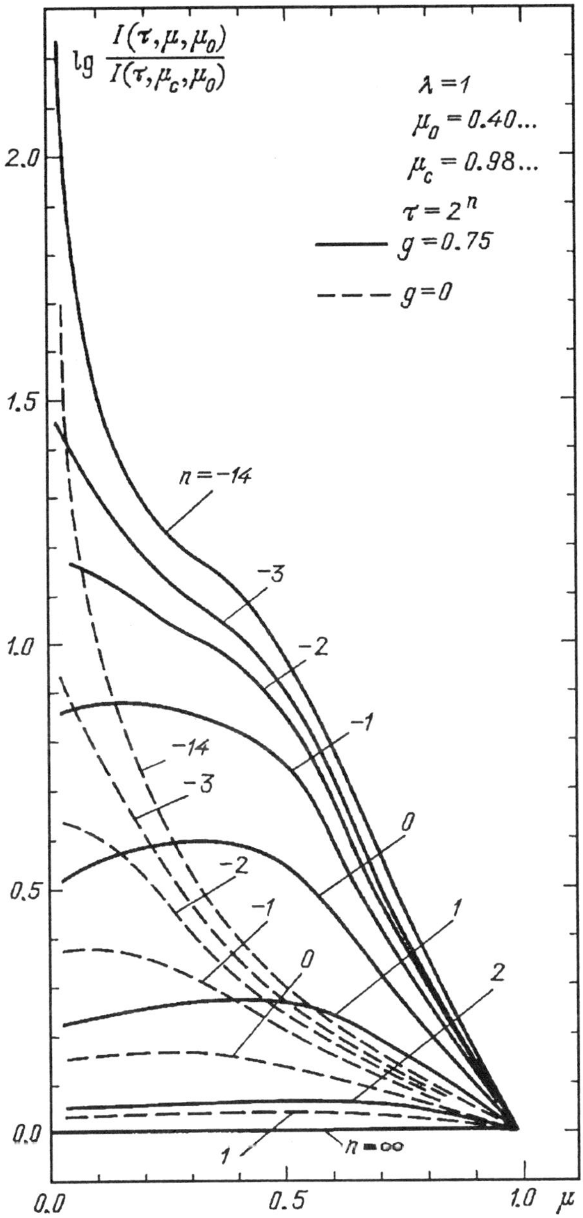

Fig. 3.2. Logarithm of relative variation of the intensity of descending ($\mu > 0$) radiation as a function of μ at different optical depths for $\mu_0 = 0.40\ldots$. Broken line corresponds to isotropic scattering ($g = 0$) and solid line to anisotropic that ($g = 0.75$)

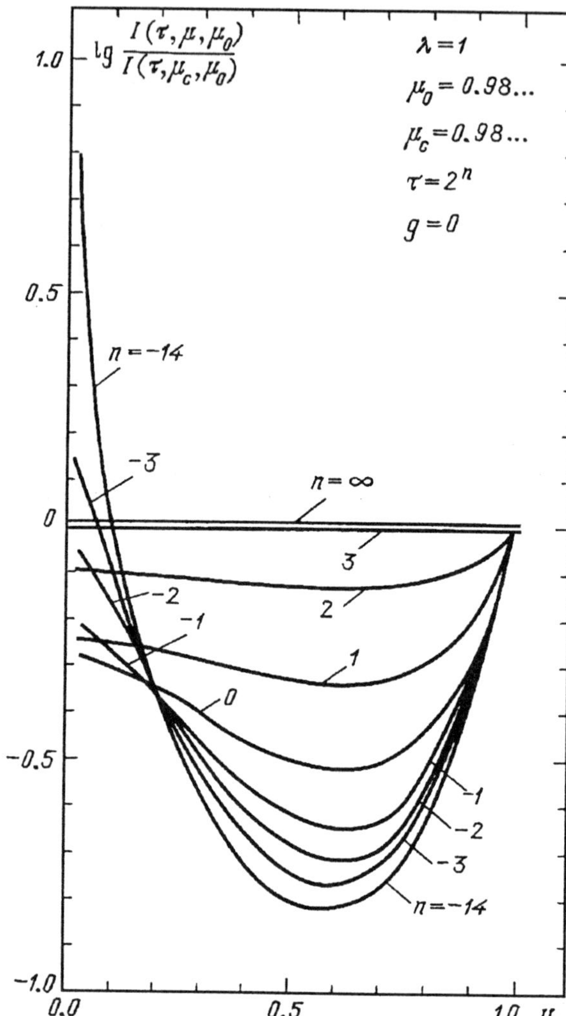

Fig. 3.3. As Fig. 3.2, with $\mu_0 = 0.98\ldots$ and $g = 0$

that the most important contribution to the ascending radiation is due to the photons that undergo multiple scattering. Therefore, the impact of the form of the phase functions is comparatively weak. Of course, this impact becomes slightly stronger with the increase in anisotropy. The asymptotic regime is reached at the values of τ that approximately equal those for the descending radiation.

The following four figures give us an idea about different ways to reach the asymptotic regime. Figures 3.6 ($\lambda = 1$) and 3.7 ($\lambda = 0.9$) show how $\lg I(\tau, \mu_c, \mu_0)$ depends on τ at $\mu_c = 0.980\ldots$. The solid lines correspond to the case when $g = 0.75, \mu_0 = 0.408\ldots$, the dotted ones to $\mu_0 = 0.980\ldots$, and the dashed line to $g = 0, \mu_0 = 0.408\ldots$. Similar graphs are provided in Fig. 3.8 ($\lambda = 1$) and Fig. 3.9 ($\lambda = 0.9$) for $\mu_c = -0.980\ldots$ (the ascending radiation).

These figures allow us to draw the following conclusions:

1. For the same phase function, the increase of absorption results in an increase of the optical depth where the asymptotic regime is reached.
2. If λ remains unchanged and g rises, the value of τ increases when the asymptotic regime is reached.
3. An increase of the angle at which radiation falls on the atmospheric boundary results in a decrease of the optical depth where the asymptotic regime is reached.

All the points mentioned above can be easily understood if we take into account the that fact the most of the photons that reach the depth of the asymptotic regime have undergone multiple scattering.

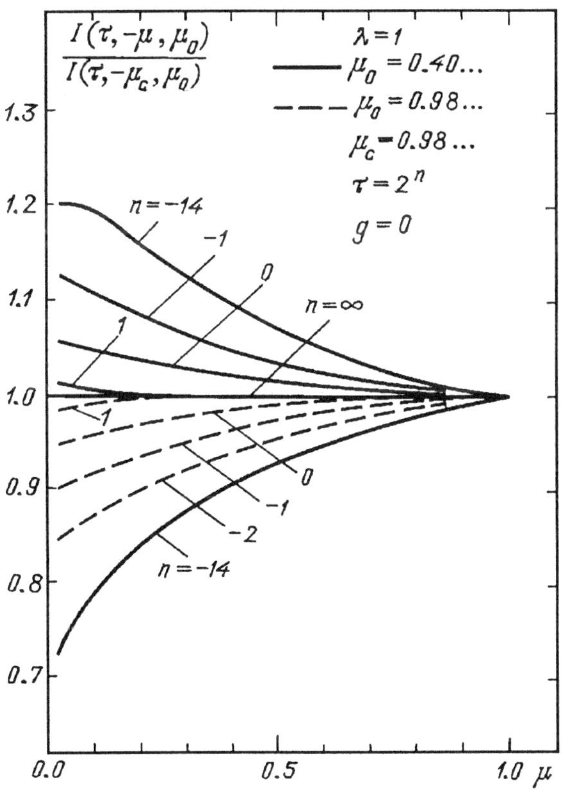

Fig. 3.4. Relative variation of the zeroth harmonic of the intensity of ascending ($\mu < 0$) radiation as a function of μ at different optical depths for isotropic scattering. Solid line: $\mu_0 = 0.40\ldots$; dotted line: $\mu_0 = 0.98\ldots$

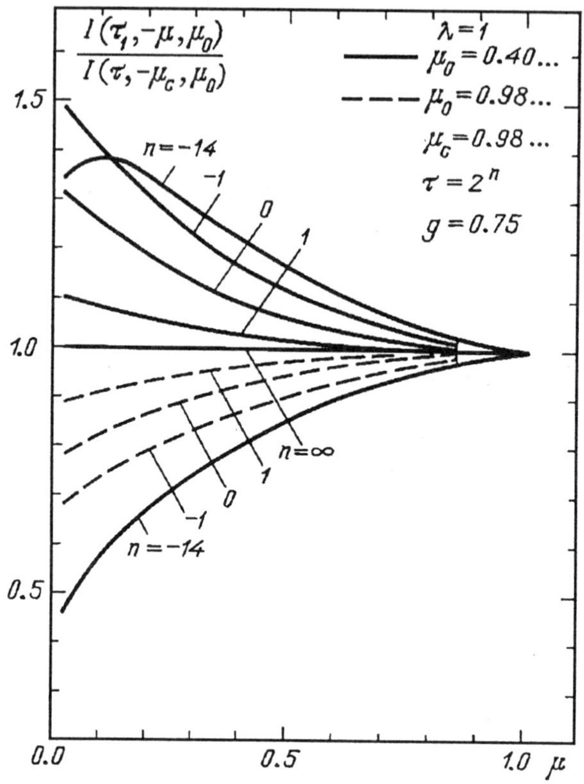

Fig. 3.5. As Fig. 3.4 for the case of anisotropic scattering, $g = 0.75$

b. Azimuthal harmonics of higher orders $(m \geq 1)$**.** Figure 3.10 reflects some results of the calculation of harmonics of higher order, showing the way $I^m(\tau, \mu, \mu_0)/I(\tau, \mu, \mu_0)$ depends on μ at three optical depths with $\mu_0 = 0.408\ldots$ for conservative scattering. The value of $g = 0.75$ was assumed.

The figure suggests that at small optical depths, the contribution of azimuthal harmonics to $I(\tau, \mu, \varphi - \varphi_0)$ is substantial with the contribution for descending radiation being much greater then that for ascending radiation. This may be due to single scattering, which in the main determines the intensity of the descending radiation at small τ. Calculations show that although the absolute values of the harmonics decrease as the absorption increases, their *relative* contribution to the radiation intensity becomes more valuable.

Up to $\tau \simeq 1$, the quantity $I^m(\tau, \mu, \mu_0)/I(\tau, \mu, \mu_0)$ very slowly decreases with the increase of τ; after that the harmonics drop abruptly. At the depth $\tau \simeq 8$ they can be neglected since, for instance at $g = 0.75$, the regime becomes asymptotic and the radiation intensity is sufficiently well determined by the asymptotic formula (3.57).

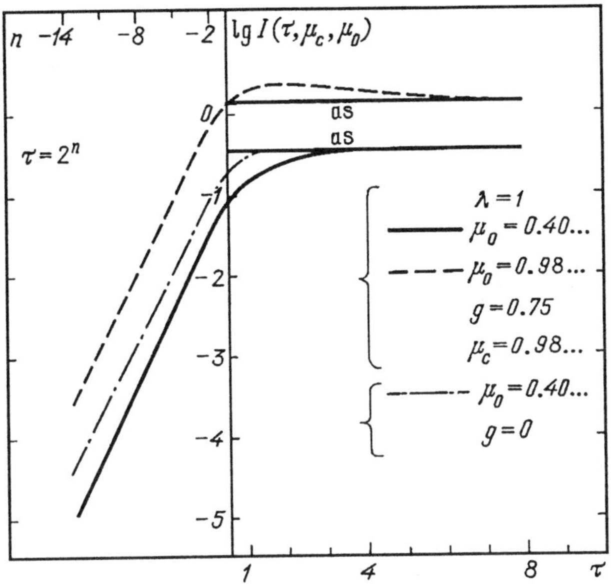

Fig. 3.6. Logarithm of zeroth azimuthal harmonic of intensity of descending ($\mu_c > 0$) radiation as a function of optical depths for conservative scattering. Dashed line corresponds to isotropic scattering ($g = 0$) with $\mu_0 = 0.40\ldots$; solid line to anisotropic scattering ($g = 0.75$) with $\mu_0 = 0.40\ldots$ and broken line to anisotropic scattering with $\mu_0 = 0.98\ldots$. The quantity n is plotted as abscissa to the left of ordinate in place of optical depth τ ($\tau = 2^n$)

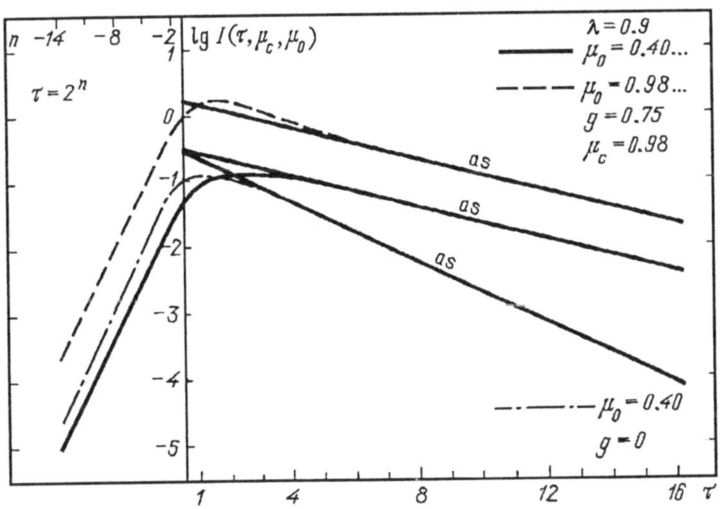

Fig. 3.7. As Fig. 3.6 for $\lambda = 0.9$

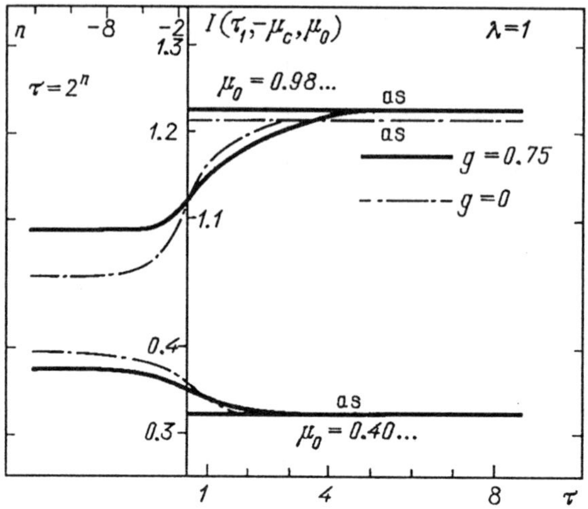

Fig. 3.8. As Fig. 3.6 for zeroth azimuthal harmonics of intensity of ascending ($\mu_c < 0$) radiation

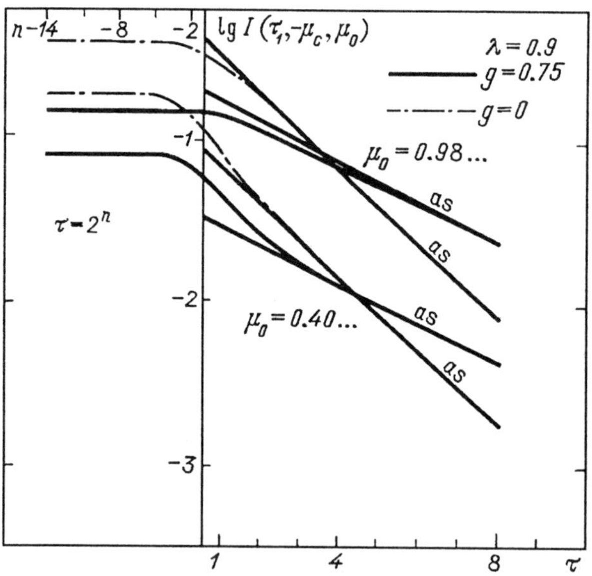

Fig. 3.9. As Fig. 3.8 for the case of non-conservative scattering with $\lambda = 0.9$

Calculations show that the increase of the angle of incidence of illuminating radiation brings about an increase in the contribution of the azimuthal harmonics to the total radiation intensity. Due to the behavior of the harmonics of the phase function, this is exactly what should be expected since single scattering becomes more important with the decrease of μ_0.

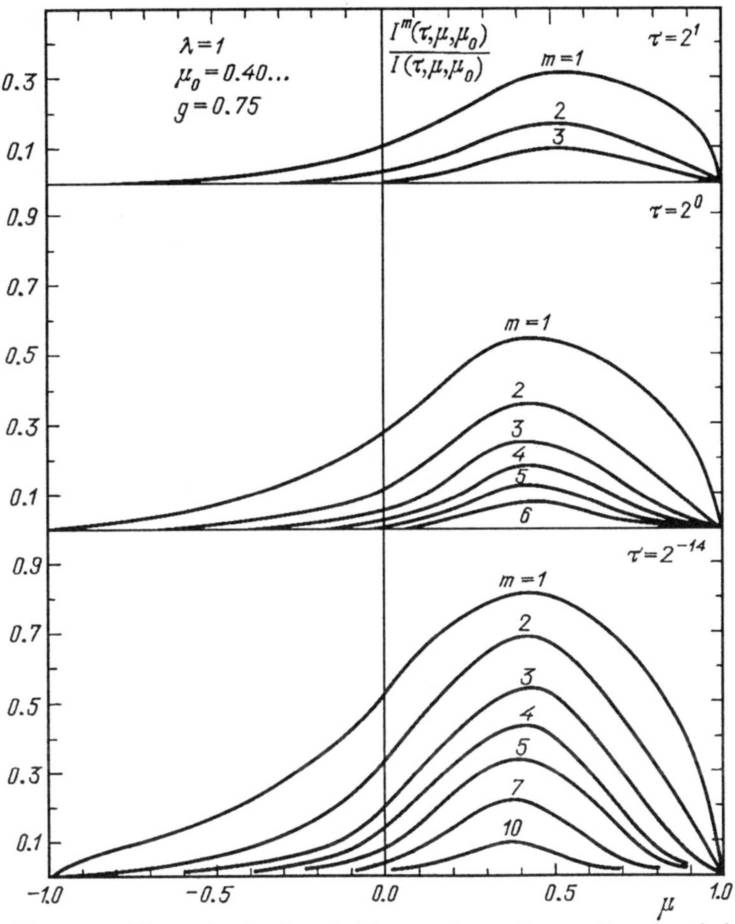

Fig. 3.10. The ratio of azimuthal harmonics to the zeroth azimuthal harmonic for different optical depths and $\mu_0 = 0.40\ldots$, in conservative scattering

3.11 Atmosphere with Nearly Conservative Scattering

In Section 2.6 we have already mentioned the practical importance of this case. For instance, if we have planetary atmospheres in mind, then the situation when $(1 - \lambda) \ll 1$ is encountered for the visible part of the spectrum of the atmospheres of the Earth, Venus and giant planets. At the same time, it is the case of weak true absorption that causes the greatest problems in calculations.

There is, however, another very important aspect of this problem. Some of the quantities characterizing the radiation field in a nearly conservatively scattering media can be represented by very simple rigorous formulas, some of which have an explicitly analytical form. Obviously, such formulas are of great significance, especially for the solution of inverse problems.

The present paragraph deals with the following problem. For a given semi-infinite atmosphere, one needs to express $I(\tau,\mu,\mu_0)$ and $I(\tau,\mu)$ in terms of conservative scattering (i.e., in terms of $I_0(\tau,\mu,\mu_0)$ and $I_0(\tau,\mu)$). The subscript "0" is, as a rule, used to denote quantities relating to conservative scattering ($\lambda = 1$). Of course, we consider only the zeroth azimuthal harmonic of the intensity in the parallel external flux problem.

3.11.1 Initial Relations

We denote

$$J(\tau,\mu,\mu_0) = I_0(\tau,\mu,\mu_0) - I(\tau,\mu,\mu_0), \tag{3.64}$$

$$J(\tau,\mu) = I_0(\tau,\mu) - I(\tau,\mu). \tag{3.65}$$

Then the following relations can be derived from the transfer equation for corresponding problems (see (1.49) and (1.50) at $m = 0$ as well as (1.67) and (1.68))

$$\mu\frac{\mathrm{d}J(\tau,\mu,\mu_0)}{\mathrm{d}\tau} + J(\tau,\mu,\mu_0) = \frac{1}{2}\int_{-1}^{+1} \chi(\mu,\mu')J(\tau,\mu',\mu_0)\,\mathrm{d}\mu'$$
$$+\frac{1-\lambda}{2}\left[\int_{-1}^{+1} \chi(\mu,\mu')I(\tau,\mu',\mu_0)\,\mathrm{d}\mu' + \frac{1}{2}\chi(\mu,\mu_0)\mathrm{e}^{-\tau/\mu_0}\right], \tag{3.66}$$

$$\mu\frac{\mathrm{d}J(\tau,\mu)}{\mathrm{d}\tau} + J(\tau,\mu) = \frac{1}{2}\int_{-1}^{+1} \chi(\mu,\mu')J(\tau,\mu')\,\mathrm{d}\mu'$$
$$+\frac{1-\lambda}{2}\int_{-1}^{+1} \chi(\mu,\mu')I(\tau,\mu')\,\mathrm{d}\mu'. \tag{3.67}$$

Let $\lambda \to 1$. We use subscript "1" to denote the solutions of (3.66) and (3.67) after the terms proportional to $1-\lambda$ have been discarded. Then, instead of (3.66) and (3.77), we obtain

$$\mu\frac{\mathrm{d}J_1(\tau,\mu,\mu_0)}{\mathrm{d}\tau} + J_1(\tau,\mu,\mu_0) = \frac{1}{2}\int_{-1}^{+1} \chi(\mu,\mu')J_1(\tau,\mu',\mu_0)\,\mathrm{d}\mu', \tag{3.68}$$

$$\mu\frac{\mathrm{d}J_1(\tau,\mu)}{\mathrm{d}\tau} + J_1(\tau,\mu) = \frac{1}{2}\int_{-1}^{+1} \chi(\mu,\mu')J_1(\tau,\mu')\,\mathrm{d}\mu'. \tag{3.69}$$

In other words, (3.68) and (3.69) involve (implicitly) only the terms proportional to $k_0 = [(1-\lambda)(3-x_1)]^{1/2}$. Equations (3.68) and (3.69) are easily seen to coincide with the Milne equation. There are four linearly independent solutions of (3.69) and $I_1(\tau,\mu)$, $I_0(\tau,\mu)$, as well as $I_1^{\mathrm{as}}(\tau,\mu)$ and $I_0^{\mathrm{as}}(\tau,\mu)$. Therefore, allowing for (3.65) and (3.69) one can write

$$I_0(\tau,\mu) - I_1(\tau,\mu) = c_1 I_1(\tau,\mu) + c_2 I_0(\tau,\mu) + c_3 I_1^{\mathrm{as}}(\tau,\mu) + c_4 I_0^{\mathrm{as}}(\tau,\mu), \quad (3.70)$$

where $c_j (j = \overline{1,4})$ are constants to be found with the help of the boundary conditions. This will turn the formal solution (3.70) into the required physical solution.

We assume that $\tau \to \infty$ in (3.70). Then $I_0(\tau,\mu) \to I_0^{\mathrm{as}}(\tau,\mu)$ and $I_1(\tau,\mu) \to I_0^{\mathrm{as}}(\tau,\mu)$. As a result, one finds $c_1 + c_3 = -1$; $c_2 + c_4 = 1$; i.e., instead of (3.70), we have ($\tau \geq 0$)

$$I_1(\tau,\mu) - I_1^{\mathrm{as}}(\tau,\mu) = -\frac{c_4}{c_3}[I_0(\tau,\mu) - I_0^{\mathrm{as}}(\tau,\mu)]. \quad (3.71)$$

As one can see, only one constant, c_4/c_3, remains to be determined, for which we have the boundary condition $J_1(0,\mu) = 0$ for $\mu \in (0,1]$.

Let us leave the Milne problem for a while and proceed to the parallel external flux problem. It follows from (1.49) and (1.50) that for $\tau \to \infty$ these equations also turn into the Milne equation. Therefore, two more solutions, $I_0^{\mathrm{as}}(\tau,\mu,\mu_0)$ and $I_1^{\mathrm{as}}(\tau,\mu,\mu_0)$, should be added to the four linearly independent solutions mentioned above. Therefore, allowing for (3.64), a formal solution of (3.68) can be written as follows:

$$\begin{aligned} I_0(\tau,\mu,\mu_0) - I_1(\tau,\mu,\mu_0) &= a_1 I_1^{\mathrm{as}}(\tau,\mu,\mu_0) + a_2 I_0^{\mathrm{as}}(\tau,\mu,\mu_0) \\ &+ a_3 I_1(\tau,\mu) + a_4 I_0(\tau,\mu) + a_5 I_1^{\mathrm{as}}(\tau,\mu) + a_6 I_0^{\mathrm{as}}(\tau,\mu), \end{aligned} \quad (3.72)$$

where $a_j (j = \overline{1,6})$ are constants to be determined with the help of the boundary conditions. Assume $\tau \to \infty$ in (3.72). As a result one finds $a_1 = -1$, $a_2 = 1$; $a_3 + a_5 = 0$; $a_4 + a_6 = 0$. Using (3.71) one can rewrite (3.72) in the following way:

$$\begin{aligned} &I_1(\tau,\mu,\mu_0) - I_1^{\mathrm{as}}(\tau,\mu,\mu_0) \\ &= I_0(\tau,\mu,\mu_0) - I_0^{\mathrm{as}}(\tau,\mu,\mu_0) + \alpha(\mu_0)[I_0(\tau,\mu) - I_0^{\mathrm{as}}(\tau,\mu)], \end{aligned} \quad (3.73)$$

where $\alpha(\mu_0) = a_6 + a_3 c_4/c_3$. Functions $I_0^{\mathrm{as}}(\tau,\mu,\mu_0)$ and $I_0^{\mathrm{as}}(\tau,\mu)$ are provided by (3.60) and (3.61), respectively, and it follows from (3.57) and (2.43) that

$$I_1^{\mathrm{as}}(\tau,\mu,\mu_0) = \left(1 + \frac{3k_0}{3 - x_1}\mu\right) \mathrm{e}^{-k_0\tau} u_1(\mu_0)\mu_0. \quad (3.74)$$

Although simple in structure, (3.71) and (3.73) deserve to be discussed in detail. The point is that they can be derived at once with the help of physical reasoning. Let us consider (3.71) for the Milne problem first.

Function $F_1(\tau,\mu) = I_1(\tau,\mu) - I_1^{\mathrm{as}}(\tau,\mu)$ is a certain characteristic of the radiation field in a *surface* atmospheric layer. By this we mean a layer of optical thickness $\tau^{\mathrm{as}} \gg 1$, for which the quantity $F_1(\tau^{\mathrm{as}},\mu)$ is infinitesimal at the boundary $\tau = \tau^{\mathrm{as}}$. The atmosphere considered is a nearly conservative one, which means that *most photons* perish at depths $\tau > \tau^{\mathrm{as}}$. Hence, any characteristic of the radiation field in the surface layer (F_1 inclusive) is *asymptotically identical* to that in a conservatively scattering one within a constant factor which takes into account the fact that the number of photons falling onto the lower boundary is less than that in a conservative medium. Thus, $0 < (-c_4/c_3) \leq 1$ and the equality is reached when $\lambda = 1$.

Consider now the parallel external flux problem. Here the surface layer is illuminated by parallel rays, from one side, and from the other side photons leave the layer entering the semi-infinite medium with $\tau > \tau^{\rm as}$. Let us again consider a characteristic of the surface layer of the medium $F_1(\tau, \mu, \mu_0) = I_1(\tau, \mu, \mu_0) - I^{\rm as}(\tau, \mu, \mu_0)$. Photons perish mostly in the half-space $\tau > \tau^{\rm as}$. Therefore, after the addition of a small term which takes the outflow of photons through the lower boundary of the layer into account, this characteristic can be considered equal to $F_0(\tau, \mu, \mu_0)$. At the same time, the outflow can be considered as a "negative source" of photons coming out of the large depths $\tau > \tau^{\rm as} \gg 1$. Such a source can, as we have already established, be described by the function $F_0(\tau, \mu)$ within a constant factor. As a result we arrive at the asymptotic formula (3.73), where $0 \leq -\alpha(\mu_0) \ll 1$. This is one more example of asymptotic separation (see Sect. 3.8) of angular variables (the third term on the right hand side of (3.73)), which arises in the study of so-called *nearly conservative asymptotics*.

3.11.2 Radiation Intensity at the Boundary of an Atmosphere

We assume that $\tau = 0$ in (3.73). Taking into account the boundary condition $I(0, \mu, \mu_0) = 0$ for $\mu > 0$ as well as (3.60), (3.61), and (3.74), we obtain

$$\left[u_0(\mu_0) - u_1(\mu_0)\left(1 + \frac{3k_0}{3 - x_1}\mu\right)\right]\mu_0 + \frac{3}{4}\alpha(\mu_0)(\gamma_0 - \mu) = 0$$

which can, by putting $\mu = 0$, be rewritten as follows

$$[u_0(\mu_0) - u_1(\mu_0)]\,\mu_0 + \frac{3}{4}\gamma_0\alpha(\mu_0) = 0. \tag{3.75}$$

On the other hand, for the ascending radiation, instead of (3.73) we have

$$\begin{aligned}&[\rho_0(\mu, \mu_0) - \rho_1(\mu, \mu_0)]\mu_0 \\ &= u_0(\mu_0) - u_1(\mu_0)i_1(-\mu) - \alpha(\mu_0)\left[u_0(\mu_0) - \frac{3}{4}(\gamma_0 + \mu)\right],\end{aligned} \tag{3.76}$$

where $i_1(\mu)$ is determined by (2.43). Multiply (3.76) by $i_1(\mu)$ and integrate it with respect to μ from 0 to 1. As a result, using (2.43), (3.11), and (3.14), we obtain

$$u_1(\mu_0) = \left(1 - \frac{2k_0}{3 - x_1}\right)u_0(\mu_0) - \frac{\alpha(\mu_0)}{2\mu_0}\left(1 - \frac{3}{2}\gamma_0\right). \tag{3.77}$$

From (3.75) and (3.77), one derives

$$\alpha(\mu_0) = -\frac{4k_0}{3 - x_1} u_0(\mu_0)\mu_0, \tag{3.78}$$

$$u_1(\mu_0) = \left(1 - \frac{3k_0\gamma_0}{3 - x_1}\right) u_0(\mu_0) + O(k_0^2). \tag{3.79}$$

The last formula suggests that for weak true absorption the *relative* distribution of the radiation intensity at the atmospheric boundary in the Milne problem does not depend on the direction of the departure of photons. This can be ascribed to the fact that photons almost never perish in the optically thick surface layer, where the angular structure of the function $u(\mu)$ is formed. The loss of a comparatively small number of photons cannot result in considerable changes in the relative dependence of the radiation intensity.

The substitution of (3.78) and (3.79) into (3.76) yields

$$\rho_1(\mu, \mu_0) = \rho_0(\mu, \mu_0) - \frac{4k_0}{3 - x_1} u_0(\mu)u_0(\mu_0) + O(k_0^2), \tag{3.80}$$

which, together with (3.14), enables us to find the function $\rho_1(\mu, \mu_0)$, even if we only know the function $\rho_0(\mu, \mu_0)$.

Formulas (3.79) and (3.80) provide expressions of the intensity at the boundary of a semi-infinite atmosphere for nearly conservative scattering. They play a very important role in the whole asymptotic theory of weakly absorbing atmospheres and are one more example of the asymptotic separation of angular variables and parameters (see Sect. 3.8), which occur when light passes through a layer of large optical thickness.

Now we can get down to the primary objective of the present section, namely, to the derivation of explicit formulas for $I_1(\tau, \mu, \mu_0)$ and $I_1(\tau, \mu)$. Before doing this, one should however notice that (3.71) involves the function $I_1^{\mathrm{as}}(\tau, \mu)$, the general expression for which is provided by (3.58). The left-hand side of this formula includes M, which is of the order of magnitude of k_0 (see (2.44)). Hence, to find $I_1^{\mathrm{as}}(\tau, \mu)$ within the accuracy of the order of magnitude of k_0, we need to know all the quantities within the accuracy of the order of magnitude of k_0^2. First of all, this is applicable to N, which is determined by (3.40).

3.11.3 Asymptotic Formulas for N and C

We will find the asymptotic expression for

$$C = 2\int_0^1 u(\mu)\mu \, \mathrm{d}\mu, \tag{3.81}$$

simultaneously with the study of the asymptotic behavior of N when $\lambda \to \infty$. It is clear that C is the radiation flux at the boundary in the Milne problem.

Let us introduce $i(\mu)$, which is provided by (2.40) into (3.12) and (3.40). This is necessary to derive the asymptotic formulas mentioned above. The addition and subtraction of the expression obtained yields

$$1+N=2C+\frac{4k_0^2}{3-x_1}\int_0^1 u_0(\mu)\left[1+\frac{10P_2(\mu)}{5-x_2}\right]\mu\,\mathrm{d}\mu+O(k_0^4),$$

$$1-N=\frac{12k_0}{3-x_1}\int_0^1 u_1(\mu)\mu^2\,\mathrm{d}\mu+O(k_0^3).$$

Substituting (3.79) into this formula, we find

$$N=1-2\frac{3k_0\gamma_0}{3-x_1}+2\left(\frac{3k_0\gamma_0}{3-x_1}\right)^2+O(k_0^3), \tag{3.82}$$

$$C=1-\frac{3k_0\gamma_0}{3-x_1}+\left(\frac{9\gamma_0^2}{3-x_1}-\frac{5\epsilon_0-x_2}{5-x_2}\right)\frac{k_0^2}{3-x_1}+O(k_0^3), \tag{3.83}$$

where

$$\varepsilon_0=6\int_0^1 u_0(\mu)\mu^3\,\mathrm{d}\mu. \tag{3.84}$$

Numerical values of ε_0, which, like γ_0, depend on the phase function very weakly, are given in Table XXI. The values of C are presented in Table XVII.

3.11.4 Radiation Intensity at an Arbitrary Optical Depth

Introduce M, $i(\mu)$, and N into (3.58). The expressions for these quantities are provided by (2.42), (2.40) (where the term proportional to k_0^3 should be discarded), and (3.82), respectively. As a result one finds

$$I_1^{\mathrm{as}}(\tau,\mu)=\frac{3}{4}\left\{\frac{3-x_1}{3k_0}\sinh k_0\tau-\mu\cosh k_0\tau+\gamma_0\mathrm{e}^{-k_0\tau}\right.$$
$$\left.+\frac{3k_0\gamma_0(\mu-\gamma_0)}{3-x_1}\mathrm{e}^{-k_0\tau}+\frac{k_0}{3}\left[\frac{10P_2(\mu)-6}{5-x_2}-\frac{3}{2}\frac{2-x_1}{3-x_1}\right]\sinh k_0\tau\right\}+O(k_0^2). \tag{3.85}$$

Now we know all the formulas needed to derive the asymptotic expression sought.

Taking (3.78) into account, one can rewrite (3.73) as follows

$$I_1(\tau,\mu,\mu_0)=I_0(\tau,\mu,\mu_0)+I_1^{\mathrm{as}}(\tau,\mu,\mu_0)$$
$$-I_0^{\mathrm{as}}(\tau,\mu,\mu_0)-\frac{4k_0}{3-x_1}[I_0(\tau,\mu)-I_0^{\mathrm{as}}(\tau,\mu)]u_0(\mu_0)\mu_0, \tag{3.86}$$

where the expressions for functions $I_0^{\mathrm{as}}(\tau,\mu,\mu_0)$ and $I_0^{\mathrm{as}}(\tau,\mu)$ are provided by (3.60) and (3.61), respectively. Using (3.74) and (3.79) we get the following expression for I_1^{as}:

$$I_1^{\mathrm{as}}(\tau,\mu,\mu_0)=\left(1+3k_0\frac{\mu-\gamma_0}{3-x_1}\right)\mathrm{e}^{-k_0\tau}u_0(\mu_0)\mu_0. \tag{3.87}$$

With the help of (3.15), one can rewrite (3.86) in the expanded form

$$I_1(\tau,\mu,\mu_0) = I_0(\tau,\mu,\mu_0) - \frac{3k_0}{3-x_1}\Bigg[2\int_0^1 I_0(\tau,\mu,\mu')\mu'\,\mathrm{d}\mu'$$
$$-\left(\gamma_0 - \frac{3-x_1}{3k_0}\right)\left(1-\mathrm{e}^{-k_0\tau}\right) + \mu\left(\mathrm{e}^{-\tau/\mu}\theta(\mu) - \mathrm{e}^{-k_0\tau}\right)\Bigg] u_0(\mu_0)\mu_0 . \quad (3.88)$$

For the Milne problem, we put $\tau = 0$ in (3.71). Then using the boundary condition $I(0,\mu) = 0$ for $\mu \in (0,1]$, from (3.61) and (3.85) we have

$$-\frac{c_4}{c_3} = 1 - \frac{3k_0\gamma_0}{3-x_1}.$$

Therefore we ultimately find

$$I_1(\tau,\mu) = I_1^{\mathrm{as}}(\tau,\mu) + \left[I_0(\tau,\mu) - I_0^{\mathrm{as}}(\tau,\mu)\right]\left(1 - \frac{3k_0\gamma_0}{3-x_1}\right), \quad (3.89)$$

where functions $I_1^{\mathrm{as}}(\tau,\mu)$ and $I_0^{\mathrm{as}}(\tau,\mu)$ are determined by (3.85) and (3.61), respectively. After having been expanded, this formula takes the very bulky form

$$I_1(\tau,\mu) = I_0(\tau,\mu)\left(1 - \frac{3k_0\gamma_0}{3-x_1}\right) + \frac{3}{4}\Bigg\{\frac{3-x_1}{3}\left(\frac{\sinh k_0\tau}{k_0} - \tau\right)$$
$$-\mu(\cosh k_0\tau - 1) - \gamma_0\left(1-\mathrm{e}^{-k_0\tau}\right) - \frac{3k_0\gamma_0}{3-x_1}\Bigg[(\mu-\gamma_0)\left(1-\mathrm{e}^{-k_0\tau}\right)$$
$$-\frac{3-x_1}{3}\tau\Bigg] + \frac{k_0}{3}\left[2\frac{5P_2(\mu)-3}{5-x_2} - \frac{3}{2}\frac{2-x_1}{3-x_1}\right]\sinh k_0\tau\Bigg\}. \quad (3.90)$$

Even if we know only $I_0(\tau,\mu,\mu_0)$ then (3.88) and (3.90) enable us to find the radiation intensities $I_1(\tau,\mu,\mu_0)$ and $I_1(\tau,\mu)$ at an *arbitrary* depth in a weakly absorbing medium. It can be shown that for the higher harmonics ($m \geq 1$), the following expansion holds: $I^m(\tau,\mu,\mu_0) = I_0^m(\tau,\mu,\mu_0) + O(k_0^2)$. Hence, the intensity in the parallel external flux problem is

$$I_1(\tau,\mu,\mu_0;\varphi-\varphi_0) = I_1(\tau,\mu,\mu_0) + 2\sum_{m=1}^{\infty} I_0^m(\tau,\mu,\mu_0)\cos m(\varphi-\varphi_0) + O(k_0^2). \quad (3.91)$$

A similar relation is obviously valid for the reflection coefficient, namely, taking (3.80) into account, one gets

$$\rho_1(\mu,\mu_0;\varphi-\varphi_0) = \rho_0(\mu,\mu_0;\varphi-\varphi_0) - \frac{4k_0}{3-x_1}u_0(\mu_0)u_0(\mu_0) + O(k_0^2), \quad (3.91a)$$

where

$$\rho_0(\mu,\mu_0;\varphi-\varphi_0) = \rho_0(\mu,\mu_0) + 2\sum_{m=1}^{\infty}\rho_0^m(\mu,\mu_0)\cos m(\varphi-\varphi_0).$$

3.11.5 Radiation Flux and K Integral

Using the definition (1.87) of the projection of the total radiation flux onto the τ-axis and (3.88), one can easily derive the following simple asymptotic formula

$$\frac{H_\tau(\tau,\mu_0)}{4\pi} = \frac{k_0}{3-x_1} u_0(\mu_0) \mathrm{e}^{-k_0\tau}\mu_0 + O(k_0^2). \tag{3.92}$$

This formula is valid for *any* optical depth. The corresponding expression for the flux in the Milne problem takes the form

$$\frac{H_\tau(\tau)}{\pi} = \frac{3k_0\gamma_0}{3-x_1}\mathrm{e}^{-k_0\tau} - \cosh k_0\tau + O(k_0^2). \tag{3.93}$$

The asymptotic expressions for the K integral can be found similarly, namely,

$$2\int_{-1}^{+1} I(\tau,\mu,\mu_0)\mu^2\,\mathrm{d}\mu + \mu_0^2\mathrm{e}^{-\tau/\mu_0} = \frac{4}{3}\left(1-\frac{3k_0\gamma_0}{3-x_1}\right) u_0(\mu_0)\mu_0\mathrm{e}^{-k_0\tau} + O(k_0^2), \tag{3.94}$$

$$2\int_{-1}^{+1} I(\tau,\mu)\mu^2\,\mathrm{d}\mu = \left(1-\frac{x_1}{3}\right)\frac{\sinh k_0\tau}{k_0} + \gamma_0\mathrm{e}^{-k_0\tau}$$
$$-\frac{3k_0\gamma_0}{3-x_1}\mathrm{e}^{-k_0\tau} - \frac{k_0}{3}\left(\frac{2}{5-x_2}+\frac{3}{2}\frac{2-x_1}{3-x_1}\right)\sinh k_0\tau + O(k_0^2). \tag{3.95}$$

It follows from (3.94) and (3.95), that if γ_0 is considered to be known (note that γ_0 depends very weakly on phase function), then for the Milne problem with weak true absorption both the flux integral and K integral can be presented as *explicit analytical expressions* at *all depths.*

It is important here that in both problems, asymptotic expressions of higher accuracy can be derived. We give these expressions without proof:

$$\frac{H_\tau(\tau,\mu_0)}{4\pi} = k_0\frac{u_0(\mu_0)\mu_0}{3-x_1}\mathrm{e}^{-k_0\tau}$$
$$-\frac{3}{4}\left[\frac{4\gamma_0\mathrm{e}^{-k_0\tau}}{3-x_1}u_0(\mu_0)\mu_0 + \frac{5R_0(\tau,\mu_0)}{5-x_2}\right]\frac{k_0^2}{3-x_1} + O(k_0^3), \tag{3.95a}$$

where function

$$R_0(\tau,\mu_0) = \mathrm{e}^{-\tau/\mu_0}\mu_0^3 + 2\int_{-1}^{+1} I_0(\tau,\mu,\mu_0)\mu^3\,\mathrm{d}\mu$$

tends to zero when $\tau\to\infty$ and

$$R_0(0,\mu_0) = v_0(\mu_0)\mu_0.$$

Similarly,

$$\frac{H_\tau(\tau)}{\pi} = \frac{3k_0\gamma_0}{3-x_1}\mathrm{e}^{-k_0\tau} - \left[1 - \frac{(9-x_2)k_0^2}{(3-x_1)(5-x_2)}\right]\cosh k_0\tau$$
$$-\frac{9k_0^2}{3-x_1}\left[\frac{\gamma_0^2}{3-x_1}\mathrm{e}^{-k_0\tau} + \frac{1}{4}R_0(\tau)\right] + O(k_0^3).$$

Function

$$R_0(\tau) = \frac{4}{5-x_2}\left[\frac{10}{3}\int_{-1}^{+1} I_0(\tau,\mu)\mu^3\,\mathrm{d}\mu + 1\right]$$

also tends to zero when $\tau \to \infty$ and at the boundary

$$R_0(0) = \frac{4}{9}\frac{9-5\varepsilon_0}{5-x_2}.$$

Function $v_0(\mu)$ and constant ε_0 are determined by (3.98) and (3.84), respectively. Their numerical values are given in Table XXI.

3.11.6 Albedo of Atmosphere

As we know, the plane $A(\mu_0)$ and spherical A_s albedos are the important characteristics of the reflective properties of atmosphere, determined by (1.104) and (1.016). What first comes to mind is to substitute the expression (3.80) for $\rho_1(\mu,\mu_0)$ into these formulas and thus obtain the corresponding asymptotic expansions within the order of k_0. One should however, realize that more accurate rigorous asymptotic formulas can be derived, namely, for $A(\mu)$ it can be done within the order of k_0^2 and for A_s even within the order of k_0^3.

Let us make a use of (3.10) and (3.11). By analogy with finding the asymptotic expansions for N and C (see Sect. 3.11.3), we substitute $i(\mu)$ into the right-hand side of (3.10) and (3.11) with the asymptotic expansion of $i_2(\mu)$ given by (2.40) where the last term is discarded. Adding the resultant expressions yields

$$1 + \left[1 + \frac{10P_2(\mu)}{5-x_2}\right]\frac{k_0^2}{3-x_1} = A(\mu) + \frac{1}{2}Mu(\mu)$$
$$+\frac{k_0^2}{3-x_1}2\int_0^1 \rho(\mu,\mu')\left[1 + \frac{10P_2(\mu')}{5-x_2}\right]\mu'\,\mathrm{d}\mu'. \tag{3.96}$$

Then, with the help of (2.44), (3.14), (3.62), and (3.79), one finds

$$A(\mu) = 1 - \frac{4\kappa_0}{3-x_1}u_0(\mu) + \frac{3k_0^2}{3-x_1}\left[\frac{5}{5-x_2}v_0(\mu) + \frac{4\gamma_0}{3-x_1}u_0(\mu)\right] + O(k_0^3), \tag{3.97}$$

where

$$v_0(\mu) = \mu^2 - 2\int_0^1 \rho(\mu,\mu')\mu'^3\,\mathrm{d}\mu' \tag{3.98}$$

is given in Table XXI for various phase functions. It can be easily shown that for isotropic scattering

$$v_0(\mu) = \frac{4}{3}(\mu - \gamma_0)u_0(\mu).$$

It is interesting to note that this equality approximately holds for *any* phase function, which the reader can easily verify with the tables of the Appendix.

If we multiply (3.96) by 2μ, then integrating the product with respect to μ from 0 to 1 and taking (1.106) and (3.81) into account, we derive

$$1 + \frac{(15 - 2x_2)k_0^2}{(3 - x_1)(5 - x_2)} = A_s + \frac{1}{2}MC + \frac{k_0^2}{3 - x_1} 2 \int_0^1 A(\mu) \left[1 + \frac{10P_2(\mu)}{5 - x_2}\right] \mu \, d\mu.$$

The substitution of (2.42), (3.83), and (3.97) into the above formula gives the following asymptotic formula:

$$A_s = 1 - \frac{4k_0}{3 - x_1} + \frac{12\gamma_0 k_0^2}{(3 - x_1)^2}$$
$$- \frac{6k_0^3}{(3 - x_1)^3} \left[2 - x_1 + 6\gamma_0^2 + \frac{2(11 + x_2 - 10\varepsilon_0)(3 - x_1)}{3(5 - x_2)}\right] + O(k_0^4). \quad (3.99)$$

As γ_0 and ε_0 depend on the phase function very weakly, (3.99) can be regarded as an *explicit analytic dependency* of A_s on the optical properties of the atmosphere which, in turn, are characterized by the particles albedo λ and the parameters of the phase function x_1 and x_2.

The following formulas, which follow from (1.107), (3.91a), are asymptotic expressions for geometric albedo A_g

$$A_g = A_{g0} - \frac{4k_0\zeta_0}{3 - x_1} + O(k_0^2),$$

where

$$A_{g0} = 2 \int_0^1 \rho_0(\mu, \mu, \pi)\mu^2 \, d\mu,$$

$$\zeta_0 = 2 \int_0^1 u_0^2(\mu)\mu^2 \, d\mu.$$

The numerical values of A_{g0} and ζ_0 are given in Tables XX and XXXVI.

It should not be difficult for the reader to get an idea about the accuracy of the asymptotic formulas derived for scattering close to conservative case. For this, one should use the results of calculations provided in the Appendix.

3.12 Q Form of the Transfer Equation and Solution to the General Problem

In Sects. 3.5 and 3.10 we found the algorithm for the solution to the parallel external flux problem. Under certain circumstances, the solution to the general problem can be reduced to the solution to the parallel external flux problem. For this, it is more convenient to use the so-called Q form of the transfer equation instead of the standard one.

3.12.1 Function $Q(\mu, \mu_0, \varphi)$ and Its Physical Meaning

We define a certain auxiliary function $Q(\mu, \mu_0, \varphi)$ from

$$Q(\mu, \mu_0, \varphi - \varphi_0) = \chi(\mu, \mu_0, \varphi - \varphi_0) + \frac{\lambda}{4\pi} \int_0^{2\pi} \mathrm{d}\varphi' \int_{-1}^{+1} Q(\mu, \mu', \varphi - \varphi')\chi(\mu_0, \mu', \varphi' - \varphi_0)\, \mathrm{d}\mu'. \qquad (3.100)$$

Then Q can always be derived for $\lambda \neq 1$ for known albedo and phase function. It is sufficient to iterate (3.100) with $Q(\mu, \mu_0, \varphi) = \chi(\mu, \mu_0, \varphi)$ as the zeroth approximation.

If the phase function can be expanded as a series with respect to azimuth (1.55), then an analogous representation follows from (3.100):

$$Q(\mu, \mu_0, \varphi - \varphi_0) = Q(\mu, \mu_0) + \sum_{m=1}^{\infty} Q^m(\mu, \mu_0) \cos m(\varphi - \varphi_0), \qquad (3.101)$$

with each azimuthal harmonic of Q defined by

$$Q^m(\mu, \mu_0) = \chi^m(\mu, \mu_0) + \frac{\lambda}{2} \int_{-1}^{+1} Q^m(\mu, \mu')\chi^m(\mu', \mu_0)\, \mathrm{d}\mu'. \qquad (3.102)$$

If the representation (1.56) holds for χ^m, then from (3.102) we obtain the following explicit expression for any azimuthal harmonic of Q:

$$Q^m(\mu, \mu_0) = \sum_{i=m}^{\infty} \frac{(2i+1)x_i^m}{2i+1-\lambda x_i} P_i^m(\mu) P_i^m(\mu_0). \qquad (3.103)$$

It is obvious that Q can be expanded in Legendre polynomials (cf. (1.53)):

$$Q(\cos\gamma) = \sum_{n=0}^{\infty} \frac{(2n+1)x_n}{2n+1-\lambda x_n} P_n(\cos\gamma), \qquad (3.103a)$$

where $\cos\gamma$ is determined by (1.43). Comparison of (1.56) and (3.103) shows that (3.103) converges more slowly than (1.56) and $Q(\mu, \mu_0, \varphi)$ ceases to exist for $\lambda = 1$ (the first term in (3.103) is equal to $(1-\lambda)^{-1}$ for $m = 0$).

We now consider the physical significance of Q.

a. Consider a homogeneous infinite medium whose optical parameters at each point are characterized by the phase function $\chi(\gamma)$ and the particle albedo $\lambda \neq 1$. Let this medium be uniformly filled by unidirectional emitters of infinite output radiating in the direction $\boldsymbol{\omega}_0 = (\mu_0, \varphi_0)$. The transfer equation for this medium will clearly lack the derivative with respect to the spatial coordinate (the radiation field in such an atmosphere will be spatially homogeneous) and can be put as

$$J^*(\mu, \mu_0, \varphi - \varphi_0) = \frac{\lambda}{4\pi}\int_0^{2\pi} \mathrm{d}\varphi' \int_{-1}^{+1} J^*(\mu', \mu_0, \varphi' - \varphi_0)\chi(\mu', \mu, \varphi' - \varphi)\,\mathrm{d}\mu' + 2\pi\delta(\mu - \mu_0)\delta(\varphi - \varphi_0), \quad (3.104)$$

where $J^*(\mu, \mu_0, \varphi - \varphi_0)$ is the intensity of the total radiation (direct + diffuse). We distinguish the diffuse intensity by means of

$$J^*(\mu, \mu_0, \varphi - \varphi_0) = J(\mu, \mu_0, \varphi - \varphi_0) + 2\pi\delta(\mu - \mu_0)\delta(\varphi - \varphi_0). \quad (3.105)$$

We substitute (3.105) into (3.104) to get

$$J(\mu, \mu_0, \varphi - \varphi_0) = \frac{\lambda}{2}\chi(\mu, \mu_0, \varphi - \varphi_0) + \frac{\lambda}{4\pi}\int_0^{2\pi} \mathrm{d}\varphi' \int_{-1}^{+1} J(\mu', \mu_0, \varphi' - \varphi_0)\chi(\mu', \mu, \varphi' - \varphi)\,\mathrm{d}\mu'. \quad (3.106)$$

Then the comparison of (3.106) with (3.100) shows that $\frac{\lambda}{2}Q(\mu, \mu_0, \varphi - \varphi_0)$ is the intensity of the diffuse radiation in an infinite homogeneous nonconservatively scattering medium containing uniformly distributed unidirectional δ-radiators of constant output radiating in the direction $\boldsymbol{\omega}_0 = (\mu_0, \varphi_0)$.

b. We continue to consider the same homogeneous infinite medium. It follows from (3.100) that Q can be represented as a Neumann series:

$$Q(\mu, \mu_0, \varphi - \varphi_0) = \chi(\mu, \mu_0, \varphi - \varphi_0) + \sum_{n=1}^{\infty} \lambda^n V_n(\mu, \mu_0, \varphi - \varphi_0), \quad (3.107)$$

where for $n = 1, 2, 3, \ldots$

$$V_n(\mu, \mu_0, \varphi - \varphi_0) = \frac{1}{4\pi}\int_0^{2\pi} \mathrm{d}\varphi' \int_{-1}^{+1} V_{n-1}(\mu, \mu', \varphi - \varphi')\chi(\mu', \mu_0, \varphi - \varphi_0)\,\mathrm{d}\mu', \quad (3.108)$$

and by definition $V_0(\mu, \mu_0, \varphi) = \chi(\mu, \mu_0, \varphi)$.

Then let there be N photons travelling in the direction $\boldsymbol{\omega}_0 = (\mu_0, \varphi_0)$, reemitted by an elementary volume in the direction $\boldsymbol{\omega} = (\mu, \varphi)$ with probability $\lambda\chi(\boldsymbol{\omega}\boldsymbol{\omega}_0)\mathrm{d}\omega/4\pi$. Clearly, this quantity is the probability of the mean number of acts of scattering experienced by N photons after reemission in direction $\boldsymbol{\omega}$.

We subsequently track these photons. Each of them sooner or later collides with a scattering particle and is reemitted in a certain direction $\chi\omega$ with probability $\lambda^2\chi(\boldsymbol{\omega}'\boldsymbol{\omega}_0)\chi(\omega\omega')\mathrm{d}\omega'/4\pi$. We now conceptually combine all scattering points into one. The elementary volume containing this point will as it were be irradiated from all directions $\boldsymbol{\omega}' = (\mu', \varphi')$ and will reemit the photons in direction $\boldsymbol{\omega} = (\mu, \varphi)$. Hence, the quantity

$$\frac{\lambda^2}{4\pi}\mathrm{d}\omega \int_0^{2\pi} \mathrm{d}\varphi' \int_{-1}^{+1} \chi(\mu', \mu_0, \varphi' - \varphi_0)\chi(\mu', \mu, \varphi' - \varphi_0)\, \mathrm{d}\mu'$$

will represent the probability of the mean number of acts of scattering for those photons in direction $\boldsymbol{\omega}$ after a second collision. The subsequent arguments are analogous. We draw the following conclusion from (3.107) and (3.108).

The function $(\lambda/4\pi)Q(\mu, \mu_0, \varphi-\varphi_0)$ is the probability density of the mean number of acts of scattering occurring in an infinite homogeneous medium by a photon that began its motion in the direction $\boldsymbol{\omega}_0 = (\mu_0, \varphi_0)$ and is reemitted in collisions on each occasion in the same direction $\boldsymbol{\omega} = (\mu, \varphi)$. In particular, we have an obvious result for isotropic scattering: $Q = (1-\lambda)^{-1} = 1 + \lambda + \lambda^2 + \ldots$.

Similarly, it follows from (1.16) and (3.103a) that for Henyey-Greenstein phase function $\chi_H(\gamma) = \chi_H(\gamma, g)$ we have

$$Q_H(\gamma) = \sum_{n=0}^{\infty} \lambda^n \chi_H(\gamma, g^{n+1}). \tag{3.108a}$$

3.12.2 Q Form of the Transfer Equation

For simplicity, we consider the zeroth azimuthal harmonic of the intensity. We multiply both parts of (1.50) with $m = 0$ by $\frac{\lambda}{2}Q(\mu, \mu_1)$ and integrate with respect to μ from -1 to +1. Then using (3.102) and replacing μ_1 with μ, we obtain

$$\frac{\lambda}{2}\int_{-1}^{+1} B(\tau, \mu', \mu_0)Q(\mu', \mu)\, \mathrm{d}\mu' = \frac{\lambda}{2}\int_{-1}^{+1} I(\tau, \mu', \mu_0)Q(\mu', \mu)\, \mathrm{d}\mu'$$
$$-\frac{\lambda}{2}\int_{-1}^{+1} I(\tau, \mu', \mu_0)\chi(\mu', \mu)\, \mathrm{d}\mu' + \frac{\lambda}{4}[Q(\mu, \mu_0) - \chi(\mu, \mu_0)]\mathrm{e}^{-\tau/\mu_0}.$$

We use (1.50) and transfer equation (1.49) (with $\mu = 0$) to get finally

$$B(\tau, \mu, \mu_0) = -\frac{\lambda}{4}\frac{\mathrm{d}}{\mathrm{d}\tau}\left[Q(\mu, \mu_0)\mathrm{e}^{-\tau/\mu_0}\mu_0 + 2\int_{-1}^{+1} I(\tau, \mu', \mu_0)Q(\mu', \mu)\mu'\, \mathrm{d}\mu'\right]. \tag{3.109}$$

Then the source function can be represented in *differential form*. In particular, taking into account (3.103) and (1.87), we get from (3.109) with $m = 0$ for isotropic scattering

$$B(\tau, \mu_0) = -\frac{1}{4\pi}\frac{\lambda}{1-\lambda}\frac{\mathrm{d}H_\tau(\tau, \mu_0)}{\mathrm{d}\tau}.$$

From (1.49) ($m = 0$), instead of (3.109) we get

$$I(\tau, \mu, \mu_0) = -\frac{\lambda}{4}\frac{\mathrm{d}}{\mathrm{d}\tau}\left[Q(\mu, \mu_0)\mathrm{e}^{-\tau/\mu_0}\mu_0 + \frac{4}{\lambda}I(\tau, \mu, \mu_0)\mu \right.$$
$$\left. +2\int_{-1}^{+1} I(\tau, \mu', \mu_0)Q(\mu', \mu)\mu' \, \mathrm{d}\mu'\right]. \tag{3.110}$$

Then (3.109) and (3.110) take the same form for the other azimuthal harmonics, so in general in place of (3.110) we have finally

$$I(\tau, \mu, \varphi - \varphi_0) = -\frac{\lambda}{4}\frac{\mathrm{d}}{\mathrm{d}\tau}\left[Q(\mu, \mu_0, \varphi - \varphi_0)\mathrm{e}^{-\tau/\mu_0}\mu_0\right.$$
$$\left. +\frac{4}{\lambda}I(\tau, \mu, \varphi - \varphi_0)\mu + \frac{1}{\pi}\int_0^{2\pi} \mathrm{d}\varphi' \int_{-1}^{+1} I(\tau, \mu', \varphi' - \varphi_0)Q(\mu', \mu, \varphi' - \varphi)\mu' \, \mathrm{d}\mu'\right]. \tag{3.111}$$

This is the desired form for the transfer equation. For simplicity, (3.111) is called the ***Q form*** or ***Q representation of the transfer equation.***

We now enumerate the main features of the Q representation.

1. It follows from (1.49) and (1.50) that the standard form of the transfer equation can be written as follows:

$$\frac{\mathrm{d}I}{\mathrm{d}\tau} = \mathcal{L}_0[I], \tag{3.112}$$

   where $\mathcal{L}_0$ is an appropriate linear integral operator, while it follows from (3.111) that

$$I = \frac{\mathrm{d}}{\mathrm{d}\tau}\mathcal{L}_1[I], \tag{3.113}$$

   where $\mathcal{L}_1$ is also a certain linear integral operator. Hence, there is a difference from the standard form of (3.112) in that the Q form represents the intensity as a derivative with respect to τ for a certain linear integral intensity operator.
2. While (1.50) contains the phase function χ, in the integral, which characterizes an act of *single* scattering, (3.111) has instead Q, which describes *multiple* scattering.
3. Formula (3.113) allows the following generalization ($n = 1, 2, 3, \ldots$):

$$I = \frac{\mathrm{d}^{n+1}}{\mathrm{d}\tau^{n+1}} \mathcal{L}_1^{n+1}[I].$$

In other words, the intensity can be represented by a derivative with respect to τ of *any* order for a certain linear integral intensity operator.

We now note some particular cases of the Q form. It is clear that the intensity $I(\tau, \mu)$ in the Milne problem will also satisfy the transfer equation in the Q representation:

$$I(\tau, \mu) = -\frac{\lambda}{4} \frac{\mathrm{d}}{\mathrm{d}\tau} \left[\frac{4}{\lambda} I(\tau, \mu)\mu + 2 \int_{-1}^{+1} I(\tau, \mu')Q(\mu', \mu)\mu' \, \mathrm{d}\mu' \right]. \tag{3.114}$$

We use the asymptotic formula (3.57) for the radiation intensity in a semi-infinite atmosphere. Substituting this expression into (3.110) we find the following Q form for $i(\mu)$:

$$i(\mu)(1 - k\mu) = \frac{\lambda k}{2} \int_{-1}^{+1} i(\mu')Q(\mu', \mu)\mu' \, \mathrm{d}\mu'. \tag{3.115}$$

3.12.3 Conservative Scattering

The function Q does not exist for $\lambda = 1$, so (3.110) for the zeroth azimuthal harmonic goes over to the standard expression for the flux integral. More precisely, an indeterminacy of $0/0$ arises on the right in (3.110) for $\lambda \to 1$. To resolve this, we use the fact that the asymptotic formula (3.95a) applies for $\lambda \to 1$:

$$2 \int_{-1}^{+1} I(\tau, \mu, \mu_0)\mu \, \mathrm{d}\mu + \mathrm{e}^{-\tau/\mu_0}\mu_0 = \frac{4k_0 \mathrm{e}^{-k_0\tau}}{3 - x_1}(1 - 3k_0\gamma_0)u_0(\mu_0)\mu_0$$
$$- \frac{15(1-\lambda)}{5 - x_2} \left[\mathrm{e}^{-\tau/\mu_0}\mu_0^3 + 2 \int_{-1}^{+1} I(\tau, \mu, \mu_0)\mu^3 \, \mathrm{d}\mu \right] + O(k_0^3). \tag{3.116}$$

Now taking into account (3.103) with $m = 0$ and expressions (3.116), (3.94), we obtain a Q representation for the transfer equation in the conservative case:

$$I_0(\tau, \mu, \mu_0) = u_0(\mu_0)\mu_0 - \frac{1}{4} \frac{\mathrm{d}}{\mathrm{d}\tau} \left[4 I_0(\tau, \mu, \mu_0)\mu \right.$$
$$\left. + Q_0^*(\mu, \mu_0)\mathrm{e}^{-\tau/\mu_0}\mu_0 + 2 \int_{-1}^{+1} I_0(\tau, \mu', \mu_0)Q_0^*(\mu, \mu')\mu' \, \mathrm{d}\mu' \right], \tag{3.117}$$

where we have put

$$Q_0^*(\mu, \mu_0) = \sum_{i=3}^{\infty} \frac{(2i+1)x_i}{2i+1-x_i} P_i(\mu)P_i(\mu_0) - \frac{15\mu_0^2}{5 - x_2} \left[1 - \frac{1}{2} x_2 P_2(\mu) \right].$$

As already noted, the Q form of the transfer equation remains unaltered for the higher azimuthal harmonics ($m = 1, 2, 3, \ldots$) in conservative scattering: it is sufficient to replace Q in (3.110) by Q^m and put $\lambda = 1$.

3.12.4 Q Representation of the Green Function for the Transfer Equation in a Plane Atmosphere

Consider a plane, semi-infinite homogeneous atmosphere. For convenience, we write down once more the basic relations for the Green function which are written in Sect. 1.11 for an inhomogeneous atmosphere. The Green function obeys the following equation:

$$\mu\frac{\partial G(\tau,\omega;\tau_1,\omega_1)}{\partial\tau}+G(\tau,\omega;\tau_1,\omega_1)$$
$$=\frac{\lambda}{4\pi}\int\limits_{(4\pi)}\chi(\omega\omega')G(\tau,\omega';\tau_1,\omega_1)\,\mathrm{d}\omega'+2\pi\delta(\tau-\tau_1)\delta(\omega-\omega_1). \quad (3.118)$$

We supplement this with the boundary conditions

$$\left.\begin{array}{lll} G(0,\omega;\tau_1,\omega_1)=0 & \text{for} & \tau_1,\mu>0, \\ G(\tau,\omega;\tau_1,\omega_1) & \text{is limited for} & |\tau-\tau_1|\to\infty, \end{array}\right\} \quad (3.119)$$

and the reciprocity relation

$$G(\tau,\omega;\tau_1,\omega_1)=G(\tau_1,-\omega_1;\tau,-\omega) \quad (3.120)$$

which expresses the fact that the radiation field is unaltered when the photon source and detector are interchanged.

Equation (3.118) corresponds to the standard form of the transfer equation. It is clear that (3.118) can be put in Q form. We now perform transformations analogous to those in Sect 3.12.2 to get

$$G(\tau,\omega;\tau_1,\omega_1)=-\frac{\partial}{\partial\tau}\left\{\mu G(\tau,\omega;\tau_1,\omega_1)+\frac{\lambda}{4\pi}\int\limits_{(4\pi)}Q(\omega\omega')G(\tau,\omega';\tau_1,\omega_1)\mu'\,\mathrm{d}\omega'\right.$$
$$\left.-\left[\frac{\lambda}{2}Q(\omega\omega_1)+2\pi\delta(\omega-\omega_1)\right]\theta(\tau-\tau_1)\right\}, \quad (3.121)$$

which is the Q representation for the Green function.

One can use the surface Green function $G(\tau,\omega;0,\omega_0)$ instead of the radiation intensity $I(\tau,\omega,\omega_0)$. They are related by the following relation

$$G(\tau,\omega;0,\omega_0)\mu_0=2I(\tau,\omega,\omega_0)+2\pi\delta(\mu-\mu_0)\delta(\varphi-\varphi_0)\mathrm{e}^{-\tau/\mu}. \quad (3.122)$$

We have then from (3.121)

$$G(\tau,\omega;0,\omega_0)=-\frac{\mathrm{d}}{\mathrm{d}\tau}\left[\mu G(\tau,\omega;0,\omega_0)+\frac{\lambda}{4\pi}\int\limits_{(4\pi)}G(\tau,\omega';0,\omega_0)Q(\omega'\omega)\mu'\,\mathrm{d}\omega'\right]. \quad (3.123)$$

This formula can also be obtained by substituting (3.122) into (3.111). We have to emphasize that (3.121–3.123) are also valid for a layer of finite thickness.

3.12.5 Solution to the General Problem

We are coming now to the main aim of the present section, namely, solving the general problem in transfer theory for a plane atmosphere which was formulated at the end of Sect. 1.4 (see also Sect. 1.11). Recall that by the general problem one usually means deriving the intensity in a medium containing internal sources whose output is dependent only on the spatial coordinate τ.

We denote by $g(\tau, \mu, \varphi)\mathrm{d}\tau\, \mathrm{d}\omega$ the amount of radiation energy emitted by the internal sources in the atmosphere at an optical depth τ in the direction $\boldsymbol{\omega}(\mu, \varphi)$ in unit time. If we recall the definition of the Green function the problem is reduced to finding the quantity

$$I_{\mathrm{s}}(\tau, \boldsymbol{\omega}) = \frac{1}{2\pi} \int_0^\infty \mathrm{d}\tau_1 \int_{(4\pi)} g(\tau_1, \boldsymbol{\omega}') G(\tau, \boldsymbol{\omega}; \tau_1, \boldsymbol{\omega}')\, \mathrm{d}\omega', \tag{3.124}$$

which is the intensity propagating at depth τ in direction $\boldsymbol{\omega}$.

We write the Green function as

$$G(\tau, \boldsymbol{\omega}; \tau_1, \boldsymbol{\omega}_1)\mu_1 = \tilde{G}(\tau, \boldsymbol{\omega}; \tau_1, \boldsymbol{\omega}_1)\mu_1$$
$$+2\pi\delta(\boldsymbol{\omega} - \boldsymbol{\omega}_1)[\theta(\mu_1)\theta(\tau - \tau_1) - \theta(-\mu_1)\theta(\tau_1 - \tau)]\mathrm{e}^{-\frac{\tau-\tau_1}{\mu_1}}. \tag{3.125}$$

Substituting (3.125) into (3.124), we obtain

$$I_{\mathrm{s}}(\tau, \boldsymbol{\omega}) = I_{\mathrm{d}}(\tau, \boldsymbol{\omega}) + \theta(\mu) \int_0^\tau g(\tau_1, \boldsymbol{\omega})\mathrm{e}^{-\frac{\tau-\tau_1}{\mu}} \frac{\mathrm{d}\tau_1}{\mu} - \theta(-\mu) \int_\tau^\infty g(\tau_1, \boldsymbol{\omega})\mathrm{e}^{-\frac{\tau-\tau_1}{\mu}} \frac{\mathrm{d}\tau_1}{\mu}. \tag{3.126}$$

The latter two terms in (3.126) describe the radiation arriving directly from the sources. These terms can be taken as known. The quantity

$$I_{\mathrm{d}}(\tau, \boldsymbol{\omega}) = \frac{1}{2\pi} \int_0^\infty \mathrm{d}\tau_1 \int_{(4\pi)} g(\tau_1, -\boldsymbol{\omega}') \tilde{G}(\tau_1, \boldsymbol{\omega}'; \tau, -\boldsymbol{\omega})\, \mathrm{d}\omega' \tag{3.127}$$

is the unknown one and defines the radiation field set up by photons scattered once or more. To write (3.127), we have used the reciprocity relation (3.120).

We now represent

$$\tilde{G}(\tau, \mu, \varphi; \tau_1, \mu_1, \varphi_1) = \tilde{G}(\tau, \mu; \tau_1, \mu_1) + 2\sum_{m=1}^\infty \tilde{G}^m(\tau, \mu; \tau_1, \mu_1) \cos m(\varphi - \varphi_1), \tag{3.128}$$

and assume that g can be represented closely by a segment of a Fourier series with respect to azimuth, i.e.,

$$g(\tau, \mu, \varphi) = g(\tau, \mu) + 2\sum_{m=1}^M [g^m(\tau, \mu) \cos m\varphi + s^m(\tau, \mu) \sin m\varphi]. \tag{3.129}$$

We can then write down

$$I_{\rm d}(\tau,\omega) = I_{\rm d}(\tau,\mu) + 2\sum_{m=1}^{M}[I_{\rm dc}^m(\tau,\mu)\cos m\varphi + I_{\rm ds}^m(\tau,\mu)\sin m\varphi], \qquad (3.130)$$

in which for all $\mu \in [-1,+1]$

$$\left.\begin{matrix} I_{\rm dc}^m(\tau,\mu) \\ I_{\rm ds}^m(\tau,\mu) \end{matrix}\right\} = \int_0^\infty {\rm d}\tau_1 \int_{-1}^{+1} \left.\begin{matrix} g^m(\tau_1,-\mu') \\ s^m(\tau_1,-\mu') \end{matrix}\right\} \tilde{G}^m(\tau_1,\mu';\tau,-\mu)\,{\rm d}\mu'. \qquad (3.131)$$

We assume that the functions $g^m(\tau,\mu)$ and $s^m(\tau,\mu)$ can be represented closely by the sums

$$\left.\begin{matrix} g^m(\tau,\mu) \\ s^m(\tau,\mu) \end{matrix}\right\} = \sum_{n=0}^{N} \left.\begin{matrix} g_n^m(\mu) \\ s_n^m(\mu) \end{matrix}\right\} L_n(\tau), \qquad (3.132)$$

where $L_n(\tau)$ are Laguerre polynomials of order n. Then in place of (3.131) we have

$$\left.\begin{matrix} I_{\rm dc}^m(\tau,\mu) \\ I_{\rm ds}^m(\tau,\mu) \end{matrix}\right\} = \sum_{n=0}^{N} \int_{-1}^{+1} \left.\begin{matrix} g_n^m(-\mu) \\ s_n^m(-\mu) \end{matrix}\right\} R_n^m(\tau,\mu',\mu)\,{\rm d}\mu', \qquad (3.133)$$

where

$$R_n^m(\tau,\mu_1,\mu) = \int_0^\infty L_n(\tau_1)\tilde{G}^m(\tau_1,\mu_1;\tau,-\mu)\,{\rm d}\tau_1. \qquad (3.134)$$

The problem is then that of determining R_n^m. We use the equation for the Fourier component $\tilde{G}^m$ for $\tilde{G}$, which follows from (3.125), (3.121) and (3.128):

$$\begin{aligned} \tilde{G}^m(\tau_1,\mu_1;\tau,-\mu) = -\frac{\partial}{\partial\tau_1}\Bigg[&\mu_1\tilde{G}^m(\tau_1,\mu_1;\tau,-\mu) \\ &+\frac{\lambda}{2}\int_{-1}^{+1} Q^m(\mu_1,\mu')\tilde{G}^m(\tau_1,\mu';\tau,-\mu)\mu'\,{\rm d}\mu\Bigg] \\ +\frac{\lambda}{2\mu}Q^m(\mu_1,-\mu)[\theta(\mu)\theta(\tau-\tau_1) &- \theta(-\mu)\theta(\tau_1-\tau)]{\rm e}^{-\frac{\tau-\tau_1}{\mu}}. \end{aligned} \qquad (3.135)$$

We use (3.134) and (3.135) with the recursion relation for the Laguerre polynomials

$$\frac{\rm d}{{\rm d}\tau}[L_n(\tau) - L_{n+1}(\tau)] = L_n(\tau), \qquad (3.136)$$

to get the recursion formula

$$\begin{aligned} R_{n+1}^m(\tau,\mu_1,\mu) &= (1-\mu_1)R_n^m(\tau,\mu_1,\mu) \\ -\frac{\lambda}{2}\int_{-1}^{+1} Q^m(\mu_1,\mu')R_n^m(\tau,\mu',\mu)\mu'\,{\rm d}\mu' &+ \frac{\lambda}{2}Q^m(\mu_1,-\mu)W_n(\tau,\mu), \end{aligned} \qquad (3.137)$$

where the function $W_n(\tau, \mu)$ is defined by the recursion relation

$$W_{n+1}(\tau, \mu) = (1+\mu)W_n(\tau, \mu) + L_{n+2}(\tau) - 2L_{n+1}(\tau) + L_n(\tau), \qquad (3.138)$$

where

$$W_0(\tau, \mu) = \mu\left(1 - \theta(\mu)\mathrm{e}^{-\tau/\mu}\right) - \tau. \qquad (3.139)$$

Then (3.135) gives the following, on the basis that (3.125) and (3.122) taken with (3.119) and (3.120) imply that $\mu_1 \tilde{G}^m(0, \mu_1; \tau, -\mu) = -2I^m(\tau, \mu, -\mu_1) \times \theta(-\mu_1)$:

$$R_0^m(\tau, -\mu_1, \mu) = -2I^m(\tau, \mu, \mu_1)\theta(\mu_1) + \frac{\lambda}{2}Q^m(\mu_1, \mu)\left[1 - \theta(\mu)\mathrm{e}^{-\tau/\mu}\right]$$
$$-\lambda \int_0^1 Q^m(\mu_1, \mu')I^m(\tau, \mu, \mu')\,\mathrm{d}\mu'. \qquad (3.140)$$

This completes the solution. Similarly, one can solve the problem for a layer of finite optical thickness if the representation (3.132) can be applied to a finite range in τ.

Then the solution to the general problem is reduced to that of a parallel external flux problem. Indeed, if function $g(\tau, \mu, \varphi)$ can be represented as the sum (3.129), where the quantities $g^m(\tau, \mu)$ and $s^m(\tau, \mu)$ can be determined by the sums (3.132) (these are sufficient conditions for reducing the solution of the general problem to that of the parallel external flux problem) then one should act as follows:

1. One solves the parallel external flux problem for the specified set of optical depths. This can be done with the help of the method described in Sect. 3.10. As a result, one obtains the set of the functions $I^m(\tau, \mu, \mu_0)$ for these optical depths.
2. One determines the set of the functions $R_n^m(\tau, \mu_1, \mu)$ ($n = 0, 1, 2, \ldots, N$) with the help of (3.140) and (3.137).
3. One substitutes the functions R_n^m into (3.133) and integrates the result with respect to μ_1. This gives the desired set of the functions $I_{dc}^m(\tau, \mu)$ and $I_{ds}^m(\tau, \mu)$.

As a result, formulas (3.126) and (3.130) give the complete solution to the general problem.

At the boundary of the atmosphere ($\tau = 0$), instead of (3.124) we have ($\mu \in [0, 1]$)

$$I_\mathrm{s}(0, -\omega) = I_\mathrm{d}(0, -\omega) + \int_0^\infty g(\tau, -\omega)\mathrm{e}^{-\tau/\mu}\frac{\mathrm{d}\tau}{\mu}. \qquad (3.141)$$

As regards the relations (3.137) and (3.140), we introduce the notation $R_n^m(0, \mu_1, -\mu) = 2R_n^m(\mu_1, \mu)$ to rewrite them as follows ($\mu \in [-1, +1]$, $\mu_0 \in [0, 1]$):

$$R_{n+1}^m(\mu,\mu_0) = (1-\mu)R_n^m(\mu,\mu_0) - \frac{\lambda}{4}\mu_0 Q^m(\mu,\mu_0)(1-\mu_0)^n\mu_0$$
$$-\frac{\lambda}{2}\int_{-1}^{+1} Q^m(\mu',\mu)R_n^m(\mu',\mu_0)\mu'\,\mathrm{d}\mu', \tag{3.142}$$

where

$$R_0^m(\mu,\mu_0) = \frac{\lambda}{4}Q^m(\mu,\mu_0) + \rho^m(-\mu,\mu_0)\theta(-\mu)\mu$$
$$-\frac{\lambda}{2}\int_0^1 \rho^m(\mu',\mu_0)Q^m(-\mu',\mu)\mu'\,\mathrm{d}\mu', \tag{3.143}$$

in which $\rho^m(\mu,\mu_0)$ is m th azimuthal harmonic of the reflection coefficient (one can use the effective numerical method of Sect. 3.5 to calculate it).

The following condition is necessary for conservative scattering: $g(\tau,\boldsymbol{\omega})$ should be such that $\int_0^\infty g(\tau,\boldsymbol{\omega})\,\mathrm{d}\tau$ converges. We represent the expression for $I_\mathrm{d}(0,-\omega)$ as the sum of two terms:

$$I_\mathrm{d}(0,-\boldsymbol{\omega}) = \frac{1}{\pi}\int\limits_{(2\pi)} I(\tau,\boldsymbol{\omega}',\boldsymbol{\omega})\frac{\mathrm{d}\omega'}{\mu}\left[\int\limits_0^{\tau_\mathrm{as}} g(\tau,-\boldsymbol{\omega}')\,\mathrm{d}\tau + \int\limits_{\tau_\mathrm{as}}^{\infty} g(\tau,-\boldsymbol{\omega}')\,\mathrm{d}\tau\right], \tag{3.144}$$

where τ_as is the optical depth at which the asymptotic formula (3.57) applies with a given accuracy, which for $\lambda = 1$ goes over to $I_0^\mathrm{as}(\tau,\mu,\mu_0) = u_0(\mu_0)\mu_0$. Consequently the second term on the right in (3.144) can be taken as known. As regards the first term, one can readily obtain the appropriate computational algorithm from (3.117) for $m = 0$, which is analogous to the one given above. We will not dwell further on the solution to the general problem with conservative scattering. This will be done in Sect. 4.10 below.

Formulas of similar type will also apply for an atmosphere of finite optical thickness τ_0, but then it is necessary to calculate not only the reflection coefficient $\rho(\mu,\mu_0,\varphi)$ but also the transmission coefficient $\sigma(\mu,\mu_0,\varphi)$, which is not difficult if one uses, for example, the doubling method (see Sect. 4.5). This case is considered in detail in Sect. 4.10. The Q form of the equation for the Green function for conservative scattering is also derived in the same section.

Thus, the Q form of the equation for the Green function provides a solution to the general problem in terms of the parallel external flux problem, which on the one hand confirms that the latter is of fundamental significance and on the other shows that the Green function, in a certain sense is "not necessary". As regards the parallel external flux problem, it is already noted above that the numerical solution for a semi-infinite medium can be obtained for example by the method expounded in Sect. 3.10.

We have used the solution to the general problem to demonstrate the convenience of the Q form, and it may be that there are other problems for which the Q representation is the most suitable for solving them.

4. Atmosphere of Finite Optical Thickness

The present chapter deals with the problem of finding the radiation field in a layer of an arbitrary thickness τ_0, which is, in other words, the parallel external flux problem.

The starting point is the general invariance relation for a layer of finite thickness. All other results, including well-known Chandrasekhar's invariance relations, are corollaries of this. The doubling method is described, which is based on the above relation and intended to calculate the reflection and transmission coefficients. Next we deal with the method of calculating the radiation field in a layer. It is of importance that this method does not require the integration on the spatial coordinate τ. We then derive some integrals of the transfer equation for a layer of a finite thickness.

An atmosphere with a large optical thickness ($\tau_0 \gg 1$) is treated separately. The asymptotic formulas are obtained for both the zeroth azimuthal harmonic of intensity and the reflection and transmission coefficients. Special attention is paid to nearly conservative scattering, when $1 - \lambda \ll 1$.

The chapter ends by analyzing the simplifications that take place in calculating some quantities in the nearly conservatively scattering atmosphere of *arbitrary* optical thickness. For further completeness, we finally give detailed consideration to the solution of the general problem, based on the Q representation of the transfer equation (see Sect. 3.12.2).

4.1 Invariance Relation

Let us consider a standard problem of light scattering in a plane atmospheric layer of optical thickness τ_1 which is illuminated by parallel rays, namely, the parallel external flux problem.

Let an atmospheric layer of optical thickness $\tau_0 \leq \tau_1$ whose upper boundary is at the depth t ($t \in [0, \tau_1 - \tau_0]$) (see Fig. 4.1) be defined. An arbitrary azimuthal harmonic with the intensity of the diffuse radiation spreading over a layer of thickness τ_1 at depth $t + \tau$ ($\tau \in [0, \tau_0]$) [i.e., $I^m(t + \tau, \mu, \mu_0, \tau_1)$] evidently equals the intensity in a layer of thickness τ_0 at depth τ whose upper boundary is illuminated both by parallel rays attenuated by a layer of thickness t and by diffuse radiation, and whose lower boundary is illuminated only by diffuse radiation. So the generalized invariance principle yields

$$I^m(\tau+t,\mu,\mu_0;\tau_1) = I^m(\tau,\mu,\mu_0;\tau_0)\mathrm{e}^{-t/\mu_0} + I^m(t,\mu,\mu_0;\tau_1)\mathrm{e}^{-\tau/\mu}\theta(\mu)$$
$$+I^m(\tau_0+t,\mu,\mu_0;\tau_1)\mathrm{e}^{\frac{\tau_0-\tau}{\mu}}\theta(-\mu) + 2\int_0^1 I^m(\tau,\mu,\mu';\tau_0)I^m(t,\mu',\mu_0;\tau_1)\,\mathrm{d}\mu'$$
$$+2\int_0^1 I^m(\tau_0-\tau,-\mu,\mu';\tau_0)I^m(\tau_0+t,-\mu',\mu_0;\tau_1)\,\mathrm{d}\mu'. \tag{4.1}$$

Of course, this formula can alternatively be derived from the equation of radiative transfer. However, we will not show it here: it is quite cumbersome.

Equation (4.1) can be rewritten in terms of the source function. Differentiating it with respect to τ and using (1.49), we obtain

$$B^m(\tau+t,\mu,\mu_0;\tau_1) = B^m(\tau,\mu,\mu_0;\tau_0)\mathrm{e}^{-t/\mu_0}$$
$$+2\int_0^1 B^m(\tau,\mu,\mu';\tau_0)I^m(t,\mu',\mu_0;\tau_1)\,\mathrm{d}\mu'$$
$$+2\int_0^1 B^m(\tau_0-\tau,-\mu,\mu';\tau_0)I^m(\tau_0+t,-\mu',\mu_0;\tau_1)\,\mathrm{d}\mu'. \tag{4.2}$$

It is clear that the large number of parameters in (4.1) and (4.2) provides considerable flexibility. This, in turn, means that one can derive many consequences, including those already known. For example, taking $\tau_1, \tau_0 \to \infty$ in (4.1) and (4.2), we arrive at the relationships (3.2) and (3.3), that have been already obtained by studying the radiation fields in a semi-infinite atmosphere.

Let us consider one of these consequences leaving the rest of them for later. Set $\tau_1 = \infty$ and direct parameter t to infinity. Figure 4.1 shows that this implies the placing of a layer of optical thickness τ_0 into the deep layers of a semi-infinite medium. Putting $m = 0$ in (4.1), (4.2) and using the asymptotic formula (3.57), we obtain

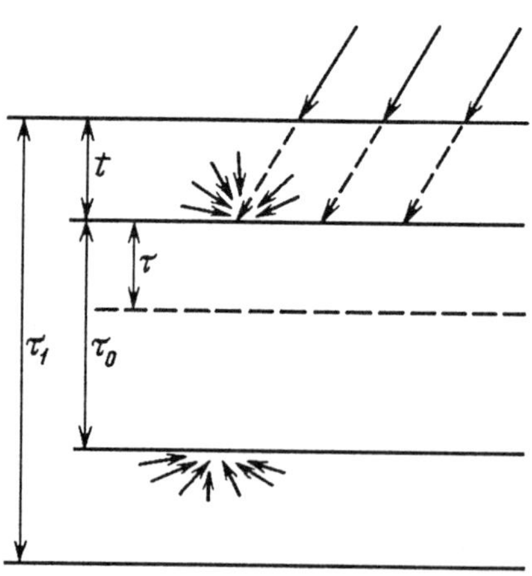

Fig. 4.1. On derivation of the invariance relation (4.1)

$$i(\mu)\mathrm{e}^{-k\tau} = i(\mu)\left[\theta(\mu) + \theta(-\mu)\mathrm{e}^{-(k-1/\mu)\tau_0}\right]\mathrm{e}^{-\tau/\mu}$$
$$+2\int_0^1 I(\tau,\mu,\mu';\tau_0)i(\mu')\,\mathrm{d}\mu' + 2\mathrm{e}^{-k\tau_0}\int_0^1 I(\tau_0-\tau,-\mu,\mu';\tau_0)i(-\mu')\,\mathrm{d}\mu' \quad (4.3)$$

The above formula expressed in terms of the source function takes then the form

$$(1-k\mu)i(\mu)\mathrm{e}^{-k\tau} = 2\int_0^1 B(\tau,\mu,\mu';\tau_0)i(\mu')\,\mathrm{d}\mu'$$
$$+2\mathrm{e}^{-k\tau_0}\int_0^1 B(\tau_0-\tau,-\mu,\mu';\tau_0)i(-\mu')\,\mathrm{d}\mu'. \quad (4.4)$$

The formulas (4.3) and (4.4) can be derived from the physical arguments relying on the generalized invariance principle. Indeed, if we take into account that the radiation intensity in an infinite medium at depth $\tau \in [0,\tau_0]$ is, at the same time, the intensity at the same depth in the layer with optical thickness τ_0 whose boundaries are illuminated by radiation spreading at the corresponding levels, then we instantly arrive at the relation (4.3).

For the conservative limit ($\lambda = 1$, $k = 0$, $i(\mu) \equiv 1$) we have, instead of (4.3) and (4.4), the following:

$$1-\left[\theta(\mu)+\theta(-\mu)\mathrm{e}^{-\tau_0/\mu}\right]\mathrm{e}^{-\tau/\mu} = 2\int_0^1 [I_0(\tau,\mu,\mu';\tau_0)+I_0(\tau_0-\tau,-\mu,\mu';\tau_0)]\,\mathrm{d}\mu', \quad (4.5)$$

$$2\int_0^1 [B_0(\tau,\mu,\mu';\tau_0) + B_0(\tau_0-\tau,-\mu,\mu';\tau_0)]\,\mathrm{d}\mu' = 1. \quad (4.6)$$

At $\lambda = 1$, one can obtain one more relation. If we take into consideration that (2.43) is valid for $\lambda \to 1$ and subtract the expression obtained from (4.3) by the substitution of μ with $-\mu$ and τ with $\tau_0 - \tau$ from (4.3), then, making the allowance for (4.5), we get

$$\mu - \frac{3-x_1}{3}\tau = \mu\left[\theta(\mu) + \theta(-\mu)\left(1 - \frac{3-x_1}{3\mu}\tau_0\right)\mathrm{e}^{\tau_0/\mu}\right]\mathrm{e}^{-\tau/\mu}$$
$$+2\int_0^1 I_0(\tau,\mu,\mu';\tau_0)\mu'\,\mathrm{d}\mu' - 2\int_0^1 I_0(\tau_0-\tau,-\mu,\mu';\tau_0)\left(\mu' + \frac{3-x_1}{3}\tau_0\right)\mathrm{d}\mu'. \quad (4.7)$$

Thus, the zeroth harmonic of radiation intensity must satisfy both the transfer equation and the integral relations (4.3), (4.5), and (4.7). The respective formulas for a semi-infinite medium have been obtained in the previous chapter ((3.7) and (3.16)). It can be easily proved that (4.7) turns into to (3.15) as $\tau_0 \to \infty$.

The invariance relation can also be written for a layer of finite optical thickness submerged in a semi-infinite atmosphere for the Milne problem (i.e., when sources of unlimited power are at infinitely large depth). This establishes the relationship between the radiation intensity (or, to be more exact, its zeroth azimuthal harmonic) in a layer of finite optical thickness and that in the Milne problem (see [51] for further details).

4.2 Equation for Radiation Intensity in Medium

First of all, one should notice that from the radiation transfer equation (1.49), the boundary condition (1.52), and the condition

$$\mu \in [0,1] \tag{4.8}$$

it follows that

$$\left.\frac{\mathrm{d}I^m(\tau,\mu,\mu_0;\tau_0)}{\mathrm{d}\tau}\right|_{\tau=0} = \frac{1}{\mu}B^m(0,\mu,\mu_0;\tau_0), \tag{4.9}$$

$$\left.\frac{\mathrm{d}I^m(\tau,-\mu,\mu_0;\tau_0)}{\mathrm{d}\tau}\right|_{\tau=\tau_0} = -\frac{1}{\mu}B^m(\tau_0,-\mu,\mu_0;\tau_0). \tag{4.10}$$

Differentiating (4.1) with respect to t, putting $t = 0$ and $\tau_1 = \tau_0$, and using (1.49), we have

$$\begin{aligned}\left(\frac{1}{\mu_0}-\frac{1}{\mu}\right) I^m(\tau,\mu,\mu_0;\tau_0) &= -\frac{1}{\mu}B^m(\tau,\mu,\mu_0;\tau_0)\\ +\frac{1}{\mu}B^m(0,\mu,\mu_0;\tau_0)\mathrm{e}^{-\tau/\mu}\theta(\mu) &+ \frac{1}{\mu}B^m(\tau_0,\mu,\mu_0;\tau_0)\mathrm{e}^{\frac{\tau_0-\tau}{\mu}}\theta(-\mu)\\ +2\int_0^1 I^m(\tau,\mu,\mu';\tau_0)&B^m(0,\mu',\mu_0;\tau_0)\frac{\mathrm{d}\mu'}{\mu'}\\ -2\int_0^1 I^m(\tau_0-\tau,-\mu,\mu';\tau_0)&B^m(\tau_0,-\mu',\mu_0;\tau_0)\frac{\mathrm{d}\mu'}{\mu'}\end{aligned} \tag{4.11}$$

for all $\mu \in [-1,+1]$. This equation generalizes (3.26) in the case of a layer of finite optical thickness.

From equations (1.50), (1.95), and (1.96) we derive

$$B^m(0,\mu,\mu_0;\tau_0) = \frac{\lambda}{4}\left[\chi^m(\mu,\mu_0)+2\int_0^1 \chi^m(\mu,-\mu')\rho^m(\mu',\mu_0;\tau_0)\,\mathrm{d}\mu'\right], \tag{4.12}$$

$$B^m(\tau_0,\mu,\mu_0;\tau_0) = \frac{\lambda}{4}\left[\chi^m(\mu,\mu_0)\mathrm{e}^{-\tau_0/\mu_0}+2\int_0^1 \chi^m(\mu,\mu')\sigma^m(\mu',\mu_0;\tau_0)\,\mathrm{d}\mu'\right]. \tag{4.13}$$

Note that the azimuthal harmonics of the reflection coefficient ρ^m and transmission coefficient σ^m depend on the parameter τ_0. Now, if we take (1.50) into consideration, one may treat (4.11) as a linear integral equation to determine the radiation intensity $I^m(\tau,\mu,\mu_0;\tau_0)$ for all depths $\tau \in (0,\tau_0)$, provided the reflection coefficient $\rho^m(\mu, \mu_0; \tau_0)$ and transmission coefficient $\sigma^m(\mu, \mu_0; \tau_0)$ have been found beforehand. It is worth noticing here that no integration over the spatial coordinate τ is required. Now we may proceed with the derivation of the corresponding equations for the aforementioned transmission and reflection coefficients.

4.3 Radiation Intensity at Atmosphere Boundaries

Equations (1.95) and (1.96) establish the relationship between $\rho(\mu, \mu_0, \varphi)$, $\sigma(\mu, \mu_0, \varphi)$, and the intensity at the boundaries of atmosphere. The system of equations determining azimuthal harmonics ρ^m and σ^m can easily be derived from (4.11). Namely, setting $\tau = 0$ in (4.11) and replacing μ by $-\mu$, one obtains

$$\begin{aligned}\rho^m(\mu, \mu_0; \tau_0)(\mu + \mu_0) &= B^m(0, -\mu, \mu_0; \tau_0) - B^m(\tau_0, -\mu, \mu_0; \tau_0)\mathrm{e}^{-\tau_0/\mu} \\ &+2\mu \int_0^1 B^m(0, \mu', \mu_0; \tau_0)\rho^m(\mu, \mu'; \tau_0)\, \mathrm{d}\mu' \\ &-2\mu \int_0^1 B^m(\tau_0, -\mu', \mu_0; \tau_0)\sigma^m(\mu, \mu'; \tau_0)\, \mathrm{d}\mu' \end{aligned} \tag{4.14}$$

for $\mu \in [0, 1]$. Similarly, putting $\tau = \tau_0$ into (4.11), we find

$$\begin{aligned}\sigma^m(\mu, \mu_0; \tau_0)(\mu - \mu_0) &= B^m(0, \mu, \mu_0; \tau_0)\mathrm{e}^{-\tau_0/\mu} - B^m(\tau_0, \mu, \mu_0; \tau_0) \\ &+2\mu \int_0^1 B^m(0, \mu', \mu_0; \tau_0)\sigma^m(\mu, \mu'; \tau_0)\, \mathrm{d}\mu' \\ &-2\mu \int_0^1 B^m(\tau_0, -\mu', \mu_0; \tau_0)\rho^m(\mu, \mu'; \tau_0)\, \mathrm{d}\mu' \end{aligned} \tag{4.15}$$

for $\mu \in [0, 1]$. The relationships given by (4.12–15) can be treated as a system of four integral equations to be used for determining the unknown functions $B^m(0, \mu, \mu_0; \tau_0)$, $B^m(\tau_0, \mu, \mu_0; \tau_0)$, $\rho^m(\mu, \mu_0; \tau_0)$, and $\sigma^m(\mu, \mu_0; \tau_0)$. If the last two functions are defined, one can determine the radiation field in a layer by solving (4.11).

The optical thickness τ_0 plays the role of parameter in (4.12–15), therefore, all the quantities involved in these equations depend on two variables, namely, μ and μ_0. One can, however, express these quantities in terms of the functions $\varphi_i^m(\mu, \tau_0)$, $\psi_i^m(\mu, \tau_0)$, called the ***Ambartsumian functions*** which depend on a single variable. For further details regarding these functions, see LSPA, Chap. 2, Sects. 1,2 and also Chaps. 6, 7, where one can find the general Sobolev theory for the radiation field in a layer of finite optical thickness. We take another approach to finding ρ^m and σ^m related to the Chandrasekhar invariance relationships and the doubling method of van de Hulst.

4.4 Further Consequences of the Basic Invariance Relation

Putting $\tau = 0$ in (4.1) and denoting $t = \tau$, one has ($\mu \in [0,1]$):

$$I^m(\tau, -\mu, \mu_0; \tau_1) = \rho^m(\mu, \mu_0; \tau_0)\mathrm{e}^{-\tau/\mu_0}\mu_0 + I^m(\tau_0 + \tau, -\mu, \mu_0; \tau_1)\mathrm{e}^{-\tau_0/\mu} + 2\int_0^1 \rho^m(\mu, \mu'; \tau_0) I^m(\tau, \mu', \mu_0; \tau_1)\mu' \,\mathrm{d}\mu' + 2\int_0^1 \sigma^m(\mu, \mu'; \tau_0) I^m(\tau_0 + \tau, -\mu', \mu_0; \tau_1)\mu' \,\mathrm{d}\mu' \quad (4.16)$$

Similarly, by putting $\tau = \tau_0$ in (4.1) and denoting $t = \tau$ one gets

$$I^m(\tau + \tau_0, \mu, \mu_0; \tau_1) = \sigma^m(\mu, \mu_0; \tau_0)\mathrm{e}^{-\tau/\mu_0}\mu_0 + I^m(\tau, \mu, \mu_0; \tau_1)\mathrm{e}^{-\tau_0/\mu} + 2\int_0^1 \sigma^m(\mu, \mu'; \tau_0) I^m(\tau, \mu', \mu_0; \tau_1)\mu' \,\mathrm{d}\mu' + 2\int_0^1 \rho^m(\mu, \mu'; \tau_0) I^m(\tau_0 + \tau, -\mu', \mu_0; \tau_1)\mu' \,\mathrm{d}\mu'. \quad (4.17)$$

Note that $\tau \in [0, \tau_1 - \tau_0]$, $\tau_1 \geq \tau_0$ in (4.16) and (4.17). The physical meaning of the above formulas can easily be understood with the help of Fig. 4.1.

Thus, we have obtained the system of integral equations that establishes the relationship between the radiation intensity in a plane layer of optical thickness τ_1 and the transmission and reflection coefficients for a plane layer of thickness τ_0. At $\tau_0 = 0$, both equations become identities.

The ***Chandrasekhar relationships*** (or ***principles***) ***of invariance*** are the consequence of relationships (4.16) and (4.17). Putting $\tau = 0$ in (4.16) and (4.17) and renaming $\tau_0 = \tau$ and $\tau_1 = \tau_0$, one obtains

$$\rho^m(\mu, \mu_0; \tau_0)\mu_0 = \rho^m(\mu, \mu_0; \tau)\mu_0 + I^m(\tau, -\mu, \mu_0; \tau_0)\mathrm{e}^{-\tau/\mu} + 2\int_0^1 \sigma^m(\mu, \mu'; \tau) I^m(\tau, -\mu', \mu_0; \tau_0)\mu' \,\mathrm{d}\mu', \quad (4.18)$$

$$I^m(\tau, \mu, \mu_0; \tau_0) = \sigma^m(\mu, \mu_0; \tau)\mu_0 + 2\int_0^1 \rho^m(\mu, \mu'; \tau) I^m(\tau, -\mu', \mu_0; \tau_0)\mu' \,\mathrm{d}\mu'. \quad (4.19)$$

On the other hand, setting $\tau = \tau_1 - \tau_0$ in (4.16) and (4.17) and renaming as above, we arrive at the following equations

$$I^m(\tau, -\mu, \mu_0; \tau_0) = \rho^m(\mu, \mu_0; \tau_0 - \tau)\mathrm{e}^{-\tau/\mu_0}\mu_0 + 2\int_0^1 \rho^m(\mu, \mu'; \tau_0 - \tau) I^m(\tau, \mu', \mu_0; \tau_0)\mu' \,\mathrm{d}\mu', \quad (4.20)$$

$$\sigma^m(\mu,\mu_0;\tau_0)\mu_0 = \sigma^m(\mu,\mu_0;\tau_0-\tau)\mathrm{e}^{-\tau/\mu_0}\mu_0 + I^m(\tau,\mu,\mu_0;\tau_0)\mathrm{e}^{-\frac{\tau_0-\tau}{\mu}}$$
$$+2\int_0^1 \sigma^m(\mu,\mu';\tau_0-\tau)I^m(\tau,\mu',\mu_0;\tau_0)\mu'\,\mathrm{d}\mu'. \quad (4.21)$$

Equations (4.18–21) are the third, second, first, and fourth Chandrasekhar invariance principles respectively (see RT, Sect. 50).

We know that the zeroth azimuthal harmonic of the radiation intensity satisfies the integral relations (4.3), (4.5), and (4.7). Assuming first that $\tau = 0$ and then that $\tau = \tau_0$, one obtains the integral relations for the transmission and reflection coefficients. Namely, it follows from (4.3) that

$$2\int_0^1 \left[\rho(\mu,\mu';\tau_0)i(\mu') + \mathrm{e}^{-k\tau_0}\sigma(\mu,\mu';\tau_0)i(-\mu')\right]\mu'\,\mathrm{d}\mu' = \left[1-\mathrm{e}^{-(k+\frac{1}{\mu})\tau_0}\right]i(-\mu), \quad (4.22)$$

$$2\int_0^1 \left[\mathrm{e}^{-k\tau_0}\rho(\mu,\mu';\tau_0)i(-\mu') + \sigma(\mu,\mu';\tau_0)i(\mu')\right]\mu'\,\mathrm{d}\mu' = \left[\mathrm{e}^{-k\tau_0} - \mathrm{e}^{-\tau_0/\mu}\right]i(\mu). \quad (4.23)$$

If the scattering is conservative ($\lambda = 1$), (4.22) and (4.23) reduce to the single equation

$$2\int_0^1 [\rho_0(\mu,\mu';\tau_0) + \sigma_0(\mu,\mu';\tau_0)]\mu'\,\mathrm{d}\mu' = 1 - \mathrm{e}^{-\tau_0/\mu}. \quad (4.24)$$

Instead of (4.7), we have

$$\mu\left(1-\mathrm{e}^{-\tau_0/\mu}\right) = \left(1-\frac{x_1}{3}\right)\tau_0\left[1 - 2\int_0^1 \rho_0(\mu,\mu';\tau_0)\mu'\,\mathrm{d}\mu'\right]$$
$$-2\int_0^1 [\rho_0(\mu,\mu';\tau_0) - \sigma_0(\mu,\mu';\tau_0)]\mu'^2\,\mathrm{d}\mu'. \quad (4.25)$$

Equations (4.24) and (4.23) could certainly be derived from physical arguments, of course. The reasoning would be the same as that provided after (4.4). Here one should add that (4.24) and (4.25) are consequences of the flux integral and K integral (see Sect. 4.7).

4.5 Doubling Method for Calculation of Transmission and Reflection Coefficients

Let us put $\tau = \tau_0$ into (4.18–21) and substitute τ_0 with $2\tau_0$. We then have

$$\rho^m(\mu,\mu_0;2\tau_0)\mu_0 = \rho^m(\mu,\mu_0;\tau_0)\mu_0 + I^m(\tau_0,-\mu,\mu_0;2\tau_0)\mathrm{e}^{-\tau_0/\mu}$$
$$+2\int_0^1 \sigma^m(\mu,\mu';\tau_0)I^m(\tau_0,-\mu',\mu_0;2\tau_0)\mu'\,\mathrm{d}\mu', \quad (4.26)$$

$$\sigma^m(\mu,\mu_0;2\tau_0)\mu_0 = \sigma^m(\mu,\mu_0;\tau_0)\mathrm{e}^{-\tau_0/\mu_0}\mu_0 + I^m(\tau_0,\mu,\mu_0;2\tau_0)\mathrm{e}^{-\tau_0/\mu} + 2\int_0^1 \sigma^m(\mu,\mu';\tau_0)I^m(\tau_0,\mu',\mu_0;2\tau_0)\mu'\,\mathrm{d}\mu', \tag{4.27}$$

$$I^m(\tau_0,\mu,\mu_0;2\tau_0) = \sigma^m(\mu,\mu_0;\tau_0)\mu_0 + 2\int_0^1 \rho^m(\mu,\mu';\tau)I^m(\tau_0,-\mu',\mu_0;2\tau_0)\mu'\,\mathrm{d}\mu', \tag{4.28}$$

$$I^m(\tau,-\mu,\mu_0;2\tau_0) = \rho^m(\mu,\mu_0;\tau_0)\mathrm{e}^{-\tau_0/\mu_0}\mu_0 + 2\int_0^1 \rho^m(\mu,\mu';\tau_0)I^m(\tau_0,\mu',\mu_0;2\tau_0)\mu'\,\mathrm{d}\mu'. \tag{4.29}$$

Equations (4.26) and (4.27) imply the following: if one knows the transmission and reflection coefficients for a layer of optical thickness τ_0 as well as the intensities of ascending and descending radiation in the middle of a layer of optical thickness $2\tau_0$, then the above coefficients are also determined for a layer of optical thickness $2\tau_0$.

The explicit expressions for the transmission and reflection coefficients determined by single scattering in a layer are commonly used as a first approximation. It follows from (4.12–15) that these quantities are

$$\rho^m(\mu,\mu_0;\tau_0) = \frac{\lambda}{4}\chi^m(-\mu,\mu_0)\frac{1-\mathrm{e}^{-\tau_0\left(\frac{1}{\mu}+\frac{1}{\mu_0}\right)}}{\mu+\mu_0}, \tag{4.30}$$

$$\sigma^m(\mu,\mu_0;\tau_0) = \frac{\lambda}{4}\chi^m(\mu,\mu_0)\frac{\mathrm{e}^{-\tau_0/\mu}-\mathrm{e}^{-\tau_0/\mu_0}}{\mu-\mu_0}. \tag{4.31}$$

We choose a small enough τ_0 for (4.30) and (4.31) to provide ρ^m and σ^m as accurately as required. Then we find the transmission and reflection coefficients for the layer of *double* optical thickness by substituting ρ^m and σ^m into (4.28) and (4.29), which produce the system of integral equations that determine $I^m(\tau_0,\mu,\mu_0;2\tau_0)$ and $I^m(\tau_0,-\mu,\mu_0;2\tau_0)$ and can be solved by the method of successive approximations. The whole procedure is repeated until $\rho^m(\mu,\mu_0,\tau_0)$ and $\sigma^m(\mu,\mu_0,\tau_0)$ are found for a layer of the necessary optical thickness.

The above method for calculating the transmission and reflection coefficients is known as the ***van de Hulst doubling method***. The method shows its great efficiency in numerous computer simulations (for further details, see MLS, Chap. 4 and [52]).

Concluding the section, it should be noted that in the case of a layer of considerable optical thickness ($\tau_0 \gg 1$), one encounters certain problems in applying the doubling method. Namely, the accuracy of the calculation of zeroth azimuthal harmonic, which can be verified with the help of (4.22–4.25), decreases dramatically. In these cases, one should use the asymptotic formulas of Sect. 4.8.1. They express $\rho(\mu,\mu_0,\tau_0)$ and $\sigma(\mu,\mu_0,\tau_0)$ in terms of the reflection coefficient $\rho(\mu, \mu_0)$ of a semi-infinite atmosphere. This method of evaluating the reflection coefficient was described in Sect. 3.5.

4.6 Radiation Field in a Layer

Let us now assume that we are to calculate not only the reflection and transmission coefficients but also the radiation intensity in a layer. It turns out that it can be done easily by using the invariance relationship (4.1) and the doubling method. Indeed, putting $t = \tau_0$, $\tau_1 = 2\tau_0$ and $t = 0$, $\tau_1 = 2\tau_0$, we obtain ($\mu \in [-1, +1]$)

$$I^m(\tau_0 + \tau, \mu, \mu_0; 2\tau_0) = I^m(\tau, \mu, \mu_0; \tau_0)\mathrm{e}^{-\tau_0/\mu_0} + I^m(\tau_0, \mu, \mu_0; 2\tau_0)\mathrm{e}^{-\tau/\mu}\theta(\mu) + 2\int_0^1 I^m(\tau, \mu, \mu'; \tau_0) I^m(\tau_0, \mu', \mu_0; 2\tau_0)\,\mathrm{d}\mu', \tag{4.32}$$

$$I^m(\tau, \mu, \mu_0; 2\tau_0) = I^m(\tau, \mu, \mu_0; \tau_0) + I^m(\tau_0, \mu, \mu_0; 2\tau_0)\mathrm{e}^{\frac{\tau_0 - \tau}{\mu}}\theta(-\mu) + 2\int_0^1 I^m(\tau_0 - \tau, -\mu, \mu'; \tau_0)\, I^m(\tau_0, -\mu', \mu_0; 2\tau_0)\,\mathrm{d}\mu'. \tag{4.33}$$

Now set $\tau = \tau_0/2$ in (4.32) and (4.33). As $I^m(\tau_0/2, \mu, \mu_0, \tau_0)$ and $I^m(\tau_0, \mu, \mu_0, 2\tau_0)$ are known by the doubling method, equations (4.32) and (4.33) provide two more points of the layer of thickness $2\tau_0$, namely $\tau = \tau_0/2$ and $\tau = 3\tau_0/2$, where the intensity is known (note that the intensity at $\tau = \tau_0$ is known). After the doubling method has been applied one more time, we can calculate the radiation intensity at seven depths, namely at $\tau = 1/2\tau_0$, τ_0, $3\tau_0/2$, $2\tau_0$, $5\tau_0/2$, $3\tau_0$, $7\tau_0/2$, etc.

The initial optical thickness of the layer used in doubling scheme is obviously the minimal step of "discretization" in depths. However, one hardly needs to calculate the internal field so elaborately. The calculation by (4.32) and (4.33) can be incorporated into the general scheme of the doubling method at any stage depending on the number of depths we want to know the field at. It is extremely important that one does not need to solve any equations or apply any iterative procedure to calculate the internal field using (4.32) and (4.33). This remarkable property is known to be characteristic of the doubling method which is exploited to calculate the internal radiation in a semi-infinite homogeneous atmosphere (see Sect. 3.10).

From a formal point of view, the proposed scheme for the calculation of the field of internal radiation is not an iterative one, like that for a semi-infinite atmosphere. Indeed, van de Hulst's doubling method — that certainly includes an iterative procedure — only calculates the transmission and reflection coefficients, as well as the radiation intensity in the middle of the layer as a byproduct. For a semi-infinite atmosphere (see Sect. 3.10), the iterative process is part of calculation of the reflection coefficient.

Dlugach applies the above scheme to construct an algorithm and calculate the radiation field in plane layers of finite optical thickness. This program shows the efficiency of the above method. For instance, incorporating the doubling method into the program results in only a 20% rise in the calculation time for the initial step of the optical depth discretization $\Delta\tau = 0.125$. The internal radiation is calculated in layers of optical thicknesses $\tau_0 = 0.5, 1.0, 2.0$.

4.7 Integrals of the Transfer Equation for a Layer of Finite Thickness

Control of the accuracy of calculations is very important to determine the internal radiation field. Of course regarding the "integral" accuracy, one can use (4.3) and (4.4), or (4.5) and (4.7) for the conservative scattering. We consider here only the zeroth azimuthal harmonic of the intensity for the angular variable $\mu_0 \in [0,1]$. It is known from Sect. 3.7, however, that in a semi-infinite medium there exists a number of integrals of the radiative transfer equation which are quadratic with respect to the intensity. Now we intend to establish similar relationships for a layer of finite thickness.

Let us multiply (4.16) by $2I^m(\tau', \mu, \mu_1; \tau_2)\mu$ and (4.17) by $2I^m(\tau_0 + \tau', -\mu, \mu_1; \tau_2)\mu$, and then integrate the resultant formulas with respect to μ from 0 to 1. Further, summing the expressions obtained and using (4.16) and (4.17), after simple manipulations we find

$$\begin{aligned}
&2\int_{-1}^{+1} I^m(\tau', \mu', \mu_1; \tau_2)\, I^m(\tau, -\mu', \mu_0; \tau_1)\mu' \,\mathrm{d}\mu' \\
&+2\int_{-1}^{+1} I^m(\tau_0 + \tau', -\mu', \mu_1; \tau_2) I^m(\tau_0 + \tau, \mu', \mu_0; \tau_1)\mu' \,\mathrm{d}\mu' \\
&= I^m(\tau', -\mu_0, \mu; \tau_2)\mathrm{e}^{-\tau/\mu_0}\mu_0 - I^m(\tau, -\mu, \mu_0; \tau_1)\mathrm{e}^{-\tau'/\mu}\mu \\
&+I^m(\tau_0 + \tau, -\mu, \mu_0; \tau_1)\mathrm{e}^{-\frac{\tau_0+\tau'}{\mu}}\mu - I^m(\tau_0 + \tau', -\mu_0, \mu; \tau_2)\mathrm{e}^{-\frac{\tau_0+\tau}{\mu_0}}\mu_0 . \qquad (4.34)
\end{aligned}$$

This formula is the first integral of the transfer equation for a layer of finite optical thickness. The second can be found in a similar way, namely by multiplying (4.16) and (4.17) by $2I^m(\tau'+\tau_0, -\mu, \mu_1; \tau_2)\mu$ and $2I^m(\tau', -\mu, \mu_1; \tau_2)\mu$, respectively, and performing similar manipulations, we derive

$$\begin{aligned}
&2\int_{-1}^{+1} I^m(\tau_0 + \tau', -\mu', \mu_1; \tau_2) I^m(\tau, -\mu', \mu_0; \tau_1)\mu' \,\mathrm{d}\mu' \\
&+2\int_{-1}^{+1} I^m(\tau', \mu', \mu_1; \tau_2) I^m(\tau_0 + \tau, \mu', \mu_0; \tau_1)\mu' \,\mathrm{d}\mu' \\
&= I^m(\tau_0 + \tau', \mu_0, \mu; \tau_2)\mathrm{e}^{-\tau/\mu_0}\mu_0 - I^m(\tau_0 + \tau, \mu, \mu_0; \tau_1)\mathrm{e}^{-\tau'/\mu}\mu \\
&+I^m(\tau, \mu, \mu_0; \tau_1)\mathrm{e}^{-\frac{\tau_0+\tau'}{\mu}}\mu - I^m(\tau', \mu_0, \mu; \tau_2)\mathrm{e}^{-\frac{\tau_0+\tau}{\mu_0}}\mu_0 . \qquad (4.35)
\end{aligned}$$

Note that $\tau \in [0, \tau_1 - \tau_0]$ and $\tau' \in [0, \tau_2 - \tau_0]$ in (4.34) and (4.35).

The integrals (4.34) and (4.35) are nonlinear relationships that include the intensities of radiation at *different* depths in layers of *different* optical thickness. The first integral establishes the relationship between the intensities of radiations propagating in *opposite directions*, and the second one between the radiations propagating in *the same direction*. The right-hand sides of (4.34) and (4.35) involve the intensities of only ascending and descending radiation, respectively.

Let us consider some particular cases of (4.34) and (4.35). It can be seen easily that for $\tau_1, \tau_2 \to \infty$, (4.34) can be transformed into the relation (3.46). Under the same conditions, one can add (4.35), which involves the intensities at four optical depths $(\tau, \tau', \tau_0 + \tau', \tau_0 + \tau)$, to the integral (3.46). Here the integral relationship involves the intensities of radiations propagating in the same direction (cf. (3.46)). Simple relations linking the intensities of radiation in layers with different optical thicknesses can easily be derived from (4.35). Namely, putting $\tau' = 0$ and $\tau = 0$, then renaming $\tau_0 = \tau$ one finds

$$\begin{aligned} &2\mu_0 \int_0^1 I^m(\tau, -\mu', \mu_1; \tau_2)\rho^m(\mu', \mu_0; \tau_1)\mu' \, d\mu' \\ &-2\mu \int_0^1 I^m(\tau, -\mu', \mu_0; \tau_1)\rho^m(\mu', \mu_1; \tau_2)\mu' \, d\mu' \\ &= I^m(\tau, \mu_0, \mu; \tau_2)\mu_0 - I^m(\tau, \mu, \mu_0; \tau_1)\mu. \end{aligned} \tag{4.36}$$

Other particular cases of (4.33) and (4.34) are not treated here.

Now we proceed with the derivation of two important relations that couple the zeroth azimuthal harmonic of the intensity in a layer of finite optical thickness with the intensity $i(\mu)$ in an infinite medium. Putting $\tau_2 = \infty$, $\tau = 0$, and $\tau' \to \infty$ in (4.34) and (4.35), and using the asymptotic formula (3.57), one obtains

$$\begin{aligned} \left[1 - e^{-\left(k+\frac{1}{\mu_0}\right)\tau}\right] i(-\mu_0)\mu_0 &= 2\mu_0 \int_0^1 \rho(\mu', \mu_0; \tau_0) i(\mu')\mu' \, d\mu' \\ &+2e^{-k\tau} \int_{-1}^{+1} I(\tau, \mu', \mu_0; \tau_0) i(-\mu')\mu' \, d\mu', \end{aligned} \tag{4.37}$$

$$\begin{aligned} \left(e^{-k\tau} - e^{-\tau/\mu_0}\right) i(\mu_0)\mu_0 &= 2 \int_{-1}^{+1} I(\tau, \mu', \mu_0; \tau_0) i(\mu')\mu' \, d\mu' \\ &+2\mu_0 e^{-k\tau} \int_0^1 \rho(\mu', \mu_0; \tau_0) i(-\mu')\mu' \, d\mu'. \end{aligned} \tag{4.38}$$

Together with (4.3) and (4.5), (4.37) and (4.38) are the auxiliary conditions that the solution of the transfer equation has to meet. At the boundaries of the layer, these conditions transform into the familiar ones, (4.22) and

(4.23), for the reflection and transmission coefficients. In the conservative limit ($\lambda = 1$), (4.37) and (4.38) yield

$$H_{0\tau}(\tau, \mu_0; \tau_0) = \pi\mu_0[1 - A_0(\mu_0, \tau_0)], \tag{4.39}$$

$$2\int_{-1}^{+1} I_0(\tau, \mu', \mu_0; \tau_0)\mu'^2 \,\mathrm{d}\mu' = \mu_0^2 \left(1 - \mathrm{e}^{-\tau/\mu_0}\right)$$
$$+2\mu_0 \int_0^1 \rho_0(\mu_0, \mu'; \tau_0)\mu'^2 \,\mathrm{d}\mu' - \left(1 - \frac{x_1}{3}\right) \tau\mu_0[1 - A_0(\mu_0, \tau_0)]. \tag{4.40}$$

Here

$$H_{0\tau}(\tau, \mu_0; \tau_0) = \pi\mu_0 \mathrm{e}^{-\tau/\mu_0} + 2\pi \int_{-1}^{+1} I_0(\tau, \mu', \mu_0; \tau_0)\mu' \,\mathrm{d}\mu' \tag{4.41}$$

is a projection of the vector of total flux of the radiation onto the axis τ (see (1.87)), and

$$A_0(\mu_0, \tau_0) = 2\int_0^1 \rho_0(\mu, \mu_0; \tau_0)\mu \,\mathrm{d}\mu. \tag{4.42}$$

is the plane albedo of the atmosphere (see (1.104)). The subscript "0" in the function H_τ refers to the conservative scattering. The relations (4.39) and (4.40) are the flux integral and K integral, respectively. The method used to derive (4.40) is similar to that used for (4.7). Setting $\tau = \tau_0$ in (4.39) and (4.40) results in the known formulas (4.24) and (4.25).

Thus, any harmonic of the radiation intensity $I^m(\tau, \mu, \mu_0; \tau_0)$ must satisfy the transfer equation and the integral relationships (4.34) and (4.35) as well. They may play a very important role for a couple of reasons. First, they may be used to test a computer algorithm aimed at evaluating the internal field of radiation. Second, these relations may also be used to control the "integral" accuracy of calculations.

With regard to the zeroth azimuthal harmonic $I(\tau, \mu, \mu_0; \tau_0)$, one sees more opportunities. For example, if $i(\mu)$ is known, one can then control the accuracy by using both angular variables, μ and μ_0. For this purpose (4.3), (4.37) and (4.38) for $\lambda \neq 1$ and (4.5), (4.7), (4.40) and (4.41) for conservative scattering can be applied. Furthermore these formulas can be used, first, as "balance" relations in solving the transfer equation by the convergence method (for details, see Sect. 3.5, where this method is used to evaluate the reflection coefficient for a semi-infinite medium), and second, to derive approximate formulas. The latter ones determine explicitly the zeroth harmonic of the radiation intensity (for general approach to this problem see LSPA, Sect. 8.1).

4.8 Atmosphere with Large Optical Thickness

Almost all the formulas derived in the previous sections of this chapter are valid for an atmosphere of arbitrary optical thickness. However, difficulties in solving the transfer equation grow with increasing optical thickness τ_0 of the layer. This mostly concerns the zeroth azimuthal harmonic of the radiation intensity. Other harmonics attenuate more rapidly as light passes through a layer of substantial optical thickness. Nevertheless, for $\tau_0 \gg 1$ it is possible to derive the asymptotic formulas that for all optical depths express $I(\tau, \mu, \mu_0; \tau_0)$ in terms of the radiation intensity in a semi-infinite medium.

To derive such a formula, we start with the invariance relation (4.1). For $m = 0$ we set $t = 0$ and $\tau_1 = \infty$ in (4.1). This results in

$$I(\tau, \mu, \mu_0) = I(\tau, \mu, \mu_0; \tau_0) + I(\tau_0, \mu, \mu_0) e^{\frac{\tau_0 - \tau}{\mu}} \theta(-\mu)$$
$$+2 \int_0^1 I(\tau_0 - \tau, -\mu, \mu'; \tau_0) I(\tau_0, -\mu', \mu_0) \, d\mu'. \qquad (4.43)$$

Assuming that the radiation intensity $I(\tau, \mu, \mu_0)$ in a semi-infinite medium is known, (4.43) can be used in determining $I(\tau, \mu, \mu_0; \tau_0)$. For large τ_0, this equation can be solved explicitly. This becomes possible owing to the asymptotic separation of angular variables in the function $I(\tau_0, \mu', \mu_0)$ in the integrand of (4.43) at large τ_0. Using (3.57) and multiplying the resultant expression by $2i(-\mu_0)$, one gets the formula which, when integrated over μ_0, from 0 to 1 yields

$$2 \int_0^1 I(\tau, -\mu, \mu'; \tau_0) i(-\mu') \, d\mu' + 2N e^{-k\tau_0} \int_0^1 I(\tau_0 - \tau, \mu, \mu'; \tau_0) i(-\mu') \, d\mu'$$
$$= -N i(-\mu) e^{-\left(k\tau_0 + \frac{\tau_0 - \tau}{\mu}\right)} \theta(\mu) + 2 \int_0^1 I(\tau, -\mu, \mu') i(-\mu') \, d\mu'. \qquad (4.44)$$

N is an asymptotic constant determined by (3.40). Replacing τ by $\tau_0 - \tau$ in (4.44), one obtains the equation that, together with (4.44), provides the method to find the integrals on the left-hand side. Solving this and performing certain manipulations, one arrives at the asymptotic formula sought

$$I(\tau, \mu, \mu_0; \tau_0) = I(\tau, \mu, \mu_0) - \{I(\tau, \mu) r(\tau_0)$$
$$-[I(\tau_0 - \tau, -\mu) - I^{\mathrm{as}}(\tau_0 - \tau, -\mu)] t(\tau_0)\} u(\mu_0) \mu_0, \qquad (4.45)$$

where

$$r(\tau_0) = N e^{-k\tau_0} t(\tau_0) = \frac{M N e^{-2k\tau_0}}{1 - N^2 e^{-2k\tau_0}}. \qquad (4.46)$$

Notice that (3.6) and (3.58) are used to derive the above formulas. One remembers that in (4.45) $\tau_0 \gg 1$, the optical depth τ is arbitrary ($\tau \in [0, \tau_0]$), and $I(\tau, \mu)$ is the Milne intensity.

The physical meaning of (4.45) is straightforward, shown in a similar way to that used in analyzing the physical meaning of (3.71) and (3.73). It is worth noticing here that $r(\tau_0)$ and $t(\tau_0)$ are usually called the ***reflection*** and ***transmission factors***, respectively. The origin of these terms is discussed later. Now we deal with the particular cases of the basic formula (4.45):

1. Let $\tau \gg 1$, i.e., we are dealing with a layer near the lower boundary of the medium. Using (3.40) and (3.58), we get

$$I(\tau,\mu,\mu_0;\tau_0) = I(\tau_0-\tau,-\mu)u(\mu_0)\mu_0 t(\tau_0). \tag{4.47}$$

2. Let $\tau_0 - \tau \gg 1$, then similar considerations yield

$$I(\tau,\mu,\mu_0;\tau_0) = I(\tau,\mu,\mu_0) - I(\tau,\mu)u(\mu_0)\mu_0 r(\tau_0). \tag{4.48}$$

3. Equation (4.45) takes a simple form far from both boundaries of the medium, i.e., when $\tau \gg 1$ and $\tau_0 - \tau \gg 1$ simultaneously, namely

$$I(\tau,\mu,\mu_0;\tau_0) = I^{\mathrm{as}}(\tau_0-\tau,-\mu)u(\mu_0)\mu_0 t(\tau_0). \tag{4.49}$$

   Hence, to calculate intensity we need to know the functions $i(\mu)$, $u(\mu)$, and the related quantities k, M, and N.

4.8.1 Reflection and Transmission Coefficients and Other Quantities

Put $\tau = 0$ in (4.45). We then get

$$\rho^{\mathrm{as}}(\mu,\mu_0;\tau_0) = \rho(\mu,\mu_0) - u(\mu)u(\mu_0)r(\tau_0). \tag{4.50}$$

Similarly, for $\tau = \tau_0$ the transmission coefficient of an optically thick atmosphere is

$$\sigma^{\mathrm{as}}(\mu,\mu_0;\tau_0) = u(\mu)u(\mu_0)t(\tau_0). \tag{4.51}$$

Equations (4.50) and (4.51) demonstrate why $r(\tau_0)$ and $t(\tau_0)$ are called the factors of reflection and transmission, respectively. The superscript "as" means that $\tau_0 \gg 1$.

The physical reasoning of such a representation is quite simple. Let us discuss it taking (4.51) as an example. The photons passing through a layer of large optical thickness contribute solely to the transmission coefficient. As we know, these photons "forget" their initial directions of motion, and their number must be proportional to the escape function $u(\mu)$. The function $\sigma^{\mathrm{as}}(\mu,\mu_0;\tau_0)$ is symmetrical with respect to permutation of μ and μ_0. Therefore, its structure should be similar to that given by (4.51). We will leave it to the reader to explain the structure of (4.50).

Now we are to give more details about the factors $r(\tau_0)$ and $t(\tau_0)$ (see (4.46)). Sometimes it seems more convenient to rewrite them in another way. Let us represent N as follows (note that $0 \le N \le 1$):

$$N = \mathrm{e}^{-2\tau^{\mathrm{e}}k}, \tag{4.52}$$

where τ^e is a so called ***extrapolated length***. Then, instead of (4.46), we have

$$r(\tau_0) = \frac{M}{2N}[\coth k(\tau_0 + 2\tau^e) - 1], \tag{4.53}$$

$$t(\tau_0) = \frac{M}{2N} \frac{1}{\sinh k(\tau_0 + 2\tau^e)}. \tag{4.54}$$

This representation is suitable to study the nearly conservative scattering of light (for more details, see Sect. 4.8.3).

If we know the reflection and transmission coefficients, it is easy to calculate the atmosphere albedo and the relative illumination of the lower boundary of the layer. Using (1.104) and (1.105) and substituting the expressions obtained into (4.50) and (4.51), we obtain

$$A^{as}(\mu_0, \tau_0) = A(\mu_0) - u(\mu_0) C r(\tau_0), \tag{4.54a}$$

$$V^{as}(\mu_0, \tau_0) = u(\mu_0) C t(\tau_0), \tag{4.54b}$$

where C is determined by (3.81).

The function

$$U(\mu_0, \tau_0) = 1 - A(\mu_0, \tau_0) - V(\mu_0, \tau_0). \tag{4.54c}$$

is an important characteristic of the absorbing properties of an atmosphere. The quantity $U(\mu_0, \tau_0) E_0$, where $E_0 = \pi \mu_0$ is the illumination of the boundary of the atmosphere by parallel rays of the Sun, is the total radiative energy absorbed by an atmospheric column of unit section. For $\tau_0 \gg 1$, (4.53–54c) give

$$U^{as}(\mu_0, \tau_0) = 1 - A(\mu_0) + \frac{CM}{2N}\left[\tanh \frac{k}{2}(\tau_0 + 2\tau^e) - 1\right] u(\mu_0). \tag{4.54d}$$

Finally, from (1.107) and (4.50) we have for geometric albedo

$$A_g^{as}(\tau_0) = A_g - \zeta r(\tau_0),$$

where

$$\zeta = 2 \int_0^1 u^2(\mu) \mu^2 \, d\mu. \tag{4.54e}$$

Numerical values of A_g and ζ are given in Tables XX and XXVI, respectively.

4.8.2 Conservative Scattering

To consider conservative scattering, we use (2.44) and (3.82). Taking the limit $\lambda \to 0$ in (4.53) and (4.54), one gets

$$r_0(\tau_0) = t_0(\tau_0) = \frac{4}{(3-x_1)(\tau_0 + 2\tau_0^{\mathrm{e}})}. \tag{4.55}$$

where as seen from comparing (3.82) with (4.52)

$$\tau_0^{\mathrm{e}} = \frac{3\gamma_0}{3-x_1}. \tag{4.56}$$

Therefore, the expanded form of (4.55) takes on the following appearance,

$$r_0(\tau_0) = t_0(\tau_0) = \frac{4}{(3-x_1)\tau_0 + 6\gamma_0}. \tag{4.57}$$

As a result, (4.50) and (4.51) become

$$\rho_0^{\mathrm{as}}(\mu, \mu_0; \tau_0) = \rho(\mu, \mu_0) - \sigma_0^{\mathrm{as}}(\mu, \mu_0; \tau_0), \tag{4.58}$$

$$\sigma_0^{\mathrm{as}}(\mu, \mu_0; \tau_0) = \frac{4u_0(\mu)u_0(\mu_0)}{(3-x_1)\tau_0 + 6\gamma_0}. \tag{4.59}$$

The formula (4.58) suggests that the sum of the reflection and transmission coefficients for conservative scattering in the optically thick layer gives the reflection coefficient of the corresponding semi-infinite atmosphere.

The general formula (4.45) for the radiation intensity takes a very simple form, namely

$$I_0(\tau, \mu, \mu_0; \tau_0) = I_0(\tau, \mu, \mu_0) - \Big\{(1 - x_1/3)\tau - \mu + \gamma_0 + 2\int_0^1 [I(\tau, \mu, \mu') - I(\tau_0 - \tau, -\mu, \mu')]\mu' \,\mathrm{d}\mu' + \Big[\mathrm{e}^{-\frac{\tau}{\mu}}\theta(\mu) - \mathrm{e}^{\frac{\tau_0-\tau}{\mu}}\theta(-\mu)\Big]\mu\Big\} \frac{u_0(\mu_0)\mu_0}{(3 - x_1/3)\tau_0 + 2\gamma_0}, \tag{4.60}$$

where (3.15) and (3.61) are applied.

Using (4.54a) and (4.57) one finds that the plane albedo of atmosphere is

$$A_0^{\mathrm{as}}(\mu_0, \tau_0), = 1 - \frac{4u_0(\mu_0)}{(3-x_1)\tau_0 + 6\gamma_0}. \tag{4.61}$$

To derive (4.61) the condition

$$C_0 = 2\int_0^1 u_0(\mu)\mu \,\mathrm{d}\mu = 1, \tag{4.62}$$

was used, which follows from (3.12) at $\lambda = 1$. Inserting (4.61) into (4.39) gives the following expression for the radiation flux

$$H_{0\tau}(\tau, \mu_0; \tau_0) = \frac{4\pi u_0(\mu)\mu_0}{(3 - x_1)\tau_0 + 6\gamma_0}. \tag{4.63}$$

Naturally the same expression for $H_{0\tau}$ can be obtained by substituting (4.60) into (4.41).

4.8.3 Nearly Conservative Scattering in an Optically Thick Layer

In fact, the solution of this problem can be found very easily; it suffices to substitute the expressions (3.88) and (3.89) for the radiation intensity in the nearly conservatively scattering semi-infinite atmosphere into (4.45) and to expand the result into a series of parameter k. Notice that some care is needed because these functions depend separately on the *small* parameter k and the parameter $k\tau$, which may take an *arbitrary* value. Therefore, it is convenient to use the representations (4.53), (4.54) for the reflection and transmission factors, which depend on the above parameters explicitly and separately. Taking this into account, we have

$$\begin{aligned} I_1(\tau, \mu, \mu_0; \tau_0) &= I_1(\tau, \mu, \mu_0) - \Bigg\{ \sinh k_0\tau + \frac{3k_0(\gamma_0 - \mu)}{3 - x_1} \cosh k_0\tau \\ &+ \frac{4k_0}{3 - x_1}[I_0(\tau, \mu) - I_0^{\mathrm{as}}(\tau, \mu)] \Bigg\} [\coth k_0(\tau_0 + 2\tau_0^{\mathrm{e}}) - 1] u_0(\mu_0)\mu_0 \\ &+ \frac{4k_0[I_0(\tau_0 - \tau, -\mu) - I_0^{\mathrm{as}}(\tau_0 - \tau, -\mu)]}{(3 - x_1)\sinh k_0(\tau_0 + 2\tau_0^{\mathrm{e}})} u_0(\mu_0)\mu_0 + O(k_0^2). \end{aligned} \tag{4.64}$$

Here, the function $I_1(\tau, \mu, \mu_0)$ is given by (3.86) or (3.88), and $k_0 = [(1 - \lambda)(3 - x_1)]^{1/2}$.

Equation (4.64) results in the following simple expression for the flux:

$$H_\tau(\tau, \mu_0; \tau_0) = \frac{4\pi k_0 \cosh k_0(\tau_0 - \tau + 2\tau_0^{\mathrm{e}})}{(3 - x_1)\sinh k_0(\tau_0 + 2\tau_0^{\mathrm{e}})} u_0(\mu_0)\mu_0 + O(k_0^2), \tag{4.65}$$

which transforms into (4.63) at $\lambda = 1$. At $\tau = \tau_0$ (4.64) gives the following asymptotic formula for the transmission coefficient:

$$\sigma^{\mathrm{as}}(\tau, \mu_0; \tau_0) = \frac{4\pi k_0 u_0(\mu) u_0(\mu_0)}{(3 - x_1)\sinh k_0(\tau_0 + 2\tau_0^{\mathrm{e}})} + O(k_0^3). \tag{4.66}$$

Using (3.61) and (3.80) at $\tau = 0$, we have

$$\rho^{\mathrm{as}}(\mu_0; \tau_0) = \rho_0(\mu, \mu_0) - \frac{4k_0 u_0(\mu) u_0(\mu_0)}{3 - x_1} \coth k_0(\tau_0 + 2\tau_0^{\mathrm{e}}) + O(k_0^2). \tag{4.66a}$$

One should note that (4.66) is valid within the accuracy of terms of the order of k_0^2. This can be verified by using (4.51) and (4.54), which are valid for each $\lambda > 0$, and the asymptotic expansions (2.42), (3.79), and (3.82). Then, from

(1.105) and (4.66), we get the following formula for the relative illumination from above of the lower atmospheric boundary ($\tau_0 \gg 1$):

$$V^{\rm as}(\mu_0, \tau_0) = \frac{4k_0 u_0(\mu_0)}{(3 - x_1)\sinh k_0(\tau_0 + 2\tau_0^{\rm e})} + O(k_0^3). \tag{4.67}$$

It is straightforward to derive the corresponding asymptotic formula for the plane albedo of a layer, namely,

$$A^{\rm as}(\mu_0, \tau_0) = 1 - \frac{4k_0 u_0(\mu)}{3 - x_1} \coth k_0(\tau_0 + 2\tau_0^{\rm e})$$
$$+ \left[\frac{15}{5 - x_2} v_0(\mu_0) + 4\tau_0^{\rm e} u_0(\mu_0)\right] \frac{k_0^2}{3 - x_1} + O(k_0^3). \tag{4.68}$$

Here the function $v_0(\mu)$ is given by (3.98), and

$$\int_0^1 v_0(\mu)\mu \, {\rm d}\mu = 0. \tag{4.69}$$

Hence, using (4.68) one arrives at the following formula for the spherical albedo of atmosphere

$$A_{\rm s}^{\rm as}(\tau_0) = 1 - \frac{4k_0}{3 - x_1} \coth k_0(\tau_0 + 2\tau_0^{\rm e}) + \frac{4\tau_0^{\rm e} k_0^2}{3 - x_1} + O(k_0^3). \tag{4.70}$$

Equation (4.70) is a noteworthy one because it simultaneously expresses the important characteristic of an optically thick layer in terms of elementary functions and remains the asymptotically exact one. We should stress also that the asymptotic constant γ_0 in (4.56) depends weakly on the phase function. Hence, it can be assumed to be equal to 0.71 without any substantial inaccuracy in calculations.

Finally, from (1.107) and (4.66a) and taking (3.91a) into account, we have for the geometric albedo

$$A_{\rm g}^{\rm as}(\tau_0) = A_{\rm g0} - \frac{4k_0\zeta_0}{3 - x_1} \coth k_0(\tau_0 + 2\tau_0^{\rm e}) + O(k_0^2).$$

The parameter ζ_0 is determined by (4.54e). Numerical values of $A_{\rm g0}$ and ζ_0 are presented in the second columns of Tables XX and XXVI, respectively.

4.8.4 Estimation of the Accuracy of Asymptotic Formulas

Using the above asymptotic formulas, we encounter the problem of finding the minimum value of τ_0 after which the use of asymptotic formulas does not result in substantial errors. The formulas under consideration involve a lot of variables and parameters, and, therefore, it seems that the problem is difficult to solve. A deeper analysis reveals, however, that this is not the case.

Let us consider first the simplest case, namely, the Milne problem. Introduce the quantity

$$\delta(\tau, \mu, \mu_0) = 1 - \frac{I(\tau, \mu, \mu_0)}{I^{\mathrm{as}}(\tau, \mu, \mu_0)}, \tag{4.70a}$$

where the function $I^{\mathrm{as}}(\tau, \mu, \mu_0)$ is determined by the Sobolev formula (3.57). Equations (3.6) and (3.7) then give

$$\begin{aligned} MI(\tau, \mu) &= i(-\mu)\mathrm{e}^{k\tau} - i(\mu)\mathrm{e}^{-k\tau} + [i(\mu) - i(-\mu)]\mathrm{e}^{-\tau/\mu}\theta(\mu) \\ &\quad -2\int_0^1 I^{\mathrm{as}}(\tau, \mu, \mu')[i(-\mu') - i(\mu')]\,\mathrm{d}\mu' \\ &\quad +2\int_0^1 I^{\mathrm{as}}(\tau, \mu, \mu')\delta(\tau, \mu, \mu')[i(-\mu') - i(\mu')]\,\mathrm{d}\mu'. \end{aligned} \tag{4.70b}$$

Obviously, the function $\delta(\tau, \mu, \mu_0)$ allows us to determine the error in the asymptotic formula (3.57). Denote

$$\delta(\tau, \lambda) = \max_{\mu,\mu_0} |\delta(\tau, \mu, \mu_0)|,$$

i.e., $\delta(\tau, \lambda)$ is the value of the quantity $|\delta(\tau, \mu, \mu_0)|$ for those μ and μ_0 where it reaches maximum. Then, with the use of (4.70b), (3.57) and (3.58) we find the following inequality:

$$M|I(\tau, \mu) - I^{\mathrm{as}}(\tau, \mu)| \le (1 - N)\mathrm{e}^{-k\tau}\delta(\tau, \lambda)i(\mu) + |i(\mu) - i(-\mu)|\mathrm{e}^{-\tau/\mu}\theta(\mu). \tag{4.70c}$$

For the conservative limit, (4.70c) gives

$$|I_0(\tau, \mu) - I_0^{\mathrm{as}}(\tau, \mu)| \le \frac{3}{4}\left[\gamma_0\delta_0(\tau, 1) + \mu\mathrm{e}^{-\tau/\mu}\theta(\mu)\right], \tag{4.70d}$$

where $I_0^{\mathrm{as}}(\tau, \mu)$ is given by (3.61).

Thus, if function $\delta(\tau, \lambda)$ is known (for any particular phase function it depends only on two parameters, namely, τ and λ), then one can estimate the error of asymptotic formulas (3.58) and (3.61) with the help of (4.70c) and (4.70d). This appears to be possible owing to the fact that (4.70c) and (4.70d) contain the same functions and parameters as the original asymptotic formulas (3.58) and (3.61).

Let us consider now a noticeably more difficult problem, namely, the problem of estimating errors in the asymptotic formula (4.45). For this, we return to the original formula (4.43) that has been used in deriving all asymptotic formulas in the present section.

Using (4.70a), we rewrite (4.43) as follows

$$\begin{aligned} \Delta I(\tau, \mu, \mu_0; \tau_0) &= I(\tau, \mu, \mu_0; \tau_0) - I^{\mathrm{as}}(\tau, \mu, \mu_0; \tau_0) \\ &= I^{\mathrm{as}}(\tau_0, \mu, \mu_0)\delta(\tau, \mu, \mu_0)\mathrm{e}^{\frac{\tau_0 - \tau}{\mu}}\theta(-\mu) \\ &\quad -2\int_0^1 \Delta I(\tau_0 - \tau, -\mu, \mu'; \tau_0)I^{\mathrm{as}}(\tau_0, -\mu', \mu_0)\,\mathrm{d}\mu' \end{aligned}$$

$$+2\int_0^1 \Delta I(\tau_0-\tau,-\mu,\mu';\tau_0) I^{\mathrm{as}}(\tau_0,-\mu',\mu_0)\delta(\tau_0,-\mu',\mu_0)\,\mathrm{d}\mu'$$
$$+2\int_0^1 I^{\mathrm{as}}(\tau_0-\tau,-\mu,\mu';\tau_0) I^{\mathrm{as}}(\tau_0,-\mu',\mu_0)\delta(\tau_0,-\mu',\mu_0)\,\mathrm{d}\mu'.$$

From now on we consider the values of τ_0 for which $\delta(\tau_0,\mu,\mu_0)$ is sufficiently small compared with one. In other words, discarding the third term on the right-hand side of the above formula, we use (3.57) to rewrite it in the following way

$$\Delta I(\tau,\mu,\mu_0;\tau_0) = \Bigg[i(\mu)\mathrm{e}^{\frac{\tau_0-\tau}{\mu}}\theta(-\mu)\delta(\tau,\mu,\mu_0)$$
$$+2\int_0^1 I^{\mathrm{as}}(\tau_0-\tau,-\mu,\mu';\tau_0)\delta(\tau_0,-\mu',\mu_0) i(-\mu')\,\mathrm{d}\mu'$$
$$-2\int_0^1 \Delta I(\tau_0-\tau,-\mu,\mu';\tau_0) i(-\mu')\,\mathrm{d}\mu'\Bigg]\mathrm{e}^{-k\tau_0}u(\mu_0)\mu_0.$$

Find in this expression the last term, as we did when deriving (4.45). As a result, we have

$$\Delta I(\tau,\mu,\mu_0;\tau_0) = \Bigg[i(\mu)\mathrm{e}^{\frac{\tau_0-\tau}{\mu}}\theta(-\mu)\delta(\tau,\mu,\mu_0)$$
$$+2\int_0^1 I^{\mathrm{as}}(\tau_0-\tau,-\mu,\mu';\tau_0)\delta(\tau_0,-\mu',\mu_0) i(-\mu')\,\mathrm{d}\mu'$$
$$+\frac{N\mathrm{e}^{-k\tau_0}F(\tau_0-\tau,-\mu;\tau_0) - F(\tau,\mu;\tau_0)}{1-N^2\mathrm{e}^{-2k\tau_0}}\Bigg]\mathrm{e}^{-k\tau_0}u(\mu_0)\mu_0, \tag{4.70e}$$

where

$$F(\tau,\mu;\tau_0) = 2\mathrm{e}^{-k\tau_0}\Bigg[i(-\mu)\mathrm{e}^{-\tau/\mu}\theta(\mu)\int_0^1 \delta(\tau_0,-\mu,\mu') i(-\mu')u(\mu')\mu'\,\mathrm{d}\mu'$$
$$+2\int_0^1 I^{\mathrm{as}}(\tau,\mu,\mu';\tau_0) i(-\mu')\,\mathrm{d}\mu' \int_0^1 \delta(\tau_0,-\mu',\mu'') i(-\mu'')u(\mu'')\mu''\,\mathrm{d}\mu''\Bigg].$$

Using (4.70a), (3.57) and taking (3.6) into consideration, we find

$$\int_0^1 \delta(\tau_0,\mu,\mu') i(-\mu')u(\mu')\mu'\,\mathrm{d}\mu'$$
$$= N - \frac{\mathrm{e}^{k\tau_0}}{i(\mu)}\left[i(-\mu)\mathrm{e}^{k\tau_0} - MI(\tau_0,\mu) - i(-\mu)\mathrm{e}^{-\tau_0/\mu}\theta(\mu)\right].$$

Equations (4.45) and (3.58) then result in

$$2\int_0^1 I^{\mathrm{as}}(\tau,\mu,\mu';\tau_0) I^{\mathrm{as}}(\tau_0,-\mu')\,\mathrm{d}\mu' = I(\tau_0-\tau,-\mu) - I^{\mathrm{as}}(\tau_0,-\mu)\mathrm{e}^{-\tau/\mu}\theta(\mu).$$

Rather tiresome manipulations of the last two formulas lead us to the expression

$$
\begin{aligned}
&N\mathrm{e}^{-k\tau_0}F(\tau_0-\tau,-\mu,\tau_0)-F(\tau,\mu,\tau_0)\\
&=2MN\mathrm{e}^{-k\tau_0}\int_0^1 I^{\mathrm{as}}(\tau_0-\tau,-\mu,\mu';\tau_0)\,[I(\tau_0,-\mu')-I^{\mathrm{as}}(\tau_0,-\mu')]\,\mathrm{d}\mu'\\
&-2M\int_0^1 I^{\mathrm{as}}(\tau,\mu,\mu';\tau_0)\,[I(\tau_0,-\mu')-I^{\mathrm{as}}(\tau_0,-\mu')]\,\mathrm{d}\mu'\\
&+N\mathrm{e}^{(\tau_0-\tau)/\mu-k\tau_0}\theta(-\mu)M[I(\tau_0,\mu)-I^{\mathrm{as}}(\tau_0,\mu)]\\
&-\mathrm{e}^{-\tau/\mu}\theta(\mu)M[I(\tau_0,-\mu)-I^{\mathrm{as}}(\tau_0,-\mu)].
\end{aligned}
$$

Substituting the above formula into (4.70e), using the inequalities (4.70c) and neglecting the terms proportional to $\mathrm{e}^{-\tau_0/\mu}$, we ultimately find

$$
\begin{aligned}
|\Delta I(\tau,\mu,\mu_0;\tau_0)|\le &\Big\{M^{-1}\left[i(\mu)\mathrm{e}^{k(\tau_0-\tau)}-(2N-1)\mathrm{e}^{-k(\tau_0-\tau)}i(-\mu)\right]\\
&+(2N-1)\mathrm{e}^{-k\tau_0}I(\tau,\mu)-I(\tau_0-\tau,-\mu)\Big\}\,t(\tau_0)\delta(\tau_0,\lambda)u(\mu_0)\mu_0.
\end{aligned}
$$

If the scattering is conservative, the above formula takes the following simple form

$$
\begin{aligned}
|\Delta I_0(\tau,\mu,\mu_0;\tau_0)|\le &\Big\{I_0(\tau,\mu)-I_0(\tau_0-\tau,-\mu)\\
&+\frac{3}{4}\left[(1-x_1/3)(\tau_0-\tau)+\mu+2\gamma_0\right]\Big\}t_0(\tau_0)u_0(\mu_0)\mu_0\delta(\tau_0,1).
\end{aligned}
$$

Therefore, we have the following estimates for the reflection and transmission coefficients

$$
\begin{aligned}
&|\Delta\rho|=|\rho(\mu,\mu_0;\tau_0)-\rho^{\mathrm{as}}(\mu,\mu_0;\tau_0)|\\
&\le\left[\frac{u(\mu)(2N-1)M-i(\mu)(1-N)}{1-N^2\mathrm{e}^{-2k\tau_0}}\right]\delta(\tau_0,\lambda)\mathrm{e}^{-2k\tau_0}u(\mu_0),
\end{aligned}
$$

$$
\begin{aligned}
&|\Delta\sigma|=|\sigma(\mu,\mu_0;\tau_0)-\sigma^{\mathrm{as}}(\mu,\mu_0;\tau_0)|\\
&\le\left[2i(\mu)-\frac{(1-N\mathrm{e}^{-2k\tau_0})i(\mu)+Mu(\mu)}{1-N^2\mathrm{e}^{-2k\tau_0}}\right]\delta(\tau_0,\lambda)\mathrm{e}^{-k\tau_0}u(\mu_0),
\end{aligned}
$$

and for conservative scattering,

$$
|\Delta\rho_0|\le\frac{4u_0(\mu)+3\gamma_0}{(3-x_1)\tau_0+6\gamma_0}u_0(\mu_0)\delta_0(\tau_0,1),
$$

$$
|\Delta\sigma_0|\le\left[1-\frac{4u_0(\mu)-3\gamma_0}{(3-x_1)\tau_0+6\gamma_0}\right]u_0(\mu_0)\delta_0(\tau_0,1).
$$

Hence, for sufficiently large τ_0, the accuracy of the asymptotic formula for the reflection coefficient is higher then that for the transmission one.

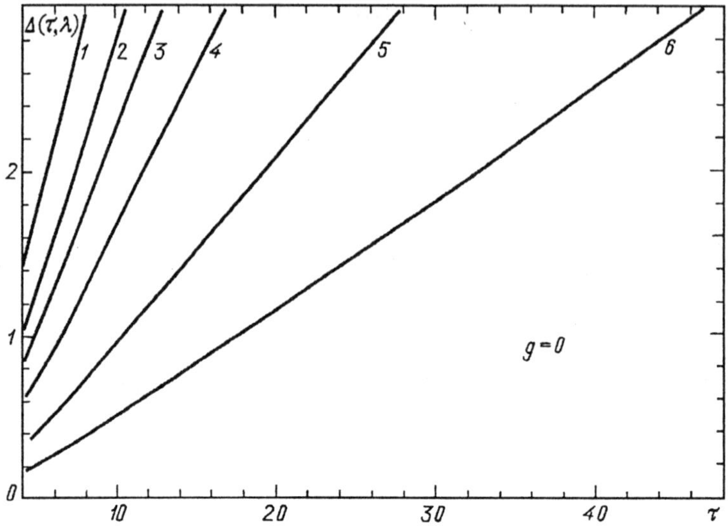

Fig. 4.2. Function $\Delta(\tau, \lambda) = -\lg |\delta(\tau, \lambda)|$ (see text) defining the error of depth asymptotics of radiation intensity in a semi-infinite atmosphere with isotropic scattering for different values of λ. Lines: 1 — $\lambda = 1$; 2 — $\lambda = 0.98$; 3 — $\lambda = 0.95$; 4 — $\lambda = 0.9$; 5 — $\lambda = 0.8$; 6 — $\lambda = 0.7$

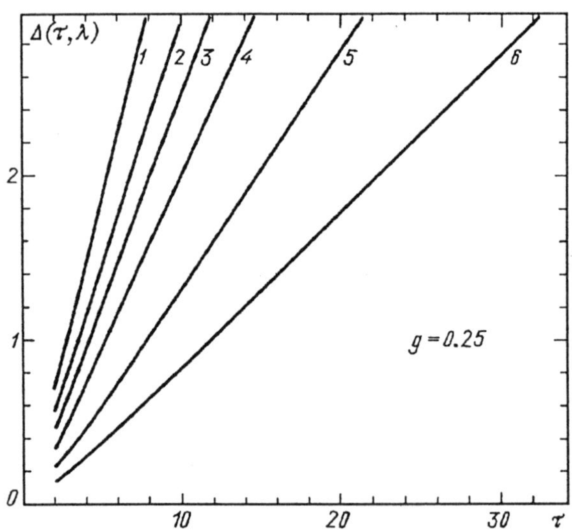

Fig. 4.3. As Fig. 4.2 for Henyey-Greenstein phase function with $g = 0.25$

Thus, we have shown that if function $\delta(\tau, \lambda)$ is known, then the functions and parameters of original asymptotics are used to evaluate the accuracy of the respective asymptotics in the Milne problem and in the parallel external flux problem as well.

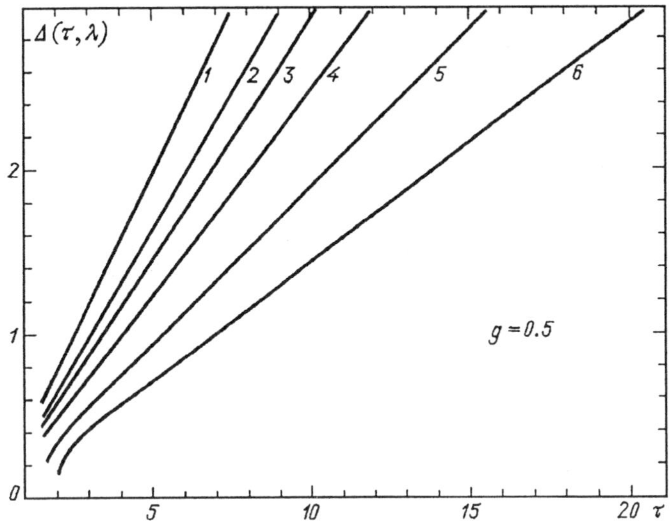

Fig. 4.4. As Fig. 4.2 for Henyey-Greenstein phase function with $g = 0.5$

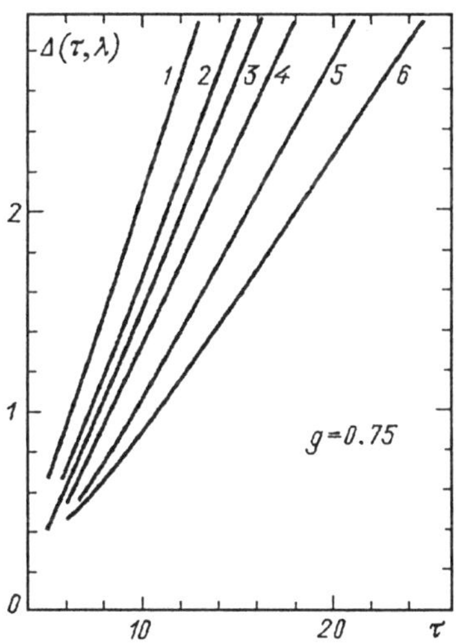

Fig. 4.5. As Fig. 4.2 for Henyey-Greenstein phase function with $g = 0.75$

Figures 4.2–4.6 show the profiles of $\Delta(\tau, \lambda) = -\lg \delta(\tau, \lambda)$ calculated for the Henyey-Greenstein phase function with different values of λ and g. Function $\delta(\tau, \lambda)$ is calculated via the Ivanov doubling method (see Sect.3.10). Using these figures and tables of functions $u(\mu)$ and $i(\mu)$, as well as constants k, M, N, and γ_0 presented in the Appendix, it is easy to estimate the accuracy of the respective asymptotics. With regard to the dependence of profiles, shown in Figs. 4.2–4.6, on g and λ, it is worth noticing that it is closely re-

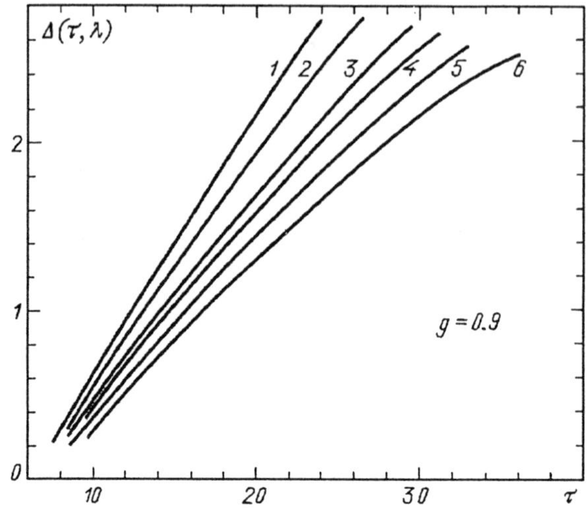

Fig. 4.6. As Fig. 4.2 for Henyey-Greenstein phase function with $g = 0.9$. Lines: 1 — $\lambda = 1$; 2 — $\lambda = 0.995$; 3 — $\lambda = 0.95$; 4 — $\lambda = 0.9$; 5 — $\lambda = 0.8$; 6 – $\lambda = 0.7$

lated to the values of several first roots of the characteristic equation (2.21) (for further details, see [53]–[55]).

4.9 Illumination of the Boundary and Albedo of Atmospheres of Arbitrary Optical Thickness for Nearly Conservative Scattering

At the beginning of Sects. 2.6 and 3.11, we pointed out the significance of the study of light scattering in an atmosphere with $(1-\lambda) \ll 1$. We have derived there some rather simple relations for a semi-infinite atmosphere (Sect. 3.11) and for a layer of large optical thickness (Sect. 4.8.3). However, by restricting ourselves to the study of the plane albedo of an atmosphere and the illumination of its lower boundary (i.e., to the study of functions $A(\mu_0, \tau_0)$ and $V(\mu_0, \tau_0)$), it seems easy to derive some comparatively simple formulas, which are valid for a layer of *arbitrary optical thickness*. These formulas are important for the physics of atmospheres so we proceed to derive them.

Pick the integral relations (4.22) and (4.23) as the starting point. Substituting function $i(\mu)$ with its asymptotic representation (2.43), we arrive at the formulas which, after replacing ρ and σ in the terms proportional to k_0 by their conservative values ρ_0 and σ_0, take the form

$$A(\mu, \tau_0) + e^{-k_0\tau_0} V(\mu, \tau_0) = 1 - \frac{4k_0}{3-x_1} u_0(\mu, \tau_0) + \frac{4k_0 e^{-k_0\tau_0}}{3-x_1} w_0(\mu, \tau_0), \quad (4.71)$$

$$A(\mu,\tau_0)\mathrm{e}^{-k_0\tau_0}+V(\mu,\tau_0)=\mathrm{e}^{-k_0\tau_0}+\frac{4k_0\mathrm{e}^{-k_0\tau_0}}{3-x_1}u_0(\mu,\tau_0)-\frac{4k_0}{3-x_1}w_0(\mu,\tau_0), \tag{4.72}$$

where

$$u_0(\mu,\tau_0)=\frac{3}{4}\left[\mu+2\int_0^1\rho_0(\mu,\mu';\tau_0)\mu'^2\,\mathrm{d}\mu'\right], \tag{4.73}$$

$$w_0(\mu,\tau_0)=\frac{3}{4}\left[\mu\mathrm{e}^{-\tau_0/\mu}+2\int_0^1\sigma_0(\mu,\mu';\tau_0)\mu'^2\,\mathrm{d}\mu'\right]. \tag{4.74}$$

Equations (1.104) and (1.105) are used to derive (4.71) and (4.72). Within the chosen approximation, we find

$$A(\mu,\tau_0)=1-\frac{4k_0\sinh^{-1}k_0\tau_0}{3-x_1}\left[u_0(\mu,\tau_0)\cosh k_0\tau_0-w_0(\mu,\tau_0)\right], \tag{4.75}$$

$$V(\mu,\tau_0)=\frac{4k_0\sinh^{-1}k_0\tau_0}{3-x_1}\left[u_0(\mu,\tau_0)-w_0(\mu,\tau_0)\cosh k_0\tau_0\right]. \tag{4.76}$$

Thus, if the scattering is nearly conservative, the illumination of the boundaries in a homogeneous atmosphere of arbitrary thickness is determined by two functions, namely, $u_0(\mu,\tau_0)$ and $w_0(\mu,\tau_0)$. For each phase function, these functions depend only on the parameters μ and τ_0.

Unfortunately, one is unable to derive an asymptotic formula that will be valid for arbitrary τ_0 for the function $U(\mu_0,\tau_0)$ determined by (4.54c). However, one may find the relations that become valid for either $\tau_0\lesssim 1$ or $\tau_0\gg 1$. For $\tau_0\lesssim 1$, using (2.40) after discarding the term proportional to k_0^3, with (4.54c), (4.22) and (4.23) we find

$$U(\mu,\tau_0)=(1-\lambda)\left\{2\tau_0[u_0(\mu,\tau_0)+w_0(\mu,\tau_0)]-\frac{15}{5-x_2}v_0(\mu,\tau_0)\right\}, \tag{4.77}$$

where

$$v_0(\mu,\tau_0)=\mu^2\left(1-\mathrm{e}^{-\tau_0/\mu}\right)-2\int_0^1[\rho_0(\mu,\mu';\tau_0)+\sigma_0(\mu,\mu';\tau_0)]\mu'^3\,\mathrm{d}\mu'. \tag{4.78}$$

In the case of $\tau_0\gg 1$, (4.54d), (3.79), (3.82), (3.83) and (3.97) lead to

$$U(\mu,\tau_0)=\frac{4k_0u_0(\mu)}{3-x_1}\tanh\frac{k_0(\tau_0+2\tau_0^{\mathrm{e}})}{2}$$
$$-\left[\frac{15}{5-x_2}v_0(\mu)+4\tau_0^{\mathrm{e}}u_0(\mu)\right](1-\lambda)+O(k_0^3). \tag{4.79}$$

Note that the introduced functions $u_0(\mu,\tau_0)$ and $v_0(\mu,\tau_0)$ transform into the known ones $u_0(\mu)$ and $v_0(\mu)$ for $\tau_0=\infty$.

Together with (4.79), (4.77) determines $U(\mu,\tau_0)$ for almost the whole range of variation of τ_0. For example, at $\lambda=0.999$ and the Henyey-Greenstein phase function (1.15) at $g = 0.75$ and $\tau_0 = 5$, (4.79) gives $U = 0.93 \cdot 10^{-2}$ at $\mu = 1$ whereas (4.77) gives $U = 0.97 \cdot 10^{-2}$.

We now introduce the following notation

$$p_0(\tau_0) = 2\int_0^1 u_0(\mu, \tau_0)\mu \, d\mu, \qquad q_0(\tau_0) = 2\int_0^1 w_0(\mu, \tau_0)\mu \, d\mu. \tag{4.80}$$

Then (4.73) and (4.74) give

$$p_0(\tau_0) + q_0(\tau_0) = 1. \tag{4.81}$$

For further considerations (Sect. 5.3) we need the spherical albedo $A_s(\tau_0)$, as well as the quantities

$$V_s(\tau_0) = 2\int_0^1 V(\mu, \tau_0)\mu \, d\mu, \qquad U_s(\tau_0) = 2\int_0^1 U(\mu, \tau_0)\mu \, d\mu. \tag{4.82}$$

Using (4.75), (4.76) and (4.81), we find

$$A_s(\tau_0) = 1 - \frac{4k_0 \sinh^{-1} k_0\tau_0}{3 - x_1}[p_0(\tau_0)(\cosh k_0\tau_0 + 1) - 1], \tag{4.83}$$

$$V_s(\tau_0) = \frac{4k_0}{3 - x_1}\left[p_0(\tau_0) \coth \frac{k_0\tau_0}{2} - \coth k_0\tau_0\right], \tag{4.84}$$

Taking into account (4.69), we finally have

$$U_s(\tau_0) = \frac{4k_0}{3 - x_1} \tanh \frac{k_0\tau_0}{2}. \tag{4.85}$$

The last formula is a noteworthy one because for *arbitrary* τ_0 *and phase function it explicitly* determines the radiative energy absorbed by a hemisphere of a planet whose albedo is equal to zero.

In the next chapter the formulas derived will be generalized to the case of an atmosphere adjacent to a reflecting surface. Tables XXIII–XXXI contain the values of functions $u_0(\mu, \tau_0)$, $\omega_0(\mu, \tau_0)$, and $v_0(\mu, \tau_0)$ as well as the quantity $p_0(\tau_0)$ for the Rayleigh and Henyey-Greenstein phase functions ($g = 0.25, 0.5, 0.75$) at various μ and τ_0.

Note that the use of (4.75) and (4.76) only provides the sufficiently accurate results for values of λ which are close to unity, whereas (4.77) gives adequate results for $\lambda = 0.95$ (for instance, at $\tau_0 = 1$ and $g = 0.75$ the inaccuracy does not exceed 5%). Nonetheless, the formulas derived for albedo and illumination give us the *strict* asymptotic dependence of the above quantities on λ, which may be useful in a number of cases; for instance in a study of the nonstationary glow of an atmosphere of arbitrary optical thickness (for details, see [14], Chap. 6 and review by Nagirner [56]).

In conclusion, we give the important formula for the projection of the flux in an atmosphere of arbitrary optical thickness at *arbitrary* depth $\tau \in [0, \tau_0]$,

$$\frac{H_\tau(\tau,\mu_0;\tau_0)}{\pi\mu_0} = \frac{4k_0 \sinh^{-1} k_0\tau_0}{3-x_1}[u_0(\mu_0,\tau_0)\cosh k_0(\tau_0-\tau) - w_0(\mu_0,\tau_0)\cosh k_0\tau]. \tag{4.86}$$

As in the case of deriving (4.75) and (4.76) we find (4.86) from (4.37) and (4.38). Equations (4.75) and (4.76) follow from (4.86) at $\tau = 0$ and $\tau = \tau_0$; at $\tau_0 = \infty$, we get (3.92). Using (4.86) and taking (4.81) into account, we find

$$\frac{1}{2\pi}\int_0^1 [H_\tau(\tau,\mu_0;\tau_0) - H_\tau(\tau_0-\tau,\mu_0;\tau_0)]\,\mathrm{d}\mu_0 = \frac{k_0 \sinh \frac{1}{2}k_0(\tau_0-2\tau)}{(3-x_1)\cosh \frac{1}{2}k_0\tau_0}. \tag{4.87}$$

This formula looks quite surprising, because it appears that an asymptotically exact integral expressed in terms of elementary functions exists in a layer of *arbitrary optical depth for any phase function and any optical depth.* Equation (4.85) follows from (4.87) at $\tau = 0$.

4.10 Algorithm for Solving the General Problem

To conclude this chapter we return to solving the general problem that has been treated in detail in Sect. 3.12 for the semi-infinite atmosphere. Therefore, it seems desirable to review the content of this section once again.

To express the problem we consider a plane homogeneous layer in the atmosphere, its optical thickness being τ_0. This layer contains sources with power depending only on the single spatial coordinate $\tau \in [0,\tau_0]$. The corresponding function $g(\tau,\boldsymbol{\omega})$ that determines this power is assumed to have the following polynomial representation:

$$g(\tau,\boldsymbol{\omega}) = \sum_{n=0}^{N} \tau^n g_n(\boldsymbol{\omega}). \tag{4.88}$$

It is necessary to determine the radiation intensity $I_\mathrm{s}(\tau,\boldsymbol{\omega})$ for all $\tau \in [0,\tau_0]$. The corresponding parallel external flux problem is assumed to be solved, i.e., the function $I(\tau,\boldsymbol{\omega},\boldsymbol{\omega}_0)$ is known. The method of finding this function is discussed in detail in Sect. 4.6.

In the case under consideration, instead of the formula (3.124), we have

$$I_\mathrm{s}(\tau,\boldsymbol{\omega}) = \frac{1}{2\pi}\int_0^{\tau} g(t,-\boldsymbol{\omega}')\,dt \int_{(4\pi)} G(\tau,\boldsymbol{\omega}';\tau,-\boldsymbol{\omega})\,d\omega', \tag{4.89}$$

where $G(\tau,\boldsymbol{\omega};\tau_1,\boldsymbol{\omega}_1)$ is the corresponding Green function. In Q representation it obeys (3.121) under the boundary conditions (1.119). This equation can be simplified by defining

$$\tilde{Q}(\boldsymbol{\omega}\boldsymbol{\omega}_1) = Q(\boldsymbol{\omega}\boldsymbol{\omega}_1) + \frac{4\pi}{\lambda}\delta(\boldsymbol{\omega}-\boldsymbol{\omega}_1). \tag{4.90}$$

As a result we obtain

$$G(\tau,\omega;t,\omega_1) = \frac{\lambda}{4\pi}\frac{\partial}{\partial\tau}\left[2\pi\,\tilde{Q}(\omega\omega_1)\theta(\tau-t) - \int\limits_{(4\pi)} \tilde{Q}(\omega'\omega)G(\tau,\omega';t,\omega_1)\mu'\,d\omega'\right]. \tag{4.91}$$

This equation will be used hereafter for non-conservative scattering. In the conservative case we need the corresponding analogue of (4.91), which will be derived below.

4.10.1 Q Form Equation for the Green Function: Conservative Scattering

Let us recall that, as follows from (3.103), the zeroth azimuthal harmonic ($m = 0$) of the Q function does not exist for the conservative scattering ($\lambda = 1$). The Q form equation for the Green function should therefore be modified correspondingly. That is why we must deal only with the zeroth azimuthal harmonics of the Green function.

Using (4.90), we rewrite (3.115) for the function $i(\mu)$,

$$i(\mu) = \frac{\lambda k}{2}\int_{-1}^{+1} \tilde{Q}(\mu,\mu')i(\mu')\mu'\,\mathrm{d}\mu'. \tag{4.92}$$

Then, instead of (4.91), for the zeroth azimuthal harmonics of the Green function, we have

$$G(\tau,\mu;t,\mu_1) = \frac{\lambda}{2}\tilde{Q}(\mu,\mu_1)\delta(\tau-t) - \frac{\lambda}{2}\frac{\partial}{\partial\tau}\int_{-1}^{+1} \tilde{Q}(\mu,\mu')G(\tau,\mu';t,\mu_1)\mu'\,\mathrm{d}\mu'. \tag{4.93}$$

From (4.92) and (4.93) it is easy to obtain

$$\frac{\partial}{\partial\tau}\int\limits_{-1}^{+1} G(\tau,\mu';t,\mu_1)i(\mu')\mu'\,\mathrm{d}\mu' = i(\mu_1)\delta(\tau-t) - k\int\limits_{-1}^{+1} G(\tau,\mu';t,\mu_1)i(\mu')\mu'\,\mathrm{d}\mu', \tag{4.94}$$

$$\frac{\partial}{\partial\tau}\int_{-1}^{+1} G(\tau,\mu';t,\mu_1)i(-\mu')\mu'\,\mathrm{d}\mu' = i(-\mu_1)\delta(\tau-t) + k\int_{-1}^{+1} G(\tau,\mu';t,\mu_1)i(-\mu')\mu'\,\mathrm{d}\mu'. \tag{4.95}$$

As usual, we denote the functions relating to the conservative scattering ($\lambda = 1$) by the subscript "0". Then, from expressions (4.94) and (4.95) and with the use of (2.40), we find

$$\frac{\partial}{\partial \tau}\int_{-1}^{+1} G(\tau,\mu;t,\mu_1)\mu \,\mathrm{d}\mu = \delta(\tau - t)$$

$$-3(1-\lambda)\int_{-1}^{+1} G_0(\tau,\mu;t,\mu_1)\mu^2 \,\mathrm{d}\mu + \frac{10(1-\lambda)}{5-x_2}\left[\delta(\tau-t)P_2(\mu_1)\right.$$

$$\left.-\frac{\partial}{\partial \tau}\int_{-1}^{+1} G_0(\tau,\mu;t,\mu_1)P_2(\mu)\mu \,\mathrm{d}\mu\right] + O\left[(1-\lambda)^2\right]. \qquad (4.96)$$

We also introduce

$$\tilde{Q}_0^*(\mu,\mu_0) = \sum_{n=2}^{\infty}\frac{(2n+1)x_n}{2n+1-x_n} - \frac{10P_2(\mu_0)}{5-x_2} + 2\delta(\mu-\mu_0), \qquad (4.97)$$

$$H_0(\tau;t,\mu_1) = \frac{1}{2}\int_{-1}^{+1} G_0(\tau,\mu;t,\mu_1)\mu \,\mathrm{d}\mu, \qquad (4.98)$$

$$K_0(\tau;t,\mu_1) = \frac{1}{2}\int_{-1}^{+1} G_0(\tau,\mu;t,\mu_1)\mu^2 \,\mathrm{d}\mu. \qquad (4.99)$$

Therefore we obtain the required Q form equation for the Green function in the conservative limit:

$$G_0(\tau,\mu;t,\mu_1) = \frac{1}{2}\left[\tilde{Q}_0^*(\mu,\mu_1) + \frac{3x_1\mu\mu_1}{3-x_1}\right]\delta(\tau-t)$$

$$-\frac{1}{2}\frac{\partial}{\partial \tau}\int_{-1}^{+1} G_0(\tau,\mu';t,\mu_1)\tilde{Q}_0^*(\mu,\mu')\mu' \,\mathrm{d}\mu'$$

$$-\frac{3x_1\mu}{3-x_1}\frac{\partial K_0(\tau;t,\mu_1)}{\partial \tau} + 3K_0(\tau;t,\mu_1). \qquad (4.100)$$

For the flux integral H_0, we have at once from (3.118),

$$\frac{\partial H_0(\tau;t,\mu_1)}{\partial \tau} = \frac{1}{2}\delta(\tau-t). \qquad (4.101)$$

For the K integral, we also have

$$\frac{\partial K_0(\tau;t,\mu_1)}{\partial \tau} = -\left(1-\frac{x_1}{3}\right)H_0(\tau;t,\mu_1) + \frac{1}{2}\mu_1\delta(\tau-t). \qquad (4.102)$$

Since it follows from (4.101) that

$$H_0(\tau;t,\mu_1) = H_0(0;t,\mu_1) + \frac{1}{2}\theta(\tau-t), \qquad (4.103)$$

we find from (4.102)

$$K_0(\tau;t,\mu_1) = K_0(0;t,\mu_1) - \left(1-\frac{x_1}{3}\right)H_0(0;t,\mu_1)\tau$$

$$-\frac{1}{2}\left[\left(1-\frac{x_1}{3}\right)(\tau-t) + \mu_1\right]\theta(\tau-t). \qquad (4.104)$$

Hence, (4.100) can finally be written in the form

$$G_0(\tau,\mu_1;t,-\mu)=\frac{1}{2}\tilde{Q}_0^*(\mu_1,-\mu)\delta(\tau-t)$$
$$-\frac{1}{2}\frac{\partial}{\partial\tau}\int_{-1}^{+1}G_0(t,\mu';\tau,-\mu)\tilde{Q}_0^*(\mu_1,\mu')\mu'\,d\mu'$$
$$+3K_0(0;\tau,-\mu)-[(3-x_1)t-x_1\mu_1]H_0(0;\tau,-\mu)$$
$$-\frac{1}{2}[(3-x_1)(t-\tau)-x_1\mu_1-3\mu]\theta(t-\tau). \tag{4.105}$$

Functions K_0 and H_0 are expressed directly in terms of the zeroth azimuthal harmonics of the radiation intensity $I(\tau,\mu,\mu_0)$ for the parallel external flux problem. Indeed, from (4.98), (4.99), and (1.121) and in view of (3.120), we find

$$H_0(0;\tau,-\mu)=-\int_0^1 I_0(\tau,\mu,\mu')\,d\mu'-\frac{1}{2}\theta(\mu)e^{-\tau/\mu}, \tag{4.106}$$

$$K_0(0;\tau,-\mu)=\int_0^1 I_0(\tau,\mu,\mu')\mu'\,d\mu'+\frac{1}{2}\mu\theta(\mu)e^{-\tau/\mu}. \tag{4.107}$$

These expressions should be substituted in (4.105).

Finally, we give two extra relations, which may be helpful when (4.105–107) are used. Namely, instead of (4.39) and (4.40), we have

$$1+2H_0(0;\tau,-\mu)+2H_0(0;\tau_0-\tau,\mu)=0, \tag{4.108}$$

$$\mu-\frac{3-x_1}{3}\tau=2K_0(0;\tau,-\mu)-2K_0(0;\tau_0-\tau,\mu)+\frac{2}{3}(3-x_1)\tau_0H_0(0;\tau_0-\tau,\mu). \tag{4.109}$$

4.10.2 Solution of the General Problem: Conservative Scattering

Let us introduce the designation

$$R_n^{c}(\mu_1;\tau,\mu)=\int_0^{\tau_0}t^nG_0(t,\mu_1;\tau,-\mu)\,dt, \tag{4.110}$$

where superscript c in R_n indicates conservative scattering.

Multiplying both parts of (4.105) by t^n and integrating with respect to t from 0 to τ_0, we obtain ($n\geq 1$)

$$R_n^{\rm c}(\mu_1;\tau,\mu)=\frac{1}{2}\Bigg[\tilde{Q}_0^*(\mu_1,-\mu)(\tau^n-\tau_0^n{\rm e}^{\frac{\tau_0-\tau}{\mu}}\theta(-\mu))$$
$$-2\tau_0^n\int\limits_0^1 I_0(\tau_0-\tau,-\mu,\mu')\tilde{Q}_0^*(\mu_1,\mu')\,{\rm d}\mu'\Bigg]+\frac{n}{2}\int\limits_{-1}^{+1}R_{n-1}^{\rm c}(\mu';\tau,\mu)\tilde{Q}_0^*(\mu_1,\mu')\mu'\,{\rm d}\mu'$$
$$+\frac{\tau_0^{n+1}}{n+1}(3K_0+x_1\mu_1H_0)-\frac{(3-x_1)\tau_0^{n+2}}{n+2}-\frac{1}{2}(3-x_1)\frac{\tau_0^{n+2}-\tau^{n+2}}{n+2}$$
$$+\frac{1}{2}[(3-x_1)\tau+x_1\mu_1-3\mu]\frac{\tau_0^{n+1}-\tau^{n+1}}{n+1}. \tag{4.111}$$

Taking $n=0$, we find from (4.105), (1.121) and (1.122)

$$R_0^{\rm c}(\mu_1;\tau,\mu)=\frac{1}{2}\tilde{Q}_0^*(\mu_1,-\mu)\left[1-{\rm e}^{-\tau/\mu}\theta(\mu)-{\rm e}^{\frac{\tau_0-\tau}{\mu}}\theta(-\mu)\right]$$
$$-\int_0^1 I_0(\tau_0-\tau,\mu,\mu')\tilde{Q}_0^*(\mu_1,\mu')\,{\rm d}\mu'-\int_0^1 I_0(\tau,\mu,\mu')\tilde{Q}_0^*(\mu_1,-\mu')\,{\rm d}\mu'$$
$$+\tau_0(3K_0+x_1\mu_1H_0)-\tfrac{1}{2}(3-x_1)\tau_0^2H_0$$
$$+\frac{1}{2}(\tau_0-\tau)\left[\frac{1}{2}(3-x_1)(\tau_0-\tau)-3\mu+x_1\mu_1\right]. \tag{4.112}$$

H_0 and K_0 in (4.111) and (4.112) are given by (4.106) and (4.107).

Therefore, the general problem solution for the zeroth azimuthal harmonic of radiation intensity reduces to successive applications of the recurrence relation (4.111) with the initial value determined by (4.112). The problem for higher azimuthal harmonics is solved by applying the formulas given below in Sect. 4.10.3.

For the intensity of radiation emerging from the layer, we introduce ($\mu\geq 0$)

$$\left.\begin{aligned}S_n^{\rm c}(\mu,\mu_1)&=\frac{1}{2}R_n^{\rm c}(\mu_1;0,-\mu),\\ T_n^{\rm c}(\mu,\mu_1)&=\frac{1}{2}R_n^{\rm c}(\mu_1;\tau_0,\mu).\end{aligned}\right\} \tag{4.113}$$

Moreover, instead of (4.106) and (4.107), we have

$$H_0(0;0,\mu)=-\int_0^1\rho_0(\mu,\mu')\mu'\,{\rm d}\mu', \tag{4.114}$$

$$H_0(0;\tau_0,-\mu)=-\int_0^1\sigma_0(\mu,\mu')\mu'\,{\rm d}\mu'-\frac{1}{2}{\rm e}^{-\tau_0/\mu}, \tag{4.115}$$

$$K_0(0;0,\mu)=\int_0^1\rho_0(\mu,\mu')\mu'^2\,{\rm d}\mu', \tag{4.116}$$

$$K_0(0;\tau_0,-\mu)=\int_0^1\sigma_0(\mu,\mu')\mu'^2\,{\rm d}\mu'+\frac{1}{2}\mu{\rm e}^{-\tau_0/\mu}, \tag{4.117}$$

for the layer boundaries. Introducing

$$A_0(\mu, \tau_0) = 2\int_0^1 \rho_0(\mu, \mu')\mu' \, d\mu', \tag{4.118}$$

(plane albedo of the atmosphere) and the quantities determined by (4.73) and (4.74), we obtain the recurrence relations for S_n^c and T_n^c (expressions (4.108) and (4.109) should be used, assuming $\tau = 0$):

$$\begin{aligned} S_n^c(\mu, \mu_1) = &-\frac{\tau_0^n}{4}\left[\tilde{Q}_0^*(\mu_1, \mu)e^{-\tau_0/\mu} + 2\int_0^1 \sigma_0(\mu, \mu')\tilde{Q}_0^*(\mu_1, \mu')\mu' \, d\mu'\right] \\ &+\frac{n}{2}\int_{-1}^{+1} S_{n-1}^c(\mu, \mu')\tilde{Q}_0^*(\mu_1, \mu')\mu' \, d\mu' + \frac{\tau_0^{n+1}}{n+1}u_0(\mu, \tau_0) \\ &-\frac{\tau_0^{n+1}}{4}\left[\frac{(3-x_1)\tau_0}{n+2} - \frac{x_1\mu_1}{n+1}\right][1 - A_0(\mu, \tau_0)], \qquad (n \geq 1) \end{aligned} \tag{4.119}$$

$$\begin{aligned} S_0^c(\mu, \mu_1) = &\frac{1}{4}\left[\left(1 - e^{-\tau_0/\mu}\right)\tilde{Q}_0^*(\mu_1, \mu) - 2\int_0^1 \sigma_0(\mu, \mu')\tilde{Q}_0^*(\mu_1, \mu')\mu' \, d\mu'\right. \\ &\left.-2\int_0^1 \rho_0(\mu, \mu')\tilde{Q}_0^*(\mu_1, -\mu')\mu' \, d\mu'\right] + \tau_0 u_0(\mu, \tau_0) \\ &-\frac{\tau_0}{8}[(3-x_1)\tau_0 - 2x_1\mu_1][1 - A_0(\mu, \tau_0)], \end{aligned} \tag{4.120}$$

$$\begin{aligned} T_n^c(\mu, \mu_1) = &\frac{\tau_0^n}{4}\left[\tilde{Q}_0^*(\mu_1, -\mu) - 2\int_0^1 \rho_0(\mu, \mu')\tilde{Q}_0^*(\mu_1, \mu')\mu' \, d\mu'\right] \\ &+\frac{n}{2}\int_0^1 T_{n-1}^c(\mu, \mu')\tilde{Q}_0^*(\mu_1, \mu')\mu' \, d\mu' + \frac{\tau_0^{n+1}}{n+1}u_0(\mu, \tau_0) \\ &-\frac{\tau_0^{n+1}}{4(n+1)}\left[\frac{(3-x_1)\tau_0}{n+2} + x_1\mu_1\right][1 - A_0(\mu, \tau_0)], \qquad (n \geq 1) \end{aligned} \tag{4.121}$$

$$T_0^c(\mu, -\mu_1) \equiv S_0^c(\mu, \mu_1). \tag{4.122}$$

Physically speaking, the quantities T_0^c and S_0^c in (4.122) determine the intensity of the radiation leaving the plane layer with the optical thickness τ_0, internal sources of radiation being distributed uniformly in the layer. Radiation intensities at the layer boundaries should apparently be the same in this case.

We note that the plane albedo $A_0(\mu, \tau_0)$ can be represented as

$$A_0(\mu, \tau_0) = 1 - \frac{4}{(3-x_1)\tau_0}[u_0(\mu, \tau_0) - w_0(\mu, \tau_0)]. \tag{4.123}$$

Functions $u_0(\mu, \tau_0)$ and $w_0(\mu, \tau_0)$ are presented in Tables XXII–XXVI for different μ and τ_0 for the Rayleigh and Henyey-Greenstein phase functions.

4.10.3 Solution Algorithm for Nonconservative Scattering

Let us denote

$$R_n(\boldsymbol{\omega}_1;\tau,\boldsymbol{\omega}) = \int\limits_0^{\tau} t^n G(t,\boldsymbol{\omega}_1;\tau,-\boldsymbol{\omega})\,dt, \tag{4.124}$$

and substitute (4.88) into (4.89). As a result, we have, instead of (4.89),

$$I^s(\tau,\boldsymbol{\omega}) = \frac{1}{2\pi}\sum_{n=0}^{N}\int\limits_{(4\pi)} g_n(-\boldsymbol{\omega}')R_n(\boldsymbol{\omega}';\tau,\boldsymbol{\omega})\,d\omega'. \tag{4.125}$$

Therefore, to solve the problem, we must find $N+1$ functions $R_n(\boldsymbol{\omega}';\tau,\boldsymbol{\omega})$ $(n=0,1,\ldots,N)$.

We multiply both sides of (4.91) by t^n and integrate with respect to t from 0 to τ_0. Using (1.122), we obtain as a result $(n \geq 1)$:

$$R_n(\boldsymbol{\omega}_1;\tau,\boldsymbol{\omega}) = \frac{\lambda}{2}\Bigg\{\tilde{Q}(-\boldsymbol{\omega}\boldsymbol{\omega}_1)\left[\tau^n - \tau_0^n \mathrm{e}^{\frac{\tau_0-\tau}{\mu}}\theta(-\mu)\right]$$
$$-\frac{\tau_0^n}{\pi}\int\limits_{(2\pi)} \tilde{Q}(\boldsymbol{\omega}'\boldsymbol{\omega}_1)I(\tau_0-\tau,-\boldsymbol{\omega},\boldsymbol{\omega}')\,d\omega'$$
$$+\frac{n}{2\pi}\int\limits_{(4\pi)} \tilde{Q}(\boldsymbol{\omega}'\boldsymbol{\omega}_1)R_{n-1}(\boldsymbol{\omega}';\tau,\boldsymbol{\omega})\mu'\,d\omega'\Bigg\}. \tag{4.126}$$

For $n=0$, with (1.121) and (1.122) taken into account, we find from (4.91):

$$R_0(\boldsymbol{\omega}_1;\tau,\boldsymbol{\omega}) = \frac{\lambda}{2}\Bigg\{\tilde{Q}(-\boldsymbol{\omega}\boldsymbol{\omega}_1)\left[1 - \mathrm{e}^{-\tau/\mu}\theta(\mu) - \mathrm{e}^{\frac{\tau_0-\tau}{\mu}}\theta(-\mu)\right]$$
$$-\frac{1}{\pi}\int\limits_{(2\pi)} \tilde{Q}(\boldsymbol{\omega}'\boldsymbol{\omega}_1)I(\tau_0-\tau,-\boldsymbol{\omega},\boldsymbol{\omega}')\,d\omega'$$
$$+\frac{1}{\pi}\int\limits_{(2\pi)} \tilde{Q}(-\boldsymbol{\omega}'\boldsymbol{\omega}_1)I(\tau,\boldsymbol{\omega},\boldsymbol{\omega}')\,d\omega'\Bigg\}. \tag{4.127}$$

Thus, once the parallel external flux problem is solved, i.e., the function $I(\tau,\boldsymbol{\omega},\boldsymbol{\omega}_1)$ is known, we find R_0 with the use of (4.127). By applying the recurrence relation (4.126), we determine all functions R_n $(n=1,2,\ldots,N)$, and as follows from (4.125), the general problem is thus solved.

Let us find the radiation intensity at the layer boundaries. For this purpose, we introduce $(\mu \geq 0)$

$$\left.\begin{aligned} S_n(\boldsymbol{\omega},\boldsymbol{\omega}_1) &= \frac{1}{2}R_n(\boldsymbol{\omega}_1;0,-\boldsymbol{\omega}), \\ T_n(\boldsymbol{\omega},\boldsymbol{\omega}_1) &= \frac{1}{2}R_n(\boldsymbol{\omega}_1;\tau_0,\boldsymbol{\omega}). \end{aligned}\right\} \tag{4.128}$$

As a result, we obtain from (4.125) and (4.128) for the layer boundaries,

$$I^s(0,-\omega) = \frac{1}{\pi}\sum_{n=0}^{N}\int_{(4\pi)} g_n(-\omega')S_n(\omega,\omega')\,d\omega', \tag{4.129}$$

$$I^s(\tau_0,\omega) = \frac{1}{\pi}\sum_{n=0}^{N}\int_{(4\pi)} g_n(-\omega')T_n(\omega,\omega')\,d\omega'. \tag{4.130}$$

For the functions S_n and T_n we have from (4.126–128) ($n \geq 1$):

$$S_n(\omega,\omega_1) = -\frac{\lambda}{4}\Bigg\{\tau_0^n e^{-\tau_0/\mu}\tilde{Q}(\omega\omega_1) + \frac{\tau_0^n}{\pi}\int_{(2\pi)} \tilde{Q}(\omega'\omega_1)\sigma(\omega,\omega')\mu'\,d\omega'$$
$$-\frac{n}{\pi}\int_{(4\pi)} \tilde{Q}(\omega'\omega_1)S_{n-1}(\omega,\omega')\mu'\,d\omega'\Bigg\}, \tag{4.131}$$

$$S_0(\omega,\omega_1) = -\frac{\lambda}{4}\Bigg\{\tilde{Q}(\omega\omega_1)\left(e^{-\tau_0/\mu}-1\right) + \frac{1}{\pi}\int_{(2\pi)} \tilde{Q}(\omega'\omega_1)\sigma(\omega,\omega')\mu'\,d\omega'$$
$$+\frac{1}{\pi}\int_{(2\pi)} \tilde{Q}(-\omega'\omega_1)\rho(\omega,\omega')\mu'\,d\omega'\Bigg\}, \tag{4.132}$$

$$T_n(\omega,\omega_1) = \frac{\lambda}{4}\Bigg\{\tau_0^n\tilde{Q}(\omega\omega_1) - \frac{\tau_0^n}{\pi}\int_{(2\pi)} \tilde{Q}(\omega'\omega_1)\rho(\omega,\omega')\mu'\,d\omega'$$
$$+\frac{n}{\pi}\int_{(4\pi)} \tilde{Q}(\omega'\omega_1)T_{n-1}(\omega,\omega')\mu'\,d\omega'\Bigg\}, \tag{4.133}$$

$$T_0(\omega,\omega_1) = \frac{\lambda}{4}\Bigg\{\tilde{Q}(-\omega\omega_1)\left(1-e^{-\tau_0/\mu}\right) - \frac{1}{\pi}\int_{(2\pi)} \tilde{Q}(\omega'\omega_1)\rho(\omega,\omega')\mu'\,d\omega'$$
$$-\frac{1}{\pi}\int_{(2\pi)} \tilde{Q}(-\omega'\omega_1)\sigma(\omega,\omega')\mu'\,d\omega'\Bigg\} = S_0(\omega,-\omega_1). \tag{4.134}$$

Thus, in order to find the radiation intensity at the boundaries of the layer with internal sources whose power in the medium is given by (4.88), it is sufficient to solve the problem of diffuse reflection and transmission of light, or, in other words, to determine $\rho(\omega,\omega_0)$ and $\sigma(\omega,\omega_0)$. This can be done by the method outlined in Sect. 4.5.

5. Atmosphere Above a Reflecting Surface

The model of an atmosphere considered in the previous chapter is hardly realized in nature because an atmosphere is usually adjacent to a solid planetary surface, which reflects light. The percentage of the radiation reflected can be rather high – up to 80% for snow covered ground. Having been scattered in the atmosphere the reflected photons undergo further reflections. Hence, there occurs the process of multiple exchange of photons between the surface and the atmosphere.

Generally, it is not difficult to take into account the reflection of light from the underlying surface. This can be reduced to changing the boundary condition at the lower boundary of atmosphere $\tau = \tau_0$. For this, it is sufficient to replace the second boundary condition in (1.44).

In this chapter, we will write a general invariance relation, which will let us construct an algorithm for the solution of the problem with surface, for an arbitrary reflection law. We will also study the case of a Lambertian surface, i.e. the case when there is no dependency of the intensity of the reflected radiation on direction (the orthotropic reflection, also the isotropic reflection).

The approximation to a Lambertian surface is a very rough one. On the other hand, it simplifies the final formulas substantially, which is the reason why this approximation is often used in practice. However, for an atmosphere of large optical thickness ($\tau_0 \gg 1$), the reflection of radiation from the surface can be taken into consideration without making any special assumption about the optical properties of the surface. This can be done by a minor modification of the standard asymptotic theory of scattering in an optically thick atmosphere with free borders, which was expounded in Sect. 4.8. The present chapter deals with the above generalization.

5.1 Radiation Field in Atmospheres

Let the reflection coefficient of the surface $\rho_s(\mu, \mu_0, \varphi)$ be known. Assume that ρ_s can be represented by a finite number l of terms of a Fourier expansion in azimuth φ, i.e.

$$\rho_s(\mu,\mu_0,\varphi) = \rho_s(\mu,\mu_0) + 2\sum_{m=1}^{l} [\rho_s^m(\mu,\mu_0)\cos m\,\varphi + \sigma_s^m(\mu,\mu_0)\sin m\varphi]. \quad (5.1)$$

We use overbar notation for the intensity of radiation in the atmosphere adjacent to the reflecting surface. By the analogy of (1.47), we have (for $\varphi_0 = 0$)

$$\bar{I}(\tau,\mu,\mu_0,\varphi;\tau_0) = \bar{I}(\tau,\mu,\mu_0;\tau_0) + 2\sum_{m=1}^{\infty} \bar{I}^m(\tau,\mu,\mu_0;\tau_0)\cos m\varphi. \quad (5.2)$$

For simplicity we will assume in further considerations that $\sigma_s^m(\mu,\mu_0) = 0$, i.e. the law of reflection is symmetrical with respect to the vertical plane whose azimuth $\varphi_0 = 0$. Therefore the following expansion can be used instead of (5.1)

$$\rho_s(\mu,\mu_0,\varphi) = \rho_s(\mu,\mu_0) + 2\sum_{m=1}^{l} \rho_s^m(\mu,\mu_0)\cos m\,\varphi. \quad (5.3)$$

The atmosphere is illuminated from above by a bunch of parallel rays and from below by the radiation reflected by the surface. According to the notation used, the intensity of the reflected radiation is $\bar{I}(\tau_0,-\mu,\mu_0,\varphi;\tau_0)$ ($\mu \in [0,1]$). Therefore, using the generalized invariance principle, we find

$$\begin{aligned}\bar{I}^m(\tau,\mu,\mu_0;\tau_0) &= I^m(\tau,\mu,\mu_0;\tau_0)\\ &+2\int_0^1 I^m(\tau_0-\tau,-\mu,\mu';\tau_0)\bar{I}^m(\tau_0,-\mu',\mu_0;\tau_0)\,\mathrm{d}\mu'\\ &+\bar{I}^m(\tau_0,\mu,\mu_0;\tau_0)\mathrm{e}^{\frac{\tau_0-\tau}{\mu}}\theta(-\mu),\end{aligned} \quad (5.4)$$

for all $\tau \in [0,\tau_0]$ and $\mu \in [-1,+1]$, $m = 0,1,2,\ldots,l$. For $m > l$ $\bar{I}^m(\tau,\mu,\mu_0;\tau_0) = I^m(\tau,\mu,\mu_0;\tau_0)$. By simple physical reasoning for the lower boundary of atmosphere, one obtains ($\mu \in [0,1]$)

$$\bar{I}^m(\tau_0,-\mu,\mu_0;\tau_0) = \rho_s^m(\mu,\mu_0)\mathrm{e}^{-\tau_0/\mu}\mu_0 + 2\int_0^1 \bar{I}^m(\tau_0,\mu',\mu_0;\tau_0)\rho_s^m(\mu,\mu')\mu'\,\mathrm{d}\mu'. \quad (5.5)$$

Here the first term on the right-hand side takes into account direct solar radiation attenuated by the atmosphere and reflected by the surface, and the second one allows for the diffuse radiation of the atmosphere, which is also reflected by the surface.

By putting $\tau = \tau_0$ in (5.4), one finds ($\mu \in [0,1]$)

$$\begin{aligned}\bar{I}^m(\tau_0,\mu,\mu_0;\tau_0) &= \sigma^m(\mu,\mu_0;\tau_0)\mu_0\\ &+2\int_0^1 \rho^m(\mu,\mu';\tau_0)\bar{I}^m(\tau_0,-\mu',\mu_0;\tau_0)\mu'\,\mathrm{d}\mu'.\end{aligned} \quad (5.6)$$

The system of linear integral equations (5.5), (5.6) is used to find the intensity of the radiation at the lower boundary for all $\mu \in [-1, +1]$. If the intensity and the radiation field in an atmosphere without reflective surface are known (see Sect. 4.6), (5.4) allows us to calculate the radiation field at arbitrary optical depth. The system of integral equations (5.5), (5.6) can easily be solved by the method of successive approximations.

5.2 Reflection and Transmission Coefficients

Using the usual notations, one has ($\mu \in [0, 1]$)

$$\left.\begin{aligned} \bar{\rho}^m(\mu, \mu_0; \tau_0)\mu_0 &= \bar{I}^m(0, -\mu, \mu_0; \tau_0), \\ \bar{\sigma}^m(\mu, \mu_0; \tau_0)\mu_0 &= \bar{I}^m(\tau_0, \mu, \mu_0; \tau_0), \end{aligned}\right\} \tag{5.7}$$

It follows from (5.4) that

$$\begin{aligned} \bar{\rho}^m(\mu, \mu_0; \tau_0)\mu_0 &= \rho^m(\mu, \mu_0; \tau_0)\mu_0 + \bar{I}^m(\tau_0, -\mu, \mu_0; \tau_0)\mathrm{e}^{-\tau_0/\mu} \\ &\quad + 2\int_0^1 \sigma^m(\mu, \mu'; \tau_0)\bar{I}^m(\tau_0, -\mu', \mu_0; \tau_0)\mu' \,\mathrm{d}\mu'. \end{aligned} \tag{5.8}$$

$$\bar{\sigma}^m(\mu, \mu_0; \tau_0)\mu_0 = \sigma^m(\mu, \mu_0; \tau_0)\mu_0 + 2\int_0^1 \rho^m(\mu, \mu'; \tau_0)\bar{I}^m(\tau_0, -\mu', \mu_0; \tau_0)\mu' \,\mathrm{d}\mu'. \tag{5.9}$$

The latter is nothing but (5.6) rewritten in new notation.

As can easily be seen, if the function $\bar{I}^m(\tau_0, -\mu, \mu_0; \tau_0)$ is known for all $m = 0, 1, 2, \ldots, l$ and $\mu \in [0, 1]$, the problem of diffuse reflection and transmission of light by a plane homogeneous atmosphere in the presence of a reflecting surface is solved. It is assumed that the above problem in the absence of a surface is already solved, of course.

5.3 The Case of a Lambertian Surface

If we suppose, which is in reality very common, that the planetary surface reflects light by the Lambert law then the above formulas simplify substantially. Then $\rho_\mathrm{s}(\mu, \mu_0) = A = \mathrm{const}$, where A is the albedo of surface and, what is more, $\rho_\mathrm{s}^m \equiv 0$ for $m \geq 1$. As the presence of an isotropically reflecting surface influences only the azimuth-averaged intensity of radiation, for $m \geq 1$ we have $\bar{I}^m(\tau_0, \mu, \mu_0; \tau_0) = I^m(\tau_0, \mu, \mu_0; \tau_0)$.

So the following formula holds instead of (5.5)

$$\bar{I}(\tau_0, -\mu, \mu_0; \tau_0) = A\left[e^{-\tau_0/\mu_0}\mu_0 + 2\int_0^1 \bar{I}(\tau_0, \mu', \mu_0; \tau_0)\mu' \, d\mu'\right]. \tag{5.10}$$

Then (5.6) can be rewritten as follows

$$\bar{I}(\tau_0, \mu, \mu_0; \tau_0) = \sigma(\mu, \mu_0; \tau_0)\mu_0$$
$$+AA(\mu, \tau_0)\left[e^{-\tau_0/\mu_0}\mu_0 + 2\int_0^1 \bar{I}(\tau_0, \mu', \mu_0; \tau_0)\mu' \, d\mu'\right], \tag{5.11}$$

where $A(\mu, \tau_0)$ is the plane albedo of atmosphere (see (1.104)). One can easily derive from (5.11) that

$$2\int_0^1 \bar{I}(\tau_0, \mu', \mu_0; \tau_0)\mu' \, d\mu' = \frac{AA_s(\tau_0)e^{-\tau_0/\mu_0}\mu_0 + 2\int\limits_0^1 \sigma^m(\mu', \mu_0; \tau_0)\mu' \, d\mu'}{1 - AA_s(\tau_0)}, \tag{5.12}$$

where $A_s(\tau_0)$ is the spherical albedo of atmosphere at $A = 0$. The substitution of (5.12) into (5.10) with allowance for (1.105) yields

$$\bar{I}(\tau_0, -\mu, \mu_0; \tau_0) = \frac{AV(\mu_0, \tau_0)\mu_0}{1 - AA_s(\tau_0)}. \tag{5.13}$$

Equation (5.11) can be rewritten as follows

$$\bar{I}(\tau_0, \mu, \mu_0; \tau_0) = \sigma(\mu, \mu_0; \tau_0)\mu_0 + \frac{AA(\mu, \tau_0)V(\mu_0, \tau_0)\mu_0}{1 - AA_s(\tau_0)}. \tag{5.14}$$

Equations (5.13) and (5.14) determine the intensity of radiation on the surface for all $\mu \in [-1, +1]$. Substituting (5.13) into (5.4) ($m = 0$) one arrives at the final result

$$\bar{I}^m(\tau, \mu, \mu_0; \tau_0) = I^m(\tau, \mu, \mu_0; \tau_0)$$
$$+AV(\mu_0, \tau_0)\mu_0 \frac{e^{\frac{\tau_0 - \tau}{\mu}}\theta(-\mu) + 2\int\limits_0^1 I(\tau_0 - \tau, -\mu, \mu'; \tau_0) \, d\mu'}{1 - AA_s(\tau_0)}. \tag{5.15}$$

So, if we take into consideration the notation of (5.7), Equation (5.15) results in the following formulas

$$\bar{\rho}(\mu, \mu_0; \tau_0) = \rho(\mu, \mu_0; \tau_0) + \frac{AV(\mu, \tau_0)V(\mu_0, \tau_0)}{1 - AA_s(\tau_0)}, \tag{5.16}$$

$$\bar{\sigma}(\mu, \mu_0; \tau_0) = \sigma(\mu, \mu_0; \tau_0) + \frac{AA(\mu, \tau_0)V(\mu_0, \tau_0)}{1 - AA_s(\tau_0)}, \tag{5.17}$$

which are known as Sobolev-van de Hulst formulas. One sees that if there is an isotropically reflecting surface, the reflection coefficient $\bar{\rho}$ is symmetrical with respect to μ and μ_0, whereas the transmission coefficient $\bar{\sigma}$ does not have this property.

5.4 Albedo of Atmospheres and Illumination of Surfaces

Using the notation introduced in Sects. 1.8, 4.8.1, and 4.9 and equations (5.16) and (5.17), we can easily find it that if a surface reflects radiation according to the Lambert law, then the plane albedo of the system consisting of an atmosphere and the surface is given by the following formula

$$\bar{A}(\mu_0, \tau_0) = A(\mu_0, \tau_0) + \frac{AV_s(\tau_0)V(\mu_0, \tau_0)}{1 - AA_s(\tau_0)}. \tag{5.18}$$

At the same time, the relative illumination of the surface is determined as follows

$$\bar{V}(\mu, \tau_0) = \frac{V(\mu_0, \tau_0)}{1 - AA_s(\tau_0)}. \tag{5.19}$$

Thus, it follows from (5.19) that the presence of the reflecting surface does not influence the relative changes in the illumination when the zenith angle of the sun (i.e. μ_0) varies.

The absorption function $U(\mu_0, \tau_0)$ is given by (4.54c). Physically it is clear that the corresponding function is

$$\bar{U}(\mu_0, \tau_0) = 1 - \bar{A}(\mu_0, \tau_0) - (1 - A)\bar{V}(\mu_0, \tau_0). \tag{5.20}$$

Substituting (5.18) and (5.19) into (5.20) and using (4.54c), we obtain

$$\bar{U}(\mu_0, \tau_0) = U(\mu_0, \tau_0) + \frac{AU_s(\tau_0)}{1 - AA_s(\tau_0)} V(\mu_0, \tau_0), \tag{5.21}$$

where $U_s(\tau_0)$ is determined by (4.82). Ultimately, it follows from (5.18) that the spherical albedo

$$\bar{A}_s(\tau_0) = A_s(\tau_0) + \frac{AV_s^2(\tau_0)}{1 - AA_s(\tau_0)}. \tag{5.22}$$

Thus, all the quantities characterizing the glowing of atmosphere for $A \neq 0$ are expressed in terms of respective quantities at $A = 0$.

5.5 Optically Thick Atmosphere Above a Reflective Surface

It follows from the results of Sect. 4.8 that if an atmosphere is of large optical thickness, then the calculation of the radiation field in the medium becomes considerably simpler. The zeroth azimuthal harmonic of the radiation intensity in an optically thick layer is expressed in terms of the corresponding harmonic of intensity in a semi-infinite medium and the intensity of the Milne problem (see (4.45)). Considerations reveal that similar simplifications take place in the presence of a reflective surface if ***the Milne problem with reflection*** has been solved beforehand. Now we are to formulate this problem.

5.5.1 The Milne Problem with Reflection

The problem is formulated just as for the ordinary Milne problem in Sect. 1.6. The only difference is that we place the reflective surface, whose law of reflection is known, at the boundary of the atmosphere. In other words, we consider the function $\rho_s(\mu, \mu_0)$ to be known. Then the following formula has to be used instead of the boundary condition (1.70), ($\mu \in [0, 1]$):

$$\bar{I}(0, \mu) = 2 \int_0^1 \rho_s(\mu, \mu')\bar{I}(0, -\mu')\mu' \, d\mu'. \tag{5.23}$$

The overbar above I and other quantities means that they relate to the problem with a reflecting surface. If $\rho_s = 0$, then all quantities will relate to either the usual parallel external flux problem or the Milne problem, whose solutions are assumed to be known.

If the radiation intensity $\bar{I}(0, \mu)$ at the boundary is known, then the problem is, in fact solved, because the solution is superposition of the Milne problem and the parallel external flux problem. Denoting $\bar{u}(\mu) = \bar{I}(0, -\mu)$, one can rewrite (5.23) as follows:

$$\bar{u}(-\mu) = 2 \int_0^1 \rho_s(\mu, \mu')\bar{u}(\mu')\mu' \, d\mu'. \tag{5.24}$$

Besides that, we have

$$\bar{u}(\mu) = u(\mu) + 2 \int_0^1 \rho(\mu, \mu')\bar{u}(-\mu')\mu' \, d\mu', \tag{5.25}$$

where $\rho(\mu, \mu')$ is the usual zeroth azimuthal harmonic of the reflection coefficient for a semi-infinite medium with $\rho_s = 0$. This relation means that one can separate the two parts of the radiation descending towards the reflecting boundary. These parts are the radiation that approaches the boundary for the first time (term $u(\mu)$) and the radiation which, moving from the boundary, gets diffusely reflected back by the medium (the integral term).

We normalize the escape function $u(\mu)$ according to the condition (3.12). Using (3.11) one can easily derive from (5.25) that

$$2 \int_{-1}^{+1} \bar{u}(\mu) i(\mu) \mu \, d\mu = 1. \tag{5.26}$$

Function $\bar{u}(\mu)$ is a certain standard angular distribution of the intensity, which plays the same role in the asymptotic theory of scattering in optically thick media with reflecting boundaries as the escape function $u(\mu)$ does at $\rho_s = 0$. The quantity $\bar{u}(\mu)$ can be found from the system of equations (5.24), (5.25) with the help of a number of methods, for instance, by iteration. Therefore, in future considerations we will assume this function to be known.

Let $I(\tau, \mu, \mu_0)$ be (as it has been) the zeroth azimuthal harmonic of radiation intensity in the parallel external flux problem for a semi-infinite medium. Also, let $I(\tau, \mu)$ be the intensity for the Milne problem without reflection. The only difference between $\bar{I}(\tau, \mu)$ and $I(\tau, \mu)$ is that caused by the additional illumination produced by the radiation reflected by the boundary. Hence, the generalized invariance principle enables us to write

$$\bar{I}(\tau, \mu) = I(\tau, \mu) + 2\int_0^1 I(\tau, \mu, \mu')\bar{u}(-\mu')\,\mathrm{d}\mu' + \bar{u}(-\mu)\mathrm{e}^{-\tau/\mu}\theta(\mu) \qquad (5.27)$$

for $\tau \geq 0$ and $\mu \in [-1, +1]$. This formula provides the solution of the problem, i.e., the solution of the Milne problem with reflection.

For deep layers of medium (when $\tau \gg 1$), using asymptotic formulas (3.57) and (3.58), we derive from (5.27)

$$M\bar{I}^{\mathrm{as}}(\tau, \mu) = i(-\mu)\mathrm{e}^{k\tau} - \bar{N}i(\mu)\mathrm{e}^{-k\tau}, \qquad (5.28)$$

where

$$\bar{N} = 2\int_{-1}^{+1} \bar{u}(\mu)i(-\mu)\mu\,\mathrm{d}\mu. \qquad (5.29)$$

To derive (5.28) we have also used (3.10) and (5.25). Using (3.10), (3.40) and (5.29) one can easily obtain from (5.25) one more formula for $\bar{N}$. Namely,

$$\bar{N} = N - 2M\int_0^1 \bar{u}(\mu)u(-\mu)\mu\,\mathrm{d}\mu. \qquad (5.30)$$

The physical meaning of (5.28) can be understood with the help of considerations similar to those given in Sect. 3.9 after (3.58). Here one should note the following. The second term on the right-hand side of (5.28) is caused by the outflow of photons through the boundary. Therefore in (3.58) this term is negative, whereas it may become positive in (5.28) (i.e. $\bar{N} < 0$); the latter may happen if the boundary is a good reflector.

For $\lambda \to 1$ one encounters a 0/0 type indeterminacy. To evaluate it, we use the expression (2.43) for $i(\mu)$ and (5.26). As a result, from (5.29), we find

$$\bar{N} = 1 - \frac{6}{3 - x_1}\bar{\gamma}_0 k_0 + O(k_0^2), \qquad (5.31)$$

where

$$\bar{\gamma}_0 = 2\int_{-1}^{+1} \bar{u}_0(\mu)\mu^2\,\mathrm{d}\mu, \qquad (5.32)$$

and $\bar{u}_0(\mu)$ is the conservative limit ($\lambda = 1$) of the function $\bar{u}(\mu)$. Using (5.31), (2.43) and (2.44) with $\lambda = 1$ we have, instead of (5.28),

$$\bar{I}_0^{\mathrm{as}}(\tau, \mu) = \frac{3}{4}\left[\left(1 - \frac{x_1}{3}\right)\tau + \bar{\gamma}_0 - \mu\right]. \qquad (5.33)$$

At $\rho_s = 0$, one easily sees that $\bar{\gamma}_0$ becomes γ_0, which is given by (3.62), and (5.33) becomes (3.61). Note that it follows from (5.30), (5.31), (2.44) and (3.82) that $\bar{\gamma}_0$ can also be determined as follows (cf. (5.32)):

$$\bar{\gamma}_0 = \gamma_0 + \frac{8}{3}\int_0^1 u_0(\mu)\bar{u}_0(-\mu)\mu\,\mathrm{d}\mu. \tag{5.33a}$$

The above results are valid for an arbitrary law for reflection of light from a surface, i.e., for an arbitrary function $\rho_s(\mu,\mu_0)$. Now let us proceed to the consideration of an important special case when the boundary reflects light isotropically, i.e., $\rho_s(\mu,\mu_0) = A = \text{const}$. Then function $\bar{u}(\mu)$ can be expressed in terms of $u(\mu)$, $A(\mu)$ and A_s explicitly. Using (5.24) and (5.25), one can easily derive

$$\bar{u}(\mu) = u(\mu) + \frac{AC}{1-AA_s}A(\mu); \qquad \bar{u}(-\mu) = \frac{AC}{1-AA_s}, \tag{5.34}$$

for $\mu \in [0,1]$; C is determined by (3.81). In particular, if the atmosphere is conservative, we have

$$\bar{u}_0(\mu) = u_0(\mu)\theta(\mu) + \frac{A}{1-A}. \tag{5.35}$$

Thus, the radiation intensity at the boundary becomes closer to isotropic when the surface albedo becomes greater.

Proceeding from (5.30) and (5.32) with the help of (5.34) and (5.35) we readily find

$$\bar{N} = N - \frac{AMC^2}{1-AA_s}, \tag{5.36}$$

$$\bar{\gamma}_0 = \gamma_0 + \frac{4}{3}\frac{A}{1-A}. \tag{5.37}$$

5.5.2 Radiation Field in Atmospheres

Firstly we are to consider the radiation intensity at the boundaries $\tau = 0$ and $\tau = \tau_0$. We have

$$\bar{I}(0,-\mu,\mu_0;\tau_0) = \rho(\mu,\mu_0)\mu_0 - u(\mu)u(\mu_0)\mu_0\bar{r}(\tau_0), \quad (\mu \in [0,1]), \tag{5.38}$$

$$\bar{I}(\tau_0,\mu,\mu_0;\tau_0) = \bar{u}(\mu)u(\mu_0)\mu_0\bar{t}(\tau_0), \quad (\mu \in [-1,+1]), \tag{5.39}$$

where transmission and reflection factors, $\bar{r}(\tau_0)$ and $\bar{t}(\tau_0)$, respectively, are involved in the following relation:

$$\bar{r}(\tau_0) = \bar{N}\mathrm{e}^{-k\tau_0}\bar{t}(\tau_0) = \frac{M\bar{N}\mathrm{e}^{-2k\tau_0}}{1-N\bar{N}\mathrm{e}^{-2k\tau_0}}. \tag{5.40}$$

On the other hand, for the conservative limit,

$$\bar{r}_0(\tau_0) = \bar{t}_0(\tau_0) = \frac{4}{3}\left[\left(1 - \frac{x_1}{3}\right)\tau_0 + \gamma_0 + \bar{\gamma}_0\right]^{-1}. \tag{5.41}$$

The overbars omitted, formulas (5.38) and (5.39) become the usual asymptotic relations for the radiation intensity at the boundaries of an optically thick layer ((4.50) and (4.51)), whereas (5.40) and (5.41) become (4.46) and (4.57), respectively.

Equation (5.39) can be proved by putting $m = 0$ and substituting (5.39) into (5.5) and (5.6) and taking (5.24), (5.25) and relations (4.50), (4.51) into account. Equation (5.38) can easily be verified with the help of (5.4).

Now we are able to write an asymptotic formula that generalizes (4.45) to the case when there is a reflecting surface. Namely, for all $\mu \in [-1, +1]$, $\tau \in [0, \tau_0]$, we have

$$\begin{aligned} \bar{I}(\tau_0, \mu, \mu_0; \tau_0) = I(\tau, \mu, \mu_0) - \{I(\tau, \mu)\bar{r}(\mu_0) \\ -[\bar{I}(\tau_0 - \tau, -\mu) - \bar{I}^{\mathrm{as}}(\tau_0 - \tau, -\mu)]\bar{t}(\tau_0)\}u(\mu_0)\mu_0. \end{aligned} \tag{5.42}$$

This formula can easily be verified by substituting it into (5.4) and taking (4.45), (5.27), (5.28) and (5.40) into consideration.

Equation (5.42) takes a simpler form near the boundaries. For example, near the upper boundary, where $(\tau_0 - \tau) \gg 1$, one finds

$$\bar{I}(\tau, \mu, \mu_0; \tau_0) = I(\tau, \mu, \mu_0) - I(\tau, \mu)\bar{r}(\tau_0)u(\mu_0)\mu_0, \tag{5.43}$$

and near the surface (where $\tau \gg 1$)

$$\bar{I}(\tau, \mu, \mu_0; \tau_0) = \bar{I}(\tau_0 - \tau, -\mu)\bar{t}(\tau_0)u(\mu_0)\mu_0. \tag{5.44}$$

If both equalities hold simultaneously, i.e., in the middle atmosphere, we have the following expression:

$$M\bar{I}(\tau, \mu, \mu_0; \tau_0) = \left[i(\mu)e^{k(\tau_0 - \tau)} - i(-\mu)e^{-k(\tau_0 - \tau)}\bar{N}\right]\bar{t}(\tau_0)u(\mu_0)\mu_0, \tag{5.45}$$

which takes a very simple form for conservative scattering

$$\bar{I}_0(\tau, \mu, \mu_0; \tau_0) = \frac{\tau_0 - \tau + 3(\mu + \bar{\gamma}_0)}{(3 - x_1)\tau_0 + 3(\gamma_0 + \bar{\gamma}_0)}u(\mu_0)\mu_0. \tag{5.46}$$

Thus, the radiation field becomes very close to an isotropic one at a point which is far from both boundaries. In other words, the radiation field hardly depends on μ (note that $(\tau_0 - \tau) \gg 1$). In particular, if the surface albedo $A \to 1$, then it follows from (5.37) that $\bar{\gamma}_0 \to \infty$ and we have $\bar{I}_0(\tau, \mu, \mu_0; \tau_0) = u_0(\mu_0)\mu_0$ instead of (5.46). This means that the radiation field is completely isotropic and does not differ from that found in deep layers of a semi-infinite and conservatively scattering atmosphere (see (3.60)).

5.5.3 Atmosphere with Nearly Conservative Scattering

We have already known that $\bar{N}$, which is determined by (5.29) or (5.30), may be either positive or negative. The sign depends on the reflective capacity of the surface. Let us use (5.36) to find critical values $A_{\rm cr}$ of the albedo of surface such that $\bar{N} \geq 0$ for $A \leq A_{\rm cr}$. Equating the right-hand side of (5.36) to zero, one finds

$$A_{\rm cr} = \frac{N}{MC^2 + NA_{\rm s}}. \tag{5.47}$$

Thus, one sees that $A_{\rm cr}$ is determined only by the properties of the atmosphere and can easily be calculated with the aid of tables provided in the Appendix. When the scattering is conservative (i.e., for pure scattering), $A_{\rm cr} = 1$ and $\bar{N}_0 = 1$. As the absorption increases, the $A_{\rm cr}$ decreases rather rapidly. For example, $A_{\rm cr} = 0.214$ at $\lambda = 0.9$ and the Henyey-Greenstein phase function with $g = 0.75$. The following asymptotic formula is valid for small values of $k_0 = \sqrt{(1-\lambda)(3-x_1)}$:

$$A_{\rm cr} = 1 - \frac{4k_0}{3-x_1} + \frac{4k_0^2}{3-x_1}(4 - 3\gamma_0) + O(k_0^3). \tag{5.48}$$

This expression can easily be derived from (5.47) with the aid of (2.44), (3.82), (3.83) and (3.99).

Consider a surface such that $\bar{N} \geq 0$ for a given atmosphere. By analogy with (4.52), one can write

$$\bar{N} = \mathrm{e}^{-2\bar{\tau}^{\rm e} k}, \tag{5.49}$$

where $\bar{\tau}^{\rm e}$ is the extrapolated length for the problem with reflecting surface. Then proceeding from (5.40) one derives the following formulas

$$\bar{r}(\tau_0) = \frac{M}{2N}[\coth k(\tau_0 + \tau^{\rm e} + \bar{\tau}^{\rm e}) - 1], \tag{5.50}$$

$$\bar{t}(\tau_0) = \frac{M}{2\sqrt{N\bar{N}}} \sinh^{-1} k(\tau_0 + \tau^{\rm e} + \bar{\tau}^{\rm e}). \tag{5.51}$$

which are analogous to (4.53) and (4.54), respectively. As a result, it turns out to be easy to write a more general analog of the relation (4.64), which determines the azimuth-averaged intensity in an optically thick atmosphere at arbitrary depth $\tau \in [0, \tau_0]$. Namely,

$$\begin{aligned}\bar{I}_1(\tau,\mu,\mu_0;\tau_0) = I_1(\tau,\mu,\mu_0) - \Bigg\{ \sinh k_0\tau + \frac{3k_0(\gamma_0-\mu)}{3-x_1}\cosh k_0\tau \\ + \frac{4k_0}{3-x_1}\left[I_0(\tau,\mu) - I_0^{\rm as}(\tau,\mu)\right]\Bigg\}(\coth k_0\tau^* - 1)u(\mu_0)\mu_0 \\ + \frac{4k_0}{3-x_1}\left[I_0(\tau_0-\tau,-\mu) - I_0^{\rm as}(\tau_0-\tau,-\mu)\right]u(\mu_0)\mu_0 \sinh^{-1}k_0\tau^* + O(k_0^2),\end{aligned} \tag{5.52}$$

where

$$\bar{\tau}_0^{\mathrm{e}} = \frac{3\bar{\gamma}_0}{3 - x_1}, \tag{5.53}$$

$$\tau^* = \tau_0 + \tau_0^{\mathrm{e}} + \bar{\tau}_0^{\mathrm{e}}. \tag{5.54}$$

Here $I_1(\tau, \mu, \mu_0)$ is the radiation intensity in a semi-infinite atmosphere, which is determined by (3.86) or (3.88), and the intensity in the Milne problem $I_0(\tau, \mu)$ is defined by (3.15). The corresponding asymptotic of $I_0^{\mathrm{as}}(\tau, \mu)$ is described by (3.61). If the surface reflects light isotropically, then we have the following relations instead of (5.53) and (5.54):

$$\bar{\tau}_0^{\mathrm{e}} = \frac{3}{3 - x_1}\left(\gamma_0 + \frac{4}{3}\frac{A}{1 - A}\right), \tag{5.55}$$

$$\tau^* = \tau_0 + \frac{6}{3 - x_1}\left(\gamma_0 + \frac{2}{3}\frac{A}{1 - A}\right). \tag{5.56}$$

To derive the above two formulas we took (4.56) and (5.37) into consideration.

By substitution of (5.52) into (1.87), we arrive at the following simple formula for the projection of the flux vector onto axis τ:

$$\frac{\bar{H}_\tau(\tau, \mu_0; \tau_0)}{4\pi} = \frac{k_0}{3 - x_1}\frac{\cosh k_0(\tau^* - \tau)}{\sinh k_0\tau^*}u_0(\mu_0)\mu_0 + O(k_0^2). \tag{5.57}$$

Setting $\tau_0 = 0$ in (5.57), one finds the plane albedo

$$\bar{A}^{\mathrm{as}}(\mu_0, \tau_0) = 1 - \frac{4k_0 u_0(\mu_0)}{3 - x_1}\coth k_0\tau^* + O(k_0^2). \tag{5.58}$$

In the conservative limit, (5.57) takes the form

$$\bar{H}_{0\tau}(\tau, \mu_0, \tau_0) = \frac{u_0(\mu_0)\mu_0}{(3 - x_1)\tau_0 + 6\gamma_0 + 4A/(1 - A)}. \tag{5.59}$$

Equations (5.52), (5.57) and (5.58) are found within the accuracy of the order of magnitude of term $k_0 = [(1 - \lambda)(3 - x_1)^{1/2}$. For a number of functions which are important in the optics of planetary atmospheres it is possible, however, to obtain the same relations which are accurate to k_0^2. Similar formulas were found in Sect. 4.8.3 for the case when there is no reflecting surface ($A = 0$). Now we are to do this for a more general problem.

First of all, one should note that (5.24) and (5.25) imply the result that if the scattering is nearly conservative then the following analogy of (3.79) holds for all $\mu \in [-1, +1]$:

$$\bar{u}(\mu) = \bar{u}_0(\mu)\left(1 - \frac{3k_0\bar{\gamma}_0}{3 - x_1}\right) + O(k_0^2), \tag{5.60}$$

where $\bar{u}_0(\mu)$ stands, just as it did earlier, for function $\bar{u}(\mu)$ for conservative scattering. Equation (5.33a) was used to derive (5.60). Now, taking into account (5.51), (3.79), (2.44) and (5.54), it is easy to obtain the following analogy of (4.66) from (5.39):

$$\bar{\sigma}^{\mathrm{as}}(\mu, \mu_0; \tau_0) = \frac{4k_0 \bar{u}_0(\mu) u_0(\mu_0)}{(3 - x_1)\sinh k_0\tau^*} + O(k_0^3). \tag{5.61}$$

If the surface reflects light isotropically, one finds the following expression for illumination:

$$\bar{V}^{\mathrm{as}}(\mu_0, \tau_0) = \frac{4k_0 u_0(\mu_0)}{(3 - x_1)(1 - A)\sinh k_0\tau^*} + O(k_0^3). \tag{5.62}$$

Using (3.97), (3.79), (3.82), (3.83), (2.44), (5.50) and (5.54) and proceeding from (5.38) we obtain a formula for the plane albedo of atmosphere:

$$\bar{A}^{\mathrm{as}}(\mu_0, \tau_0) = 1 - \frac{4k_0 u_0(\mu_0)}{3 - x_1}\coth k_0\tau^*$$
$$+3\left[\frac{5}{5 - x_2} v_0(\mu_0) + \frac{4\gamma_0}{3 - x_1} u_0(\mu_0)\right]\frac{k_0^2}{3 - x_1} + O(k_0^3). \tag{5.63}$$

After having substituted (5.62) and (5.63) into (5.20) one arrives at the expression for the absorption function

$$\bar{U}^{\mathrm{as}}(\mu_0, \tau_0) = \frac{4k_0 u_0(\mu_0)}{3 - x_1}\tanh\frac{k_0\tau^*}{2}$$
$$-3\left[\frac{5}{5 - x_2} v_0(\mu_0) + \frac{4\gamma_0}{3 - x_1} u_0(\mu_0)\right]\frac{k_0^2}{3 - x_1} + O(k_0^3). \tag{5.64}$$

For the spherical albedo, it follows from (5.63) that

$$\bar{A}_{\mathrm{s}}^{\mathrm{as}}(\tau_0) = 1 - \frac{4k_0}{3 - x_1}\coth k_0\tau^* + \frac{12\gamma_0 k_0^2}{(3 - x_1)^2} + O(k_0^3). \tag{5.65}$$

The superscript "as" in the above formulas, as usual, means that we are dealing with an optically thick atmosphere layer, i.e. $\tau_0 \gg 1$. If there is no reflecting surface, the formulas derived become the corresponding formulas which were found in Sect. 4.8.3.

And a final remark. Some of these formulas depend on the albedo of the surface only through the parameter τ^*, which is determined by (5.54) or (5.56). This means that if the albedo of the surface is such that $\bar{\tau}_0^{\mathrm{e}} \ll (\tau_0 + \tau_0^{\mathrm{e}})$, then there are cases when the influence of the surface can be neglected. For example, if we apply the above consideration to (5.64), we come to the conclusion that it is the atmosphere, not the planetary surface, that absorbs most of the photons.

It should be noted in conclusion that a considerable simplification in the general formulas of Sects. 5.1, 5.2, and 5.5.1 occurs if the azimuth-averaged reflection coefficient $\rho_{\mathrm{s}}(\mu, \mu_0) = A\rho(\mu)\rho(\mu_0)$, where $\rho(\mu)$ is a given function, or even in a more general form

$$\rho_{\mathrm{s}}(\mu, \mu_0, \varphi) = \sum_{m=0}^{l}\sum_{i=0}^{k} a_i^m \rho_i^m(\mu)\rho_i^m(\mu_0)\cos m\varphi.$$

The simplest of the cases mentioned is considered in the book by Minin [14], Sect. 3.4.

Bibliographical Comments and Additions to Part I

The content of Sects. 1.1–1.3 and 1.8 is mostly based on Ivanov's exposition of these problems in [13], Sects. 1.1, 2.1.

The Mie theory of light scattering on homogeneous spheres is expounded in the monographs by Shifrin [57], van de Hulst [58], Bohren and Huffman [59]; a brief discussion of the problem can also be found in the books by Born and Wolf [60], Chap. 13 and Newton [61], Chaps. 1–3.

The monograph by Deirmendjian [62] gives a clear explanation of the Mie theory and reduces it to a form convenient for the solution of particular problems and for computer calculations. This monograph also contains numerous tables showing results of calculations. A convenient algorithm for calculations is provided in the article by Loskutov [63]; however, the latest program realization of the Mie theory is given in the work by Wiscombe [64]. As to the calculation of the coefficients x_n of the expansion of phase function in terms of Legendre's polynomials, it should be noted that this problem has already been solved for a more general case, namely, for the scattering of polarized radiation, when the coefficients are sought of the expansion of the phase matrix in terms of generalized spherical functions. The corresponding algorithms are described in the works by Bugaenko [65], Domke [66], de Rooij and van der Stap [67]. The work by Mishchenko [68] contains a comparison of efficiency between the methods described in [65] and [66]. The algorithm for the calculation of x_n in the scalar case is given in the work by Dlugach [69].

In Sect. 1.4, the transfer equation was used in the form neglecting the phenomenon of refraction. Its impact on transfer processes was elaborately studied by Minin (see [14], Sects. 1.8 and 4.1). He showed that in most cases the refraction can be neglected.

It should be noted that the title of Sect. 1.7 "The problem for two-sided infinity" was taken from the work by Ambartsumian [3], p.181. To all appearances it was the author's work [70] where the formulation of this problem, as in this section was given for the first time. Equation (1.107) was derived in [71] (see, also, LSPA Sect. 9.2). Detailed consideration of (1.110) is given in the work by Fymat and Lenoble [72], where this relation is studied for an atmosphere whose optical properties depend on coordinates in an arbitrary way. The author of the present book used (1.110) to estimate the distribution of the absorption coefficient by height in the Venusian atmosphere [73], and Gavrilova [74] in the Earth's atmosphere (also, see Sect. 14.2.1 of the present book).

The exposition of Sect. 1.11 is based on the ideas of Chap. 2 of the book by Case and Zweifel [45] and the article by Ivanov and Volkov [75]. The invariance principle was introduced by Ambartsumian in [76] in 1943 (see, also, [3]). Later, Ambartsumian developed the method of addition of layers for a one-dimensional model [77, 78]. This idea anticipated Chandrasekhar's method for deriving invariance relations for a plane layer, which establishes a non-linear relationship between the radiation intensity in the atmosphere at an arbitrary depth, and that on the boundary (RT, Chap. 4.7). For details, see article by Pikichian [79].

In the middle of the 1970s, Engibarian and Mnatsakanian [80], Mnatsakanian [81] and Ivanov [82] each independently found that the invariance principle for homogeneous atmospheres can be formulated in a way yielding non-linear relationships between intensities at different depths. This was a further development of the application of the invariance principles to radiation transfer theory.

Section 1.12 contains the formulation of the generalized invariance principle proposed by the author in [83]. The mathematical aspects of this formulation are discussed in the article by Rogovtsov [84] (also, see [85]). The term "generalized invariance principle" is taken from the work by Ivanov [86]. It is worth noting here that the generalized invariance principle is, in fact, a corollary of the uniqueness theorem for the transfer equation (see [45], Sect. 2.3). The same idea can be found in the method of surface pseudosources, which was proposed by Laletin [87] (see, also, the monograph by Marchuk and Lebedev [46], Sect. 6.5).

Further applications of the generalized invariance principle to the study of radiation fields in media of various configurations were reported in works by Rogovtsov [88, 89] and Pikichian [90]–[92]. Additional information on the development of invariance principles as well as on the results achieved can be found in the Proceedings of the symposium commemorating the 40th anniversary of the introduction of invariance principles into radiation transfer theory [93].

Chapter 2 contains the results that can, in most cases, be found in LSPA, Sect. 2.1 and MLS, Sect. 6.2 as well as in Loskutov's article [94]. The mathematical aspects of the characteristic equation are more elaborately studied in the work by Maslennikov [40], articles by Germogenova [95, 96] and works of other authors. In particular, the method of determining the number of roots of the characteristic equation (Sect. 2.3) is presented for the first time in [95], the numerical method for the solution of the characteristic equation is outlined in accordance with [48]. The program for finding all the roots of the equation can be found in the work by Konovalov [97] and the formulas for nearly conservative scattering were derived in the author's work [98]. However, the quadratic forms for the asymptotic constant M are provided here for the first time.

Sections 3.1–3.4 are expounded according to work by Ivanov [82]. The basic invariance relation (3.1) was also independently found in [80]. Shimizu [99] was, to all appearances, the first person to find the basic relation of invariance; later it was exploited to study the transfer of neutrons [22]. Relations (3.10, 3.11) were established by Sobolev [100] and van de Hulst [101]. Equation (3.14) is Chandrasekhar's relation of invariance (see RT, Sect. 29.3). For the remaining bibliographical information on the results of the above sections, see [82].

A somewhat different method for the derivation of Ambartsumian's equation (3.32) can be found in LSPA, Sect. 2.2. The method for numerical solving of Ambartsumian's equation is expounded according to [49]. Konovalov [97, 53] also used relation (3.11) to solve (3.32). Having tested many different methods, he came to the conclusion that for a semi-infinite medium this method for the calculating the zeroth harmonic of the reflection coefficients is the fastest and most convenient.

It turned out that the method used for the calculation of the reflection coefficient was not, in fact a new one. It was actually a modification of the known method for accelerating the convergence of the iterative process (see, for instance, [23], Sect. 18 and [46], Sect. 10.26). This method was first used in transfer theory by Takahashi [102] as long ago as 1959. However, to the best of the author's knowledge, the first application of this method to the problem of diffuse reflection was first reported in [49, 53]. De Rooij and Domke [103] elaborately studied the convergence of the iterative procedure in solving Ambartsumian's equation for azimuthal harmonics of the reflection coefficient. They also considered the problem of the selection of a "physical" solution.

The integral relations of Sect. 3.6 are taken from [101] (equation (3.41)) and [82]. The integrals of the transfer equation which were provided in Sect. 3.7 were derived in the author's article [104]. For a homogeneous atmosphere, there exist integrals of an other type, which were established by Rybicki in [105]. These new integrals establish quadratic relationships between intensities at the same depth. Ivanov [106] generalized the results of [105] to the case of different optic depths.

The physical ideas of Sect. 3.8 are expounded according to the article by Ivanov [107]. The asymptotic formula (3.57) was found by Sobolev and (3.58) by van de Hulst [101]; both derivations are different from those exploited by the author.

The doubling formula (3.63) and methods for its application are provided in [82]. Dlugach [50] described how this method can be realized in practice. Sect. 3.10 is based on this work. Earlier, the calculation of the internal radiation field was done by another method; the results of calculation can be found in the article by Kattawar and Plass [108].

The way provided in Sect. 3.11 to derive the main formulas (3.89) and (3.90) have never been specified before. It was another idea that the author of this book exploited in order to obtain these formulas in [109]. The physical considerations explaining the structure of (3.71) and (3.73) in Sect. 3.11.1 are close to those given by van de Hulst in MLS, Chap. 5. The important formulas (3.79), (3.80) were independently derived by Sobolev and van de Hulst in [110, 101] respectively (see, also, LSPA, Sect. 2.6 and MLS, Sects. 5.4.1, 5.4.2). The expansions (3.82), (3.83) were found in MLS, Sect. 5.41 (see, also, [111]). The asymptotic formulas for flux and K integral were obtained in [104] (see, also, [109]). For the asymptotics provided in Sect. 3.11.6, see LSPA, Sect. 2.5 and MLS, Sect. 5.41 and also [112] and [98]. The latter contains a derivation of (3.99) and estimations of albedo of particles in the Venusian atmosphere. The approximate formula for spherical albedo at high frequencies was obtained by van de Hulst in [113].

Sect. 3.12 expounds the content of the author's work [114]. More details on the representation of the Green function can also be found in [75]. As to the results of Sect. 3.12.5, one should add that for isotropic scattering, the problem of radiation emerging from a layer was considered by Sobolev for different internal sources in [5], Sect. 4.3. In particular, the case was considered when the function describing internal sources can be represented as an expansion in optical depth. As a result, the problem was — just as in Sect. 3.12.5 — reduced to the usage of recurrence relations. This problem was considered by Minin [115] for anisotropic scattering.

The main results of Sect. 4.1–4.7 can be found in the author's article [51] (see, also, LSPA, Sects. 3.1, 3.2). The doubling method was pioneered by Peebles and Plesset [116] as early as 1951. This method became very popular after having been re-discovered by van de Hulst [117] and exploited in numerous numerical calculations (see, for instance, [118, 52]) including application to the transfer of polarized radiation [119, 120]. A detailed study of the accuracy of this method was given in the works by Wiscombe [121, 122].

Another approach to the calculation of the radiation field can be based on Sobolev's general theory (LSPA, Chaps. 5, 6). Numerical realizations of this method are elaborately described in the monograph by Viik, Ryym and Heinlo [123].

The content of Sect. 4.8 corresponds to that of Sect. 1 of the article by Ivanov [124]. Equations (4.47–49) were earlier derived by Germogenova [125] (see, also, [126, 111]).

The asymptotic formulas (4.50), (4.51) were obtained by different methods at different times [125, 100, 101]. The representation of these formulas in the form of (4.53) and (4.54) is taken from [127].

The formulas of Sect. 4.8.3 were mainly obtained in [109, 128] (see, also, the articles by Minin [129, 111]). The most important results of Sect. 4.8.4 can be found in the author's works [54, 55]. The main idea of Sect. 4.9 can be found in [128] and [130]. It is worth noticing here that the asymptotic theory has been used successfully to construct various numerical algorithms (see, for instance, [131]). Sect. 4.10 grounds on the article by the author and Dlugach [132].

The content of Chap. 5 is based on the results of LSPA, Chap. 4 and the article by Ivanov [133]. Equations (5.16), (5.17) were derived in 1948 independently of each other by Sobolev [134] and van de Hulst [135]. Formulas (5.42), (5.52), (5.57) were found in the work by the author [109] whereas it is probable that (5.62–65) have never been provided before. The expression (5.52) for the radiation intensity was used by Fomin [136] to study the non-stationary glowing of an atmosphere.

Part II

MULTILAYER ATMOSPHERE

* * *

It has been stated in Part I that the theory of transfer of monochromatic unpolarized radiation in a homogeneous plane atmosphere is now complete. The analytical forms and numerical methods have been elaborated, so that any problem in this field can in principle be solved. One may naturally expect that the next step is a study of light scattering in a plane inhomogeneous medium whose optical properties vary with depth.

A multilayer medium that consists of *any* number of plane homogeneous layers with an arbitrary optical depth is the simplest example of a plane inhomogeneous atmosphere. It is of practical importance to investigate such a model. One reason for this is that stellar and planetary atmospheres as well as other media encountered in nature and in laboratories can very often be treated as multilayer media. Furthermore, if the layers are thin enough, one then approximates the plane inhomogeneous atmosphere, with λ and $\chi(\gamma)$ depending *arbitrarily* on the optical depth. One sees that the accuracy of such an approximation depends solely on the number of layers: the greater this number, the better the approximation. Finally in this part we will show that the algorithm of evaluating the radiation field in a multilayer medium reduces generally speaking, to a "superposition" of algorithms for calculating the radiation field in homogeneous layers treated separately. The latter problem has already been solved above.

Based on the above-stated reasons, we present the main results of the theory of light scattering in multilayer atmospheres in this separate part.

6. Parallel External Flux Problem and the Milne Problem

The key problem studied in this chapter is to determine a radiation field in a plane multilayer atmosphere illuminated by parallel rays. The geometry of this problem is given in Fig. 6.1.

6.1 Formulation of the Problem

We are dealing with a multilayer atmosphere consisting of n ($n \geq 2$) homogeneous layers. It is illuminated by parallel rays incident at a zenith angle $\cos^{-1}\mu_0$ and azimuthal angle φ_0 and creating illumination π of a unit area perpendicular to the rays. We denote the optical thickness of each layer by τ_j ($j = 1, 2, 3, \dots, n$). Let $I_{1,n}(\tau, \mu, \mu_0, \varphi - \varphi_0)$ be the intensity of the diffuse radiation in the multilayer atmosphere propagating at optical depth τ in the direction $\cos^{-1}\mu$ at azimuth φ (see Fig. 6.1; $\tau \in [0, \tau_0]$) with

$$\tau_0 = \sum_{j=1}^{n} \tau_j.$$

The first subscript in $I_{j,n}$ corresponds hereafter to the j-th layer illuminated directly by parallel rays.

The optical properties of each homogeneous layer are assumed given. In other words, λ_j and $\chi_j(\gamma)$ are known. We also assume that the solution of the parallel external flux problem for each homogeneous layer treated separately is known. This means that the intensities of radiation $I_j(\tau, \mu, \mu_0, \varphi - \varphi_0)$ ($\tau \in [0, \tau_j]$) are given. The procedure of determining them is outlined in Part I.

We need to calculate the radiation field in a multilayer atmosphere, or equivalently, to find the quantity $I_{1,n}(\tau, \mu, \mu_0, \varphi - \varphi_0)$ for any $\tau \in [0, \tau_0]$.

6.2 A Two-Layer Atmosphere

We begin naturally with the simplest case of the medium that consists of a couple of arbitrarily chosen layers, say, the layer $n - 1$ and the layer n. We make a common assumption that the phase functions $\chi_n(\gamma)$ and $\chi_{n-1}(\gamma)$ can

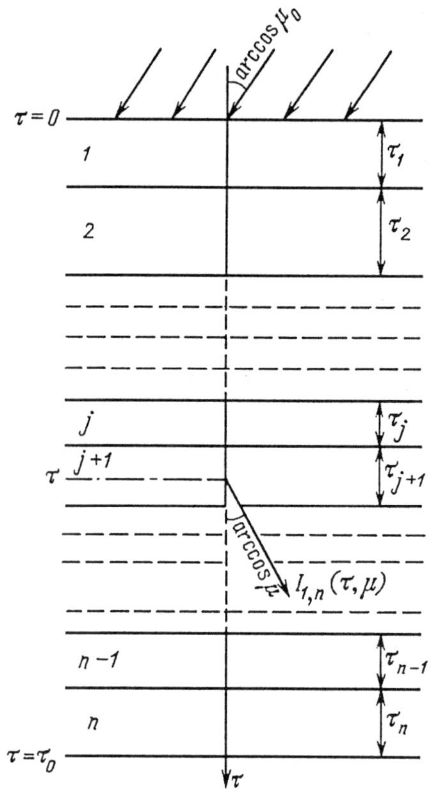

Fig. 6.1. The parallel external flux problem for a multilayer atmosphere

be represented by a sum of terms of their expansions in azimuthal angle; of N_n and N_{n-1} terms respectively. Similar formulas hold for the quantities I_n and I_{n-1} (see (1.47) and (1.55)). We can then write,

$$I_{n-1,n}(\tau,\mu,\mu_0,\varphi-\varphi_0) = I_{n-1,n}(\tau,\mu,\mu_0) + 2\sum_{m=1}^{N_{n-1,n}} I^m_{n-1,n}(\tau,\mu,\mu_0)\cos m(\varphi-\varphi_0), \tag{6.1}$$

where $N_{n-1,n} = \max\{N_{n-1}, N_n\}$.

We consider first the zeroth azimuthal harmonic of the radiation intensity. Applying the generalized invariance principle enables us to obtain the invariance relations that determine the radiation field in a two-layer medium. Indeed, the layer $n-1$ is illuminated from above by a parallel beam and from the interlayer boundary by a radiation of intensity $I_{n-1,n}(\tau_{n-1},-\mu,\mu_0)$ $(\mu \in [0,1])$. The n-th layer is illuminated only from the interlayer boundary by the attenuated direct radiation and by diffuse radiation of intensity $I_{n-1,n}(\tau_{n-1},\mu,\mu_0)$. Then, applying similar arguments to those used in deriving the invariance relations (4.1), we find $(\mu \in [-1,+1])$,

$$I_{n-1,n}(\tau,\mu,\mu_0) = I_{n-1}(\tau,\mu,\mu_0) + I_{n-1,n}(\tau_{n-1},\mu,\mu_0)\mathrm{e}^{\frac{\tau_{n-1}-\tau}{\mu}}\theta(-\mu)$$
$$+2\int_0^1 I_{n-1,n}(\tau_{n-1},-\mu',\mu_0)I_{n-1}(\tau_{n-1}-\tau,-\mu,\mu')\,\mathrm{d}\mu' \quad (\tau\in[0,\tau_{n-1}]), \tag{6.2}$$

$$I_{n-1,n}(\tau_{n-1}+\tau,\mu,\mu_0) = I_n(\tau,\mu,\mu_0)\mathrm{e}^{-\frac{\tau_{n-1}}{\mu_0}} + I_{n-1,n}(\tau_{n-1},\mu,\mu_0)\mathrm{e}^{-\frac{\tau}{\mu}}\theta(\mu)$$
$$+2\int_0^1 I_{n-1,n}(\tau_{n-1},\mu',\mu_0)I_n(\tau,\mu,\mu')\,\mathrm{d}\mu' \quad (\tau\in[0,\tau_n]). \tag{6.3}$$

Put $\tau=\tau_{n-1}$ and $\tau=0$ respectively into (6.2) and (6.3). Taking also into account (1.95) and (1.96), we obtain the system of equations to determine the radiation intensity on the interlayer boundary,

$$I_{n-1,n}(\tau_{n-1},\mu,\mu_0) = \sigma_{n-1}(\mu,\mu_0)\mu_0$$
$$+2\int_0^1 \rho_{n-1}(\mu,\mu')I_{n-1,n}(\tau_{n-1},-\mu',\mu_0)\mu'\,\mathrm{d}\mu', \tag{6.4}$$

$$I_{n-1,n}(\tau_{n-1},-\mu,\mu_0) = \rho_n(\mu,\mu_0)\mathrm{e}^{-\frac{\tau_{n-1}}{\mu_0}}\mu_0$$
$$+2\int_0^1 \rho_n(\mu,\mu')I_{n-1,n}(\tau_{n-1},\mu',\mu_0)\mu'\,\mathrm{d}\mu'. \tag{6.5}$$

With regard to the reflection and transmission coefficients for the two-layer atmosphere (we denote them by $\rho_{n-1,n}$ and $\sigma_{n-1,n}$), setting, respectively, $\tau=0$ and $\tau=\tau_n$ in (6.2) and (6.3) we obtain the following expressions:

$$\rho_{n-1,n}(\mu,\mu_0)\mu_0 = \rho_{n-1}(\mu,\mu_0)\mu_0 + I_{n-1,n}(\tau_{n-1},-\mu,\mu_0)\mathrm{e}^{-\frac{\tau_{n-1}}{\mu}}$$
$$+2\int_0^1 \sigma_{n-1}(\mu,\mu')I_{n-1,n}(\tau_{n-1},-\mu',\mu_0)\mu'\,\mathrm{d}\mu', \tag{6.6}$$

$$\sigma_{n-1,n}(\mu,\mu_0)\mu_0 = \sigma_n(\mu,\mu_0)\mathrm{e}^{-\frac{\tau_{n-1}}{\mu}}\mu_0 + I_{n-1,n}(\tau_{n-1},\mu,\mu_0)\mathrm{e}^{-\frac{\tau_n}{\mu}}$$
$$+2\int_0^1 \sigma_n(\mu,\mu')I_{n-1,n}(\tau_{n-1},\mu',\mu_0)\mu'\,\mathrm{d}\mu'. \tag{6.7}$$

For given ρ_n and σ_{n-1}, the quantity $I_{n-1,n}(\tau_{n-1},\pm\mu,\mu_0)$ can be found by solving the system of equations (6.4) and (6.5) by successive approximations. The speed of convergence is mainly determined by λ_n and λ_{n-1}, as well as by the optical thicknesses of the layers; the thinner the layers, the faster the convergence process. In the case of two adjacent half-spaces ($\tau_n=\tau_{n-1}=\infty$) the proof of the convergence of the corresponding iterative procedure is given in [137]. If the system of equations (6.4) and (6.5) for the zeroth azimuthal harmonic is solved, the complete solution to the parallel external flux problem for the two-layer atmosphere is then given by (6.2) and (6.3).

It is obvious that these equations remain unchanged for all higher azimuthal harmonics for which the functions $\chi^m_{n-1}(\mu,\mu_0)$ and $\chi^m_n(\mu,\mu_0)$ are both non-zero in both layers. If $N_{n-1} > N_n$, then for all $m = N_n+1, N_n+2, \ldots, N_{n-1}$ functions $I^m_n(\tau,\mu,\mu_0) \equiv 0$, i.e., in the lower layer considered separately the diffuse radiation is absent for these harmonics. From (6.2–5) we have the trivial result:

$$I^m_{n-1,n}(\tau,\mu,\mu_0) = I^m_{n-1}(\tau,\mu,\mu_0) \qquad (\tau \in [0,\tau_{n-1}]), \tag{6.8}$$

$$I^m_{n-1,n}(\tau_{n-1}+\tau,\mu,\mu_0) = 0 \qquad (\tau \in [0,\tau_n]), \tag{6.9}$$

i.e., the two-layer atmosphere in this case does not differ from the one-layer one. If $N_{n-1} < N_n$, then for all $m = N_{n-1}+1, N_{n-1}+2, \ldots, N_n$ the quantity $I^m_{n-1}(\tau,\mu,\mu_0) \equiv 0$, and hence

$$I^m_{n-1,n}(\tau,\mu,\mu_0) = 0 \qquad (\tau \in [0,\tau_{n-1}]), \tag{6.10}$$

$$I^m_{n-1,n}(\tau_{n-1}+\tau,\mu,\mu_0) = I_n(\tau,\mu,\mu_0)\mathrm{e}^{-\frac{\tau_{n-1}}{\mu_0}} \qquad (\tau \in [0,\tau_n]). \tag{6.11}$$

The influence of the upper layer manifests itself only in attenuating the direct radiation incident on the lower layer. The complete solution of the problem is given by (6.1).

6.3 Choosing the Direction to Add Layers

It seems clear at this moment how to obtain the algorithm for calculating the radiation field in a multilayer medium. One just needs to add a third layer (say with properties j) to a two-layer atmosphere and to write down the corresponding invariance relations similar to (6.2) and (6.3). Afterwards one should add a fourth layer to the three-layer atmosphere and so on.

One can make up a three-layer atmosphere in two ways. One way is to attach the layer j on the side of the layer $n-1$. In this case the multilayer medium is illuminated on the side of the layer j. Another way is to add the layer j on the side of the layer n. In this case the layer $n-1$ is illuminated by parallel rays; see Fig. 6.2. These ways seem to have no essential difference. However this is not true, the reason that a multilayer atmosphere is an *inhomogeneous* medium. This means that the *radiation field in this medium depends on which side of the medium (upper or lower) is illuminated* (it will be different for each case[1]). Thus, the addition of the third layer to the two-layer atmosphere from below (Fig. 6.2b) would require us to define a new function for this atmosphere – the radiation intensity in the inhomogeneous layer illuminated from below (in addition to the already introduced intensity in the layer illuminated from above). On the other hand, adding a homo-

[1] Any rule always has exclusions. For example, if the optical properties of the layer are symmetrical with respect to its midpoint, the illumination of the upper side and the lower side of the medium is equivalent, i.e., the radiation field remains unchanged.

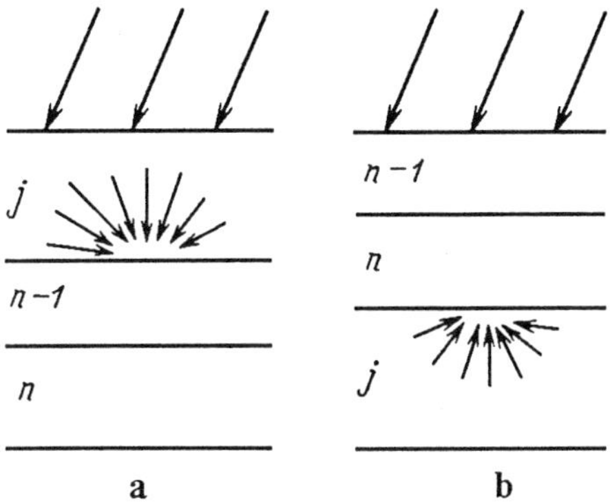

Fig. 6.2. Choosing the direction to add layers in a multilayer atmosphere

geneous layer to the top (Fig. 6.2a) creates additional illumination in the two-layer atmosphere *from above*. However, the upper layer, while it is illuminated from below, is homogeneous. Thus the introduction of the additional radiation intensity is not necessary.

That is why we will add the layers from the top upwards under the assumption that the upper layer is illuminated by parallel rays.

6.4 Radiation Field in a Multilayer Atmosphere

Let us add layer $n-2$ to a two-layer atmosphere consisting of layers $n-1$ and n. Let us further add a layer $n-3$ and so on. This results in the following invariance relations for the zeroth azimuthal harmonic of the radiation intensity in the multilayer medium consisting of $n-j+1$ layers, $j = n-1, n-2, \ldots, 1$

$$I_{j,n}(\tau,\mu,\mu_0) = I_j(\tau,\mu,\mu_0) + I_{j,n}(\tau_j,\mu,\mu_0)\mathrm{e}^{\frac{\tau_j-\tau}{\mu}}\theta(-\mu)$$
$$+2\int_0^1 I_{j,n}(\tau_j,-\mu',\mu_0)I_j(\tau_j-\tau,-\mu,\mu')\,\mathrm{d}\mu' \qquad (\tau\in[0,\tau_j]), \quad (6.12)$$

$$I_{j,n}(\tau_j+\tau,\mu,\mu_0) = I_{j+1,n}(\tau,\mu,\mu_0)\mathrm{e}^{-\frac{\tau_j}{\mu_0}} + I_{j,n}(\tau_j,\mu,\mu_0)\mathrm{e}^{-\frac{\tau}{\mu}}\theta(\mu)$$
$$+2\int_0^1 I_{j,n}(\tau_j,\mu',\mu_0)I_{j+1,n}(\tau,\mu,\mu')\,\mathrm{d}\mu' \quad (\tau\in[0,\tau_{j+1}^s = \sum_{i=j+1}^{n}\tau_i]). \quad (6.13)$$

Hence, we have for $\mu\in[0,1]$

$$I_{j,n}(\tau_j,\mu,\mu_0) = \sigma_j(\mu,\mu_0)\mu_0 + 2\int_0^1 \rho_j(\mu,\mu')I_{j,n}(\tau_j,-\mu',\mu_0)\mu'\,\mathrm{d}\mu', \quad (6.14)$$

$$I_{j,n}(\tau_j,-\mu,\mu_0) = \rho_{j+1,n}(\mu,\mu_0)\mathrm{e}^{-\frac{\tau_j}{\mu_0}}\mu_0 + 2\int_0^1 \rho_{j+1,n}(\mu,\mu')I_{j,n}(\tau_j,\mu',\mu_0)\mu'\,\mathrm{d}\mu'. \quad (6.15)$$

Separate formulas for the reflection $\rho_{j,n}$ and transmission $\sigma_{j,n}$ coefficients can be readily obtained. Setting $\tau = 0$ and $\tau = \tau^s_{j+1}$ in (6.12) and (6.13), respectively, we obtain

$$\rho_{j,n}(\mu,\mu_0)\mu_0 = \rho_j(\mu,\mu_0)\mu_0 + I_{j,n}(\tau_j,-\mu,\mu_0)\mathrm{e}^{-\frac{\tau_j}{\mu}} \\ +2\int_0^1 \sigma_j(\mu,\mu')I_{j,n}(\tau_j,-\mu',\mu_0)\mu'\,\mathrm{d}\mu', \tag{6.16}$$

$$\sigma_{j,n}(\mu,\mu_0)\mu_0 = \sigma_{j+1,n}(\mu,\mu_0)\mathrm{e}^{-\frac{\tau_j}{\mu_0}}\mu_0 + I_{j,n}(\tau_j,\mu,\mu_0)\mathrm{e}^{-\frac{\tau^s_{j+1}}{\mu}} \\ +2\int_0^1 \sigma_{j+1,n}(\mu,\mu')I_{j,n}(\tau_j,\mu',\mu_0)\mu'\,\mathrm{d}\mu'. \tag{6.17}$$

The relations (6.13), (6.15) and (6.17) are of recursive type with respect to j with the initial conditions $I_{n,n} \equiv I_n$, $\rho_{n,n} \equiv \rho_n$ and $\sigma_{n,n} \equiv \sigma_n$. Using (6.12–17) for $j = n-1, n-2, \ldots, 2, 1$, the problem posed may be completely solved. This implies that for the multilayer atmosphere we can determine $I_{1,n}(\tau,\mu,\mu_0)$ at any optical depth. The iterative procedure is used only for solving the system of equations (6.14) and (6.15) which gives the boundary radiation intensity.

Equations (6.12–17) are obviously valid for any azimuthal harmonic. One point worth emphasizing is that as m increases, the harmonic of the phase function may vanish in some of the layers, so that the corresponding values of $I_j^m(\tau,\mu,\mu_0)$ become equal to zero. This results in the disappearance of the terms in (6.12–17) that correspond to these values of j and m.

Thus, to determine the radiation field in the multilayer medium, one needs:

1. to calculate the harmonics of the reflection and transmission coefficients for all separate layers, as well as the radiation intensities for the chosen values of the optical depth;
2. using the relations (6.12–17) to calculate successively all harmonics of the radiation intensity for all values of j from $j = n-1$ to $j = 1$ at the depths at which they are known for the separate layers;
3. to substitute the values $I^m_{1,n}(\tau,\mu,\mu_0)$ into the equation

$$I_{1,n}(\tau,\mu,\mu_0,\varphi-\varphi_0) = I_{1,n}(\tau,\mu,\mu_0) + 2\sum_{m=1}^{N_{1,n}} I^m_{1,n}(\tau,\mu,\mu_0)\cos m(\varphi-\varphi_0), \tag{6.17a}$$

where $N_{1,n} = \max\{N_1, N_2, \ldots, N_n\}$, and N_j is the number of azimuthal harmonics in the expansion (1.55) of the phase function of the layer j.

It is worth noticing that in a stratified atmosphere for all harmonics of the radiation intensities at the interface boundaries, the following integral relations hold:

$$\int_{-1}^{+1} I_{j,n}^m(\tau_j,\mu',\mu_0) I_{j,n}^m(\tau_j,-\mu',\mu_1)\mu'\,\mathrm{d}\mu'$$
$$= \mu_0 \int_0^1 \sigma_j^m(\mu',\mu_0) I_{j,n}^m(\tau_j,-\mu',\mu_1)\mu'\,\mathrm{d}\mu'$$
$$-\mu_1 \int_0^1 \sigma_j^m(\mu',\mu_1) I_{j,n}^m(\tau_j,-\mu',\mu_0)\mu'\,\mathrm{d}\mu'$$
$$= \mu_1 \mathrm{e}^{-\frac{\tau_j}{\mu_1}} \int_0^1 \rho_{j+1,n}^m(\mu',\mu_1) I_{j,n}^m(\tau_j,\mu',\mu_0)\mu'\,\mathrm{d}\mu'$$
$$-\mu_0 \mathrm{e}^{-\frac{\tau_j}{\mu_0}} \int_0^1 \rho_{j+1,n}^m(\mu',\mu_0) I_{j,n}^m(\tau_j,\mu',\mu_1)\mu'\,\mathrm{d}\mu'. \tag{6.18}$$

These are supplemented with the following:

$$\mu_0 \int_0^1 \left[\rho_{j,n}^m(\mu',\mu_0) - \rho_j^m(\mu',\mu_0)\right] I_{j,n}^m(\tau_j,-\mu',\mu)\mu'\,\mathrm{d}\mu'$$
$$= \mu \int_0^1 \left[\rho_{j,n}^m(\mu',\mu) - \rho_j^m(\mu',\mu)\right] I_{j,n}^m(\tau_j,-\mu',\mu_0)\mu'\,\mathrm{d}\mu'. \tag{6.18a}$$

They can be derived from (6.14), (6.15), and (6.16) using the method applied earlier to obtain the expression (4.34). For $\mu \neq \mu_0$ (at $\mu = \mu_0$ (6.18) and (6.18a) are identities) these expressions can, in particular, be employed to test the accuracy of our calculations. Now we go on to some particular cases.

6.5 A Semi-Infinite Multilayer Atmosphere

Let the lower, n-th layer be semi-infinite. It is clear that the method proposed for defining the radiation field remains valid. In this case, it is expedient to calculate the reflection coefficient for the lower layer and the radiation field in this layer by using the method described in Sects. 3.5 and 3.10. It is of interest also to obtain an asymptotic formula that determines the radiation field at a large depth in the lower semi-infinite layer, i.e., with the condition $(\tau - \tau_j^*) \gg 1$, where

$$\tau_j^* = \sum_{i=j}^{n-1} \tau_i. \tag{6.19}$$

It is clear from physical considerations that owing to the angular photon relaxation, the radiation intensity is proportional to $i_n(\mu)$, and the attenuation of the radiation flux with depth obeys the law $\mathrm{e}^{-k_n(\tau-\tau_j^*)}$. Therefore, we have

$$I_{j,n}^{\mathrm{as}}(\tau,\mu,\mu_0) = i_n(\mu)\mathrm{e}^{-k_n(\tau-\tau_j^*)} U_{j,n}(\mu_0). \tag{6.20}$$

Substituting (6.20) into (6.13) we obtain the following recursion relation for the function $U_{j,n}(\mu_0)$

$$U_{j,n}(\mu_0) = U_{j+1,n}(\mu_0)\mathrm{e}^{-\frac{\tau_j}{\mu_0}} + 2\int_0^1 U_{j+1,n}(\mu')I_{j,n}(\tau_j,\mu',\mu_0)\,\mathrm{d}\mu' \qquad (6.21)$$

with the initial condition $U_{n,n}(\mu_0) = u_n(\mu_0)\mu_0$. It is seen that, in order to calculate the radiation intensity in deep layers of the n-th semi-infinite layer, we only need to know the radiation intensities at the interlayer boundaries.

6.6 A Multilayer Atmosphere Above a Reflecting Surface

It is readily seen that this case reduces to the algorithm already outlined. Indeed, let the n-th layer be semi-infinite with the known reflection coefficient $\rho_s(\mu,\mu_0,\varphi-\varphi_0)$, then it determines the reflection law for a surface adjacent to a multilayer atmosphere that consists of $(n-1)$ layers. Let us expand the function ρ_s in a series (5.3) with respect to the azimuthal angle. Replacing ρ_n with ρ_s in (6.5), we obtain a system of equations (6.4) and (6.5) to determine $\bar{I}_{n-1,n}(\tau_{n-1},\mu,\mu_0)$. This function defines directly the corresponding azimuthal harmonics of the radiation intensity on the boundary between the reflecting surface and the atmosphere (usually, the overbar refers to a medium with a reflecting surface). Further, the entire process of calculating the radiation field reduces to successive uses of the relations (6.12–15) for $j = n-2, n-3, \ldots, 1$.

In the simplest case of a surface that reflects light isotropically, i.e., when $\rho_s(\mu,\mu_0,\varphi-\varphi_0) = A = \text{const}$, the surface affects only the radiation intensity averaged over the azimuth. In this case, from (6.4) and (6.5), we find for the intensity at the atmosphere–surface boundary ($\mu \in [0,1]$),

$$\bar{I}_{n-1,n}(\tau_{n-1},\mu,\mu_0) = \sigma_{n-1}(\mu,\mu_0)\mu_0 + \frac{AA(\mu,\tau_{n-1})V(\mu_0,\tau_{n-1})}{1 - AA_s(\tau_{n-1})}\mu_0, \qquad (6.22)$$

where, in accordance with the notation introduced in Sect. 1.9,

$$A(\mu,\tau_{n-1}) = 2\int_0^1 \rho_{n-1}(\mu,\mu')\mu'\,\mathrm{d}\mu', \qquad (6.23)$$

$$V(\mu,\tau_{n-1}) = \mathrm{e}^{-\frac{\tau_{n-1}}{\mu}} + 2\int_0^1 \sigma_{n-1}(\mu,\mu')\mu'\,\mathrm{d}\mu', \qquad (6.24)$$

$$A_s(\tau_{n-1}) = 4\int_0^1 \mu\,\mathrm{d}\mu \int_0^1 \rho_{n-1}(\mu,\mu')\mu'\,\mathrm{d}\mu'. \qquad (6.25)$$

Other azimuthal harmonics of the radiation intensity can be calculated similarly, by analogy with the case when the reflecting surface is absent.

6.7 The Milne Problem

We now turn to solving another fundamental problem in the theory of light scattering in plane media, namely, the Milne problem.

Let us suppose that a plane atmosphere is given. It consists of n homogeneous layers whose optical thicknesses τ_j ($j = 1, 2, 3, \ldots, n$) and properties are assumed to be given for each layer. Moreover, $\tau_n = \infty$ (see Fig. 6.3). This model of the atmosphere requires us to solve not only the parallel external flux problem but also the Milne problem, (i.e., to find the radiation field in a medium in which primary sources of unlimited power are at infinity). For each layer taken separately, we assume that the diffuse reflection problem is solved, i.e., the functions $I_j(\tau, \mu, \mu_0)$ ($\tau \in [0, \tau_j]$; $\mu \in [-1, +1]$, $\mu_0 \in [0, 1]$) are known.

In the multilayer atmosphere, the parallel external flux problem is solved in the same way as explained above. As a result, we may assume that both the functions $I_{j,n}(\tau, \mu, \mu_0)$ and the functions $\rho_{j,n}(\mu, \mu_0)$, $\sigma_{j,n}(\mu, \mu_0)$ are known. It is obvious that for the Milne problem we require only the zeroth azimuthal harmonics of these functions. We now offer the solution of this problem.

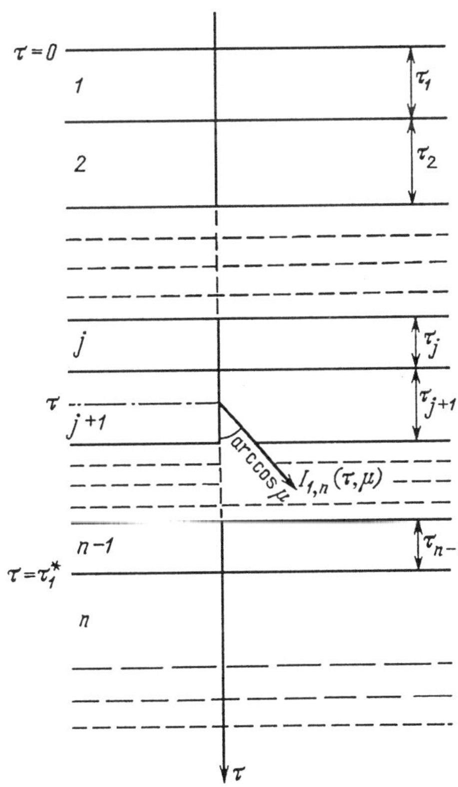

Fig. 6.3. The Milne problem for a multilayer atmosphere

Let us deal with a semi-infinite layer with properties "n". According to the results given in Sect. 3.3, when we consider the Milne problem for such an atmosphere, the properly normalized radiation intensity $I_n(\tau, \mu)$ is determined by ($\tau \in [0, \infty), \mu \in [-1, +1]$)

$$M_n I_n(\tau, \mu) = i_n(-\mu) \mathrm{e}^{k_n \tau} - i_n(-\mu) \mathrm{e}^{-\frac{\tau}{\mu}} \theta(\mu)$$
$$-2 \int_0^1 I_n^\infty(\tau, \mu, \mu') i_n(-\mu') \, \mathrm{d}\mu' \qquad (\lambda_n \neq 1). \tag{6.26}$$

In the conservative case ($\lambda_n = 1$)

$$I_n(\tau, \mu) = \frac{3}{4} \left[\left(1 - \frac{x_{1n}}{3}\right) \tau - \mu + 2 \int_0^1 I_n^\infty(\tau, \mu, \mu') \mu' \, \mathrm{d}\mu' + \mu \mathrm{e}^{-\frac{\tau}{\mu}} \theta(\mu) \right]. \tag{6.27}$$

Here, as usual, the function $i_n(\mu)$ is the radiation intensity in the diffusion regime for an infinite atmosphere with the properties "n", k_n is the corresponding diffusion exponent, x_{1n} is the first coefficient in the expansion of the phase function $\chi_n(\gamma)$ in Legendre polynomials, and $I_n^\infty(\tau, \mu, \mu_0)$ is the zeroth azimuthal harmonic of the radiation intensity in a semi-infinite homogeneous layer as designated by superscript "∞". Since the function $I_n^\infty(\tau, \mu, \mu_0)$ is assumed known, $I_n(\tau, \mu)$, — the Milne's radiation intensity for a homogeneous atmosphere — is also known.

We now add the layer $n-1$ to the layer n. Let us refer the optical depth τ to the upper edge of the layer $n-1$. For this two-layer atmosphere, we have the following invariance relations ($\mu \in [-1, +1]$):

$$I_{n-1,n}(\tau, \mu) = I_{n-1,n}(\tau_{n-1}, \mu) \mathrm{e}^{\frac{\tau_{n-1} - \tau}{\mu}} \theta(-\mu)$$
$$+2 \int_0^1 I_{n-1,n}(\tau_{n-1}, -\mu') I_{n-1}(\tau_{n-1} - \tau, -\mu, \mu') \, \mathrm{d}\mu' \quad (\tau \in [0, \tau_{n-1}]), \tag{6.28}$$

$$I_{n-1,n}(\tau_{n-1} + \tau, \mu) = I_n(\tau, \mu) + I_{n-1,n}(\tau_{n-1}, \mu) \mathrm{e}^{-\frac{\tau}{\mu}} \theta(\mu)$$
$$+2 \int_0^1 I_{n-1,n}(\tau_{n-1}, \mu') I_n^\infty(\tau, \mu, \mu') \, \mathrm{d}\mu' \quad (\tau \geq 0), \tag{6.29}$$

where $I_{n-1,n}(\tau, \mu)$ is the intensity in the Milne problem for the two-layer atmosphere. Then, as follows from (6.28) and (6.29), the problem is completely solved (up to normalization) if we find the boundary intensity $I_{n-1,n}(\tau_{n-1}, \mu)$. To get the latter one, we obtain from (6.28), (6.29) and (6.26) the system of equations ($\mu \in [0, 1]$)

$$I_{n-1,n}(\tau_{n-1}, \mu) = 2 \int_0^1 I_{n-1,n}(\tau_{n-1}, -\mu') \rho_{n-1}(\mu, \mu') \mu' \, \mathrm{d}\mu', \tag{6.30}$$

$$I_{n-1,n}(\tau_{n-1}, -\mu) = u_n(\mu) + 2 \int_0^1 I_{n-1,n}(\tau_{n-1}, \mu') \rho_n^\infty(\mu, \mu') \mu' \, \mathrm{d}\mu', \tag{6.31}$$

which have to be solved numerically by successive approximations. Here, as usual, we denote the escape function by $u_n(\mu) = I_n(0, -\mu)$. In particular, as follows from (6.28), this function for the two-layer atmosphere is given by

$$u_{n-1,n}(\mu) = I_{n-1,n}(\tau_{n-1}, -\mu)\mathrm{e}^{-\frac{\tau_{n-1}}{\mu}} + 2\int_0^1 I_{n-1,n}(\tau_{n-1}, -\mu')\sigma_{n-1}(\mu, \mu')\mu' \, \mathrm{d}\mu'. \tag{6.32}$$

We emphasize that the function $I_{n-1,n}(\tau, \mu)$ (and, in particular, the function $u_{n-1,n}(\mu)$) must be suitably normalized. We choose the normalization in Sect. 6.9.

Continuing successively to add the layers $n-2, n-3, \ldots, j$ to the two-layer atmosphere, we find for the atmosphere consisting of $n-j+1$ layers ($\mu \in [-1, +1]$),

$$I_{j,n}(\tau, \mu) = I_{j,n}(\tau_j, \mu)\mathrm{e}^{\frac{\tau_j - \tau}{\mu}}\theta(-\mu)$$
$$+2\int_0^1 I_{j,n}(\tau_j, -\mu')I_j(\tau_j - \tau, -\mu, \mu') \, \mathrm{d}\mu' \qquad (\tau \in [0, \tau_j]), \tag{6.33}$$

$$I_{j,n}(\tau_j + \tau, \mu) = I_{j+1,n}(\tau, \mu) + I_{j,n}(\tau_j, \mu)\mathrm{e}^{-\frac{\tau}{\mu}}\theta(\mu)$$
$$+2\int_0^1 I_{j,n}(\tau_j, \mu')I_{j+1,n}(\tau, \mu, \mu') \, \mathrm{d}\mu' \qquad (\tau \geq 0). \tag{6.34}$$

The boundary intensity $I_{j,n}(\tau_j, \mu)$ is determined via the system of equations

$$I_{j,n}(\tau_j, \mu) = 2\int_0^1 I_{j,n}(\tau_j, -\mu')\rho_j(\mu, \mu')\mu' \, \mathrm{d}\mu', \tag{6.35}$$

$$I_{j,n}(\tau_j, -\mu) = u_{j+1,n}(\mu) + 2\int_0^1 I_{j,n}(\tau_j, \mu')\rho_{j+1,n}(\mu, \mu')\mu' \, \mathrm{d}\mu', \tag{6.36}$$

where

$$u_{j,n}(\mu) = I_{j,n}(\tau_j, -\mu)\mathrm{e}^{-\frac{\tau_j}{\mu}} + 2\int_0^1 I_{j,n}(\tau_j, -\mu')\sigma_j(\mu, \mu')\mu' \, \mathrm{d}\mu'. \tag{6.37}$$

Thus, the solution of our problem is given by the recursion system of equations (6.34), (6.36) and relations (6.33), (6.35) and (6.37) with $j = 1$. The initial conditions for these recursion relations are: $I_{n,n}(\tau, \mu, \mu_0) \equiv I_n^\infty(\tau, \mu, \mu_0)$; $I_{n,n}(\tau, \mu) \equiv I_n(\tau, \mu)$; $\rho_{n,n}(\mu, \mu_0) \equiv \rho_n^\infty(\mu, \mu_0)$. The following integral relations, which can be used as the control ones, must also be satisfied:

$$\int_{-1}^{+1} I_{j,n}(\tau_j, -\mu')I_{j,n}(\tau_j, \mu', \mu)\mu' \, \mathrm{d}\mu' = \mu\int_0^1 I_{j,n}(\tau_j, -\mu')\sigma_j(\mu, \mu')\mu' \, \mathrm{d}\mu'$$
$$= \int_0^1 u_{j+1,n}(\mu')I_{j,n}(\tau_j, \mu', \mu)\mu' \, \mathrm{d}\mu' - \mathrm{e}^{-\frac{\tau_j}{\mu}}\mu\int_0^1 I_{j,n}(\tau_j, \mu')\rho_{j+1,n}(\mu, \mu')\mu' \, \mathrm{d}\mu'. \tag{6.38}$$

They can be obtained from (6.35), (6.36), (6.14) and (6.15).

6.8 The Milne Intensity at a Large Depth in Layer n

It is of interest to elucidate the structure of the radiation field at a large optical depth in the lower (n-th) layer, i.e., in case when the condition $(\tau-\tau^*) \gg 1$ holds. τ^* is defined by the expression (6.19). It is clear from physical considerations that, owing to the angular photon relaxation, the general structure of the corresponding asymptotic expression for the radiation intensity in layer n must be the same as if the external layers $(n-1, n-2, \ldots, j)$ were absent. Therefore, taking into account the corresponding equation (3.58) for a homogeneous half-space, we write

$$M_n I^{\mathrm{as}}_{j,n}(\tau,\mu) = i_n(-\mu)\mathrm{e}^{k_n(\tau-\tau_j^*)} - i_n(\mu)\mathrm{e}^{-k_n(\tau-\tau_j^*)} N^*_{j,n} \tag{6.39}$$

trying to find the quantity $N^*_{j,n}$. The superscript "as" means that large depths in the layer n are considered, i.e., $(\tau-\tau^*) \gg 1$.

We substitute (6.39) into (6.34). Taking into account (6.20), we arrive at the following recursion relation:

$$N^*_{j,n} = N^*_{j+1,n} - 2M_n \int_0^1 I_{j,n}(\tau_j,\mu')U_{j+1,n}(\mu')\,\mathrm{d}\mu' \tag{6.40}$$

with the initial condition

$$N^*_{n,n} = N_n = 2\int_0^1 u_n(\mu) i_n(-\mu)\mu\,\mathrm{d}\mu. \tag{6.41}$$

Here, as in the case of calculating the deep conditions in the parallel external flux problem, the function $I^{\mathrm{as}}_{j,n}(\tau,\mu)$ is merely defined by the radiation intensity at the layer surfaces.

6.9 Normalization of the Solution of the Milne Problem

Needless to say, choosing the normalization is, generally speaking, arbitrary. However, we will follow the normalization method that we have chosen for homogeneous atmospheres. Namely, the escape function $u(\mu)$ is normalized according to the condition (3.12), where $i(\mu)$ is the radiation intensity in an inhomogeneous infinite atmosphere with sources at minus infinity. Consequently, to write the normalization of the same type for multilayer media, we must introduce some *infinite* atmosphere that corresponds to the given one. This can be done as follows (see Sect. 1.7).

Let us imbed our semi-infinite atmosphere in an infinite one, in such a way that the optical properties satisfy the relations $\lambda(\tau) = \lambda(-\tau)$ and $\chi(\gamma,\tau) = \chi(\gamma,-\tau)$; see Fig. 6.4. The radiation intensity in the multilayer infinite medium built in such manner is denoted by $I^{\infty}_{1,n}(\tau,\mu)$ $(\mu \in [-1,+1])$. First we find the radiation intensity at the boundary between two half-spaces placed symmetrically.

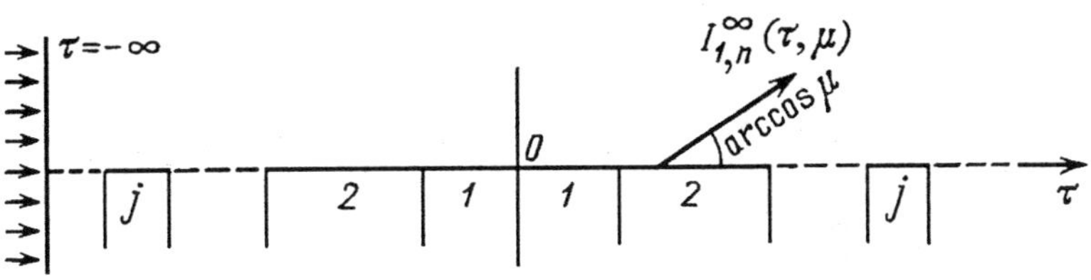

Fig. 6.4. The problem for two-sided infinity in a multilayer atmosphere

Using the same arguments as in deriving (3.10) and (3.11), we obtain the following system of equations which defines the function $I_{1,n}^{\infty}(0,\mu)$ ($\mu \in [0.1]$):

$$I_{1,n}^{\infty}(0,\mu) = M_{1,n}u_{1,n}(\mu) + 2\int_0^1 \rho_{1,n}(\mu',\mu)I_{1,n}^{\infty}(0,-\mu')\mu'\,\mathrm{d}\mu', \tag{6.42}$$

$$I_{1,n}^{\infty}(0,-\mu) = 2\int_0^1 \rho_{1,n}(\mu',\mu)I_{1,n}^{\infty}(0,\mu')\mu'\,\mathrm{d}\mu', \tag{6.43}$$

where the normalizing factor $M_{1,n}$ will be determined below. We assume that the function $I_{1,n}^{\infty}(0,\mu)$ is normalized according to the condition

$$\frac{\lambda(0)}{2}\int_{-1}^{+1} I_{1,n}^{\infty}(0,\mu)\,\mathrm{d}\mu = 1. \tag{6.44}$$

Normalization for the escape function is taken as

$$2\int_0^1 u_{1,n}(\mu)I_{1,n}^{\infty}(0,\mu)\mu\,\mathrm{d}\mu = 1. \tag{6.45}$$

This provides full consistency with the normalizations taken for a homogeneous medium (compare (6.44) and (6.45) with (2.7) and (3.12), respectively). Using (6.42), (6.43) and (6.45), we obtain

$$M_{1,n} = 2\int_{-1}^{+1} \left[I_{1,n}^{\infty}(0,\mu)\right]^2 \mu\,\mathrm{d}\mu. \tag{6.46}$$

It is obvious that the following relations are valid ($\tau \geq 0, \mu \in [-1,+1]$)):

$$I_{1,n}^{\infty}(\tau,\mu) = 2\int_0^1 I_{1,n}^{\infty}(0,\mu')I_{1,n}(\tau,\mu,\mu')\,\mathrm{d}\mu' + I_{1,n}^{\infty}(0,\mu)\mathrm{e}^{-\frac{\tau}{\mu}}\theta(\mu), \tag{6.47}$$

$$I_{1,n}^{\infty}(-\tau,\mu) = M_{1,n}I_{1,n}(\tau,-\mu)$$
$$+2\int_0^1 I_{1,n}^{\infty}(0,-\mu')I_{1,n}(\tau,-\mu,\mu')\,\mathrm{d}\mu' + I_{1,n}^{\infty}(0,\mu)\mathrm{e}^{\frac{\tau}{\mu}}\theta(-\mu). \tag{6.48}$$

Formulas (6.47) and (6.48) have the same physical meaning as (3.7) and (3.6). In particular, the relation (6.48) links the Milne to the parallel external flux problem in the case of the multilayer atmosphere. However, in this case we have an essential difference. For the homogeneous atmosphere the function $I_\infty(\tau, \mu)$ can be assumed to be given, whereas for the multilayer space the quantity $I_{1,n}^{\infty}(\tau, \mu)$ is, in general, unknown. This fact prevents us from using the relation (6.48) directly in order to obtain the solution to the Milne problem based on the known solution of the parallel external flux problem.

Thus, to solve the Milne problem in the multilayer atmosphere one needs:

1. to solve the parallel external flux problem for the chosen set of optical depths;
2. using the relations (6.33–36), to calculate successively the radiation intensities for all values of j from $j = n-1$ to $j = 1$ at the same depths (it is expedient to solve the two problems simultaneously on a computer);
3. to solve the system of equations (6.42) and (6.43) iteratively with allowance for the normalizations (6.44) and (6.45). The normalization of the escape function $U_{1,n}(\mu)$ found by means of (6.45) must be taken into account at all optical depths.

6.10 Solution of the General Problem

In order to make our presentation complete, we must also specify the algorithm for solving the general problem for a multilayer atmosphere. We remember that this algorithm for the homogeneous layer taken separately is described in Sect. 4.10.

We consider a plane atmosphere of optical thickness τ_0 which consists of an arbitrary number n of homogeneous layers ($\tau_0 = \sum_{j=1}^{n} \tau_j$). The optical properties of each layer are characterized by the optical thickness τ_j, single-scattering albedo λ_j and scattering phase function χ_j. The source function $g_j(t, \omega_1)$ ($t \in [0, \tau_j]$) is given for each separate layer. The quantity $g_j(t, \omega_1)dt$ is the radiation energy emitted by internal sources lying at the optical depth t in the layer j in the elementary volume of unit cross-section and optical thickness dt, this radiation being emitted in the direction ω_1 during the unit time interval ($t \in [0, \tau_j]$). We assume that $g_j(t, \omega_1)$ allows a polynomial representation, i.e.,

$$g_j(t, \omega_1) = \sum_{k=0}^{N_j} g_j^{(k)}(\omega_1) t^k. \tag{6.49}$$

The unknown function is the intensity $I_{1,n}^s(\tau, \omega)$ for any values of $\tau \in [0, \tau_0]$ and any direction of propagation ω.

Let $I_{j,n}^s(\tau, \omega)$ and $I_{j,n}(\tau, \omega, \omega_0)$ stand for the intensity of radiation in the general problem and in the parallel external flux problem, respectively; the radiation propagates at the depth τ in the direction ω in a multilayer medium consisting of $n - j + 1$ ($j = 1, 2, 3, \ldots, n$) homogeneous layers. In this case by definition $I_{n,n}^s(\tau, \omega) \equiv I_n^s(\tau, \omega)$ and $I_{n,n}(\tau, \omega, \omega_0) \equiv I_n(\tau, \omega, \omega_0)$.

Let us introduce the notation

$$\tau_j^* = \sum_{i=j}^{n} \tau_i, \tag{6.50}$$

i.e., τ_j^* is the total optical thickness of these layers which lie below the layer $j-1$. It is clear that $\tau_1^* = \tau_0$ is the optical thickness of the whole medium which consists of n layers, and $\tau \in [0, \tau_1^*]$.

We will start with considering, as before, a two-layer medium with internal sources consisting of the layers n and $n-1$. In such a medium, the upper layer $n-1$ is exposed from below to radiation of intensity $I^s_{n-1,n}(\tau_{n-1}, -\boldsymbol{\omega})$, and the lower layer n is exposed from above to radiation of intensity $I^s_{n-1,n}(\tau_{n-1}, \boldsymbol{\omega})$. Then, the generalized invariance principle tells us that for an arbitrary direction $\boldsymbol{\omega}$,

$$I^s_{n-1,n}(\tau, \boldsymbol{\omega}) = I^s_{n-1}(\tau, \boldsymbol{\omega}) + I^s_{n-1,n}(\tau_{n-1}, \boldsymbol{\omega}) \mathrm{e}^{\frac{\tau_{n-1}-\tau}{\mu}} \theta(-\mu)$$
$$+ \frac{1}{\pi} \int\limits_{(2\pi)} I^s_{n-1,n}(\tau_{n-1}, -\boldsymbol{\omega}') I_{n-1}(\tau_{n-1} - \tau, -\boldsymbol{\omega}, \boldsymbol{\omega}') \, \mathrm{d}\omega' \quad (\tau \in [0, \tau_{n-1}), \tag{6.51}$$

and

$$I^s_{n-1,n}(\tau_{n+1} + \tau, \boldsymbol{\omega}) = I^s_{n-1}(\tau, \boldsymbol{\omega}) + I^s_{n-1,n}(\tau_{n-1}, \boldsymbol{\omega}) \mathrm{e}^{-\frac{\tau}{\mu}} \theta(\mu)$$
$$+ \frac{1}{\pi} \int\limits_{(2\pi)} I^s_{n-1,n}(\tau_{n-1}, \boldsymbol{\omega}') I_n(\tau, \boldsymbol{\omega}, \boldsymbol{\omega}') \, \mathrm{d}\omega' \quad (\tau \in [0, \tau_n). \tag{6.52}$$

Here $\mathrm{d}\omega' = \mathrm{d}\mu' \mathrm{d}\varphi'$ ($\mu' \in [0,1]$, $\varphi' \in [0, 2\pi]$), and $\theta(\mu)$ is the unit step function: $\theta(\mu) = 1$ for $\mu \geq 0$; $\theta(\mu) = 0$ for $\mu < 0$.

Setting $\tau = \tau_{n-1}$ and $\tau = 0$ in (6.51) and (6.52) respectively, and introducing the reflection coefficient ρ_j by the relation ($\mu, \mu_0 > 0$)

$$I_j(0, -\boldsymbol{\omega}, \boldsymbol{\omega}_0) = \rho_j(\boldsymbol{\omega}, \boldsymbol{\omega}_0) \mu_0, \tag{6.53}$$

we obtain for $\mu > 0$:

$$I^s_{n-1,n}(\tau_{n-1}, \boldsymbol{\omega}) = I^s_{n-1}(\tau_{n-1}, \boldsymbol{\omega}) + \frac{1}{\pi} \int\limits_{(2\pi)} I^s_{n-1,n}(\tau_{n-1}, -\boldsymbol{\omega}') \rho_{n-1}(\boldsymbol{\omega}, \boldsymbol{\omega}') \mu' \, \mathrm{d}\omega', \tag{6.54}$$

$$I^s_{n-1,n}(\tau_{n-1}, -\boldsymbol{\omega}) = I^s_n(0, -\boldsymbol{\omega}) + \frac{1}{\pi} \int\limits_{(2\pi)} I^s_{n-1,n}(\tau_{n-1}, \boldsymbol{\omega}') \rho_n(\boldsymbol{\omega}, \boldsymbol{\omega}') \mu' \, \mathrm{d}\omega'. \tag{6.55}$$

Relations (6.54) and (6.55) form the system of integral equations with respect to $I^s_{n-1,n}(\tau_{n-1}, \boldsymbol{\omega})$ and $I^s_{n-1,n}(\tau_{n-1}, -\boldsymbol{\omega})$ (functions $I^s_{n-1}(\tau_{n-1}, \boldsymbol{\omega})$, $I^s_n(0, -\boldsymbol{\omega})$, $\rho_{n-1}(\boldsymbol{\omega}, \boldsymbol{\omega}')$ and $\rho_n(\boldsymbol{\omega}, \boldsymbol{\omega}')$ are assumed to be known). This set of equations can be solved numerically by successive approximations. Then the intensity at any optical depth in the two-layer atmosphere is found from (6.51) and (6.52).

Let us consider an arbitrary layer j. A multilayer medium that consists of $n-j$ layers and has the optical thickness τ^*_{j+1} lies below this layer. Then, in full analogy, we can obtain for the layer j ($\tau \in [0, \tau_j]$),

$$I^s_{j,n}(\tau, \boldsymbol{\omega}) = I^s_j(\tau, \boldsymbol{\omega}) + I^s_{j,n}(\tau_j, \boldsymbol{\omega})\mathrm{e}^{\frac{\tau_j - \tau}{\mu}}\theta(-\mu) + \frac{1}{\pi}\int\limits_{(2\pi)} I^s_{j,n}(\tau_j, -\boldsymbol{\omega}')I_j(\tau_j - \tau, -\boldsymbol{\omega}, \boldsymbol{\omega}')\,\mathrm{d}\omega', \tag{6.56}$$

while

$$I^s_{j,n}(\tau_j + \tau, \boldsymbol{\omega}) = I^s_{j+1,n}(\tau, \boldsymbol{\omega}) + I^s_{j,n}(\tau_j, \boldsymbol{\omega})\mathrm{e}^{-\frac{\tau}{\mu}}\theta(\mu) + \frac{1}{\pi}\int\limits_{(2\pi)} I^s_{j,n}(\tau_j, \boldsymbol{\omega}')I_{j+1,n}(\tau, \boldsymbol{\omega}, \boldsymbol{\omega}')\,\mathrm{d}\omega' \tag{6.57}$$

holds for all layers lying below ($\tau \in [0, \tau^*_{j+1}]$). It is obvious that if $j = n-1$, the expressions (6.56) and (6.57) turn into (6.51) and (6.52), respectively. From (6.56) and (6.57), we get the system of integral equations for the radiation intensity at the interlayer boundary ($\mu \geq 0$),

$$I^s_{j,n}(\tau, \boldsymbol{\omega}) = I^s_j(\tau_j, \boldsymbol{\omega}) + \frac{1}{\pi}\int\limits_{(2\pi)} I^s_{j,n}(\tau_j, -\boldsymbol{\omega}')\rho_j(\boldsymbol{\omega}, \boldsymbol{\omega}')\mu'\,\mathrm{d}\omega', \tag{6.58}$$

$$I^s_{j,n}(\tau, -\boldsymbol{\omega}) = I^s_{j+1,n}(0, -\boldsymbol{\omega}) + \frac{1}{\pi}\int\limits_{(2\pi)} I^s_{j,n}(\tau_j, \boldsymbol{\omega}')\rho_{j+1,n}(\boldsymbol{\omega}, \boldsymbol{\omega}')\mu'\,\mathrm{d}\omega', \tag{6.59}$$

where the reflection coefficient for the multilayer atmosphere, $\rho_{j,n}$, is substituted ($\mu, \mu_0 > 0$):

$$I^s_{j,n}(0, -\boldsymbol{\omega}, \boldsymbol{\omega}_0) = \rho_{j,n}(\boldsymbol{\omega}, \boldsymbol{\omega}_0)\mu_0. \tag{6.60}$$

Setting successively $j = n-1, n-2, \ldots, 1$ in (6.56–59) and taking into account that, by definition, $\rho_{n,n}(\boldsymbol{\omega}, \boldsymbol{\omega}_0) = \rho_n(\boldsymbol{\omega}, \boldsymbol{\omega}_0)$ we obtain the solution of the problem, i.e., the function $I^s_{1,n}(\tau, \boldsymbol{\omega})$ for all $\tau \in [0, \tau_0]$. At each stage, when the next layer is added, it is necessary to solve the system of integral equations (6.58–59). The number of such systems is $n-1$.

Thus, in order to apply the proposed algorithm, two problems must first be solved. The first is the parallel external flux problem for a multilayer atmosphere with $I_{j,n}(\tau, \boldsymbol{\omega}, \boldsymbol{\omega}_0)$ as its solution. The second is the determination of $I^s_j(\tau, \boldsymbol{\omega})$ for each layer. This has been done in Sect. 6.4 and Sect. 4.10.

Thus, the present chapter deals with an atmosphere that consists of an arbitrary number of homogeneous layers. We have developed here the algorithm for solving three typical problems for such a medium: the parallel external flux problem, the Milne problem, and the general problem. However, as emphasized in the preceding part, a number of problems related to these problems can be substantially simplified when optically thick layers (especially those for which scattering is close to conservative) are treated. However, before going on to this important problem we must consider an auxiliary one. The next chapter is aimed at solving this auxiliary problem.

7. Light Scattering in Two Adjacent Half-Spaces

An unbounded medium consisting of two homogeneous half-spaces is one of the simplest examples of a multilayer atmosphere. The study of the radiation transfer in such a medium is of considerable interest from the standpoint of the methods that can be used to derive asymptotic formulas for atmospheres consisting of homogeneous layers of large optical thickness. We will give detailed consideration to this subject in the next chapter, so this chapter bears, on the whole, auxiliary character.

7.1 Statement of the Problem and Main Equations

Let us consider the following generalization of the Milne problem. Let there be two adjacent homogeneous semi-spaces, i and j, the optical properties of each space determined by particles albedo and phase functions λ_i, $\chi_i(\gamma)$ and λ_j, $\chi_j(\gamma)$ respectively. The radiation field is generated by a source that is situated within the half-space i and infinitely far away from the boundary separating the half-spaces. We need to find the radiation field in this medium.

Let us measure the optical depth τ from the boundary and assume that $\tau < 0$ for the half-space i containing the source. Denote the intensity of radiation in the resulting unbounded medium by $\tilde{I}_{ij}(\tau, \mu)$ and accept that $\mu > 0$ for the radiation propagating at the direction that τ increases. The intensity of radiation at the boundary will be denoted by $i_{ij}(\mu)$, i.e.,

$$i_{ij}(\mu) = \tilde{I}_{ij}(0, \mu), \qquad (\mu \in [-1, +1]). \tag{7.1}$$

The indices i and j will be used to denote quantities relative to i and j half-spaces considered separately.

It is obvious that the following relations hold ($\tau \geq 0$, $\mu \in [-1, +1]$):

$$\tilde{I}_{ij}(-\tau, \mu) = I_i(\tau, -\mu) + i_{ij}(\mu) e^{\tau/\mu} \theta(-\mu) + 2 \int_0^1 I_i(\tau, -\mu, \mu') i_{ij}(-\mu') \, d\mu', \tag{7.2}$$

$$\tilde{I}_{ij}(\tau, \mu) = i_{ij}(\mu) e^{-\tau/\mu} \theta(\mu) + 2 \int_0^1 I_j(\tau, \mu, \mu') i_{ij}(\mu') \, d\mu'. \tag{7.3}$$

To derive (7.2) and (7.3), we use the same considerations as for (6.47) and (6.48). Note the following fact: setting the coefficient by the Milne intensity $I_i(\tau, -\mu)$ in (7.2) equal to one, we defined the normalization of $\tilde{I}_{ij}(\tau, \mu)$.

7.2 Radiation Intensity at the Boundary

Considering (7.2) and (7.3), we can see that to calculate the radiation field in two adjacent half-spaces one needs to find the intensity $i_{ij}(\mu)$ at the boundary (for each half-space considered separately, the problem is assumed to be solved). The function $i_{ij}(\mu)$ is determined by the following equations (they result from (7.2) and (7.3) at $\tau = 0$ ($\mu \in [0, 1]$)):

$$i_{ij}(\mu) = u_i(\mu) + 2\int_0^1 \rho_i^\infty(\mu, \mu')i_{ij}(-\mu')\mu' \,\mathrm{d}\mu', \tag{7.4}$$

$$i_{ij}(-\mu) = 2\int_0^1 \rho_j^\infty(\mu, \mu')i_{ij}(\mu')\mu' \,\mathrm{d}\mu'. \tag{7.5}$$

Substituting (7.5) into (7.4), we come to a single equation determining $i_{ij}(\mu)$ for $\mu \in [0, 1]$. Solving this equation, say by iteration technique and using (7.5), we can find $i_{ij}(\mu)$ for $\mu \in [-1, 0]$.

The intensity $i_{ij}(\mu)$ of radiation at the boundary can be regarded as a generalization of the intensity $u_i(\mu)$ in the standard Milne problem for the "naked" half-space i. The difference between $i_{ij}(\mu)$ and $u_i(\mu)$ is due to the fact that it is possible for photons to be reflected by the medium j. On the other hand, the function $i_{ij}(\mu)$ can also be considered as a generalization of the intensity $i_i(\mu)$ for a two layer medium. The difference between $i_{ij}(\mu)$ and $i_i(\mu)$ is due to the difference between the optical properties of the half-spaces. If these properties do not differ (i.e., $i = j$), then it follows from (7.4), (7.5), (3.10) and (3.11), that

$$i_{jj} = i_j/M_j. \tag{7.6}$$

Equations (7.4) and (7.5) lead to the following normalizing condition for $i_{ij}(\mu)$:

$$2\int_{-1}^{+1} i_{ij}(\mu)i_i(\mu)\mu \,\mathrm{d}\mu = 1. \tag{7.7}$$

To derive this, it is sufficient to multiply (7.4) by $2i_i(\mu)\mu$ and integrate it over μ from 0 to 1. Then, taking into account (3.10) and (3.11), we arrive at the condition (7.7) .

The function $i_{ij}(\mu)$ satisfies a number of useful relationships, each of them easily proved with the help of (7.4) and (7.5). Let us denote

$$\alpha_{ij} = \alpha = 2\int_0^1 i_{ji}(\mu)u_i(\mu)\mu \,\mathrm{d}\mu, \tag{7.8}$$

$$\beta_{ij} = 2\int_0^1 i_{ij}(-\mu)u_i(\mu)\mu \,\mathrm{d}\mu. \tag{7.9}$$

Permutating of the subscripts i and j in (7.4), multiplying by $2i_{ij}(\mu)\mu$ and integrating the result over μ, gives

$$\alpha = 2\int_{-1}^{+1} i_{ij}(\mu)i_{ji}(\mu)\mu \,\mathrm{d}\mu. \tag{7.10}$$

If we multiply (7.4) by $2i_i(-\mu)\mu$ and integrate it by μ, then, using (3.10), we obtain

$$M_i\beta_{ij} = N_i - N_{ij}, \tag{7.11}$$

where

$$N_{ij}=2\int_{-1}^{+1} i_{ij}(\mu)i_i(-\mu)\mu\,\mathrm{d}\mu. \tag{7.12}$$

And finally, multiplying (7.5) by $2i_j(-\mu)\mu$ and integrating over μ with use of (3.10), we have

$$2\int_{-1}^{+1} i_{ij}(\mu)i_j(\mu)\mu\,\mathrm{d}\mu = \alpha M_j. \tag{7.13}$$

The relation

$$\int_{-1}^{+1} i_{ij}(\mu)i_j(-\mu)\mu\,\mathrm{d}\mu = 0 \tag{7.14}$$

can be proved in the same way. If the properties of both half-spaces are identical (i.e., $i = j$), then it follows from (7.6), (7.8) and (7.9) that

$$\alpha_{jj} = M_j^{-1}; \qquad \beta_{jj} = N_j M_j^{-1}. \tag{7.15}$$

It should be stressed that (7.10) proves the existence of the symmetry with respect to the permutation of i and j (this is the reason why i and j are not used as indices to denote different half-spaces). The parameters α and β_{ij} are important for the problems to be considered in the next chapter.

It is not clear so far whether it is possible to obtain the function $i_{ij}(\mu)$ in the explicit form without using additional assumptions. Two cases where it can be done are to be considered below.

7.3 Isotropic Scattering

Let the phase function in both half spaces be spherical. We use well known formulas (see, for example, RT, Sects. 33, 38, LSPA, Chap. 2 and (3.36)):

$$\rho^{\infty}(\mu,\mu_0) = \frac{\lambda}{4}\frac{\varphi(\mu)\varphi(\mu_0)}{\mu+\mu_0}, \tag{7.16}$$

$$u(\mu) = \left(\frac{\lambda k N}{2M}\right)^{1/2}\frac{\varphi(\mu)}{1-k\mu}, \tag{7.17}$$

$$[\varphi(\mu)]^{-1} = (1-\lambda)^{1/2} + \frac{\lambda}{2}\int_0^1 \frac{\varphi(\mu')\mu'\,\mathrm{d}\mu'}{\mu+\mu'}. \tag{7.18}$$

We substitute (7.16) and (7.17) into (7.4) and (7.5). As a result, we find that the functions $i_{ij}(\mu)$ and $i_{ij}(-\mu)$ can be written as follows:

$$i_{ij}(\mu) = \lambda_i\varphi_i(\mu)v_{ij}(\mu), \quad i_{ij}(-\mu) = \lambda_j\varphi_j(\mu)v_{ij}(-\mu), \tag{7.19}$$

where

$$\upsilon_{ij}(\mu) = \frac{1}{\lambda_i}\left(\frac{\lambda_i k_i N_i}{2M_i}\right)^{1/2} \frac{1}{1-k_i\mu} + \frac{\lambda_j}{2}\int_0^1 \frac{\varphi_i(\mu')\varphi_j(\mu')}{\mu+\mu'}\upsilon_{ij}(-\mu')\mu'\,\mathrm{d}\mu', \tag{7.20}$$

$$\upsilon_{ij}(-\mu) = \frac{\lambda_i}{2}\int_0^1 \frac{\varphi_i(\mu')\varphi_j(\mu')}{\mu+\mu'}\upsilon_{ij}(\mu')\mu'\,\mathrm{d}\mu'. \tag{7.21}$$

Comparison of (7.20) and (7.21) with (7.18) suggests that the functions υ_{ij} may be represented in the following form

$$\upsilon_{ij}(\mu) = \frac{a_{ij}}{(1-k_i\mu)\varphi_j(\mu)}, \tag{7.22}$$

$$\upsilon_{ij}(-\mu) = \frac{a_{ij}}{(1+k_i\mu)\varphi_i(\mu)}. \tag{7.23}$$

Bearing in mind that

$$\frac{1}{(\mu+\mu')(1-k_i\mu')} = \frac{1}{1+k_i\mu}\left(\frac{1}{\mu+\mu'} + \frac{k_i}{1-k_i\mu'}\right), \tag{7.24}$$

$$\int_0^1 \varphi(\mu)\,\mathrm{d}\mu = \frac{2}{\lambda}(1-\sqrt{1-\lambda}), \tag{7.25}$$

substituting (7.22) and (7.23) into (7.21) and using (7.18), we can see that we have taken the correct form of the functions $\upsilon_{ij}(\mu)$.To find the coefficient a_{ij}, we substitute (7.22) and (7.23) into (7.20). By simple manipulations, we get, making allowance for (7.18) and (7.24), that

$$a_{ij} = \frac{1}{\lambda_i}\left(\frac{\lambda_i k_i N_i}{2M_i}\right)^{1/2} \varphi_j\left(\frac{1}{k_i}\right). \tag{7.26}$$

And, ultimately, we find that

$$i_{ij}(\mu) = c_{ij}\frac{\lambda_i\varphi_i(\mu)}{(1-k_i\mu)\varphi_j(\mu)}, \tag{7.27}$$

$$i_{ij}(-\mu) = c_{ij}\frac{\lambda_j\varphi_j(\mu)}{(1+k_i\mu)\varphi_i(\mu)}, \tag{7.28}$$

where

$$c_{ij} = \left(\frac{k_i N_i}{2\lambda_i M_i}\right)^{1/2} \varphi_j\left(\frac{1}{k_i}\right). \tag{7.29}$$

Inasmuch as the Ambartsumian function $\varphi(\mu)$ is known in explicit form and elaborately tabulated, the problem may be considered solved. Note that c_{ij} is a mean radiation intensity at the boundary between the half-spaces,

$$\frac{1}{2}\int_{-1}^{+1} i_{ij}(\mu)\,\mathrm{d}\mu = c_{ij}, \tag{7.30}$$

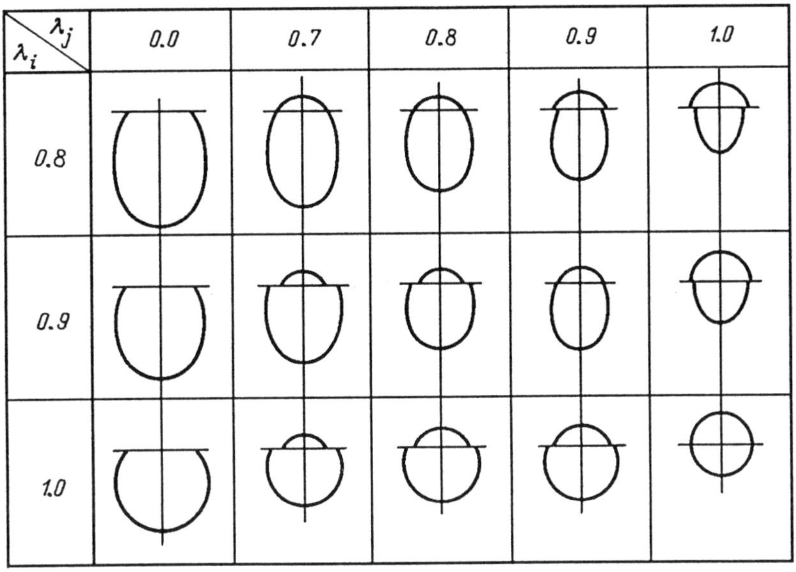

Fig. 7.1. Polar diagrams of radiation intensity at the boundary separating two adjacent half-spaces i and j for isotropic scattering

and the flux through the boundary

$$2\pi \int_{-1}^{+1} i_{ij}(\mu)\mu \,\mathrm{d}\mu = \frac{4\pi c_{ij}}{k_i}\sqrt{(1-\lambda_i)(1-\lambda_j)}. \tag{7.31}$$

The parameter α can be easily found with the help of (7.10), (7.27) and (7.28) giving

$$\alpha = 4\frac{\lambda_i - \lambda_j}{k_j - k_i} c_{ij} c_{ji}. \tag{7.32}$$

It can be shown that

$$N_{ij} = \frac{2(\lambda_i - \lambda_j)}{\lambda_i^2 k_i M_i} \frac{\varphi_j^2\left(\dfrac{1}{k_i}\right)}{\varphi_i^2\left(\dfrac{1}{k_i}\right)}, \tag{7.33}$$

which can be rewritten in the form

$$N_{ij} = N_i \frac{\lambda_i - \lambda_j}{\lambda_i} \varphi_j^2\left(\frac{1}{k_i}\right). \tag{7.34}$$

At $\lambda_j = 0$, it follows from (7.33) that

$$N_i = 2\left[\lambda_i k_i M_i \varphi_i^2\left(\frac{1}{k_i}\right)\right]^{-1}. \tag{7.35}$$

Equations (7.33) and (7.34) suggest that N_{ij} and N_{ji} will always be of opposite sign. Note, also, that

$$\alpha^2 = -\frac{4k_i k_j N_{ij} N_{ji}}{(k_i - k_j)^2 M_i M_j}. \tag{7.36}$$

This formula can be easily derived from (7.29), (7.32) and (7.33).

Fig. 7.1 shows the angular dependence of the intensity of radiation at the boundary separating the two half-spaces. In other words, it shows the polar diagrams of intensity both normalized to unit mean intensity (i.e., we took $c_{ij} = 1$ in (7.27) and (7.28)). The sources of radiation are situated in the upper half-space. When $\lambda_i \neq \lambda_j$, the intensity is discontinuous at $\mu = 0$.

7.4 Radiation Field for Nearly Conservative Scattering

Let us consider another special case where we have an explicit asymptotic expression for $i_{ij}(\mu)$. Suppose that both half-spaces are weakly absorbing ($1 - \lambda_i \ll 1,\ \ 1 - \lambda_j \ll 1$) and the phase functions in the half-spaces are arbitrary. Then,

$$i_{ij}(\mu) = \frac{a}{4}(1 + 3\kappa_j \mu + \ldots), \tag{7.37}$$

where

$$a = (\kappa_i + \kappa_j)^{-1}, \tag{7.38}$$

$$\kappa_i = \left(\frac{1 - \lambda_i}{3 - x_{1i}}\right)^{1/2}. \tag{7.39}$$

and x_{ij} stands for the i-th coefficient of the expansion of the phase function $\chi_j(\gamma)$ into series in terms of Legendre polynomials for the j-th layer of the atmosphere. The validity of the expansion can verified by substituting it into (7.4) and (7.5) making allowance for (3.13), (3.14) and (3.80).

Let us turn to the derivation of the asymptotic formulas for the constants α and N_{ij}. We rewrite (7.12) in a form

$$N_{ij} = 2\int_{-1}^{+1} i_{ij}(\mu)[1 - 3\kappa_i \mu + \kappa_i^2 i_2^{(i)}(\mu) - \kappa_i^{(3)} i_3^{(i)}(\mu) + \ldots]\mu\, \mathrm{d}\mu, \tag{7.40}$$

where, according to (2.40)

$$i_2^{(i)}(\mu) = (3 - x_{1i})\left[1 + \frac{10P_2(\mu)}{5 - x_{2i}}\right], \tag{7.41}$$

$$i_3^{(i)}(\mu) = 3\left\{\frac{2(3 - x_{1i})}{5 - x_{2i}}\left[\frac{7P_3(\mu)}{7 - x_{3i}} + \frac{\mu}{3 - x_{1i}}\right] + \frac{3}{2}(2 - x_{1i})\mu\right\}. \tag{7.42}$$

Similarly, (7.7) can be written in the form

$$1 = 2\int_{-1}^{+1} i_{ij}(\mu)[1 + 3\kappa_i\mu + \kappa_i^2 i_2^{(i)}(\mu) + \kappa_i^3 i_3^{(i)}(\mu) + \ldots]\mu\, \mathrm{d}\mu. \tag{7.43}$$

Adding and subtracting (7.40) and (7.43) term-by-term, we get

$$1 - N_{ij} = 12\kappa_i \int_{-1}^{+1} i_{ij}(\mu)\mu^2\, \mathrm{d}\mu + 4\kappa_i^3 \int_{-1}^{+1} i_{ij}(\mu) i_3^{(i)}(\mu)\mu\, \mathrm{d}\mu, \tag{7.44}$$

$$1 + N_{ij} = 4\int_{-1}^{+1} i_{ij}(\mu)\mu\, \mathrm{d}\mu + 4\kappa_i^2 \int_{-1}^{+1} i_{ij}(\mu) i_2^{(i)}(\mu)\mu\, \mathrm{d}\mu. \tag{7.45}$$

Substituting (7.37), (7.42) and (7.41) into the corresponding terms in (7.44) and (7.45) and performing integration by μ, we get

$$1 - N_{ij} = 12\kappa_i \int_{-1}^{+1} i_{ij}(\mu)\mu^2\, \mathrm{d}\mu + ab_i\kappa_i^3 + \ldots, \tag{7.46}$$

$$1 + N_{ij} = 4\int_{-1}^{+1} i_{ij}(\mu)\mu\, \mathrm{d}\mu + 2a\kappa_i^2\kappa_j c_i + \ldots, \tag{7.47}$$

where

$$b_i = 3(2 - x_{1i}) + \frac{4(3 - x_{1i})}{5 - x_{2i}}, \tag{7.48}$$

$$c_i = \frac{(3 - x_{1i})(9 - x_{2i})}{5 - x_{2i}}. \tag{7.49}$$

Proceeding the same way, the integrals (7.13) and (7.14) give us the following expressions

$$\alpha M_j = 12\kappa_j \int_{-1}^{+1} i_{ij}(\mu)\mu^2\, \mathrm{d}\mu + ab_j\kappa_j^3 + \ldots, \tag{7.50}$$

$$\alpha M_j = 4\int_{-1}^{+1} i_{ij}(\mu)\mu\, \mathrm{d}\mu + 2ac_j\kappa_j^3 + \ldots. \tag{7.51}$$

Eliminating the integrals from (7.46), (7.50) and (7.47), (7.51), we arrive at the following system to determine α and N_{ij}

$$\frac{1 - N_{ij}}{\kappa_i} - \frac{\alpha M_j}{\kappa_j} = a(b_i\kappa_i^2 - b_j\kappa_j^2) + \ldots, \tag{7.52}$$

$$1 + N_{ij} - \alpha M_j = 2a\kappa_j(\kappa_i^2 c_i - \kappa_j^2 c_j) + \ldots. \tag{7.53}$$

Solving the system of equations (7.52) and (7.53), we find

$$\alpha = \frac{a}{4}\left[1 - \frac{a}{2}(b_j\kappa_j^3 + b_i\kappa_i^3) - a(c_j\kappa_j + c_i\kappa_i)\kappa_i\kappa_j\right] + \ldots, \tag{7.54}$$

$$N_{ij} = 1 - 2a\kappa_i - a^2\kappa_i\kappa_j[(b_i\kappa_i^2 - b_j\kappa_j^2) - 2(c_i\kappa_i^2 - c_j\kappa_j^2)] + \ldots. \tag{7.55}$$

In deriving (7.54) we used the asymptotic expression (2.42) for M_j. Brief note: it follows from (7.55) and(7.38) that $N_{ij} = -N_{ji}$ in the approximation taken.

Ultimately, using (7.11) and asymptotic formulas (7.55), (2.42) and (3.82), we get

$$8\beta_{ij} = 2a - 6\gamma_i + 18\gamma_i^2\kappa_i - a^2(b_i\kappa_i^3 + b_j\kappa_j^3) - 2a^2(c_i\kappa_i^3 - c_j\kappa_j^3) - 4a^2c_i\kappa_i^2\kappa_j + \dots . \tag{7.56}$$

Note that the expansions (7.50) and (7.51) together with (7.54) make it possible to get the asymptotic formulas for the flux of radiation and K integral at the boundary separating layers.

7.5 Radiation Field Away from the Boundary

Let us obtain the intensity away from the boundary. If we consider deep layers of the half-space j, which does not include sources, the intensity can be described by the following asymptotic formula ($\tau \gg 1$):

$$\tilde{I}_{ij}^{\mathrm{as}}(\tau, \mu) = i_j(\mu)\alpha \mathrm{e}^{-k_j\tau}. \tag{7.57}$$

This expression follows from (7.3) and (7.8) when we take into account (3.57). For $\lambda_i < 1$ the intensity of radiation in the interior of the half-space i, which contains a source is

$$M_i\tilde{I}_{ij}^{\mathrm{as}}(-\tau, \mu) = i_i(\mu)\mathrm{e}^{k_i\tau} - i_i(-\mu)\mathrm{e}^{-k_i\tau}N_{ij}. \tag{7.58}$$

This formula is derived from (7.2) taking into account (7.9) and (7.11) and asymptotic formulas (3.57), (3.58). If $\lambda_i \equiv 1$, then

$$\tilde{I}_{ij}^{\mathrm{as}}(-\tau, \mu) = \frac{3}{4}\left[\left(1 - \frac{x_{1i}}{3}\right)\tau + \gamma_{ij} + \mu\right], \tag{7.59}$$

where

$$\gamma_{ij} = 2\int_{-1}^{+1} i_{ij}(\mu)\mu^2\,\mathrm{d}\mu. \tag{7.60}$$

The above expression is derived from (7.58) by the passage to the limit $k_i \to 0$ allowing for (7.46).

The relationship (7.58) has, like (3.58), a clear physical meaning. The presence of the second term in the right side of (7.58) is determined by the difference between the optical properties of the half-space i and the half-space j ($i = j, N_{jj} \equiv 0$). For $N_{ij} > 0$, the outflow of photons through the boundary $\tau = 0$ is compensated less than it would be in a homogeneous medium. For $N_{ij} < 0$, the perturbation caused by the difference between the properties of the half-spaces i and j will be positive since the reflection from the region $\tau < 0$ is more effective that in a homogeneous infinite medium. Hence, it becomes clear why the quantities N_{ij} and N_{ji} are always of opposite sign.

Note that we have already encountered a similar problem in Sect. 5.5.1 when considering the Milne problem with reflection. The surface there played the role of the half-space j and the formulas corresponding to (7.58) and (7.59) (see (5.28) and (5.33), respectively) were derived there.

8. Atmosphere Consisting of Layers with Large Optical Thickness

The equations derived in Chap. 6 are valid for layers of arbitrary optical thickness. However, when τ_j ($j = 1, 2, \ldots, n$) increases (especially if it occurs simultaneously in all layers), it becomes more difficult to solve these equations. In order to describe the internal radiation field in detail, one needs to take an enormous number of equations and, consequently, to store a lot of information on computer. Therefore, it is important that one can derive simple asymptotic formulas for the quantities that describe the radiation field in an atmosphere consisting of layers with large optical thickness ($\tau_j \gg 1$). Accuracy of these formulas grows with increasing τ_j. It turns out to be possible to express analytically these quantities in terms of functions describing the radiation field in the corresponding semi-infinite layers and intensities at interlayer boundaries. The present chapter deals with such asymptotic formulas.

8.1 Parallel External Flux Problem: General Discussion

Let us consider this problem, which was originally formulated in Sect. 6.1, for the case when all atmospheric layers are optically thick (i.e., $\tau_j \gg 1$, $j = 1, 2, \ldots, n$). It suffices to consider only the zeroth azimuthal harmonics of the radiation intensity over the whole atmosphere excepting the upper layer.

As follows from the formulas given in Sect. 6.4, the key point in solving the parallel external flux problem is to obtain the function $I_{j,n}(\tau_j, \mu, \mu_0)$, the intensity of the radiation at the boundary separating the upper layer from the rest of the atmosphere. Let us try to use physical arguments in order to shed light on the general structure of this function.

There is only one way for photons to reach the boundary at the depth $\tau = \tau_j$: they diffuse through the optically thick atmospheric layer. As result, they completely "forget" the initial directions of their velocities, and the angular variables in the expression for $I_{j,n}$ become separable. The function $u_j(\mu_0)\mu_0$ obviously has to be a proportionality factor depending on μ_0 (cf. (3.57)).

The structure of the second factor, that depends on μ, can be established with the aid of the following arguments. The boundary $\tau = \tau_j$ lies between two *optically thick* layers of property j and $j+1$. So, the photons emitted by the source at the boundary $\tau = 0$ should be asymptotically distributed over all angles, proportionally to the function $i_{j,j+1}(\mu)$. $i_{j,k}(\mu)$ is the intensity of radiation between two adjacent homogeneous half-spaces with properties j and k (see Sect. 7.1). On the other hand, we also have the "negative" sources, which are caused by the layers $j+2$, $j+3, \ldots, n$, and a vacuum. Owing to these sources, the angular structure of radiation intensity at the boundary $\tau = \tau_j$ should apparently be $i_{j+1,j}(-\mu)$. Summarizing, one concludes that the function $I^{\mathrm{as}}_{j,n}(\tau_j, \mu, \mu_0)$ is to be sought in the form

$$I_{j,n}(\tau_j, \mu, \mu_0) = [a_{j,n} i_{j,j+1}(\mu) - c_{j,n} i_{j+1,j}(-\mu)]\, u_j(\mu_0)\mu_0, \tag{8.1}$$

where the angular independent factors $a_{j,n}$ and $c_{j,n}$ are to be determined.

Now we proceed to the a-priori determination of the structure of the reflection, $\rho^{\mathrm{as}}_{j,n}(\mu, \mu_0)$, and transmission, $\sigma^{\mathrm{as}}_{j,n}(\mu, \mu_0)$, coefficients for the multilayer atmosphere. As is known (see Sect. 4.8.1), for an optically thick homogeneous layer j the following asymptotic formulas hold:

$$\rho^{\mathrm{as}}_j(\mu, \mu_0) = \rho^{\infty}_j(\mu, \mu_0) - u_j(\mu)u_j(\mu_0)r_j, \tag{8.2}$$

$$\sigma^{\mathrm{as}}_j(\mu, \mu_0) = u_j(\mu)u_j(\mu_0)t_j. \tag{8.3}$$

Here both the reflection, r_j and transmission, t_j, factors can be expressed in terms of the corresponding asymptotic constants by means of (4.46). Superscript "as" in both coefficients designates that a layer, or collection of layers, is optically thick. Superscript ∞ is used to denote the quantities that refer to a semi-infinite medium.

The reflection coefficient $\rho^{\mathrm{as}}_{j,n}$ can be divided into two parts. The first one is determined by the photons reflected by the layer j (i.e., those photons that never pass the boundary $\tau = \tau_j$ from above). It is clear that this part is equal to $\rho^{\mathrm{as}}_j(\mu, \mu_0)$. The other part, $\rho^{\mathrm{as}}_{j,n}$, is determined by the photons passing the boundary from below and diffusing through an optically thick layer. Therefore, this part is proportional to $u_j(\mu)$. Taking the symmetry of the reflection coefficient with respect to μ and μ_0 into account, one writes

$$\rho^{\mathrm{as}}_{j,n}(\mu, \mu_0) = \rho^{\mathrm{as}}_j(\mu, \mu_0) + u_j(\mu)u_j(\mu_0)r^*_{j,n}.$$

Using (8.2), we finally obtain

$$\rho^{\mathrm{as}}_{j,n}(\mu, \mu_0) = \rho^{\infty}_j(\mu, \mu_0) - u_j(\mu)u_j(\mu_0)r_{j,n}, \tag{8.4}$$

where $r_{j,n} = r_j - r^*_{j,n}$, the light reflection factor of a stratified medium, needs to be found.

The structure of the transmission coefficient can be determined with the help of the same physical arguments as those used after the derivation of (4.51) (see Sect. 4.8.1).

The transmission of light through an optically thick multilayer atmosphere can be divided into three stages. The first is the "pumping" of external radiation into deep layers. It takes place at the illuminated upper boundary.

After photons have passed through the upper j-th layer, angular relaxation occurs, and the function $\sigma^{\text{as}}_{j,n}(\mu,\mu_0)$ gains the factor $u_j(\mu_0)$ (cf. (4.51)). The second stage is the diffusion of photons through internal atmospheric layers. This results in the appearance of the transmission factor $t_{j,n}$. The third stage is a departure from the optically thick surface layer n, which results in the formation of the final angular structure of the transmitted radiation and introducing the factor $u_n(\mu)$. So, we have

$$\sigma^{\text{as}}_{j,n}(\mu,\mu_0) = u_n(\mu)u_j(\mu_0)t_{j,n}. \tag{8.5}$$

Here one should emphasize the following two points. First, upon comparing (8.3) with (8.5), one sees that the symmetry of the transmission coefficient with respect to μ and μ_0 does not result in the multilayer atmosphere having the same property: $\sigma^{\text{as}}_{j,n}(\mu,\mu_0) \neq \sigma^{\text{as}}_{j,n}(\mu_0,\mu)$. The reasons for this have already been discussed in Sect. 6.3. Second, it follows from (8.5) that the *relative* angular structure of the intensity of the radiation that leaves a multilayer medium is solely determined by the properties of the lowest layer.

Of course, the results represented by (8.1), (8.4) and (8.5) cannot be regarded as rigorous. To check their validity, one must substitute them into the exact equations which are given in Sect. 6.1. This substitution will also allow us to determine $a_{j,n}$, $c_{j,n}$, $r_{j,n}$, and $t_{j,n}$, which is the task to be undertaken in the next section.

8.2 Asymptotic Formulas for Intensity at a Boundary and for Transmission and Reflection Coefficients

Our aim is to find $a_{j,n}$ and $c_{j,n}$, which enter into (8.1) and into the reflection, $r_{j,n}$ and transmission, $t_{j,n}$ factors. Let us assume, first, that all the layers of the atmosphere are absorbing, i.e., $\lambda_j \neq 1$ ($j = 1,2,3,\ldots,n$). Substituting (8.1–4) into the rigorous relationships (6.14) and (6.15) and neglecting the first term on the right-hand side of (6.15), we obtain the system of equations determining $a_{j,n}$ and $c_{j,n}$. The solutions of the system are

$$a_{j,n} = \frac{(1 + r_{j+1,n}\beta_{j+1,j})t_j}{(1 + r_j\beta_{j,j+1})(1 + r_{j+1,n}\beta_{j+1,j}) - r_j r_{j+1,n}\alpha^2_{j,j+1}}, \tag{8.6}$$

$$c_{j,n} = \frac{r_{j+1,n}\alpha_{j,j+1}}{1 + r_{j+1,n}\beta_{j+1,j}} a_{j,n}, \tag{8.7}$$

where $\alpha_{i,k}$ and $\beta_{i,k}$ are determined by (7.8) and (7.9) respectively, and considered to be known. So, we still must find $r_{j,n}$ and $t_{j,n}$. To do this, we discard the second term on the right-hand side of (6.16) and substitute (8.1–4) into the resultant expression. Allowing for (8.6) and (8.7), and performing simple manipulations, we find the following recursion formula ($j = n-1, n-2, \ldots, 2, 1$):

$$r_{j,n} = r_j - \cfrac{t_j^2}{r_j + \cfrac{1}{\beta_{j,,j+1} - \cfrac{\alpha_{j,j+1}^2}{\beta_{j+1,,j} + r_{j+1,n}^{-1}}}} \tag{8.8}$$

with the initial condition $r_{n,n} \equiv r_n$.

Formula (8.6) can be substantially simplified with the use of (8.8). Precisely, bearing in mind (8.8), we have

$$\beta_{j,,j+1} - \alpha_{j,j+1}^2 \left(\beta_{j+1,,j} + r_{j+1,j}^{-1}\right)^{-1} = \left(r_j - r_{j,n}\right)\left(r_j r_{j,n} + t_j^2 - r_j^2\right)^{-1}. \tag{8.9}$$

Substituting (8.9) into (8.6), we find

$$a_{j,n} = \left(r_j r_{j,n} + t_j^2 - r_j^2\right) t_j^{-1}, \tag{8.10}$$

i.e., $a_{j,n}$ can be directly expressed in terms of the reflection and transmission factors only.

Now, let us proceed to determine the transmission factor $t_{j,n}$. By discarding the first two terms on the right-hand side of (6.17) and then substituting (8.1) and (8.5) into the expression obtained, we have

$$t_{j,n} = \frac{c_{j,n}}{r_{j+1,n}} t_{j+1,n}. \tag{8.11}$$

From (8.11) we easily obtain

$$t_{j,n} = t_n \prod_{i=j}^{n-1} \frac{c_{i,n}}{r_{i+1,n}}. \tag{8.12}$$

Hence, we have all the quantities necessary to find the radiation intensity, namely $I_{j,n}(\tau_j, \mu, \mu_0)$ ((8.1), (8.6), (8.7) and (8.8)), the reflection coefficient $\rho_{j,n}(\mu, \mu_0)$ ((8.4), (8.8)), and the transmission coefficient $\sigma_{j,n}(\mu, \mu_0)$ ((8.5), (8.12), (8.7), (8.6) and (8.8)). The finding of the above quantities implies that the problem of determining the radiation intensities $i_{j,j+1}(\mu)$ and $i_{j+1,j}(\mu)$ ($j = 1, 2, 3, \ldots, n$) at the boundary of two adjacent half-spaces is solved. If all layers are media which scatter light nearly conservatively ($(1 - \lambda_j) \ll 1$), the situation is trivial since the asymptotic expression for the function $i_{j,k}(\mu)$ is given by the simple formula (7.37).

Let us consider now the simplifications taking place in the evaluation of the radiation field in the medium.

8.3 Radiation Field in a Medium

To calculate the radiation intensity averaged over the azimuthal angle, instead of Eqs (6.12) and (6.13), we have

$$I_{j,n}(\tau,\mu,\mu_0) = I_j(\tau,\mu,\mu_0) + I_{j,n}(\tau_j,\mu,\mu_0)\mathrm{e}^{\frac{\tau_j-\tau}{\mu}}\theta(-\mu)$$
$$+2\int_0^1 I_{j,n}(\tau_j,-\mu',\mu_0)I_j(\tau_j-\tau,-\mu,\mu')\,\mathrm{d}\mu' \qquad (\tau\in[0,\tau_j]), \quad (8.13)$$

$$I_{j,n}(\tau_j+\tau,\mu,\mu_0) = I_{j,n}(\tau_j,\mu,\mu_0)\mathrm{e}^{-\frac{\tau}{\mu}}\theta(\mu)$$
$$+2\int_0^1 I_{j,n}(\tau_j,\mu',\mu_0)I_{j+1,n}(\tau,\mu,\mu')\,\mathrm{d}\mu' \qquad (\tau\in[0,\tau_{j+1}^s]), \quad (8.14)$$

and the initial condition $I_{n,n} = I_n(\tau,\mu,\mu_0)$. Taking into consideration (4.45) and that the layers are of the considerable optical thickness, we obtain

$$I_{j,n}(\tau,\mu,\mu_0) = I_j^{\infty}(\tau,\mu,\mu_0) - J_j(\tau,\mu)u_j(\mu_0)\mu_0, \quad (8.15)$$

where

$$J_j(\tau,\mu) = I_j(\tau,\mu)r_j - [I_j(\tau_j-\tau,-\mu) - I_j^{\mathrm{as}}(\tau_j-\tau,-\mu)]t_j. \quad (8.16)$$

Formula (8.1) can be rewritten as follows

$$I_{j,n}(\tau_j,\mu,\mu_0) = W_{j,n}(\mu)u_j(\mu_0)\mu_0, \quad (8.17)$$

where

$$W_{j,n}(\mu) = a_{j,n}i_{j,j+1}(\mu) - c_{j,n}i_{j+1,j}(-\mu). \quad (8.18)$$

Hence, one can, in place of (8.13) and (8.14), write

$$I_{j,n}(\tau,\mu,\mu_0) = I_j^{\infty}(\tau,\mu,\mu_0) - \Bigg[J_j(\tau,\mu) - W_{j,n}(\mu)\mathrm{e}^{\frac{\tau_j-\tau}{\mu}}\theta(-\mu)$$
$$-2\int_0^1 W_{j,n}(-\mu')I_j(\tau_j-\tau,-\mu,\mu')\,\mathrm{d}\mu'\Bigg]u_j(\mu_0)\mu_0 \quad (\tau\in[0,\tau_j]), \quad (8.19)$$

$$I_{j,n}(\tau_j+\tau,\mu,\mu_0) = \Bigg[W_{j,n}(\mu)\mathrm{e}^{-\frac{\tau}{\mu}}\theta(\mu)$$
$$+2\int_0^1 I_{j+1,n}(\tau,\mu,\mu')W_{j,n}(\mu')\,\mathrm{d}\mu'\Bigg]u_j(\mu_0)\mu_0, \quad (\tau\in[0,\tau_{j+1}^s]). \quad (8.20)$$

Thus, for all the layers except the upper one, the angular variables are separable, which substantially simplifies the calculation of the internal radiation field.

The above treatment is applicable only to the azimuthally averaged radiation intensity. In order to find other azimuthal harmonics ($m \geq 1$) one should deal with the upper layer only. It is suggested that the latter one is semi-infinite. Thus, the final solution for the problem of the internal radiation field is given by the following formula (cf. (6.17a)):

$$I_{1,n}(\tau, \mu, \mu_0, \varphi - \varphi_0) = I_{1,n}(\tau, \mu, \mu_0) + 2 \sum_{m=1}^{N_1} I_1^m(\tau, \mu, \mu_0) \cos m(\varphi - \varphi_0). \tag{8.21}$$

If all the layers are optically thick and the lowest layer is a semi-infinite, calculation of the radiation field in the deep layers via (6.20) is simplified. Indeed, in this case we have, instead of (6.21)

$$U_{j,n}(\mu_0) = 2 \int_0^1 U_{j+1,n}(\mu') I_{j,n}(\tau_j, \mu', \mu_0) \mu' \, \mathrm{d}\mu'. \tag{8.22}$$

Substituting (8.1) into (8.22) and performing simple manipulations we find

$$U_{j,n}(\mu_0) = u_j(\mu_0) \mu_0 T^*_{j,n}, \tag{8.23}$$

where

$$T^*_{j,n} = \prod_{i=j}^{n-1} \frac{c_{i,n}}{r_{i+1,n}}. \tag{8.24}$$

So, instead of (6.20), we can finally write ($\tau - \tau_j^* \gg 1$),

$$I^{\mathrm{as}}_{j,n}(\tau, \mu, \mu_0) = i_n(\mu) u_j(\mu_0) \mu_0 \mathrm{e}^{-k_n(\tau - \tau_j^*)} T^*_{j,n}. \tag{8.25}$$

8.4 Conservatively Scattering Atmosphere

It follows from (7.54) and (7.56), and (7.38) and (7.39) that if for two adjacent half-spaces with the properties i and k, $\lambda_i \to 1$ and $\lambda_k \to 1$, then the asymptotic constants α and β go to infinity as

$$\left[\left(\frac{1 - \lambda_i}{3 - x_{1j}} \right)^{1/2} + \left(\frac{1 - \lambda_k}{3 - x_{1k}} \right)^{1/2} \right]^{-1}.$$

Hence, calculating the reflection factor by means of the formula (8.8), we encounter indeterminacy. Handling the problem with the help of (7.54) and (7.56), we have

$$\frac{1}{r^{(0)}_{j,n}} = \frac{1}{r^{(0)}_{j+1,n}} + \frac{1}{r^{(0)}_j} - \frac{3}{4}(\gamma_j + \gamma_{j+1}). \tag{8.26}$$

Here we take into account that for conservative scattering, $r_j^{(0)} = t_j^{(0)}$. Hereafter, the superscript "0" denotes that the corresponding quantity is taken for conservative scattering. If all the layers of the medium scatter light conservatively, then, using (4.55) and the recursion relation (8.26), we find

$$r_{j,n}^{(0)} = \frac{4}{3}\left[\sum_{i=j}^{n}\left(1-\frac{x_{1i}}{3}\right)\tau_i + \gamma_j + \gamma_n\right]^{-1} . \tag{8.27}$$

We see that this formula includes the constants γ which relate to the lower and the upper layers.

Note the following consequence of (8.27). If the lower layer is semi-infinite ($\tau_n = \infty$), (8.27) then $r_{1n}^{(0)} = 0$ and, consequently, $\rho_{j,n}^{\text{as}}(\mu,\mu_0) = \rho_j^{\infty}(\mu,\mu_0)$. Thus, if the scattering is conservative, then the reflection coefficient for a semi-infinite medium consisting of layers of considerable optical thickness does not differ from that for a *homogeneous* semi-infinite atmosphere with the properties identical to those of the upper layer. Though strange at first sight, this can easily be explained on the physical level. The reflection factor $r_{j,n}$ in (8.4) can be either positive or negative. Everything depends on the degree that the downward outflux of photons through the boundary between j-th and $j+1$-th layers is compensated by the upward influx, as compared with the situation in a homogeneous half-space with properties j. We will see later that the radiation intensity in a multilayer atmosphere at the boundary specified above will be exactly the same as in the semi-infinite homogeneous atmosphere with the properties j (see (8.28) when $\tau \to \infty$). This has the result that $r_{j,n}^{(0)} = 0$.

Let us consider the connection between the reflection factor $r_{j,n}$ and transmission factor $t_{j,n}$ for the case of conservative scattering. Going to the limit of $\lambda_j \to 1$ in (8.10), we find $a_{j,n}^{(0)} = r_{j,n}^{(0)}$. It follows thus from (8.7) that $c_{j,n}^{(0)} = r_{j,n}^{(0)}$. Using (8.12), we have $t_{j,n}^{(0)} = r_{j,n}^{(0)}$. In other words, like in the case of the homogeneous and optically thick layer, the reflection factor and transmission factor coincide for conservative scattering in a multilayer, optically thick atmosphere.

The expression (8.1) also becomes simplified. Namely, using the asymptotic expressions (7.37), (7.54), (7.56) and (8.7), and taking the conservative limit, we obtain

$$I_{j,n}^{(0)}(\tau_j,\mu,\mu_0) = r_{j,n}^{(0)}\left(\frac{1}{r_{j+1,n}^{(0)}} - \frac{3}{4}\gamma_{j+1} + \frac{3}{4}\mu\right)u_j^{(0)}(\mu_0)\mu_0,$$

from where with the help of (8.27) we ultimately find:

$$I_{j,n}^{(0)}(\tau_j,\mu,\mu_0) = \frac{\sum\limits_{i=j+1}^{n}\left(1-\dfrac{x_{1i}}{3}\right)\tau_i + \gamma_n + \mu}{\sum\limits_{i=j}^{n}\left(1-\dfrac{x_{1i}}{3}\right)\tau_i + \gamma_j + \gamma_n} u_j^{(0)}(\mu_0)\mu_0 . \tag{8.28}$$

In particular, if the properties of all layers are the same (i.e., the atmosphere is homogeneous), expression (8.28) takes then the form

$$I_{j,n}^{(0)}(\tau_j,\mu,\mu_0) = \frac{\left(1-\frac{x_{1j}}{3}\right)(\tau_0 - \tau j) + \gamma_j + \mu}{\left(1-\frac{x_{1j}}{3}\right)\tau_0 + 2\gamma_j} u_j^{(0)}(\mu_0)\mu_0 .$$

It coincides with putting $\tau \gg 1$ and $(\tau_0 - \tau) \gg 1$ into (4.60).

Let us now consider (8.25). For the pure scattering, it follows from (8.24) and (8.27) that $T_{j,n}^*$ takes the form

$$T_{j,n}^* = \left[1 + \frac{\sum\limits_{i=j}^{n-1}\left(1-\frac{x_{1i}}{3}\right)\tau_i + \gamma_j - \gamma_n}{\left(1-\frac{x_{1n}}{3}\right)\tau_n + 2\gamma_n}\right]^{-1} . \tag{8.29}$$

Therefore $T_{j,n}^* \to 1$ as $\tau_n \to \infty$. Hence, we can, in place of (8.25), write

$$I_{j,n}^{\mathrm{as}}(\tau,\mu,\mu_0) = u_j^{(0)}(\mu_0)\mu_0 , \tag{8.30}$$

which coincides with the result following from (8.28) for $\tau_n \to \infty$. It can be shown that the above conclusion is valid for all $\tau > \tau_j$ and for an arbitrary number of layers, i.e., if a stratified atmosphere is semi-infinite and the scattering is pure, the radiation that occurs in deep layers of the atmosphere becomes completely isotropic.

8.5 The Milne Problem

As in the case of solving the parallel external flux problem, we intend to find the asymptotic expression for the radiation intensity at the boundary between the upper layer and the layer below. It turns out that for all $\mu \in [-1, +1]$,

$$I_{j,n}(\tau,\mu) = \left[\frac{1 + r_j\beta_{j,j+1}}{r_j\alpha_{j,j+1}} i_{j+1,j}(-\mu) - i_{j,j+1}(\mu)\right] A_{j,n} , \tag{8.31}$$

where

$$A_{j,n} = \frac{t_{j+1}}{r_{j+1}}\frac{r_j}{t_j}\frac{c_{j,n}}{r_{j+1,n}} A_{j+1,n} . \tag{8.32}$$

The validity of (8.31) can be checked by substituting it into (6.35) and (6.36) and using (7.4), (7.5), (7.8), (7.9), (8.2), (8.4), (8.6), (8.7). Substitution of (8.31) into (6.37) yields

$$u_{j,n}(\mu) = \frac{t_j}{r_j} A_{j,n} u_j(\mu) , \tag{8.33}$$

from which it follows that $t_n A_{n,n}/r_n = 1$. Using this condition, instead of (8.32), we have

$$A_{j,n} = \frac{r_j}{t_j} T^*_{j,n}, \tag{8.34}$$

where $T^*_{j,n}$ is determined by (8.24). Thus,

$$u_{j,n}(\mu) = T^*_{j,n} u_j(\mu), \tag{8.35}$$

i.e., the relative angular distribution of the intensity of the radiation emerging from the multilayer atmosphere is completely determined by the optical properties of the upper layer j ($\tau_j \gg 1$).

Formula (8.31), together with (6.33), (6.34) and asymptotic formulas (8.15), (8.16), (3.57), allow us to express $I_{1,n}(\tau, \mu)$ ($\tau \geq 0$) in terms of $I_j^{\infty}(\tau, \mu, \mu_0)$ explicitly. To do this, one should have the auxiliary quantities $i_{j,j+1}(\mu)$, $i_{j+1,j}(\mu)$ and $u_j(\mu)$, which depend only on the single angular variable ($j = 1, 2, 3, \ldots, n-1$). The final expressions are rather cumbersome, therefore we will not give them here.

Let us consider now the simplifications taking place in the case under consideration in the formula (6.39), which determines the radiation field at considerable depths in the lower layer n. Substituting (8.31) into (6.40) and using (8.23), (8.34), (7.8) and (7.9) we have

$$N^*_{j,n} = N^*_{j+1,n} - M_n T^*_{j,n} T^*_{j+1,n} p_j, \tag{8.36}$$

where

$$p_j = \frac{(1 + r_j \beta_{j,j+1})\beta_{j+1,j} - r_j \alpha^2_{j,j+1}}{t_j \alpha_{j,j+1}}, \tag{8.37}$$

with $T^*_{n,n} = 1$. Taking (6.41) into account, we easily find

$$N^*_{j,n} = N_n - M_n \sum_{i=j}^{n-1} T^*_{i,n} T^*_{i+1,n} p_i. \tag{8.38}$$

Equation (6.39) describes the intensity of radiation in the lower layer, where the quantity $N^*_{j,n}$ can be found with the help of (8.38).

Note that in obtaining the above results, the normalization of the solution of the Milne problem was not taken into account.

If the scattering is conservative and $\lambda_j = 1$ ($j = 1, 2, 3, \ldots, n$) for each layer, the formulas (8.31) and (8.32) become much more simple. In this case, as we have already seen in the previous section, $T^*_{j,n} = 1$ and we have the following formula

$$u_{j,n}(\mu) = u_j^{(0)}(\mu) \tag{8.39}$$

instead of (8.35). Moving to the conservative limit results in the following expression

$$I^{\text{as}}_{j,n}(\tau, \mu) = \frac{1}{r_j^{(0)}} - \frac{3}{4}(\gamma_j + \mu). \tag{8.40}$$

Using (4.55), we find

$$I^{\mathrm{as}}_{j,n}(\tau,\mu) = \frac{3}{4}\left[\left(1-\frac{x_{1j}}{3}\right)\tau_j + \gamma_j - \mu\right]. \tag{8.41}$$

This formula completely coincides with that in the conservative Milne problem for deep layers of homogeneous half-space (cf. (8.41) and (3.61)). Note that there is no dependence of the radiation intensity on the optical properties of the layers underneath $(j+1, j+2, \ldots, n)$.

Proceeding from (8.37), it is not difficult to derive the corresponding expression for $p_j^{(0)}$. It turns out that

$$p_j^{(0)} = \frac{1}{r_j^{(0)}} - \frac{3}{4}(\gamma_j + \gamma_{j+1}). \tag{8.42}$$

Using (4.55), we have

$$p_j^{(0)} = \frac{3}{4}\left[\left(1-\frac{x_{1j}}{3}\right) + \gamma_j - \gamma_{j+1}\right]. \tag{8.43}$$

Taking into account that $T^*_{j,n} = 1$ and using (8.38) and (8.43), and corresponding expressions for the asymptotic constants M_n, N_n and function $i_n(\mu)$, we find in the conservative limit that

$$I^{\mathrm{as}}_{j,n}(\tau,\mu) = \frac{3}{4}\left[\sum_{i=j}^{n-1}\left(1-\frac{x_{1i}}{3}\right)\tau_i + \left(1-\frac{x_{1n}}{3}\right)(\tau-\tau^*) + \gamma_j - \mu\right],$$
$$(\tau-\tau^* \gg 1), \tag{8.44}$$

at large depths in the n-th layer. Here τ^* is determined by (6.19). In this case the radiation field at large depths is, in contrast to the parallel external flux problem (see (8.30)), determined by the properties of the layers constituting the atmosphere.

8.6 Normalization of the Milne Problem

Our intention now is to normalize the solutions found in the previous section. Let us follow the approach described in Sect. 6.9 and find the radiation intensity $I^{\infty}_{1,n}(0,\mu)$ at the boundary between two adjacent multilayer half-spaces. This intensity obeys the system of equations (6.42) and (6.43). It can be easily checked that for the situation in question, i.e., when the layers are optically thick, we have

$$I^{\infty}_{1,n}(0,\mu) = \frac{M_1 + r_{1,n}N_1}{M_1(1-N_1)r_{1,n}} i_1(\mu) - \frac{r_{1,n}}{M_1(1-N_1)r_{1,n}} i_1(-\mu), \tag{8.45}$$

and

$$u_{1,n}^{\mathrm{as}}(\mu) = \left(1 - \frac{1-N_1}{M_1} r_{1,n}\right) u_1(\mu), \tag{8.46}$$

$$M_{1,n} = M_1 \frac{M_1 + (1+N_1) r_{1,n}}{M_1 - (1-N_1) r_{1,n}}. \tag{8.47}$$

Note that (8.45) does not change its form only in the case when layer "1" has a large optical thickness and the others are arbitrary. In this case, however, $r_{1,n}$ cannot be found with the help of (8.8). If the half-space is homogeneous, then $r_{1,n} \equiv 0$ and (8.45–47) yield the standard result.

Hence, the escape function $u_{1,n}^{\mathrm{as}}(\mu)$ normalized in the aforementioned manner, is determined by (8.46), where $r_{1,n}$ can be found with the use of the recursion relation (8.8). On the other hand, (8.35) provides the expression for the same quantity, but not normalized (here one should put $j = 1$). Therefore, the factor $C_{1,n}$, that normalizes the radiation intensity in the Milne problem for an atmosphere consisting of layers of large optical thickness becomes equal to

$$C_{1,n} = \left(1 - \frac{1-N_1}{M_1} r_{1,n}\right) (T_{1,n}^*)^{-1}. \tag{8.48}$$

This quantity has a clear physical meaning. Let us consider a homogeneous medium with the properties "1". Then, $r_{1,n} = 0$, $\alpha_{j,j+1} = M_1^{-1}$, and $\beta_{j,j+1} = \beta_{j+1,j} = N_1 M_1^{-1}$. Using (4.46), (8.6–8), it can be seen that (8.24) results in $T_{1,n}^* = \mathrm{e}^{-k_1 \tau^*}$. τ^* is the optical thickness of the layers above the n-th one, so we have $C_{1,n} = \mathrm{e}^{k_1 \tau^*}$. As known from Sect. 3.2, within the Milne problem, the removal of a layer of optical thickness τ^* from the surface results in increasing the intensity at all depths by $\mathrm{e}^{k_1 \tau^*}$. Here we *add* a layer of thickness τ^* to the initial layer "1". In other words, we *decrease* the radiation intensity by $\mathrm{e}^{-k_1 \tau^*}$. This is the reason why the normalizing factor appears to be equal to $\mathrm{e}^{k_1 \tau^*}$.

For a stratified atmosphere, the factor $C_{1,n}$ could be taken into account at the first stage of the solution by introducing it into (6.26) (to do this, we would have had to find $r_{1,n}$ and $T_{1,n}^*$ beforehand). As a result, the first term in the right hand side of (6.31) would have been $C_{1,n} u_n(\mu)$ instead of $u_n(\mu)$. So after solving the Milne problem we would have had the normalized result. This treatment shows that the factor $C_{1,n}$ is the number of times by which the radiation intensity increases at all optical depths (for a fixed μ) in the multilayer Milne problem after removing the $(n-1)$-th layer from the atmospheric surface.

Since $r_{j,n}^{(0)} = 0$ and $T_{j,n}^* = 1$ for conservative scattering, then the normalizing factor $C_{1,n}^{(0)} = 1$. Thus, there is no need for any normalization in this case.

8.7 A Two-Layer Atmosphere: Basic Formulas

Winding up this chapter, let us consider an atmosphere that consists of two layers of large optical thickness, which is the simplest example of stratified atmosphere. We will also pay attention to both the situation when the atmosphere is adjacent to an isotropically reflecting surface and that with conservative scattering. A more general situation with the atmosphere consisting of two layers with an arbitrary optical thickness, has already been studied in Sect. 6.2.

8.7.1 A Two-Layer Semi-Infinite Atmosphere

Let a semi-infinite atmosphere consist of a layer with the optical thickness $\tau_1 \gg 1$. Also, $\lambda = \lambda_1$ and the phase function $\chi(\gamma) = \chi_1(\gamma)$. Let this layer lie within the half-space with $\lambda = \lambda_2$ and $\chi(\gamma) = \chi_2(\gamma)$.

According to (8.4), the reflection coefficient for such an atmosphere is determined by the formula

$$\rho_{1,2}(\mu, \mu_0) = \rho_1^\infty(\mu, \mu_0) - u_1(\mu)u_1(\mu_0)r_{1,2}^\infty, \tag{8.49}$$

where the superscript "∞" refers to a semi-infinite atmosphere. Because $r_2 = 0$, with the use of (8.8), we have

$$r_{1,2}^\infty = r_1 - \frac{t_1^2\beta_{1,2}}{1 + r_1\beta_{1,2}}. \tag{8.50}$$

Using (4.46) and (7.11), we find

$$r_{1,2}^\infty = \frac{N_{1,2}}{N_1}R_{1,2}, \tag{8.51}$$

where

$$R_{1,2} = \frac{r_1}{1 + r_1\beta_{1,2}} = N_1\mathrm{e}^{-k_1\tau_1}T_{1,2} = \frac{M_1N_1\mathrm{e}^{-2k_1\tau_1}}{1 - N_1N_{1,2}\mathrm{e}^{-2k_1\tau_1}}. \tag{8.52}$$

It follows from (8.6), (8.7) and (8.52) that in the case in question, $c_1 = 0$ and $a_{1,2} = T_{1,2}$. Hence, according to (8.1), we have at the interlayer boundary

$$I_{1,2}(\tau_1, \mu, \mu_0) = i_{1,2}(\mu)u_1(\mu_0)\mu_0T_{1,2}. \tag{8.53}$$

Now we proceed with the consideration of (8.19) and (8.20) which describe the radiation field at all depths. Using (8.15–18) and the expression (7.2) that determines the radiation field in two adjacent half-spaces, we obtain, instead of (8.19), the following ($\tau \in [0, \tau_1]$):

$$I_{1,2}(\tau, \mu, \mu_0) = I_1^\infty(\tau, \mu, \mu_0) - \{I_1(\tau, \mu)r_{1,2}^\infty$$
$$-[\tilde{I}_{1,2}(\tau - \tau_1, \mu) - \tilde{I}_{1,2}^{\mathrm{as}}(\tau - \tau_1, \mu)]T_{1,2}\}u_1(\mu_0)\mu_0. \tag{8.54}$$

Here $\tilde{I}_{1,2}(\tau,\mu)$ is the intensity in two adjacent half-spaces. $\tilde{I}^{\mathrm{as}}_{1,2}(\tau,\mu)$ is determined by (7.58). One sees that the structure of (8.54) is similar to that of (8.15). At $\tau = \tau_1$, (8.54) results in (8.53). The expression (8.54) becomes substantially simplified if $\tau \gg 1$ and $(\tau_1 - \tau) \gg 1$. In this case we derive

$$I^{\mathrm{as}}_{1,2}(\tau,\mu,\mu_0) = \tilde{I}^{\mathrm{as}}_{1,2}(\tau - \tau_1,\mu)u_1(\mu_0)\mu_0 T_{1,2}. \tag{8.55}$$

Let us now consider (8.20). Using (7.3), (8.17) and (8.18), we get ($\tau \geq 0$),

$$I_{1,2}(\tau_1 + \tau,\mu,\mu_0) = \tilde{I}_{1,2}(\tau,\mu)u_1(\mu_0)\mu_0 T_{1,2}. \tag{8.56}$$

In particular, taking (7.57) into account, for large enough depths ($\tau \gg 1$) in a semi-infinite layer, we find

$$I^{\mathrm{as}}_{1,2}(\tau_1 + \tau,\mu,\mu_0) = i_2(\mu)u_1(\mu_0)\mu_0 \mathrm{e}^{-k_2\tau}\alpha T_{1,2}. \tag{8.57}$$

This expression can be derived in an alternative way proceeding from (8.25).

Thus, for the two-layer semi-infinite atmosphere, the radiation intensity at all optical depths is determined by (8.54) and (8.56).

8.7.2 Optically Thick Layers

The main arguments used to derive the basic formulas remain the same as those in Sect. 8.7.1. Using (4.46) and (7.11), from (8.8) we obtain

$$r_{1,2} = r^{\infty}_{1,2} + \frac{\alpha^2 R_{2,1} T^2_{1,2}}{1 - \alpha^2 R_{1,2} R_{2,1}}. \tag{8.58}$$

$r^{\infty}_{1,2}$ is determined by (8.51). Substituting (8.58) into (8.10) yields

$$a_{1,2} = \frac{T_{1,2}}{1 - \alpha^2 R_{1,2} R_{2,1}}. \tag{8.59}$$

On the other hand, (8.59) and (8.7) give

$$c_{1,2} = \frac{\alpha T_{1,2} R_{2,1}}{1 - \alpha^2 R_{1,2} R_{2,1}}, \tag{8.60}$$

Upon substituting (8.60) into (8.12), one finds

$$t_{1,2} = \frac{\alpha T_{1,2} T_{2,1}}{1 - \alpha^2 R_{1,2} R_{2,1}}. \tag{8.61}$$

Hence, the reflection and transmission factors for an optically thick two-layer atmosphere are determined by (8.58) and (8.61), respectively. Notice that R_{ik} and T_{ik} are determined by (8.52).

Now, by means of (8.15–19), we can find the formula determining the radiation field inside the upper layer. After rather tiresome manipulations one can obtain for $\tau \in [0,\tau_1]$,

$$I_{1,2}(\tau,\mu,\mu_0) = I_1^{\infty}(\tau,\mu,\mu_0) - \{I_1(\tau,\mu)r_{1,2} - [\tilde{I}_{1,2}(\tau-\tau_1,\mu)$$

$$-\tilde{I}_{1,2}^{\mathrm{as}}(\tau-\tau_1,\mu)]a_{1,2} + [\tilde{I}_{2,1}(\tau_1-\tau,-\mu) - \tilde{I}_{2,1}^{\mathrm{as}}(\tau_1-\tau,-\mu)]c_{1,2}\}u_1(\mu_0)\mu_0. \tag{8.62}$$

$\tilde{I}_{ik}(\tau,\mu)$ and $\tilde{I}_{ik}^{\mathrm{as}}(\tau,\mu)$ are determined by (7.2), (7.3) and (7.57), (7.58), respectively. Far enough from both boundaries (i.e., $\tau \gg 1$, $\tau_1-\tau \gg 1$), the expression (8.62) takes the very simple form:

$$\begin{aligned} I_{1,2}^{\mathrm{as}}(\tau,\mu,\mu_0) &= [a_{1,2}\tilde{I}_{1,2}^{\mathrm{as}}(\tau-\tau_1,\mu) - c_{1,2}\tilde{I}_{2,1}^{\mathrm{as}}(\tau_1-\tau,-\mu)]u_1(\mu_0)\mu_0 \\ &= \left[i_1(\mu)\mathrm{e}^{-k_1\tau_1} - I_1^{\mathrm{as}}(\tau,\mu)r_{1,2}\right]u_1(\mu_0)\mu_0. \end{aligned} \tag{8.63}$$

Let us now find the radiation field in the lower layer. Using (8.20) and taking into account (8.15–18), we obtain, for $\tau \in [0,\tau_2]$,

$$\begin{aligned} I_{1,2}(\tau_1+\tau,\mu,\mu_0) = \{&a_{1,2}\tilde{I}_{1,2}(\tau,\mu) - c_{1,2}\tilde{I}_{2,1}(-\tau,-\mu) \\ &+[I_2(\tau_2-\tau,-\mu) - I_2^{\mathrm{as}}(\tau_2-\tau,-\mu)]t_{1,2}\}u_1(\mu_0)\mu_0. \end{aligned} \tag{8.64}$$

In particular, if $\tau=\tau_2$, (8.64) results in (8.5), and for $\tau \gg 1$ and $\tau_2-\tau \gg 1$ we have

$$I_{1,2}^{\mathrm{as}}(\tau_1+\tau,\mu,\mu_0) = [a_{1,2}\tilde{I}_{1,2}^{\mathrm{as}}(\tau,\mu) - c_{1,2}\tilde{I}_{2,1}^{\mathrm{as}}(-\tau,-\mu)]u_1(\mu_0)\mu_0. \tag{8.65}$$

Expressions for the intensity in the first and second layers look quite similar (cf. (8.63–65)). The only difference between them is that the signs of the spatial argument τ in the functions $\tilde{I}_{ik}(\tau,\mu)$ at the respective coefficients $a_{1,2}$ and $c_{1,2}$ are opposite.

The formulas derived can easily be generalized to the case when the lower atmospheric layer is adjacent to an isotropically reflecting surface with a given albedo A. To do this, it is sufficient in (8.8) to substitute $r_{2,2} \equiv r_2$ with $\bar{r}_2$, which is determined by (5.40), and set $I_{2,2}(\tau,\mu,\mu') = \bar{I}_2(\tau,\mu,\mu')$ in (8.20). This function is given by (5.42). As usual, the overbar means that the corresponding quantity is related to the case when there is a light-reflecting surface.

As a result, it turns out that for the above generalization it is sufficient to substitute $R_{2,1}$ with

$$\bar{R}_{2,1} = \bar{N}_2\mathrm{e}^{-k_2\tau_2}\bar{T}_{2,1} = \frac{M_2\bar{N}_2\mathrm{e}^{-2k_2\tau_2}}{1-\bar{N}_2N_{2,1}\mathrm{e}^{-2k_2\tau_2}} \tag{8.66}$$

and $T_{2,1}$ with $\bar{T}_{2,1}$. Equation (8.64) takes the following form

$$\begin{aligned} \bar{I}_{1,2}(\tau_1+\tau,\mu,\mu_0) = \{&\bar{a}_{1,2}\tilde{I}_{1,2}(\tau,\mu) - \bar{c}_{1,2}\tilde{I}_{2,1}(-\tau,-\mu) \\ &+[\bar{I}_2(\tau_2-\tau,-\mu) - \bar{I}_2^{\mathrm{as}}(\tau_2-\tau,-\mu)]\bar{t}_{1,2}\}u_1(\mu_0)\mu_0, \end{aligned} \tag{8.67}$$

where $\bar{I}_2(\tau,\mu)$ is the radiation intensity in the Milne problem with reflection (see Sect. 5.5.1). Note that quantities $\bar{r}_{1,2}$, $\bar{a}_{1,2}$, $\bar{c}_{1,2}$, and $\bar{t}_{1,2}$ are determined by (8.58–61) where $R_{2,1}$ and $T_{2,1}$ are substituted by $\bar{R}_{2,1}$ and $\bar{T}_{2,1}$, respectively.

8.7.3 Conservative Scattering

Equations (8.58–61) are not applicable for conservative scattering since they lead to a 0/0 type indeterminacy. To avoid this problem, we have to rewrite the expressions for $R_{1,2}$ and $T_{1,2}$ as

$$\alpha R_{1,2} = \frac{r_1}{\alpha^{-1} + r_1 \beta_{1,2}\alpha^{-1}}, \tag{8.68}$$

$$\alpha T_{1,2} = \frac{t_1}{\alpha^{-1} + r_1 \beta_{1,2}\alpha^{-1}}. \tag{8.69}$$

It follows from (7.54) and (7.56) that

$$\beta_{1,2}\alpha^{-1} = 1 - 3\gamma_1(\kappa_1 + \kappa_2) + \delta_{1,2}^{(1)}(\kappa_1 + \kappa_2) + \dots, \tag{8.70}$$

where

$$\delta_{1,2}^{(1)} = 9\gamma_1^2\kappa_1 + a(c_1\kappa_1^2 - c_2\kappa_2^2). \tag{8.71}$$

As a result, (8.68–70) yield

$$(\alpha R_{1,2})^{-1} = 1 + \left(\frac{4}{r_1} - 3\gamma_1 + \delta_{1,2}^{(1)}\right)(\kappa_1 + \kappa_2) + \dots, \tag{8.72}$$

$$(\alpha T_{1,2})^{-1} = \frac{r_1}{t_1}\left[1 + \left(\frac{4}{r_1} - 3\gamma_1 + \delta_{1,2}^{(1)}\right)(\kappa_1 + \kappa_2)\right] + \dots. \tag{8.73}$$

To derive the expansions of functions $(\alpha\bar{T}_{2,1})^{-1}$ and $(\alpha\bar{R}_{2,1})^{-1}$ from (8.72) and (8.73), it suffices to make a permutation of the indices and put $r_2 = \bar{r}_2$ and $t_2 = \bar{t}_2$. Note that it follows from (8.72) and (8.71),

$$\begin{aligned}(\alpha^2 R_{1,2}\bar{R}_{2,1})^{-1} - 1 = \big[&4(r_1^{-1} + \bar{r}_2^{-1}) - 3(\gamma_1 + \gamma_2) + 9(\gamma_1^2\kappa_1 + \gamma_2^2\kappa_2) \\ &+ (4r_1^{-1} - 3\gamma_1)(4\bar{r}_2^{-1} - 3\gamma_2)(\kappa_1 + \kappa_2)\big](\kappa_1 + \kappa_2) + \dots. \end{aligned} \tag{8.74}$$

Then, passing to the conservative limit (λ_1 and $\lambda_2 \to 1$) and using (8.72) and (8.73), from (8.58) and (8.61) we find

$$\bar{r}_{1,2}^{(0)} = \bar{t}_{1,2}^{(0)} = \frac{4}{3}\left[\left(1 - \frac{x_{11}}{3}\right)\tau_1 + \left(1 - \frac{x_{12}}{3}\right)\tau_2 + \gamma_1 + \bar{\gamma}_2\right]^{-1}. \tag{8.75}$$

If we set $\gamma_n = \gamma_n$ in (8.27), then it turns out to be applicable to the case of a multilayer atmosphere being adjacent to a light reflecting surface. So (8.75) is the particular case of the expression (8.27).

It can be easily proved that $\bar{a}_{1,2}^{(0)} = \bar{c}_{1,2}^{(0)} = \bar{r}_{1,2}^{(0)}$ and

$$\bar{a}_{1,2} - \bar{c}_{1,2} = 3\bar{r}_{1,2}^{(0)}\left[\left(1 - \frac{x_{12}}{3}\right) + \bar{\gamma}_2\right](\kappa_1 + \kappa_2) + \dots. \tag{8.76}$$

As a result, from (8.67) and (7.37) we find the following expression for the radiation intensity at the interlayer boundary:

$$\bar{I}_{1,2}(\tau_1, \mu, \mu_0) = \frac{3}{4}\left[\left(1 - \frac{x_{12}}{3}\right)\tau_2 + \bar{\gamma}_2 + \mu\right]\bar{r}_{1,2}^{(0)}u_1^{(0)}(\mu_0)\mu_0. \tag{8.77}$$

With the use of (5.33), (8.77) is rewritten more simply as

$$\bar{I}_{1,2}(\tau_1, \mu, \mu_0) = \bar{r}_{1,2}^{(0)}\bar{I}_0^{\rm as}(\tau_2, -\mu)u_1^{(0)}(\mu_0)\mu_0. \tag{8.78}$$

Of course, the expressions (8.75) and (8.77) could be derived directly from the general relationships (8.27) and (8.28). However, the above way is chosen because we considered it useful to provide the asymptotic expressions (8.72–74). They may turn out to be applicable in order to derive the formulas for the intensity in a two-layer atmosphere for nearly conservative scattering. In which one should use the asymptotic formula (5.52). This determines the radiation field in an optically thick atmosphere adjacent to an isotropically reflecting surface. Although the derivation of the corresponding expressions meets with no difficulty, they appear to be rather cumbersome, so we omit them here.

Bibliographical Comments and Additions to Part II

Scientists began the study of light scattering in multilayer atmospheres with the simplest case of a two layer atmosphere. In this respect one should first mention the paper by Gutshabash [138]. Here the author studied the rigorous separation of angular variables in the problem of the radiative transfer in a two layer, semi-infinite atmosphere with isotropic scattering. It was shown for the first time that for a *homogeneous* medium, the angular variables of the intensity *inside* the medium get separated. For the particular case of an atmospheric boundary this fact was, as we know, established by Ambartsumian (see (3.36)). In a way, this work anticipated the results obtained later by Mullikin [139], Danielian and Mnatsakanian [140] and the author [104, 141]. One should also note the series of papers by Kolesov [142]–[145]. The principal aspiration of the above works is to the rigorous analytical separation of angular variables.

Takashima [146, 147] is known to be the first to exploit the idea of adding layers to study the polarized radiation in multilayer media.

The text of Chaps. 6 and 8 is mainly based on two papers by Dlugach and the author [148, 149]. The work of these chapters was considerably stimulated by the appearance of the paper by Germogenova and Konovalov [137] as well as those by Ivanov [124, 150]. The above works deal mostly with finding the intensity at the boundaries of layers whose optical thickness is assumed to be very large. The Milne problem was not considered in those works. The main content of Chap. 6 is the generalization of results from [137, 124, 150] for arbitrary optical depth. As for Chap. 8, mostly it includes the results of the works [137, 150].

The method of adding layers, in which the boundary is assigned to an arbitrary place in the multilayer medium, was exploited in the works by Lacis and Hansen [151], and Viik [152]. The latter author essentially developed this method in his subsequent work [153]. At the beginning of the algorithm for calculations in [152, 153], he used the invariance relationships derived in [51].

As for the direction of building up the layers, the author pointed out in 1963 [154] the expediency of introducing the inverse counting of optical depths. This method was also used by Viik [153].

Sect. 6.10 is written on the basis of the paper by the author and Dlugach [132]. See also the monograph by Viik [155], Sects. 2.5 and 3.8 on this matter, Chap. 5 of the book [155] contains the results of some calculations.

The explicit expressions for the radiation intensity at the boundary of two adjacent isotropically scattering half-spaces (see Chap. 7) have been known for a long time. To derive these expressions, Davison [156] used the Winer-Hopf method (also see [21] Chap. 8), while Chandrasekhar proceeded from the invariance relationships [157]. Respective formulas were also found by Romanov [158] as well as Mendelson and Summerfield [159] (also see [45], Sect. 6.5). The main formulas of Sect. 7.3 can also be found in the works [124, 137], which we have already mentioned. At the end of the paper [124] the author provides an elaborate table of quantities N_{ij}, which are determined by (7.33) or (7.34) for different λ_i and λ_j. Figure 7.1 is taken from [124].

The asymptotic formulas for constants α, β_{ik}, and N_{ik} within an accuracy to terms of order κ^2 are presented here for the first time (see Sect. 7.4).

Special attention should be paid to the monograph by Viik [155], since it sums up his data [152, 153] and provides programs as well as numerous results of calculations (including those for continuous inhomogeneity). He also considered the case of internal sources of radiation. The work [155] contains additional bibliographic sources.

Devaux et al. [160] used the F_N method to calculate the radiation field in a multilayer medium. Saad et al. also considered the radiation transfer in a multilayer medium in [161].

The asymptotic formulas for a two-layer atmosphere were also studied in [162, 163]. In the latter work the following effect of the separation of parameters for nearly conservative scattering was described: the dependence of the radiation intensity on depth in the lower layer does not depend on the optical properties of upper layers. This fact substantially facilitates the processing of corresponding measures (see also Sect. 14.2.2). The works by Konovalov [53] and [164, 165] deal with the numerical realization of the methods of the asymptotic theory of multilayer media and with estimation of errors of corresponding asymptotic formulas. The algorithm, method and results of computation of the intensity of radiation leaving a multilayer atmosphere are described in the paper by Dlugach [166].

One should add that Stamnes and Conklin [167] designed an algorithm for the calculation of the radiation field in a multilayer medium, which is based on the method of discrete ordinates.

And finally, the paper by Cogley [168] should be mentioned, where the matrix-vector method has been worked out to solve the problem of radiative transfer in a plane medium consisting of an unrestricted number of inhomogeneous layers; the possibility of the presence of internal isotropic delta-sources of heat in the medium was also taken into consideration. The fact that in [168] the equations of invariant imbedding are used for calculations in a single inhomogeneous layer allowed Sharma et al. [169] to estimate errors caused by the replacement of inhomogeneous layer with a number of an homogeneous layers and subsequent use of the addition method.

Part III

ATMOSPHERE WITH CONTINUOUSLY VARYING PARAMETERS

* * *

In the theory of radiative energy transfer, phase function $\chi(\gamma)$ (γ is the angle of scattering) and probability of photon survival are usually supposed to be independent of optical depth τ. In other words, it is supposed that the light scattering takes place in a homogeneous medium. Usually, this does not hold for planetary and stellar atmospheres. For instance, when studying radiation transfer in a stellar atmosphere (in spectral line), one should take into account the depth dependence of the ratio of scattering coefficient in line to the absorption coefficient in continuum.

The fact that concentration and dimensions of aerosol particles, their phase state, chemical composition of gases, etc. depend on height in planetary atmospheres results in the phase function and the single scattering albedo depending substantially on height. The same is the case with other natural and artificial media. Hence, the model of atmosphere that implies $\lambda = \lambda(\tau)$ and $\chi(\gamma) = \chi(\gamma, \tau)$ (the optical properties of the medium are characterized by certain functions of optical depth) is much closer to reality.

Of course, this generalization of the model of the atmosphere has the result that the theory becomes much more complicated. For example, for a homogeneous medium the coefficient of reflection of light by the atmosphere is determined by Ambartsumian's nonlinear integral equation, whereas for a non-homogeneous medium one has to consider a nonlinear integro-differential equation of first order. The derivative appears here because one needs to take into account that the optical properties vary with depth. Besides, the fact that the functions λ and χ now depend on τ (in contrast to a homogeneous atmosphere) results in a number of parameters, which, in turn, make calculations of the radiation field in media more complicated. Therefore, one would think that substantial development of radiation transfer theory in inhomogeneous media is not possible unless a particular form of functions $\lambda(\tau)$ and $\chi(\gamma, \tau)$ is chosen. This is only true to a certain extent, for the following two reasons. Firstly, the formulation of the generalized invariance principle for plane atmospheres is actually identical for both homogeneous and inhomogeneous media (for details, see Sect. 1.12), and secondly, if a photon undergoes multiple scattering in a homogeneous or inhomogeneous medium, then angular relaxation occurs. The photon "forgets" its initial direction of movement, which results in the separation of variables, which, in turn, simplifies the problem. For angular relaxation it is only important that the photon should undergo a sufficiently large number of scatterings. The other factors, for instance the geometry of the medium, have no influence on the process. So, both reasons mentioned above have the result that the transfer theory in an inhomogeneous medium sometimes turns out to be very similar to that in a homogeneous one.

As we have already said, this part of the book deals with the development of radiation transfer theory in an inhomogeneous medium as far as it is possible *without imposing too strong restrictions on the behavior of functions* $\lambda(\tau)$ *and* $\chi(\gamma, \tau)$. A more detailed idea about the content can be gained with the help of the Table of Contents.

9. Diffuse Reflection and Transmission of Light by Atmospheres

From now on we commence a systematical study of radiation fields in an inhomogeneous atmosphere whose optical properties depend on only one spatial coordinate, namely, optical depth. In the beginning we will consider the simpler problem of how to determine the transmission and reflection coefficients for a plane layer of inhomogeneous atmosphere. So, in other words, we will be interested in the intensity of the radiation diffusely transmitted and reflected by the medium (see Sect. 1.9). This problem is one of the most important in astrophysics. Observing celestial bodies, we mostly deal with either the intensity of the radiation diffusely reflected by their surface (planets, gaseous and dust nebula) or the intensity of the radiation that diffuses through their atmosphere from a source in the deep layers of the medium (stars). For a homogeneous atmosphere, both problems are known to be closely connected (see Sect. 3.3). It is important to note here that the problem of determination of the reflection and transmission coefficients can be solved without the main problem (i.e., the determination of the radiation field in the atmosphere) being solved.

In the beginning of the chapter we will exploit the method of differentiation by optical thickness τ_0 to obtain a system of integro-differential equations determining azimuthal harmonics of the reflection and transmission coefficients. These coefficients can be found by solving a system of integral equations for generalized Ambartsumian φ and ψ functions. Later, similar relations will be derived by the so-called method of "truncated" atmosphere. This method enables us to derive another system of integral equations determining the above functions. The case of a semi-infinite inhomogeneous atmosphere is considered in the last section of the chapter.

9.1 Integro-Differential Equations for the Source Function and Reflection and Transmission Coefficients

Let us consider the standard parallel external flux problem for a plane inhomogeneous atmosphere (see Sect. 1.5). Azimuthal harmonics of the source functions $B^m(\tau, \mu, \mu_0; \tau_0)$ are determined by the integral equation (1.63). We know (see Sect. 6.3) that the radiation field in a plane inhomogeneous layer depends on which boundary (the upper or lower one) is illuminated by parallel flux. Therefore in addition to function $B^m(\tau,\mu,\mu_0;\tau_0)$, we consider the function $\tilde{B}^m(\tau,\mu,\mu_0;\tau_0)$, which determines the respective azimuthal harmonic of the source function in the layer illuminated from below. From now on, a tilde is used to denote functions relating to the situation when a plane inhomogeneous layer is illuminated by parallel rays from below. The equation for $\tilde{B}^m(\tau,\mu,\mu_0;\tau_0)$ is of the following form

$$\tilde{B}^m(\tau,\mu,\mu_0;\tau_0) = \frac{\lambda(\tau)}{2}\int_0^1 \chi^m(\mu,\mu';\tau)\,\mathrm{d}\mu' \int_0^\tau \tilde{B}^m(\tau',\mu',\mu_0;\tau_0)\mathrm{e}^{-\frac{\tau-\tau'}{\mu'}}\frac{\mathrm{d}\tau'}{\mu'}$$
$$+\frac{\lambda(\tau)}{2}\int_0^1 \chi^m(\mu,-\mu';\tau)\,\mathrm{d}\mu' \int_\tau^{\tau_0} \tilde{B}^m(\tau',-\mu',\mu_0;\tau_0)\mathrm{e}^{-\frac{\tau'-\tau}{\mu'}}\frac{\mathrm{d}\tau'}{\mu'}$$
$$+\frac{\lambda(\tau)}{4}\chi^m(-\mu,\mu_0;\tau)\mathrm{e}^{-\frac{\tau_0-\tau}{\mu_0}}. \tag{9.1}$$

Differentiating (1.63) with respect to the parameter τ_0 and exploiting the principle of superposition, we obtain

$$\frac{\partial B^m(\tau,\mu,\mu_0;\tau_0)}{\partial \tau_0} = 2\int_0^1 B^m(\tau_0,-\mu',\mu_0;\tau_0)\tilde{B}^m(\tau,-\mu,\mu';\tau_0)\frac{\mathrm{d}\mu'}{\mu'}. \tag{9.2}$$

Performing the same manipulations with (9.1), we have

$$\frac{\partial \tilde{B}^m(\tau,\mu,\mu_0;\tau_0)}{\partial \tau_0} = -\frac{1}{\mu_0}\tilde{B}^m(\tau,\mu,\mu_0;\tau_0)$$
$$+2\int_0^1 \tilde{B}^m(\tau_0,-\mu',\mu_0;\tau_0)\tilde{B}^m(\tau,\mu,\mu';\tau_0)\frac{\mathrm{d}\mu'}{\mu'}. \tag{9.3}$$

Setting $\tau=\tau_0$ in (1.63) and (9.1) and having in mind that expressions for harmonics of the reflection and transmission coefficients (see (1.97–100)) are determined by the following formulas:

$$\rho^m(\mu,\mu_0;\tau_0)\mu_0 = \int_0^{\tau_0} B^m(\tau',-\mu,\mu_0;\tau_0)\mathrm{e}^{-\frac{\tau'}{\mu}}\frac{\mathrm{d}\tau'}{\mu}, \tag{9.4}$$

$$\sigma^m(\mu,\mu_0;\tau_0)\mu_0 = \int_0^{\tau_0} B^m(\tau',\mu,\mu_0;\tau_0)\mathrm{e}^{-\frac{\tau_0-\tau'}{\mu}}\frac{\mathrm{d}\tau'}{\mu}, \tag{9.5}$$

$$\tilde{\rho}^m(\mu,\mu_0;\tau_0)\mu_0 = \int_0^{\tau_0} \tilde{B}^m(\tau',\mu,\mu_0;\tau_0)\mathrm{e}^{-\frac{\tau_0-\tau'}{\mu}}\frac{\mathrm{d}\tau'}{\mu}, \tag{9.6}$$

$$\tilde{\sigma}^m(\mu,\mu_0;\tau_0)\mu_0 = \int_0^{\tau_0} \tilde{B}^m(\tau',-\mu,\mu_0;\tau_0)\mathrm{e}^{-\frac{\tau'}{\mu}}\frac{\mathrm{d}\tau'}{\mu}, \tag{9.7}$$

we find

$$B^m(\tau_0,\mu,\mu_0;\tau_0) = \frac{\lambda(\tau_0)}{4}\Bigg[\chi^m(\mu,\mu_0;\tau_0)e^{-\tau_0/\mu_0}$$
$$+2\mu_0\int_0^1 \chi^m(\mu,\mu';\tau_0)\sigma^m(\mu',\mu_0;\tau_0)\,d\mu'\Bigg], \tag{9.8}$$

$$\tilde{B}^m(\tau_0,\mu,\mu_0;\tau_0) = \frac{\lambda(\tau_0)}{4}\Bigg[\chi^m(-\mu,\mu_0;\tau_0)$$
$$+2\mu_0\int_0^1 \chi^m(\mu,\mu';\tau_0)\tilde{\rho}^m(\mu',\mu_0;\tau_0)\,d\mu'\Bigg]. \tag{9.9}$$

The system of equations (9.2) and (9.3) for the source functions deserves more elaborate consideration. If we put $\tau = \tau_0$ in (9.2) and (9.3), we obtain the system of equations determining the quantities (9.8) and (9.9). On the other hand, (9.8) and (9.9) suggest that to find the boundary values of B^m and $\tilde{B}^m$ it is sufficient to find the transmission coefficient σ^m and reflection coefficient $\tilde{\rho}^m$. It turns out to be possible to derive separate equations for these quantities.

Let us differentiate (9.4–7) with respect to τ_0. Using (9.2) and (9.3) one derives the following expressions:

$$\frac{\partial\rho^m}{\partial\tau_0}\mu\mu_0 = B^m(\tau_0,-\mu,\mu_0;\tau_0)e^{-\tau_0/\mu}$$
$$+2\mu\int_0^1 B^m(\tau_0,-\mu',\mu_0;\tau_0)\tilde{\sigma}^m(\mu,\mu';\tau_0)\,d\mu', \tag{9.10}$$

$$\frac{\partial\sigma^m}{\partial\tau_0}\mu\mu_0 + \sigma^m(\mu,\mu_0;\tau_0) = B^m(\tau_0,\mu,\mu_0;\tau_0)$$
$$+2\mu\int_0^1 B^m(\tau_0,-\mu',\mu_0;\tau_0)\tilde{\rho}^m(\mu,\mu';\tau_0)\,d\mu', \tag{9.11}$$

$$\frac{\partial\tilde{\rho}^m}{\partial\tau_0}\mu\mu_0 = -(\mu+\mu_0)\tilde{\rho}^m(\mu,\mu_0;\tau_0) + \tilde{B}^m(\tau_0,\mu,\mu_0;\tau_0)$$
$$+2\mu\int_0^1 \tilde{B}^m(\tau_0,-\mu',\mu_0;\tau_0)\tilde{\rho}^m(\mu,\mu';\tau_0)\,d\mu', \tag{9.12}$$

$$\frac{\partial\tilde{\sigma}^m}{\partial\tau_0}\mu\mu_0 + \tilde{\sigma}^m(\mu,\mu_0;\tau_0) = \tilde{B}^m(\tau_0,-\mu,\mu_0;\tau_0)e^{-\tau_0/\mu}$$
$$+2\mu\int_0^1 \tilde{B}^m(\tau_0,-\mu',\mu_0;\tau_0)\tilde{\sigma}^m(\mu,\mu';\tau_0)\,d\mu'. \tag{9.13}$$

Relations (9.10–12) together with (9.8) and (9.9) constitute a system of integro-differential equations determining the reflection and transmission coefficients. Note that (9.12) is the only equation independent of the others.

Now we proceed to represent the reflection and transmission coefficients, which depend on two angular variables μ and μ_0, in terms of auxiliary functions depending on only one variable. The azimuthal harmonic can be expressed as in (1.56), therefore (1.63) and (9.1) result in the following expressions:

$$B^m(\tau,\mu,\mu_0;\tau_0) = \sum_{i=m}^{n} x_i^m(\tau) B_i^m(\tau,\mu_0;\tau_0) P_i^m(\mu), \tag{9.14}$$

$$\tilde{B}^m(\tau,\mu,\mu_0;\tau_0) = \sum_{i=m}^{n} x_i^m(\tau) \tilde{B}_i^m(\tau,\mu_0;\tau_0) P_i^m(\mu), \tag{9.15}$$

where $x_i^m(\tau)$ is determined by (1.57). The respective systems of integral equations for functions B_i^m and $\tilde{B}_i^m$ take the form

$$B_k^m(\tau,\mu_0;\tau_0) = \frac{\lambda(\tau)}{2} \sum_{i=m}^{n} \int_0^{\tau_0} x_i^m(t) B_i^m(t,\mu_0;\tau_0) \operatorname{sign}^{i+k}(\tau - t) E_{ik}^m(|\tau - t|)\,\mathrm{d}t + \frac{\lambda(\tau)}{4} P_k^m(\mu_0) \mathrm{e}^{-\tau/\mu_0}, \tag{9.16}$$

$$\tilde{B}_k^m(\tau,\mu_0;\tau_0) = \frac{\lambda(\tau)}{2} \sum_{i=m}^{n} \int_0^{\tau_0} x_i^m(t) \tilde{B}_i^m(t,\mu_0;\tau_0) \operatorname{sign}^{i+k}(\tau - t) E_{ik}^m(|\tau - t|)\,\mathrm{d}t + \frac{\lambda(\tau)}{4} P_k^m(-\mu_0) \mathrm{e}^{-\frac{\tau_0-\tau}{\mu_0}}, \tag{9.17}$$

where function $E_{ik}^m(z)$ is determined by (1.66). Now proceeding from (9.14) and (9.15) and taking into consideration (9.8), (9.9) and the reciprocity relations (for the proof, see Sect. 1.11 and also LSPA, Sect. 3.5)

$$\left.\begin{aligned} \rho^m(\mu,\mu_0) &= \rho^m(\mu_0,\mu), \\ \tilde{\rho}^m(\mu,\mu_0) &= \tilde{\rho}^m(\mu_0,\mu), \\ \sigma^m(\mu,\mu_0) &= \tilde{\sigma}^m(\mu_0,\mu), \end{aligned}\right\} \tag{9.18}$$

we have

$$B^m(\tau_0,\mu,\mu_0;\tau_0) = \frac{\lambda(\tau_0)}{4} \sum_{i=m}^{n} x_i^m(\tau_0) \psi_i^m(\mu_0,\tau_0) P_i^m(\mu), \tag{9.19}$$

$$\tilde{B}^m(\tau_0,\mu,\mu_0;\tau_0) = \frac{\lambda(\tau_0)}{4} \sum_{i=m}^{n} x_i^m(\tau_0) \tilde{\varphi}_i^m(\mu_0,\tau_0) P_i^m(\mu), \tag{9.20}$$

where

$$\psi_k^m(\mu_0,\tau_0) = P_k^m(\mu_0) \mathrm{e}^{-\tau_0/\mu_0} + 2\mu_0 \int_0^1 \sigma^m(\mu',\mu_0;\tau_0) P_k^m(\mu')\,\mathrm{d}\mu', \tag{9.21}$$

$$\tilde{\varphi}_k^m(\mu_0, \tau_0) = P_k^m(\mu_0) + 2\mu_0 \int_0^1 \tilde{\rho}^m(\mu', \mu_0; \tau_0) P_k^m(\mu') \, \mathrm{d}\mu', \tag{9.22}$$

As a result, we have the following system instead of (9.10–9.13):

$$\frac{\partial \rho^m}{\partial \tau_0} \mu\mu_0 = \frac{\lambda(\tau_0)}{4} \sum_{i=m}^{n} (-1)^{i+m} x_i^m(\tau_0) \psi_i^m(\mu, \tau_0) \psi_i^m(\mu_0, \tau_0), \tag{9.23}$$

$$\frac{\partial \sigma^m}{\partial \tau_0} \mu\mu_0 = -\sigma^m(\mu, \mu_0; \tau_0)\mu_0 + \frac{\lambda(\tau_0)}{4} \sum_{i=m}^{n} x_i^m(\tau_0) \psi_i^m(\mu_0, \tau_0) \tilde{\varphi}_i^m(\mu, \tau_0), \tag{9.24}$$

$$\frac{\partial \tilde{\rho}^m}{\partial \tau_0} \mu\mu_0 = -\tilde{\rho}^m(\mu, \mu_0; \tau_0)(\mu + \mu_0)$$
$$+ \frac{\lambda(\tau_0)}{4} \sum_{i=m}^{n} (-1)^{i+m} x_i^m(\tau_0) \tilde{\varphi}_i^m(\mu, \tau_0) \tilde{\varphi}_i^m(\mu_0, \tau_0), \tag{9.25}$$

$$\frac{\partial \tilde{\sigma}^m}{\partial \tau_0} \mu\mu_0 = -\tilde{\sigma}^m(\mu, \mu_0; \tau_0)\mu + \frac{\lambda(\tau_0)}{4} \sum_{i=m}^{n} x_i^m(\tau_0) \tilde{\varphi}_i^m(\mu_0, \tau_0) \psi_i^m(\mu, \tau_0), \tag{9.26}$$

$$\rho^m(\mu, \mu_0; \tau_0) = \frac{1}{4} \sum_{i=m}^{n} (-1)^{i+m} \int_0^{\tau_0} \lambda(t) x_i^m(t) \psi_i^m(\mu, t) \psi_i^m(\mu_0, t) \frac{\mathrm{d}t}{\mu\mu_0}, \tag{9.27}$$

$$\sigma^m(\mu, \mu_0; \tau_0) = \frac{1}{4} \sum_{i=m}^{n} \int_0^{\tau_0} \lambda(t) x_i^m(t) \psi_i^m(\mu_0, t) \tilde{\varphi}_i^m(\mu, t) \mathrm{e}^{-\frac{\tau_0 - t}{\mu}} \frac{\mathrm{d}t}{\mu\mu_0}, \tag{9.28}$$

$$\tilde{\rho}^m(\mu, \mu_0; \tau_0)$$
$$= \frac{1}{4} \sum_{i=m}^{n} (-1)^{i+m} \int_0^{\tau_0} \lambda(t) x_i^m(t) \tilde{\varphi}_i^m(\mu_0, t) \tilde{\varphi}_i^m(\mu, t) \mathrm{e}^{-(\tau_0 - t)\left(\frac{1}{\mu} + \frac{1}{\mu_0}\right)} \frac{\mathrm{d}t}{\mu\mu_0}, \tag{9.29}$$

and, as it follows from (9.18), $\tilde{\sigma}^m(\mu, \mu_0; t) = \sigma^m(\mu_0, \mu; t)$. The substitution of (9.28) and (9.27) into (9.21) and (9.22) yields the following system determining the auxiliary functions $\tilde{\varphi}_k^m$ and ψ_k^m

$$\psi_k^m(\mu, \tau_0) = P_k^m(\mu) \mathrm{e}^{-\tau_0/\mu}$$
$$+ \frac{1}{2} \sum_{i=m}^{n} \int_0^{\tau_0} \lambda(t) x_i^m(t) \psi_i^m(\mu, t) \, \mathrm{d}t \int_0^1 \tilde{\varphi}_i^m(\mu', t) \mathrm{e}^{-\frac{\tau_0 - t}{\mu'}} P_k^m(\mu') \frac{\mathrm{d}\mu'}{\mu'}, \tag{9.30}$$

$$\tilde{\varphi}_k^m(\mu, \tau_0) = P_k^m(\mu)$$
$$+ \frac{1}{2} \sum_{i=m}^{n} (-1)^{i+k} \int_0^{\tau_0} \lambda(t) x_i^m(t) \tilde{\varphi}_i^m(\mu, t) \, \mathrm{d}t \int_0^1 \tilde{\varphi}_i^m(\mu', t) \mathrm{e}^{-(\tau_0 - t)\left(\frac{1}{\mu} + \frac{1}{\mu'}\right)} P_k^m(\mu') \frac{\mathrm{d}\mu'}{\mu'}. \tag{9.31}$$

Thus, the determination of azimuthal harmonics of the transmission and reflection coefficients, each depending on two angular variables, is reduced to the determination of auxiliary functions $\tilde{\varphi}_k^m$ and ψ_k^m, each depending on only one angular variable. Functions $\tilde{\varphi}_k^m$ and ψ_k^m are generalizations of Ambartsumian's functions φ and ψ for the inhomogeneous case. These functions were introduced to solve a similar problem for a homogeneous atmosphere.

Hence, to find functions $\rho(\mu, \mu_0, \varphi)$ and $\sigma(\mu, \mu_0, \varphi)$ for an inhomogeneous atmosphere of optical thickness τ_0, one needs to solve the system of integral equations (9.30) and (9.31) first. Then by substituting $\tilde{\varphi}_k^m$ and ψ_k^m into (9.27) and (9.28), we find $\rho^m(\mu, \mu_0; \tau_0)$ and $\sigma^m(\mu, \mu_0; \tau_0)$, which, in turn, are to be substituted into (1.99) and (1.100). Thus, the solution of the problem is found. There exists, however, another approach to solving this problem, which is expounded in the next section.

9.2 Method of Truncated Atmosphere for Determining Reflection and Transmission Coefficients

We have already met the idea of truncation of a semi-infinite atmosphere (for example see Sect. 3.2). Sobolev was the first scientist to put this idea forward for a semi-infinite inhomogeneous atmosphere, and the idea turned out to be very fruitful. The author of this book was, in all probability, the person who first suggested the term itself and its physical interpretation.

Imagine that a plane layer of optical thickness α is discarded from a plane inhomogeneous layer of optical thickness τ_0 from above. For simplicity, we will call such an atmosphere a ***truncated atmosphere*** and α ***the parameter of truncation***. Denoting $\tau_0 - \alpha = z$ (i.e. z becomes the optical thickness of the layer), we have, instead of (1.63),

$$\begin{aligned} &B^m(\tau, \mu, \mu_0; z, \alpha) \\ &= \frac{\lambda(\tau+\alpha)}{2} \int_0^1 \chi^m(\mu, \mu'; \tau+\alpha)\, \mathrm{d}\mu' \int_0^\tau B^m(\tau', \mu', \mu_0; z, \alpha) \mathrm{e}^{-\frac{\tau-\tau'}{\mu'}} \frac{\mathrm{d}\tau'}{\mu'} \\ &+ \frac{\lambda(\tau+\alpha)}{2} \int_0^1 \chi^m(-\mu, \mu'; \tau+\alpha)\, \mathrm{d}\mu' \int_\tau^z B^m(\tau', -\mu', \mu_0; z, \alpha) \mathrm{e}^{-\frac{\tau'-\tau}{\mu'}} \frac{\mathrm{d}\tau'}{\mu'} \\ &+ \frac{\lambda(\tau+\alpha)}{4} \chi^m(\mu, \mu_0; \tau+\alpha) \mathrm{e}^{-\tau/\mu_0}. \end{aligned} \tag{9.32}$$

We differentiate (9.32) with respect to τ and after that again differentiate (9.32) with respect to α bearing in mind that $z = \tau_0 - \alpha$. By subtracting from the second relation the first one, we arrive at the following formula

$$\begin{aligned} &F(\tau, \mu, \mu_0; z, \alpha) \\ &= \frac{\lambda(\tau+\alpha)}{2} \int_0^1 \chi^m(\mu, \mu'; \tau+\alpha)\, \mathrm{d}\mu' \int_0^\tau F(\tau', \mu', \mu_0; z, \alpha) \mathrm{e}^{-\frac{\tau-\tau'}{\mu'}} \frac{\mathrm{d}\tau'}{\mu'} \end{aligned}$$

$$+\frac{\lambda(\tau+\alpha)}{2}\int_0^1 \chi^m(-\mu,\mu';\tau+\alpha)\,\mathrm{d}\mu'\int_\tau^z F(\tau',-\mu',\mu_0;z,\alpha)\mathrm{e}^{-\frac{\tau'-\tau}{\mu'}}\frac{\mathrm{d}\tau'}{\mu'}$$
$$+\frac{\lambda(\tau+\alpha)}{2}\int_0^1 \chi^m(\mu,\mu';\tau+\alpha)\,B^m(0,\mu',\mu_0;z,\alpha)\mathrm{e}^{-\tau'/\mu'}\frac{\mathrm{d}\mu'}{\mu'}$$
$$-\frac{\lambda(\tau+\alpha)}{4\mu_0}\chi^m(\mu,\mu_0;\tau+\alpha)\mathrm{e}^{-\tau/\mu_0}, \qquad (9.33)$$

where

$$F(\tau,\mu,\mu_0;z,\alpha)=\frac{\partial B^m(\tau,\mu,\mu_0;z,\alpha)}{\partial\tau}-\frac{\partial B^m(\tau,\mu,\mu_0;z,\alpha)}{\partial\alpha}.$$

Comparing (9.33) with (9.32) and using the principle of superposition, we find

$$\frac{\partial B^m(\tau,\mu,\mu_0;z,\alpha)}{\partial\tau}-\frac{\partial B^m(\tau,\mu,\mu_0;z,\alpha)}{\partial\alpha}=-\frac{1}{\mu_0}B^m(\tau,\mu,\mu_0;z,\alpha)$$
$$+2\int_0^1 B^m(0,\mu',\mu_0;z,\alpha)B^m(\tau,\mu,\mu';z,\alpha)\frac{\mathrm{d}\mu'}{\mu'}. \qquad (9.34)$$

Hence, the method of truncated atmosphere yields one integro-differential equation determining the azimuthal harmonic of the source function, and does not require the layer illuminated from below to be considered. The equation similar to (9.34) for the layer illuminated from below can, however, be easily derived. Namely,

$$\frac{\partial\tilde{B}^m}{\partial\tau}-\frac{\partial\tilde{B}^m}{\partial\alpha}=2\int_0^1\tilde{B}^m(0,\mu',\mu_0;z,\alpha)B^m(\tau,\mu,\mu';z,\alpha)\frac{\mathrm{d}\mu'}{\mu'}. \qquad (9.35)$$

Here the source function of the atmosphere illuminated from below is, in contrast to the previous case, expressed via the source function of the atmosphere illuminated from above. Denoting

$$\rho^m(\mu,\mu_0;z,\alpha)\mu_0=\int_0^z B^m(\tau',-\mu,\mu_0;z,\alpha)\mathrm{e}^{-\frac{\tau'}{\mu}}\frac{\mathrm{d}\tau'}{\mu}, \qquad (9.36)$$

$$\sigma^m(\mu,\mu_0;z,\alpha)\mu_0=\int_0^z B^m(\tau',\mu,\mu_0;z,\alpha)\mathrm{e}^{-\frac{z-\tau'}{\mu}}\frac{\mathrm{d}\tau'}{\mu}, \qquad (9.37)$$

by analogy with the considerations, we find

$$(\mu+\mu_0)\rho^m(\mu,\mu_0;z,\alpha)-\frac{\partial\rho^m}{\partial\alpha}\mu\mu_0=B^m(0,-\mu,\mu_0;z,\alpha)$$
$$+2\mu\int_0^1 B^m(0,\mu',\mu_0;z,\alpha)\rho^m(\mu,\mu';z,\alpha)\,\mathrm{d}\mu', \qquad (9.38)$$

where, as follows from (9.32),

$$B^m(0,\mu,\mu_0;z,\alpha) = \frac{\lambda(\alpha)}{4}\Bigg[\chi^m(\mu,\mu_0;\alpha)$$

$$+2\mu_0 \int_0^1 \chi^m(-\mu,\mu';\alpha)\rho^m(\mu',\mu_0;z,\alpha)\,\mathrm{d}\mu'\Bigg], \tag{9.39}$$

and

$$\mu\sigma^m(\mu,\mu_0;z,\alpha) - \mu\mu_0\frac{\partial\sigma^m}{\partial\alpha} = B^m(0,\mu,\mu_0;z,\alpha)\mathrm{e}^{-z/\mu}$$

$$+2\mu\int_0^1 B^m(0,\mu',\mu_0;z,\alpha)\sigma^m(\mu,\mu';z,\alpha)\,\mathrm{d}\mu'. \tag{9.40}$$

In deducing (9.40) we used the following equation

$$\int\limits_0^z \frac{\partial B^m}{\partial\alpha}\mathrm{e}^{-\frac{z-\tau'}{\mu}}\,\mathrm{d}\tau' = \mu\frac{\partial\sigma^m}{\partial\alpha} - \sigma^m(\mu,\mu_0;z,\alpha) + B^m(z,\mu,\mu_0;z,\alpha).$$

The relations (9.38) and (9.40) together with (9.39) can be regarded as a system of two integro-differential equations for finding the transmission and reflection coefficients. The system we found earlier consisted of three equations (9.10–12) determining the same quantities and the additional quantity $\tilde{\rho}^m$. Hence, the method of truncated atmosphere provides a simpler result.

By analogy with Sect. 9.1, one can write

$$B^m(0,\mu,\mu_0;z,\alpha) = \frac{\lambda(\alpha)}{4}\sum_{i=m}^{n} x_i^m(\alpha)\varphi_i^m(\mu_0;z,\alpha)P_i^m(\mu), \tag{9.41}$$

where

$$\varphi_k^m(\mu_0,z,\alpha) = P_k^m(\mu) + 2\mu\int\limits_0^1 \rho^m(\mu,\mu';z,\alpha)P_k^m(-\mu')\,\mathrm{d}\mu'. \tag{9.42}$$

Then, instead of (9.38) and (9.40) we have

$$(\mu+\mu_0)\rho^m(\mu,\mu_0;z,\alpha) - \frac{\partial\rho^m}{\partial\alpha}\mu\mu_0$$

$$= \frac{\lambda(\alpha)}{4}\sum_{i=m}^{n}(-1)^{i+m}x_i^m(\alpha)\varphi_i^m(\mu;z,\alpha)\varphi_i^m(\mu_0;z,\alpha), \tag{9.43}$$

$$\mu\sigma^m(\mu,\mu_0;z,\alpha) - \mu\mu_0\frac{\partial\sigma^m}{\partial\alpha}$$

$$= \frac{\lambda(\alpha)}{4}\sum_{i=m}^{n} x_i^m(\alpha)\tilde{\psi}_i^m(\mu;z,\alpha)\varphi_i^m(\mu_0;z,\alpha), \tag{9.44}$$

where

$$\tilde{\psi}_i^m(\mu; z, \alpha) = P_i^m(\mu)\mathrm{e}^{-z/\mu} + 2\mu \int_0^1 \mu\sigma^m(\mu, \mu'; z, \alpha) P_i^m(\mu')\,\mathrm{d}\mu'$$
$$= P_i^m(\mu)\mathrm{e}^{-z/\mu} + 2\mu \int_0^1 \tilde{\sigma}^m(\mu', \mu; z, \alpha) P_i^m(\mu')\,\mathrm{d}\mu'. \tag{9.45}$$

After integration and appropriate change of variables, (9.43) and (9.44) yield

$$\rho^m(\mu, \mu_0; z, \alpha) = \frac{1}{4}\sum_{i=m}^{n}(-1)^{i+m}\int_0^z \lambda(z+\alpha-z')x_i^m(z+\alpha-z')$$
$$\cdot\varphi_i^m(\mu; z', \alpha)\varphi_i^m(\mu_0; z', \alpha)\mathrm{e}^{-(z-z')\left(\frac{1}{\mu}+\frac{1}{\mu_0}\right)}\frac{\mathrm{d}z'}{\mu\mu_0}, \tag{9.46}$$

$$\sigma^m(\mu, \mu_0; z, \alpha) = \frac{1}{4}\sum_{i=m}^{n}\int_0^z \lambda(z+\alpha-z')x_i^m(z+\alpha-z')$$
$$\cdot\varphi_i^m(\mu_0; z', \alpha)\tilde{\psi}_i^m(\mu; z', \alpha)\mathrm{e}^{-\frac{(z-z')}{\mu_0}}\frac{\mathrm{d}z'}{\mu\mu_0}, \tag{9.47}$$

$$\varphi_k^m(\mu; z, \alpha) = P_k^m(\mu) + \frac{1}{2}\sum_{i=m}^{n}(-1)^{i+m}\int_0^z \lambda(z+\alpha-z')x_i^m(z+\alpha-z')$$
$$\cdot\varphi_i^m(\mu; z', \alpha)\,\mathrm{d}z'\int_0^1 \varphi_i^m(\mu'; z', \alpha)\mathrm{e}^{-(z-z')\left(\frac{1}{\mu}+\frac{1}{\mu'}\right)}P_k^m(\mu')\frac{\mathrm{d}\mu'}{\mu'}, \tag{9.48}$$

$$\tilde{\psi}_k^m(\mu; z, \alpha) = P_k^m(\mu)\mathrm{e}^{-z/\mu} + \frac{1}{2}\sum_{i=m}^{n}\int_0^z \lambda(z+\alpha-z')x_i^m(z+\alpha-z')$$
$$\cdot\varphi_i^m(\mu; z', \alpha)\mathrm{e}^{-\frac{(z-z')}{\mu}}\,\mathrm{d}z'\int_0^1 \tilde{\psi}_i^m(\mu'; z', \alpha)P_k^m(\mu')\frac{\mathrm{d}\mu'}{\mu'}. \tag{9.49}$$

Equations (9.46) and (9.48) suggest the following important conclusion: in determining the reflection coefficient it is sufficient to consider *one* system of integral equations for functions φ_i^m whereas the relations for the reflection coefficient found in the previous section consist of *two* systems of integral equations for $\tilde{\varphi}_i^k$ and ψ_i^k. There exists another way to derive (9.46–49) (at $\alpha = 0$, i.e. when $z = \tau_0$). Imagine that, in contrast to the previous case, the origin of the optical depth is chosen at the lower boundary and optical depths are read in the "backward" direction. Then performing manipulations similar to those we did for the "standard" way of counting of optical depths, we derive (9.40–43), which depend on the parameter τ_0. We prefer, however, to obtain this result by the method of truncated atmosphere because this method itself is very important for the study of light scattering in inhomogeneous atmospheres.

The previous paragraph can easily be explained with the help of the reasoning of Sect. 6.3, which concerns the direction of the build-up of layers. Indeed, the differentiation with respect to τ_0 is equivalent to the addition of a layer of infinitesimal thickness to the atmosphere. For the "standard" counting of optical depth, in differentiation we add a layer of small thickness from below. This creates additional illumination of the atmosphere which comes from the non-illuminated border, which, in turn, requires an additional equation to be introduced. On the contrary in differentiating an equation with "reverse counting" of optical depths, a thin layer is added from above, which does not require any new equations.

Finally, one can derive one more kind of formulas determining the reflection and transmission coefficients. Setting $\alpha = 0$ (i.e. $z = \tau_0$) in (9.46) and (9.47) and differentiating with respect to τ_0 we find derivatives $\partial\rho^m/\partial\tau_0$ and $\partial\sigma^m/\partial\tau_0$. By substituting these derivatives into (9.23) and (9.24), we obtain

$$\begin{aligned}(\mu+\mu_0)\rho^m(\mu,\mu_0;\tau_0) = \frac{1}{4}\sum_{i=m}^{n}(-1)^{i+m}\Big\{&\lambda(0)x_i^m(0)\varphi_i^m(\mu,\tau_0)\varphi_i^m(\mu_0,\tau_0)\\ &-\lambda(\tau_0)x_i^m(\tau_0)\psi_i^m(\mu,\tau_0)\psi_i^m(\mu_0,\tau_0)\\ &+\int_0^{\tau_0}\frac{\partial}{\partial\tau_0}[\lambda(\tau_0-t)x_i^m(\tau_0-t)]\varphi_i^m(\mu,t)\varphi_i^m(\mu_0,t)\mathrm{e}^{-(\tau_0-t)\left(\frac{1}{\mu}+\frac{1}{\mu_0}\right)}\,\mathrm{d}t\Big\},\end{aligned} \tag{9.50}$$

$$\begin{aligned}(\mu-\mu_0)\sigma^m(\mu,\mu_0;\tau_0) = \frac{1}{4}\sum_{i=m}^{n}\Big\{&\lambda(0)x_i^m(0)\varphi_i^m(\mu_0,\tau_0)\tilde{\psi}_i^m(\mu,\tau_0)\\ &-\lambda(\tau_0)x_i^m(\tau_0)\psi_i^m(\mu_0,\tau_0)\tilde{\varphi}_i^m(\mu,\tau_0)\\ &+\int_0^{\tau_0}\frac{\partial}{\partial\tau_0}[\lambda(\tau_0-t)x_i^m(\tau_0-t)]\varphi_i^m(\mu_0,t)\psi_i^m(\mu,t)\mathrm{e}^{-\frac{\tau_0-t}{\mu_0}}\,\mathrm{d}t\Big\}.\end{aligned} \tag{9.51}$$

The reflection coefficient of an atmosphere illuminated from below can be found similarly, namely,

$$\begin{aligned}(\mu+\mu_0)\tilde{\rho}^m(\mu,\mu_0;\tau_0) = \frac{1}{4}\sum_{i=m}^{n}(-1)^{i+m}\Big\{&\lambda(\tau_0)x_i^m(\tau_0)\tilde{\varphi}_i^m(\mu,\tau_0)\tilde{\varphi}_i^m(\mu_0,\tau_0)\\ &-\lambda(0)x_i^m(0)\tilde{\psi}_i^m(\mu,\tau_0)\tilde{\psi}_i^m(\mu_0,\tau_0)\\ &+\int_0^{\tau_0}\frac{\mathrm{d}}{\mathrm{d}t}[\lambda(t)x_i^m(t)]\tilde{\varphi}_i^m(\mu,t)\tilde{\varphi}_i^m(\mu_0,t)\mathrm{e}^{-(\tau_0-t)\left(\frac{1}{\mu}+\frac{1}{\mu_0}\right)}\,\mathrm{d}t\Big\}.\end{aligned} \tag{9.52}$$

If we take into account that $\sigma^m(\mu,\mu_0;\tau_0) = \tilde{\sigma}^m(\mu,\mu_0;\tau_0)$ and substitute (9.50–52) into (9.42), (9.21), (9.45) and (9.22), we find a system consisting of four integral equations for functions φ_k^m, $\tilde{\varphi}_k^m$, ψ_k^m, and $\tilde{\psi}_k^m$. This system is different from those we found earlier.

Denote the integrals on the right-hand side of (9.50–52) by I_ρ, I_σ, and $I_{\tilde{\rho}}$, respectively. These integrals can be regarded as indicators of inhomogeneity of medium since they involve the derivatives of parameters characterizing optical properties of the media. If we neglect the integrals, we obtain equations that hardly differ from those for a homogeneous medium. The relative error of such an omission is determined by quantities I_ρ/ρ, I_σ/σ, and $I_{\tilde{\rho}}/\tilde{\rho}$.

9.3 A Semi-Infinite Atmosphere

If the optical thickness of the atmosphere $\tau_0 = \infty$, then the above formulas become significantly more simple. To determine the azimuthal harmonic of the reflection coefficient from (9.27), we have the following expression

$$\rho^m(\mu,\mu_0) = \frac{1}{4}\sum_{i=m}^{n}(-1)^{i+m}\int_0^\infty \lambda(t)x_i^m(t)\psi_i^m(\mu,t)\psi_i^m(\mu_0,t)\frac{\mathrm{d}t}{\mu\mu_0}, \tag{9.53}$$

with functions ψ_i^m being found from the system of equations (9.30) and (9.31).

Furthermore, renaming quantities in (9.43) and (9.42) we derive the following expression for the reflection coefficient for the truncated semi-infinite atmosphere

$$\rho^m(\mu,\mu_0;\alpha)$$
$$= \frac{1}{4}\sum_{i=m}^{n}(-1)^{i+m}\int_\alpha^\infty \lambda(t)x_i^m(t)\varphi_i^m(\mu,t)\varphi_i^m(\mu_0,t)\mathrm{e}^{-(\tau-\alpha)\left(\frac{1}{\mu}+\frac{1}{\mu_0}\right)}\frac{\mathrm{d}t}{\mu\mu_0}, \tag{9.54}$$

where

$$\varphi_k^m(\mu,t) = P_k^m(\mu) + \frac{1}{2}\sum_{i=m}^{n}(-1)^{i+m}\int_t^\infty \lambda(t')x_i^m(t')\varphi_i^m(\mu,t')\,\mathrm{d}t'$$
$$\cdot\int_0^1 \varphi_i^m(\mu',t')\mathrm{e}^{-(t'-t)\left(\frac{1}{\mu}+\frac{1}{\mu_0}\right)}P_k^m(\mu')\frac{\mathrm{d}\mu'}{\mu'}. \tag{9.55}$$

Function $\rho^m(\mu,\mu_0;\alpha)$ can also be derived from the integro-differential equation following from (9.38)

$$(\mu+\mu_0)\rho^m(\mu,\mu_0;\alpha) - \mu\mu_0\frac{\partial\rho^m}{\partial\alpha} = B^m(0,-\mu,\mu_0;\alpha)$$
$$+2\mu\int_0^1 B^m(0,\mu',\mu_0;\alpha)\rho^m(\mu,\mu';\alpha)\,\mathrm{d}\mu', \tag{9.56}$$

where

$$B^m(0,\mu,\mu_0;\alpha) = \frac{\lambda(\alpha)}{4}\left[\chi^m(\mu,\mu_0;\alpha) + 2\mu_0 \int_0^1 \chi^m(-\mu,\mu';\alpha)\rho^m(\mu',\mu_0;\alpha)\,\mathrm{d}\mu'\right]. \qquad (9.57)$$

Equation (9.56) together with (9.57) can be treated as a ***Cauchy problem*** with the initial condition $\rho^m(\mu,\mu_0;\infty) = 0$. The system of equations (9.10–12) with the initial conditions $\rho^m(\mu,\mu_0;0) = 0$, $\tilde{\rho}^m(\mu,\mu_0;0) = 0$ and $\sigma^m(\mu,\mu_0;0) = 0$, as well as (9.38) and (9.39) with the initial conditions $\rho^m(\mu,m_0;0,\tau_0) = 0$ and $\sigma^m(\mu,\mu_0;0,\tau_0) = 0$ can also be regarded as Cauchy problem. For the methods of solving this problem and their application to the transfer theory, see, for example, the book by Casti and Kalaba [9].

Here we stop our consideration of the problem of diffuse transmission and reflection of light by a plane inhomogeneous atmosphere. Note that one more kind of equation, determining the intensity of radiation leaving an isotropically scattering atmosphere, will be obtained in Sect. 13.4.

10. Basic Equations Defining the Radiation Field in a Vertically Inhomogeneous Layer

In Chap. 9 we considered the problem of calculating the reflection and transmission coefficients of a plane-parallel layer. In other words, the problem of finding the intensity of the radiation at the boundaries of the layer was tackled.

The general problem is now how to calculate the radiation field at any depth in an atmosphere whose optical properties are assumed to be given.

At the beginning of this chapter, we will obtain an equation that defines the azimuthal harmonics of the intensity of the diffuse radiation, through the corresponding harmonics of the reflection coefficient for any optical depth. Then we will derive a slightly different equation which includes the derivative of the intensity with respect to the optical depth of the layer. Further, on the basis of the general invariance principle, we will derive a new form of the transfer equation in an inhomogeneous plane layer. This form has a property necessary to determine the upward and downward diffuse intensities in the atmosphere independently. Finally, we will briefly discuss numerical methods that can be applied to determine the radiation field in an inhomogeneous plane layer.

To conclude the present chapter we will give the equations defining the radiation field in an inhomogeneous plane layer adjacent to a reflecting surface. It will be shown that the intensity of radiation can be found quite easily if the law of light reflection by the surface is given and the radiation field in an inhomogeneous layer without the reflective surface is known. The simplest case of the Lambertian surface will be discussed in detail.

10.1 Equation for the Radiation Intensity in a Plane Layer

This equation can be easily obtained from (9.34) for the truncated atmosphere. Indeed, taking into account that

$$I^m(\tau, \mu, \mu_0; z, \alpha) = \int_0^\tau B^m(\tau', \mu, \mu_0; z, \alpha) \mathrm{e}^{-\frac{\tau - \tau'}{\mu}} \frac{\mathrm{d}\tau'}{\mu}, \qquad (10.1)$$

$$I^m(\tau,-\mu,\mu_0;z,\alpha) = \int\limits_{\tau}^{z} B^m(\tau',-\mu,\mu_0;z,\alpha)\mathrm{e}^{-\frac{\tau'-\tau}{\mu}}\frac{\mathrm{d}\tau'}{\mu}, \tag{10.2}$$

where $\mu \in [0,1]$, $\tau \in [0, z = \tau_0 - \alpha]$, we find from (9.34), (10.1) and (10.2) that ($\mu \in [-1,1]$)

$$\left(\frac{1}{\mu_0} - \frac{1}{\mu}\right) I^m(\tau,\mu,\mu_0;z,\alpha) = \frac{\partial I^m(\tau,\mu,\mu_0;z,\alpha)}{\partial\alpha} - \frac{1}{\mu}B^m(\tau,\mu,\mu_0;z,\alpha)$$
$$+\frac{1}{\mu}B^m(0,\mu,\mu_0;z,\alpha)\mathrm{e}^{-\frac{\tau}{\mu}}\theta(\mu) + 2\int\limits_0^1 B^m(0,\mu',\mu_0;z,\alpha)I^m(\tau,\mu,\mu';z,\alpha)\frac{\mathrm{d}\mu'}{\mu'}. \tag{10.3}$$

Here $\theta(\mu)$ is a unit step function: $\theta(\mu) = 1$ ($\mu \geq 0$); $\theta(\mu) = 0$ ($\mu < 0$). The function $B^m(\tau,\mu,\mu_0;z,\alpha)$ is, as follows from (1.50), expressed in terms of the intensity $I^m(\tau,\mu,\mu_0;z,\alpha)$ (as usual, we say "intensity" instead of "the m-th harmonic of the intensity", "source function" instead of "the m-th harmonic of the source function" etc.) by the equation

$$B^m(\tau,\mu,\mu_0;z,\alpha) = \frac{\lambda(\tau+\alpha)}{2}\int_{-1}^{+1}\chi^m(\mu,\mu';\tau+\alpha)I^m(\tau,\mu',\mu_0;z,\alpha)\,\mathrm{d}\mu'$$
$$+\frac{\lambda(\tau+\alpha)}{4}\chi^m(\mu,\mu_0;\tau+\alpha)\mathrm{e}^{-\frac{\tau}{\mu_0}}, \tag{10.4}$$

and $B^m(0,\mu,\mu_0;z,\alpha)$ is given in terms of the reflection coefficient by (9.39). The equation for the intensity in the truncated layer illuminated from below can be written similarly. Specifically, taking into account that for $\mu \in [0,1]$

$$\tilde{I}^m(\tau,\mu,\mu_0;z,\alpha) = \int_0^{\tau} \tilde{B}^m(\tau',\mu,\mu_0;z,\alpha)\mathrm{e}^{-\frac{\tau-\tau'}{\mu}}\frac{\mathrm{d}\tau'}{\mu},$$
$$\tilde{I}^m(\tau,-\mu,\mu_0;z,\alpha) = \int_{\tau}^{z} \tilde{B}^m(\tau',-\mu,\mu_0;z,\alpha)\mathrm{e}^{-\frac{\tau'-\tau}{\mu}}\frac{\mathrm{d}\tau'}{\mu},$$

from (9.35) we have

$$\tilde{I}^m(\tau,\mu,\mu_0;z,\alpha) + \mu\frac{\partial\tilde{I}^m(\tau,\mu,\mu_0;z,\alpha)}{\partial\alpha} = \tilde{B}^m(\tau,\mu,\mu_0;z,\alpha)$$
$$-\tilde{B}^m(0,\mu,\mu_0;z,\alpha)\mathrm{e}^{-\frac{\tau}{\mu}} - 2\mu\int_0^1 \tilde{B}^m(0,\mu',\mu_0;z,\alpha)\tilde{I}^m(\tau,\mu,\mu';z,\alpha)\frac{\mathrm{d}\mu'}{\mu'}, \tag{10.4a}$$

and

$$\tilde{I}^m(\tau,-\mu,\mu_0;z,\alpha) - \mu\frac{\partial\tilde{I}^m(\tau,-\mu,\mu_0;z,\alpha)}{\partial\alpha} = \tilde{B}^m(\tau,-\mu,\mu_0;z,\alpha)$$
$$-\tilde{B}^m(z,-\mu,\mu_0;z,\alpha)\mathrm{e}^{-\frac{z-\tau}{\mu}} + 2\mu\int_0^1 \tilde{B}^m(0,\mu',\mu_0;z,\alpha)\tilde{I}^m(\tau,-\mu,\mu';z,\alpha)\frac{\mathrm{d}\mu'}{\mu'}. \tag{10.4b}$$

Thus, we have obtained the integro-differential equation which connects the intensity of radiation with the reflection coefficient for a layer of optical thickness z. The equation (10.3) does not contain the derivative with respect to the space coordinate τ (in contrast to the radiative transfer equation). This fact enables us to find the intensity at an arbitrary optical depth in an independent way, provided that the reflection coefficient is known for any value of the truncation parameter $\alpha \in [0, \tau_0]$.

One can obtain an additional equation defining the intensity in the inhomogeneous atmosphere. To do this, we will use the "reverse counting" of optical depth mentioned in Sect. 9.2. Let us assume that the reference point for the optical depth is placed at the lower boundary of the plane layer (the boundary that is not illuminated directly by the incident radiation) and the depth is measured from the bottom to the top. For simplicity, the angles defining the direction of light propagation are measured in the "usual" reference system. Then, instead of (1.63), we can write:

$$\begin{aligned}{}^{*}B^m(\tau,\mu,\mu_0;\tau_0) &= \frac{\tilde{\lambda}(\tau)}{2}\int\limits_0^1 \tilde{\chi}(-\mu,\mu';\tau)\mathrm{d}\mu'\int\limits_0^{\tau} {}^{*}B^m(\tau',-\mu',\mu_0;\tau_0)\mathrm{e}^{-\frac{\tau-\tau'}{\mu'}}\frac{\mathrm{d}\tau'}{\mu'}\\ &+\frac{\tilde{\lambda}(\tau)}{2}\int\limits_0^1 \tilde{\chi}^m(\mu,\mu';\tau)\mathrm{d}\mu'\int\limits_{\tau}^{\tau_0} {}^{*}B^m(\tau',\mu',\mu_0;\tau_0)\mathrm{e}^{-\frac{\tau'-\tau}{\mu'}}\frac{\mathrm{d}\tau'}{\mu'}\\ &+\frac{\tilde{\lambda}(\tau)}{4}\tilde{\chi}^m(\mu,\mu_0;\tau)\mathrm{e}^{-\frac{\tau_0-\tau}{\mu_0}},\end{aligned} \tag{10.5}$$

where $\tilde{\lambda}(\tau) = \lambda(\tau_0 - \tau)$, $\tilde{\chi}^m(\mu,\mu_0;\tau) = \chi^m(\mu,\mu_0,\tau_0-\tau)$, and the asterisk is used to point out the fact that the source function corresponds to the case of the "reverse counting" of optical depth.

Differentiating (10.5) with respect to τ_0 and using the superposition principle, we find:

$$\begin{aligned}\frac{\partial\,{}^{*}B^m(\tau,\mu,\mu_0;\tau_0)}{\partial\tau_0} &= -\frac{1}{\mu_0}\,{}^{*}B^m(\tau,\mu,\mu_0;\tau_0)\\ &+2\int_0^1 {}^{*}B^m(\tau_0,\mu',\mu_0;\tau_0){}^{*}B^m(\tau,\mu,\mu';\tau_0)\frac{\mathrm{d}\mu'}{\mu'},\end{aligned} \tag{10.6}$$

where

$$\begin{aligned}{}^{*}B^m(\tau_0,\mu,\mu_0;\tau_0) &= \frac{\tilde{\lambda}(\tau_0)}{4}[\tilde{\chi}^m(\mu,\mu_0;\tau_0)\\ &+2\mu_0\int_0^1 \tilde{\chi}^m(-\mu,\mu';\tau_0)\rho^m(\mu',\mu_0;\tau_0)\,\mathrm{d}\mu'].\end{aligned} \tag{10.7}$$

Now, just as above, from (10.6) we easily find the equation for the intensity in the layer considered

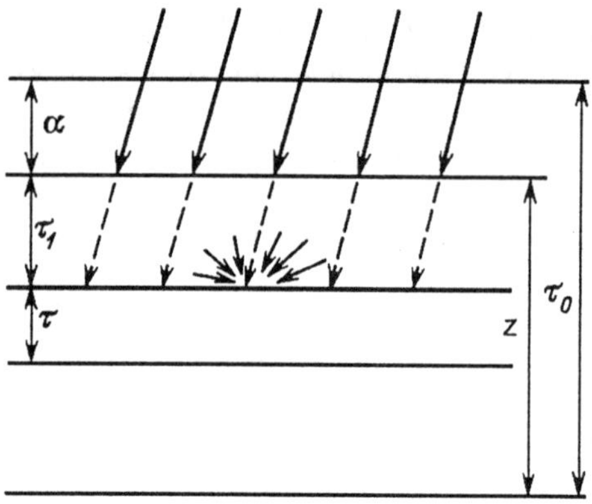

Fig. 10.1. On the derivation of invariance relation (10.9)

$$\frac{\partial\, {}^*I^m(\tau,\mu,\mu_0;\tau_0)}{\partial\tau_0} = -\frac{1}{\mu_0}\, {}^*I^m(\tau,\mu,\mu_0;\tau_0)$$

$$+2\int_0^1 {}^*B^m(\tau_0,\mu',\mu_0;\tau_0)\, {}^*I^m(\tau,\mu,\mu';\tau_0)\frac{\mathrm{d}\mu'}{\mu'} + \frac{1}{\mu}\, {}^*B^m(\tau_0,\mu,\mu_0;\tau_0)\mathrm{e}^{-\frac{\tau_0-\tau}{\mu}}\theta(\mu). \tag{10.8}$$

A characteristic and important property of (10.8) is that, in contrast to (10.3), the intensity of both the upward and downward radiation is defined by two independent equations. It will, however, be shown later that (10.3) can be put in a form which allows one to find the upward and downward intensities separately. To do that, the general invariance principle for a plane inhomogeneous atmosphere (see Sect. 1.12) will be used as a starting point.

10.2 Invariance Relation for a Plane Sublayer and Some of Its Corollaries

Let us consider a plane inhomogeneous sublayer of the truncated atmosphere with an optical depth $z = \tau_0 - \alpha$ illuminated by parallel rays from above (Fig. 10.1). Within this layer we select two levels, τ_1 and τ, measured from the new boundary. It is obvious that $\tau_1 \in [0, z]$, $\tau \in [0, z - \tau_1]$. Then, applying the general invariance principle, for each azimuthal harmonic of the source function we have

$$B^m(\tau+\tau_1,\mu,\mu_0;z,\alpha) = B^m(\tau,\mu,\mu_0;z-\tau_1,\alpha+\tau_1)\mathrm{e}^{-\tau_1/\mu_0}$$
$$+2\int_0^1 I^m(\tau_1,\mu',\mu_0;z,\alpha)B^m(\tau,\mu,\mu';z-\tau_1,\alpha+\tau_1)\,\mathrm{d}\mu'. \tag{10.9}$$

Similarly, for the same layer but illuminated from below, we have

$$\tilde{B}^m(\tau+\tau_1,\mu,\mu_0;z,\alpha)=\tilde{B}^m(\tau,\mu,\mu_0;z-\tau_1,\alpha+\tau_1)$$
$$+2\int_0^1 \tilde{I}^m(\tau_1,\mu',\mu_0;z,\alpha)B^m(\tau,\mu,\mu';z-\tau_1,\alpha+\tau_1)\,\mathrm{d}\mu'. \quad (10.10)$$

Here we have taken into account the absence of direct radiation incident on the truncated layer from above.

Thus, Sobolev's idea of the "truncation" of an inhomogeneous atmosphere permits us to use only one function where, seemingly, it should have been two (indeed, if we subtract a layer of depth α from an inhomogeneous layer then we have a layer with different optical properties). We thereby bring the theory of light scattering in inhomogeneous atmospheres very close to the analogous theory developed for homogeneous atmospheres. This will be demonstrated in detail in the following chapter for an example of an inhomogeneous atmosphere. Now we consider some particular cases of (10.9) and (10.10).

We set $\tau=0$ and designate $\tau_1=\tau$ in (10.9). As a result we find

$$B^m(\tau,\mu,\mu_0;z,\alpha)=B^m(0,\mu,\mu_0;z-\tau,\alpha+\tau)\mathrm{e}^{-\tau/\mu_0}$$
$$+2\int_0^1 I^m(\tau,\mu',\mu_0;z,\alpha)B^m(0,\mu,\mu';z-\tau,\alpha+\tau)\,\mathrm{d}\mu'. \quad (10.11)$$

Taking into account the transfer equation (1.49), for $\alpha=0$ we obtain

$$\mu\frac{\mathrm{d}I^m(\tau,\mu,\mu_0;\tau_0)}{\mathrm{d}\tau}+I^m(\tau,\mu,\mu_0;\tau_0)$$
$$=2\int_0^1 I^m(\tau,\mu',\mu_0;\tau_0)B^m(0,\mu,\mu';\tau_0-\tau,\tau)\,\mathrm{d}\mu'$$
$$+B^m(0,\mu,\mu_0;\tau_0-\tau,\tau)\mathrm{e}^{-\tau/\mu_0}. \quad (10.12)$$

Here the source function at the boundary of the truncated atmosphere with the truncation parameter $\alpha=\tau$ is directly expressed in terms of the reflection coefficient with the help of (9.39). Another nonstandard form of the transfer equation can be obtained by substituting the expression (10.11) for the function $B^m(\tau,\mu,\mu_0;z,\alpha)$ in (10.3). As a result we get

$$\left(\frac{1}{\mu_0}-\frac{1}{\mu}\right)I^m(\tau,\mu,\mu_0;z,\alpha)=\frac{\partial I^m(\tau,\mu,\mu_0;z,\alpha)}{\partial\alpha}$$
$$-\frac{1}{\mu}\left[B^m(0,\mu,\mu_0;z-\tau,\alpha+\tau)\mathrm{e}^{-\tau/\mu_0}-B^m(0,\mu,\mu_0;z,\alpha)\mathrm{e}^{-\tau/\mu}\theta(\mu)\right]$$
$$+2\int_0^1 I^m(\tau,\mu,\mu';z,\alpha)B^m(0,\mu',\mu_0;z,\alpha)\frac{\mathrm{d}\mu'}{\mu'}$$
$$-2\frac{1}{\mu}\int_0^1 I^m(\tau,\mu',\mu_0;z,\alpha)B^m(0,\mu,\mu';z-\tau,\alpha+\tau)\,\mathrm{d}\mu'. \quad (10.13)$$

With the help of (10.1) and (10.2) the invariance relation (10.9) can be rewritten in terms of the intensities (see Chap. 11). From an expression obtained in such a way (or just from the simple physical considerations analogous to those used in RT, Sect. 50) for $\mu \in [0,1]$ we obtain

$$I^m(\tau,-\mu,\mu_0;z,\alpha) = \rho^m(\mu,\mu_0;z-\tau,\alpha+\tau)\mathrm{e}^{-\tau/\mu_0}\mu_0$$
$$+2\int_0^1 I^m(\tau,\mu',\mu_0;z,\alpha)\rho^m(\mu,\mu';z-\tau,\alpha+\tau)\mu'\mathrm{d}\mu', \tag{10.14}$$

$$\sigma^m(\mu,\mu_0;z,\alpha)\mu_0 = \sigma^m(\mu,\mu_0;z-\tau,\alpha+\tau)\mu_0$$
$$+2\int_0^1 I^m(\tau,\mu',\mu_0;z,\alpha)\sigma^m(\mu,\mu';z-\tau,\alpha+\tau)\mu'\mathrm{d}\mu' + I^m(\tau,\mu,\mu_0;z,\alpha)\mathrm{e}^{-\frac{z-\tau}{\mu}}. \tag{10.15}$$

The equations (10.14) and (10.15) are generalizations of the first and fourth Chandrasekhar invariance relations[1] for an inhomogeneous case. The analogies for the second and third invariance relations take the form

$$\tilde{I}^m(\tau,-\mu,\mu_0;z,\alpha) = \tilde{\sigma}^m(\mu,\mu_0;z-\tau,\alpha+\tau)\mu_0$$
$$+2\int_0^1 \tilde{I}^m(\tau,\mu',\mu_0;z,\alpha)\rho^m(\mu,\mu';z-\tau,\alpha+\tau)\mu'\mathrm{d}\mu', \tag{10.16}$$

$$\tilde{\rho}^m(\mu,\mu_0;z,\alpha)\mu_0 = \tilde{\rho}^m(\mu,\mu_0;z-\tau,\alpha+\tau)\mu_0$$
$$+2\int_0^1 \tilde{I}^m(\tau,\mu',\mu_0;z,\alpha)\sigma^m(\mu,\mu';z-\tau,\alpha+\tau)\mu'\mathrm{d}\mu' + \tilde{I}^m(\tau,\mu,\mu_0;z,\alpha)\mathrm{e}^{-\frac{z-\tau}{\mu}}. \tag{10.17}$$

Thus, if the problem of diffuse reflection and transmission of light by a plane layer is solved, (10.12) and (10.13) enable us to find *the intensity of the downward radiation separately*. Thereafter, using (10.14), we can find the intensity of the upward radiation by simple integration. In solving the transfer equation numerically this fact is of considerable significance inasmuch as after discretization of the integrals, it permits one to reduce the angle dimensions of the required matrix I^m by a factor of two.

Besides the analogies of the Chandrasekhar invariance relations, one can also obtain some integral relations of another kind. For these we multiply both sides of (10.14) by $2I^m(\tau,\mu,\mu_1;z,\alpha)$ and integrate over μ from zero to one. As a result, using (10.14) twice, we obtain

[1] These equations are termed invariance principles in RT. However, in accordance with what has been said in Sect. 1.12, we call them relations.

$$2\int_{-1}^{+1} I^m(\tau,-\mu,\mu_0;z,\alpha)I^m(\tau,\mu,\mu_1;z,\alpha)\mu\,\mathrm{d}\mu$$
$$= I^m(\tau,-\mu_0,\mu_1;z,\alpha)\mathrm{e}^{-\tau/\mu_0}\mu_0 - I^m(\tau,-\mu_1,\mu_0;z,\alpha)\mathrm{e}^{-\tau/\mu_1}\mu_1. \quad (10.18)$$

Similarly, multiplying (10.16) by $2\tilde{I}^m(\tau,\mu,\mu_1;z,\alpha)\mu$ and integrating over μ from zero to one and also using (10.16) and (10.17), we find

$$2\int_{-1}^{+1} \tilde{I}^m(\tau,-\mu,\mu_0;z,\alpha)\tilde{I}^m(\tau,\mu,\mu_1;z,\alpha)\mu\,\mathrm{d}\mu$$
$$= \tilde{I}^m(\tau,\mu_1,\mu_0;z,\alpha)\mathrm{e}^{-\frac{z-\tau}{\mu_1}}\mu_1 - \tilde{I}^m(\tau,\mu_0,\mu_1;z,\alpha)\mathrm{e}^{-\frac{z-\tau}{\mu_0}}\mu_0. \quad (10.19)$$

Integral relations like (10.18) and (10.19) will be discussed in detail in the next chapter, where the case of a semi-infinite inhomogeneous medium is considered.

In conclusion, we note that the relations for an inhomogeneous atmosphere found here turn into the corresponding expressions obtained in Chap. 4. For example, (10.3) turns into (4.11) and (10.18) and (10.19) turn into (4.34) if we set $\tau = \tau' = 0$ and $\tau_1 = \tau_2$ in the latter equation.

10.3 On Numerical Methods to Compute Radiation Field in an Inhomogeneous Atmosphere

It is obvious that complete solution of the problem of finding the radiation field in an inhomogeneous atmosphere is, generally speaking, impossible without the use of modern computers. As one can see from the above analysis, this solution can be achieved in three steps:

1. Solving the boundary problem, i.e., finding the reflection and transmission coefficients. The system of integro-differential equations (9.38) and (9.40) serves this purpose. In general, the equations are of the following form:
$$\frac{\mathrm{d}y}{\mathrm{d}\alpha} = \mathcal{F}[y,\alpha], \quad (10.20)$$
where $\mathcal{F}$ is an appropriate integral operator. The boundary condition $y|_{\alpha=\tau_0} = 0$ must to be added to (10.20). It is obvious that to solve (10.20) one of the modern methods for solving the Cauchy problem may be used.
2. When ρ and σ are found for a sufficiently large set of discrete values of $\alpha \in [0,\tau_0]$ we can proceed by calculating the radiation field in the layer with the help, for instance, of (10.13). As we see, (10.13) is of the same type as (10.20). Hence, to solve this equation one can also use the same methods. The relationships (10.15) and (10.17–19) can be used for testing purposes at each computational step.

3. The calculation of the intensity of the upward radiation according to (10.14) and (10.16).

All the above are, of course, rather general remarks. The choice of a particular computational method depends, to a considerable extent, on the behavior of the functions $\lambda(\tau)$ and $\chi(\gamma, \tau)$ with the optical depth τ. For example, if these functions behave smoothly enough, the calculation of the radiation field can be performed with sufficient accuracy by employing the method of adding homogeneous layers and using the equations obtained in the preceding section of this book. In bibliographic notes, more details about some particular numerical implementations of other methods are given.

Now we would like to emphasize the following: the above-mentioned computational scheme can be successfully applied to an optically thin inhomogeneous medium. But it becomes much less efficient for optically thick atmospheres, especially in the case of nearly conservative scattering. Therefore another approach to the problem will be developed in Chap. 12. It is based on rigorous asymptotic formulas which will be obtained there.

10.4 An Inhomogeneous Atmosphere Above a Reflecting Surface

Until now, we have considered a plane sublayer of an inhomogeneous atmosphere in the absence of a reflecting underlying surface. However, a real planetary atmosphere may be adjacent to a reflecting surface which participates in the radiation exchange. Consequently, the radiation field depends on the law according to which light is reflected by the surface.

10.4.1 The Radiation Field in an Atmosphere

Just as in chapter 5, we assume that the reflection coefficient of the surface $\rho_s(\mu, \mu_0, \varphi)$ is given by the expansion (5.3). As before, we denote the intensity of the radiation in an inhomogeneous atmosphere adjacent to the reflecting surface by an overbar. By analogy with (5.2) we write (at $\varphi_0 = 0$)

$$\bar{I}(\tau, \mu, \mu_0, \varphi) = \bar{I}^0(\tau, \mu, \mu_0) + 2 \sum_{m=1}^{n} \bar{I}^m(\tau, \mu, \mu_0) \cos m\varphi. \tag{10.21}$$

An inhomogeneous truncated atmosphere is illuminated from above by parallel rays falling at the angle $\cos^{-1} \mu_0$ to the outward normal. From below it is illuminated by the radiation reflected from the surface. In this case, according to the general invariance principle, we can write ($\mu \in [-1, +1]$)

$$\bar{I}^m(\tau,\mu,\mu_0;z,\alpha) = I^m(\tau,\mu,\mu_0;z,\alpha)$$

$$+2\int_0^1 \tilde{I}^m(\tau,\mu,\mu';z,\alpha)\bar{I}^m(z,-\mu',\mu_0;z,\alpha)\,\mathrm{d}\mu' + \bar{I}^m(z,\mu,\mu_0;z,\alpha)\mathrm{e}^{\frac{z-\tau}{\mu}}\theta(-\mu). \tag{10.22}$$

From physical considerations, for the lower boundary of the atmosphere we have

$$\bar{I}^m(z,-\mu,\mu_0;z,\alpha) = \rho_s^m(\mu,\mu_0)\mathrm{e}^{-z/\mu_0}\mu_0 + 2\int_0^1 \bar{I}^m(z,\mu',\mu_0;z,\alpha)\rho_s^m(\mu,\mu')\mu'\mathrm{d}\mu' \tag{10.23}$$

where $\mu \in [0,1]$. If we set $\tau = z$ in (10.22), we obtain

$$\bar{I}^m(z,\mu,\mu_0;z,\alpha) = \sigma^m(\mu,\mu_0;z,\alpha)\mu_0$$

$$+2\int_0^1 \tilde{\rho}^m(\mu,\mu';z,\alpha)\bar{I}^m(z,-\mu',\mu_0;z,\alpha)\mu'\mathrm{d}\mu'. \tag{10.24}$$

The system of integral equations (10.23) and (10.24) can be used to find the intensity of radiation at the lower boundary, after which the relation (10.22) enables us to calculate the radiation field at an arbitrary optical depth, provided that the radiation field in the atmosphere without the reflecting surface is known.

Solving the system of equations (10.23) and (10.24) can be carried out with the use of the method of successive approximations.

As can easily be seen, the formulas for the reflection and transmission coefficients, which are completely analogous to (5.8) and (5.9), follow from (10.22).

One should take into account that if $l > n$ then (10.21) must contain summands with $m > n$, which are defined by the radiation that came directly from the Sun and was reflected by the surface. If $n > l$ then $\rho_s^m = 0$ at $m > l$ and hence, $\bar{I}^m(z,\mu,\mu_0;z,\alpha) = I^m(z,\mu,\mu_0;z,\alpha)$ for these m.

10.4.2 The Case of a Lambertian Surface. Reflection and Transmission Coefficients

If one assumes that planetary surfaces reflect light according to the Lambert law then the formulas derived above simplify significantly. Indeed, in this case, $\rho_s^0(\mu,\mu_0) = \mathrm{const} = A$ where A is the albedo of the surface. In so far as the existence of an isotropically reflecting surface influences only the azimuthally averaged intensity, we simply have $\bar{I}^m(\tau,\mu,\mu_0;z,\alpha) = I^m(\tau,\mu,\mu_0;z,\alpha)$ for $m \geq 1$.

We find from (10.23) (as before, we omit the superscript "0")

$$\bar{I}(z,-\mu,\mu_0;z,\alpha) = A\left[e^{-z/\mu_0} + 2\int_0^1 \bar{I}(z,\mu',\mu_0;z,\alpha)\mu'\mathrm{d}\mu'\right]. \tag{10.25}$$

In this case, the relation (10.24) can be rewritten as follows:

$$\begin{aligned}\bar{I}(z,\mu,\mu_0;z,\alpha) &= \sigma(\mu,\mu_0;z,\alpha)\mu_0 \\ +A\left[e^{-z/\mu_0}\mu_0 + 2\int_0^1 \bar{I}(z,\mu',\mu_0;z,\alpha)\mu'\mathrm{d}\mu'\right] & 2\int_0^1 \tilde{\rho}(\mu,\mu';z,\alpha)\mu'\mathrm{d}\mu',\end{aligned} \tag{10.26}$$

from which

$$2\int_0^1 \bar{I}(z,\mu',\mu_0;z,\alpha)\mu'\mathrm{d}\mu' = \frac{A\tilde{A}_s(z,\alpha)e^{-z/\mu_0} + 2\int_0^1 \sigma(\mu',\mu_0;z,\alpha)\mu'\mathrm{d}\mu'}{1 - A\tilde{A}_s(z,\alpha)}, \tag{10.27}$$

where

$$\tilde{A}_s(z,\alpha) = 4\int_0^1 \mu\mathrm{d}\mu \int_0^1 \tilde{\rho}(\mu,\mu';z,\alpha)\mu'\mathrm{d}\mu' \tag{10.28}$$

is the spherical albedo of the atmosphere illuminated from below (the reflecting surface is absent). Substituting (10.27) into (10.25) we obtain

$$\bar{I}(z,-\mu,\mu_0;z,\alpha) = \frac{AV(\mu_0;z,\alpha)\mu_0}{1 - A\tilde{A}_s(z,\alpha)}, \tag{10.29}$$

where $V(\mu;z,\alpha)$ is the total relative illumination of the surface [see (1.105)]. After substituting (10.29) into (10.22) we finally find

$$\begin{aligned}\bar{I}(\tau,\mu,\mu_0;z,\alpha) &= I(\tau,\mu,\mu_0;z,\alpha) \\ +A\mu_0 V(\mu_0;z,\alpha) & \frac{e^{\frac{z-\tau}{\mu}}\theta(-\mu) + 2\int_0^1 \tilde{I}(\tau,\mu,\mu';z,\alpha)\,\mathrm{d}\mu'}{1 - A\tilde{A}_s(z,\alpha)}.\end{aligned} \tag{10.30}$$

Introducing the notation ($\mu \in [0,1]$)

$$\bar{I}(0,-\mu,\mu_0;z,\alpha) = \bar{\rho}(\mu,\mu_0;z,\alpha)\mu_0, \tag{10.31}$$

$$\bar{I}(z,\mu,\mu_0;z,\alpha) = \bar{\sigma}(\mu,\mu_0;z,\alpha)\mu_0, \tag{10.32}$$

we obtain the following expressions for the boundaries of the atmosphere:

$$\bar{\rho}(\mu,\mu_0;z,\alpha) = \rho(\mu,\mu_0;z,\alpha) + \frac{AV(\mu_0;z,\alpha)V(\mu;z,\alpha)}{1 - A\tilde{A}_s(z,\alpha)}, \tag{10.33}$$

$$\bar{\sigma}(\mu,\mu_0;z,\alpha)=\sigma(\mu,\mu_0;z,\alpha)+\frac{A\tilde{A}(\mu;z,\alpha)V(\mu_0;z,\alpha)}{1-A\tilde{A}_s(z,\alpha)}, \tag{10.34}$$

where$\tilde{A}(\mu_0;z,\alpha)$ is the plane albedo of the atmosphere illuminated from below (without the reflecting surface) which is defined by (10.36).

If one compares the formulas

$$V(\mu_0;z,\alpha)=\mathrm{e}^{-z/\mu_0}+2\int_0^1\sigma(\mu,\mu_0;z,\alpha)\mu\,\mathrm{d}\mu, \tag{10.35}$$

$$\tilde{A}(\mu_0;z,\alpha)=2\int_0^1\tilde{\rho}(\mu,\mu_0;z,\alpha)\mu\,\mathrm{d}\mu, \tag{10.36}$$

with (9.42) and (9.21), one can see that

$$V(\mu_0;z,\alpha)=\frac{1}{\mu_0}\psi_1^0(\mu_0;z,\alpha), \tag{10.37}$$

$$\tilde{A}(\mu_0;z,\alpha)=1-\frac{1}{\mu_0}\tilde{\varphi}_1^0(\mu_0;z,\alpha). \tag{10.38}$$

Hence, taking into account (10.28), we have

$$\tilde{A}_s(z,\alpha)=1-\int_0^1\tilde{\varphi}_1^0(\mu;z,\alpha)\mathrm{d}\mu. \tag{10.39}$$

10.4.3 Albedo of the Atmosphere and Illumination of the Surface

Proceeding as in the case of a homogeneous atmosphere (see Chap. 5), one can write the corresponding formulas for an inhomogeneous atmosphere. Indeed, we already have the following relation for the absorption function in the absence of a reflecting surface (see (4.54c)):

$$U(\mu_0;z,\alpha)=1-A(\mu_0;z,\alpha)-V(\mu_0;z,\alpha). \tag{10.40}$$

If $A\neq 0$ then from (10.33) and (10.34) we easily find

$$\bar{A}(\mu_0;z,\alpha)=A(\mu_0;z,\alpha)+\frac{AV_s(z,\alpha)V(\mu_0;z,\alpha)}{1-A\tilde{A}_s(z,\alpha)}, \tag{10.41}$$

$$\bar{V}(\mu_0;z,\alpha)=\frac{V(\mu_0;z,\alpha)}{1-A\tilde{A}_s(z,\alpha)}, \tag{10.42}$$

where

$$V_s(z,\alpha) = 2\int_0^1 V(\mu; z,\alpha)\mu\,\mathrm{d}\mu. \tag{10.43}$$

It is obvious from physical considerations that the absorption function is

$$\bar{U}(\mu_0; z,\alpha) = 1 - \bar{A}(\mu_0; z,\alpha) - (1-A)\bar{V}(\mu_0; z,\alpha). \tag{10.44}$$

Substituting (10.41) and (10.42) into the last expression and taking into account (10.40) we obtain

$$\bar{U}(\mu_0; z,\alpha) = U(\mu_0; z,\alpha) + A\frac{1 - \tilde{A}_s(z,\alpha) - V_s(z,\alpha)}{1 - A\tilde{A}_s(z,\alpha)}V(\mu_0; z,\alpha). \tag{10.45}$$

It follows from (10.41) that the spherical albedo of the planet is defined, in this case, by the formula

$$\bar{A}_s(z,\alpha) = A_s(z,\alpha) + \frac{AV_s^2(z,\alpha)}{1 - A\tilde{A}_s(z,\alpha)}. \tag{10.46}$$

As should have been expected, in the case of a homogeneous atmosphere the formulas derived above turn into the corresponding expressions obtained in Chap. 5. The main difference between the relations derived here and those obtained for a homogeneous medium is that now we have a number of functions for an "inverted" atmosphere. It is obvious that the appearance of these functions is connected with the fact that the photons reflected by the surface illuminate the atmosphere from below.

In this chapter we have obtained all the principal formulas and equations describing the radiation field in an inhomogeneous plane layer of finite optical thickness. The next chapter will be devoted to a detailed study of a semi-infinite inhomogeneous atmosphere.

11. Invariance Relations and Their Corollaries for a Semi-Infinite Atmosphere

The prime objective of this chapter is to derive all the most important rigorous equations and formulas relating to the parallel external flux problem and to the Milne problem. The derivation is based on the generalized invariance principle and the concept of a truncated inhomogeneous atmosphere. In the final section of the chapter, a new important concept — the inverted semi-infinite atmosphere — is introduced, and a general approach to the solution of the main problems is discussed.

11.1 Invariance Relations

Consider an inhomogeneous truncated atmosphere illuminated by parallel rays (see Fig. 11.1). Let us use α to denote the parameter of truncation (see Sect. 9.2). In this atmosphere the radiation intensity at optical depth $\tau + \tau_1$ equals the intensity of radiation in the same layer at depth τ with the parameter of truncation $\tau_1 + \alpha$; the boundary of the layer is assumed to be illuminated by both direct radiation attenuated by the factor $e^{-\tau_1/\mu_0}$, and

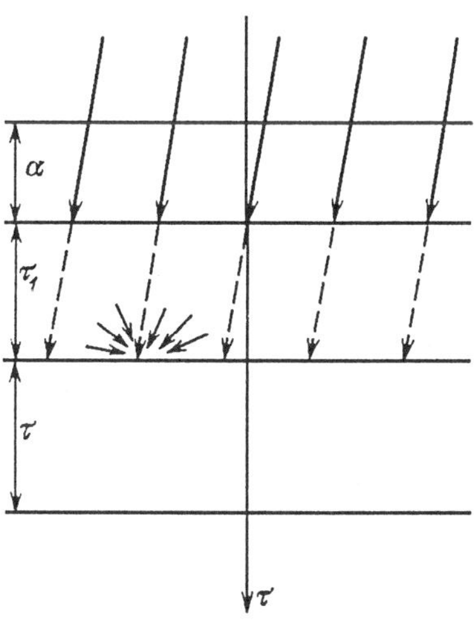

Fig. 11.1. On the derivation of invariance relation (11.1)

diffuse radiation of intensity $I^m(\tau_1, \mu, \mu_0; \alpha)$. Therefore, for any azimuthal harmonic of radiation we have ($\mu \in [-1, +1]$)

$$I^m(\tau + \tau_1, \mu, \mu_0; \alpha) = I^m(\tau, \mu, \mu_0; \tau_1 + \alpha)\mathrm{e}^{-\tau_1/\mu_0}$$
$$+I^m(\tau_1, \mu, \mu_0; \alpha)\mathrm{e}^{-\tau/\mu_0}\theta(\mu) + 2\int_0^1 I^m(\tau_1, \mu', \mu_0; \alpha)I^m(\tau, \mu, \mu'; \tau_1 + \alpha)\,\mathrm{d}\mu'. \tag{11.1}$$

We consider now the Milne problem (i.e., a semi-infinite atmosphere with sources placed at infinity) for the truncated atmosphere with the truncation parameter α. As can be inferred from Fig. 11.1, the radiation intensity at depth $\tau_1 + \tau$ equals the radiation intensity at depth τ in the same layer with a layer of optical thickness τ_1 having been removed and the boundary being illuminated by radiation of intensity $I(\tau_1, \mu; \alpha)$. Hence, for all $\mu \in [-1, +1]$, the following equality holds for the zeroth azimuthal harmonic:

$$I(\tau + \tau_1, \mu; \alpha) = I(\tau, \mu; \tau_1 + \alpha) + I(\tau_1, \mu; \alpha)\mathrm{e}^{-\tau/\mu}\theta(\mu)$$
$$+2\int_0^1 I(\tau_1, \mu'; \alpha)I(\tau, \mu, \mu'; \tau_1 + \alpha)\,\mathrm{d}\mu'. \tag{11.2}$$

Differentiating (11.1) and (11.2) with respect to τ and making use of the following transfer equations

$$\mu\frac{\mathrm{d}I^m(\tau, \mu, \mu_0; \alpha)}{\mathrm{d}\tau} + I^m(\tau, \mu, \mu_0; \alpha) = B^m(\tau, \mu, \mu_0; \alpha), \tag{11.3}$$

$$\mu\frac{\mathrm{d}I(\tau, \mu; \alpha)}{\mathrm{d}\tau} + I(\tau, \mu; \alpha) = B(\tau, \mu; \alpha), \tag{11.4}$$

we obtain the invariance relations

$$B^m(\tau + \tau_1, \mu, \mu_0; \alpha) = B^m(\tau, \mu, \mu_0; \tau_1 + \alpha)\mathrm{e}^{-\tau_1/\mu_0}$$
$$+2\int_0^1 I^m(\tau_1, \mu', \mu_0; \alpha)B^m(\tau, \mu, \mu'; \tau_1 + \alpha)\,\mathrm{d}\mu', \tag{11.5}$$

$$B(\tau+\tau_1, \mu; \alpha) = B(\tau, \mu; \tau_1+\alpha)+2\int_0^1 I(\tau_1, \mu'; \alpha)B(\tau, \mu, \mu'; \tau_1+\alpha)\,\mathrm{d}\mu', \tag{11.6}$$

which, in fact, are (11.1) and (11.2) rewritten in terms of source function. The physical meaning of (11.5) and (11.6) is as clear as that of (11.1) and (11.2); (11.5) and (11.6) can also be derived from physical reasoning (see Sect. 10.2).

The invariance relations (11.1), (11.2) and (11.5), (11.6) are generalizations of (3.2), (3.4), (3.3) and (3.5) in the case of inhomogeneous medium. As was mentioned when studying the truncation of homogeneous atmosphere in the Milne problem, $I(\tau, \mu; \alpha) = I(\tau, \mu)\mathrm{e}^{k\alpha}$.

11.2 Basic Equations Determining the Radiation Field

We denote (for $\mu \in [0,1]$)

$$I^m(0,-\mu,\mu_0;\alpha) = \rho^m(\mu,\mu_0;\alpha)\mu_0, \tag{11.7}$$

$$I(0,-\mu;\alpha) = u(\mu,\alpha). \tag{11.8}$$

The normalization of the escape function $u(\mu,\alpha)$ will be chosen later (see Sect. 11.3). By assuming $\tau = 0$ in (11.1) and (11.2) and renaming $\tau_1 = \tau$, one gets

$$I^m(\tau,-\mu,\mu_0;\alpha) = \rho^m(\mu,\mu_0;\tau+\alpha)\mathrm{e}^{-\tau/\mu_0}\mu_0$$
$$+2\int_0^1 I^m(\tau_1,\mu',\mu_0;\alpha)\rho^m(\mu,\mu';\tau+\alpha)\mu'\,\mathrm{d}\mu', \tag{11.9}$$

$$I(\tau,-\mu;\alpha) = u(\mu;\tau+\alpha) + 2\int_0^1 I(\tau,\mu';\alpha)\rho(\mu',\mu;\tau+\alpha)\mu'\,\mathrm{d}\mu'. \tag{11.10}$$

These formulas are analogous to the Chandrasekhar invariance relations (see RT, Sect. 29.1, 50). For a homogeneous atmosphere, we have $u(\mu,\alpha) = u(\mu)\mathrm{e}^{k\alpha}$.

Proceeding from (11.3), we find ($\mu \in [0,1]$)

$$\mu\frac{\mathrm{d}I(\tau,\mu,\mu_0;\alpha)}{\mathrm{d}\tau}\bigg|_{\tau=0} = B(0,\mu,\mu_0;\alpha). \tag{11.11}$$

Therefore, by differentiating (11.5) with respect to τ_1 and setting $\tau_1 = 0$, we find

$$\frac{\partial B^m(\tau,\mu,\mu_0;\alpha)}{\partial\tau} - \frac{\partial B^m(\tau,\mu,\mu_0;\alpha)}{\partial\alpha} = -\frac{1}{\mu_0}B^m(\tau,\mu,\mu_0;\alpha)$$
$$+2\int_0^1 B^m(0,\mu',\mu_0;\alpha)B^m(\tau,\mu,\mu';\alpha)\frac{\mathrm{d}\mu'}{\mu'}. \tag{11.12}$$

Similarly, one derives from (11.6) that

$$\frac{\partial B(\tau,\mu;\alpha)}{\partial\tau} - \frac{\partial B(\tau,\mu;\alpha)}{\partial\alpha} = 2\int_0^1 B(0,\mu';\alpha)B(\tau,\mu,\mu';\alpha)\frac{\mathrm{d}\mu'}{\mu'}. \tag{11.13}$$

Equations (11.12) and (11.13) can be regarded as a system of integro-differential equations determining functions $B^m(\tau,\mu,\mu_0;\alpha)$ and $B(\tau,\mu,\alpha)$ (i.e., the radiation field inside the atmosphere) providing that the boundary problem has been solved, which means that functions $B^m(0,\mu,\mu_0;\alpha)$ and $B(0,\mu,\alpha)$ are known. To find these functions one can use a separate system of equations, which is derived as follows:

Using (11.4), (11.5) and (11.7), (11.8), we find that for $\mu > 0$

$$\mu \frac{\mathrm{d}I^m(\tau, -\mu, \mu_0; \alpha)}{\mathrm{d}\tau}\bigg|_{\tau=0} = \rho^m(\mu, \mu_0; \alpha)\mu_0 - B^m(0, -\mu, \mu_0; \alpha), \qquad (11.14)$$

$$\mu \frac{\mathrm{d}I(\tau, -\mu; \alpha)}{\mathrm{d}\tau}\bigg|_{\tau=0} = u(\mu, \alpha) - B(0, -\mu; \alpha). \qquad (11.15)$$

Differentiating (11.9), (11.10) with respect to τ and setting then $\tau = 0$, we obtain

$$(\mu + \mu_0)\rho^m(\mu, \mu_0; \alpha) = \frac{\partial \rho^m(\mu, \mu_0; \alpha)}{\partial \alpha}\mu\mu_0 + B^m(0, -\mu, \mu_0; \alpha)$$
$$+2\mu \int_0^1 B^m(0, \mu', \mu_0; \alpha)\rho^m(\mu, \mu'; \alpha)\,\mathrm{d}\mu', \qquad (11.16)$$

$$u(\mu, \alpha) = \frac{\partial u(\mu, \alpha)}{\partial \alpha}\mu + B(0, -\mu; \alpha) + 2\mu \int_0^1 B(0, \mu'; \alpha)\rho(\mu, \mu'; \alpha)\,\mathrm{d}\mu'. \quad (11.17)$$

On the other hand, it is known that

$$B^m(0, \mu, \mu_0; \alpha) = \frac{\lambda(\alpha)}{4}\left[\chi^m(\mu, \mu_0; \alpha) + 2\mu_0 \int_0^1 \chi^m(\mu, -\mu'; \alpha)\rho^m(\mu', \mu_0; \alpha)\,\mathrm{d}\mu'\right], \qquad (11.18)$$

$$B(0, \mu; \alpha) = \frac{\lambda(\alpha)}{2}\int_0^1 \chi(\mu, \mu'; \alpha)u(\mu', \alpha)\,\mathrm{d}\mu'. \qquad (11.19)$$

Thus, the relations (11.16–19) is a system of integro-differential equations which determines four functions, namely $B^m(0, \mu, \mu_0; \alpha)$, $B(0, \mu, \alpha)$, $u(\mu, \alpha)$, and $\rho^m(\mu, \mu_0; \alpha)$. Equation (11.17) can be rewritten to put it into another form. The substitution of (11.19) into (11.17) yields

$$u(\mu, \alpha) = \frac{\partial u(\mu, \alpha)}{\partial \alpha}\mu + \frac{\lambda(\alpha)}{2}\int_0^1 \chi(\mu, \mu'; \alpha)u(\mu', \alpha)\,\mathrm{d}\mu'$$
$$+\lambda(\alpha)\mu \int_0^1 u(\mu''; \alpha)\,\mathrm{d}\mu'' \int_0^1 \chi(\mu', -\mu''; \alpha)\rho(\mu, \mu'; \alpha)\,\mathrm{d}\mu'. \qquad (11.20)$$

Making use of (11.18), one ultimately finds

$$u(\mu, \alpha) = \frac{\partial u(\mu, \alpha)}{\partial \alpha}\mu + 2\int_0^1 u(\mu', \alpha)B(0, \mu', \mu; \alpha)\,\mathrm{d}\mu'. \qquad (11.21)$$

With the help of (9.41), which is a representation of function B, we find the following formula instead of (11.21)

$$u(\mu, \alpha) = \frac{\partial u(\mu, \alpha)}{\partial \alpha}\mu + \frac{\lambda(\alpha)}{2}\sum_{i=0}^{n} x_i(\alpha)\varphi_i(\mu, \alpha)\int_0^1 u(\mu', \alpha)P_i(\mu')\,\mathrm{d}\mu'. \qquad (11.21a)$$

Equations (11.12), (11.13), (11.16), (11.17), (11.21) and (11.21a) are generalizations of the Ambartsumian invariance relations in the case of an inhomogeneous atmosphere.

One can derive a number of other equations also determining the radiation intensity. For (11.3) and (11.4), we know that

$$B^m(\tau,\mu,\mu_0;\alpha) = \frac{\lambda(\tau+\alpha)}{2}\int_{-1}^{+1} I^m(\tau,\mu',\mu_0;\alpha)\chi^m(\mu',\mu;\tau+\alpha)\,\mathrm{d}\mu' + \frac{\lambda(\tau+\alpha)}{4}\chi^m(\mu,\mu_0;\tau+\alpha)\mathrm{e}^{-\tau/\mu_0}, \tag{11.22}$$

$$B(\tau,\mu;\alpha) = \frac{\lambda(\tau+\alpha)}{2}\int_{-1}^{+1} I(\tau,\mu';\alpha)\chi(\mu',\mu;\tau+\alpha)\,\mathrm{d}\mu'. \tag{11.23}$$

We differentiate (11.1) with respect to τ_1 and set $\tau_1 = 0$. With the help of (11.11) we find

$$\mu_0\frac{\partial I^m(\tau,\mu,\mu_0;\alpha)}{\partial\tau} - \mu_0\frac{\partial I^m(\tau,\mu,\mu_0;\alpha)}{\partial\alpha} + I^m(\tau,\mu,\mu_0;\alpha) = 2\mu_0\int_0^1 B^m(0,\mu',\mu_0;\alpha)I^m(\tau,\mu,\mu';\alpha)\frac{\mathrm{d}\mu'}{\mu'} + \frac{\mu_0}{\mu}B^m(0,\mu,\mu_0;\alpha)\mathrm{e}^{-\tau/\mu}\theta(\mu). \tag{11.24}$$

With the help of the transfer equation (11.3) we eliminate the derivative $\partial I^m/\partial\tau$ from (11.24) and obtain one more equation

$$\left(\frac{1}{\mu_0}-\frac{1}{\mu}\right)I^m(\tau,\mu,\mu_0;\alpha) = \frac{\partial I^m(\tau,\mu,\mu_0;\alpha)}{\partial\alpha} - \frac{1}{\mu}B^m(\tau,\mu,\mu_0;\alpha) + \frac{1}{\mu}B^m(0,\mu,\mu_0;\alpha)\mathrm{e}^{-\tau/\mu}\theta(\mu) + 2\int_0^1 B^m(0,\mu',\mu_0;\alpha)I^m(\tau,\mu,\mu';\alpha)\frac{\mathrm{d}\mu'}{\mu'}, \tag{11.25}$$

where $B^m(\tau,\mu,\mu_0;\alpha)$ is determined by (11.22). And finally, by setting $\tau = 0$ in (11.5) and using (11.3), we arrive at the following equation:

$$\mu\frac{\mathrm{d}I^m(\tau,\mu,\mu_0;\alpha)}{\mathrm{d}\tau} + I^m(\tau,\mu,\mu_0;\alpha) = B^m(0,\mu,\mu_0;\tau+\alpha)\mathrm{e}^{-\tau/\mu_0} + 2\int_0^1 I^m(\tau,\mu',\mu_0;\alpha)B^m(0,\mu,\mu';\tau+\alpha)\,\mathrm{d}\mu'. \tag{11.26}$$

Now, using (11.26) to eliminate the derivative $\partial I^m/\partial\tau$ from (11.24), we find one more equation

$$\frac{\partial I^m(\tau,\mu,\mu_0;\alpha)}{\partial\alpha} + \left(\frac{1}{\mu}-\frac{1}{\mu_0}\right)I^m(\tau,\mu,\mu_0;\alpha) = 2\int_0^1 I^m(\tau,\mu',\mu_0;\alpha)B^m(0,\mu,\mu';\tau+\alpha)\frac{\mathrm{d}\mu'}{\mu}$$

$$-2\int_0^1 I^m(\tau,\mu,\mu';\alpha)B^m(0,\mu',\mu_0;\alpha)\frac{d\mu'}{\mu'}$$
$$+\frac{1}{\mu}\left[B^m(0,\mu,\mu_0;\tau+\alpha)e^{-\tau/\mu_0} - B^m(0,\mu,\mu_0;\alpha)\,e^{-\tau/\mu}\theta(\mu)\right]. \quad (11.27)$$

It follows from (11.26), (11.27) and (11.18) that if the problem of the diffuse reflection of light has been solved (i.e., $\rho(\mu,\mu_0;\alpha)$ is known for all $\alpha \geq 0$), then the intensity of descending radiation ($\mu \in [0,1]$) can be found irrespective of whether we know the intensity for $\mu \in [-1,0]$ or not. After that, the intensity of ascending radiation can be found with the help of (11.9). Equation (11.27) is a particular case of (10.13).

Now some remarks on the Milne problem. Putting $\tau = 0$ in (11.6), one can rewrite the transfer equation for the Milne problem as follows:

$$\mu\frac{dI(\tau,\mu;\alpha)}{d\tau}+I(\tau,\mu;\alpha) = B(0,\mu;\tau+\alpha)+2\int_0^1 I(\tau,\mu';\alpha)B(0,\mu,\mu';\tau+\alpha)\,d\mu'. \quad (11.28)$$

The following equation is an analogy of (11.25) for the Milne problem:

$$B(\tau,\mu;\alpha) = \mu\frac{\partial I(\tau,\mu;\alpha)}{\partial\alpha} + I(\tau,\mu;\alpha)$$
$$+2\mu\int_0^1 B(0,\mu';\alpha)I(\tau,\mu,\mu';\alpha)\frac{d\mu'}{\mu'} + B(0,\mu;\alpha)e^{-\tau/\mu}\theta(\mu). \quad (11.29)$$

Now, there is no point in discussing this problem more elaborately since we will see later that the solution of the transfer equation, for an infinite medium and a semi-infinite atmosphere illuminated by parallel rays, gives the complete solution of the Milne problem. For instance, the escape function is determined by (11.21a). It is possible, however, to find another way to find this function, and we are to do it below.

11.3 Two Integral Relations, Normalization of Escape Function and *M* Integral

Let us complete the initial semi-infinite medium evenly, so that we have an infinite medium, i.e., put $\chi(\mu,\mu_0;\tau) = \chi(\mu,\mu_0;-\tau)$ and $\lambda(\tau) = \lambda(-\tau)$. In other words, let us consider the problem for two-sided infinity (see Sect. 1.7). This problem is known to be reduced to the solution of the equation

$$\mu\frac{dI_\infty(\tau,\mu)}{d\tau} + I_\infty(\tau,\mu) = \frac{\lambda(\tau)}{2}\int_{-1}^{+1} I_\infty(\tau,\mu')\chi(\mu,\mu';\tau)\,d\mu' \quad (11.30)$$

provided the condition (1.77a) holds.

Physical considerations clearly suggest (see Fig. 1.5) that the intensity $I_\infty(-\tau,\mu)$ at the cut at $-\tau$ consists of two parts. The first is due to those photons that have passed diffusely from the source through the semi-infinite medium on the left of the cut (i.e., due to the photons that have never crossed the boundary $-\tau$). This part of the intensity is determined by the function $u(\mu,\tau)$. The other part is due to the photons that have been diffusely reflected by the half-space on the left of the cut. So, with the help of the generalized invariance principle one can write

$$I_\infty(-\tau,\mu) = Mu(\mu,\tau) + 2\int_0^1 \rho(\mu,\mu';\tau) I_\infty(-\tau,-\mu')\mu'\,\mathrm{d}\mu', \tag{11.31}$$

where M is a normalizing function of the truncation parameter τ. For the conservative case, the inhomogeneous analogue of the Chandrasekhar invariance relation follows from (11.31) (see RT, Sect. 29.3)

$$u_0(\mu,\tau) = \frac{3}{4}\left[\mu + 2\int_0^1 \rho_0(\mu,\mu';\tau)\mu'^2\,\mathrm{d}\mu'\right]. \tag{11.31a}$$

For the homogeneous case, this formula was derived in Sect. 3.3. The proof of (11.31a) will be given later, in Sect. 12.5.

Similarly for intensity $I_\infty(\tau;-\mu)$ we find

$$I_\infty(\tau,-\mu) = 2\int_0^1 \rho(\mu,\mu';\tau) I_\infty(\tau,\mu')\mu'\,\mathrm{d}\mu'. \tag{11.32}$$

Equations (11.31) and (11.32) are generalizations of the Sobolev–van de Hulst relations (3.10) and (3.11) for the inhomogeneous case. It follows from (11.31) that if the intensity $I_\infty(\tau,\mu)$ and the reflection coefficient $\rho(\mu,\mu_0,\tau)$ are known, then the escape function $u(\mu,\tau)$ is also known. However, this function still needs normalizing, for which we require that the following equality holds:

$$2\int_0^1 u(\mu) I_\infty(0,\mu)\mu\,\mathrm{d}\mu = 1. \tag{11.33}$$

Here $u(\mu) = u(\mu,0)$. This equality is an analogy of the condition (3.12). Now we are to show that in this case the following equality holds for the truncated atmosphere:

$$2\int_0^1 u(\mu,\tau) I_\infty(\tau,\mu)\mu\,\mathrm{d}\mu = 1. \tag{11.34}$$

Rename α by τ in (11.20) and multiply the result by $2I_\infty(\tau,\mu)$. Integrating the product with respect to μ from 0 to 1 and making use of (11.32), we find

$$\begin{aligned}2\int_0^1 u(\mu,\tau) I_\infty(\tau,\mu)\,\mathrm{d}\mu &= 2\int_0^1 \frac{\mathrm{d}u(\mu,\tau)}{\mathrm{d}\tau} I_\infty(\tau,\mu)\mu\,\mathrm{d}\mu \\ &+\lambda(\tau)\int_0^1 I_\infty(\tau,\mu)\chi(\mu,\mu';\tau)\,\mathrm{d}\mu\int_0^1 u(\mu',\tau)\mathrm{d}\mu' \\ &+\lambda(\tau)\int_0^1 u(\mu,\tau)\mathrm{d}\mu\int_0^1 I_\infty(\tau,-\mu')\chi(\mu',-\mu;\tau)\,\mathrm{d}\mu'.\end{aligned} \tag{11.35}$$

Making allowance for the fact that $\chi(-\mu', -\mu; \tau) = \chi(\mu', \mu, \tau)$, one can rewrite (11.35) as follows:

$$2\int_0^1 u(\mu,\tau)I_\infty(\tau,\mu)\,\mathrm{d}\mu = 2\int_0^1 \frac{du(\mu,\tau)}{\mathrm{d}\tau}I_\infty(\tau,\mu)\mu\,\mathrm{d}\mu$$
$$+\lambda(\tau)\int_0^1 u(\mu,\tau)\mathrm{d}\mu\int_{-1}^{+1}\chi(\mu',\mu;\tau)I_\infty(\tau,\mu')\,\mathrm{d}\mu'. \tag{11.36}$$

And with the help of the transfer equation (11.30) we have

$$2\frac{\mathrm{d}}{\mathrm{d}\tau}\int_0^1 u(\mu,\tau)I_\infty(\tau,\mu)\mu\,\mathrm{d}\mu = 0.$$

If (11.33) holds, then we arrive at the normalizing condition (11.34). Applying it to (11.31) and (11.32), we find

$$M = 2\int_{-1}^{+1} I_\infty(\tau,\mu)I_\infty(-\tau,\mu)\mu\,\mathrm{d}\mu. \tag{11.37}$$

The next important result follows directly from the transfer equation (11.30):

$$\frac{\mathrm{d}M}{\mathrm{d}\tau} = 0. \tag{11.38}$$

In fact, by differentiating (11.37) with respect to τ and using (11.30), we derive (11.38).

Thus, *M is invariant with respect to τ* in the problem of radiation transfer in an infinite inhomogeneous medium whose optical properties have been expanded evenly over negative depths. M we will call to be the M integral of this equation. The condition (11.38) can be used for the verification of calculations. Therefore the invariance of the M integral may be very important for the numerical solution of (11.30) for all τ.

11.4 Relationship Between the Milne Problem and the Parallel External Flux Problem

Let us consider, like we did in the previous section, an infinite inhomogeneous atmosphere with sources placed at minus infinity whose optical properties have been continued into the negative depths in the even way. Assume also that the particles albedo and phase function in the half-space on the left (or right) side of the cut $\tau = 0$ vary in such a way that there exists a solution to the Milne problem.

Consider the half-space on the left of the cut $\tau = 0$ (Fig. 1.5). For all $\tau < 0$, the intensity in an infinite medium can, at the same time, be considered as the intensity in a semi-infinite medium with one of the boundaries illuminated by sources lying at minus infinity and with the boundary $\tau = 0$ illuminated by radiation of the intensity $I_\infty(0, -\mu)$. Then we have

$$I_\infty(-\tau, -\mu) = MI(\tau, \mu) + 2\int_0^1 I(\tau, \mu, \mu')I_\infty(0, -\mu')\,\mathrm{d}\mu' + I_\infty(0, -\mu)\mathrm{e}^{-\tau/\mu}\theta(\mu), \tag{11.39}$$

where M is a normalizing factor, which in general, depends on τ. The half-space on the right of the cut $\tau = 0$ is illuminated along cut only by radiation of intensity $I_\infty(0, \mu)$. Therefore, for $\tau \geq 0$ we find

$$I_\infty(\tau, \mu) = 2\int_0^1 I(\tau, \mu, \mu')I_\infty(0, \mu')\,\mathrm{d}\mu' + I_\infty(0, \mu)\mathrm{e}^{-\tau/\mu}\theta(\mu). \tag{11.40}$$

Proceeding from the same physical ideas, one can rewrite (11.39) and (11.40) in terms of source functions as follows ($\tau \geq 0$):

$$B_\infty(-\tau, \mu) = MB(\tau, \mu) + 2\int_0^1 B(\tau, \mu, \mu')I_\infty(0, -\mu')\,\mathrm{d}\mu', \tag{11.41}$$

$$B_\infty(\tau, \mu) = 2\int_0^1 B(\tau, \mu, \mu')I_\infty(0, \mu')\,\mathrm{d}\mu'. \tag{11.42}$$

By differentiating (11.39) and (11.40) with respect to τ and making use of the transfer equation as well as (11.41) and (11.42), we arrive at the conclusion that $\mathrm{d}M/\mathrm{d}\tau = 0$. As (11.39) and (11.40) at $\tau = 0$ turn into the ready-known (11.31) and (11.32) with τ substituted with $\tau = 0$, one comes to the conclusion that M in (11.39) is the M integral, which is provided by (11.37).

Equations (11.39) and (11.40) are a generalization for an inhomogeneous medium of equalities (3.6) and (3.7). Equation (11.39) suggests that, just as was the case for a homogeneous medium, the solution to the problem for two-sided infinity together with the solution to the parallel external flux problem provides the complete solution to the Milne problem. Hence, we can state — as we could for the inhomogeneous case — that the Milne problem is not a separate case.

The following expressions are the conservative limits of (11.39) and (11.40) respectively:

$$2\int_0^1 I_0(\tau, \mu, \mu')\,\mathrm{d}\mu' - 1 - \mathrm{e}^{-\tau/\mu}\theta(\mu), \tag{11.43}$$

$$I_0(\tau, \mu) = \frac{3}{4}\left\{\left[1 - \frac{\bar{x}_1(\tau)}{3}\right]\tau - \mu + 2\int_0^1 I_0(\tau, \mu, \mu')\mu'\,\mathrm{d}\mu' + \mu\mathrm{e}^{-\tau/\mu}\theta(\mu)\right\}, \tag{11.43a}$$

where $\bar{x}_1(\tau)$ is determined according to (1.117). These formulas will be proved in Sect. 12.6. Now we would like to draw the reader's attention to the fact that (11.43) and (11.43a) are surprisingly similar to corresponding formulas for a homogeneous medium ((3.16) and (3.15) respectively). Namely, (11.43) and (3.16) just coincide, and the only difference between (11.43a) and (3.15) is that in (11.43a), x_1 is substituted with its value averaged over the interval of optical depths from 0 to τ.

11.5 Some Integral Relations

Besides the transfer equation, the intensity of radiation in an inhomogeneous atmosphere also satisfies a number of non-linear relations. Now we are to derive these integral relations.

Multiply (11.3) with $m = 0$ by $2u(\mu, \tau + \alpha)$ and integrate the product with respect to μ from 0 to 1. Allowing for the fact that

$$2\int_0^1 \frac{\mathrm{d}I(\tau, \mu, \mu_0; \alpha)}{\mathrm{d}\tau} u(\mu, \tau + \alpha)\mu \,\mathrm{d}\mu = \frac{\partial}{\partial\tau}\left[2\int_0^1 I(\tau, \mu, \mu_0; \alpha)u(\mu, \tau + \alpha)\mu \,\mathrm{d}\mu\right]$$
$$-2\int_0^1 I(\tau, \mu, \mu_0; \alpha)\frac{\partial u(\mu, \tau + \alpha)}{\partial\alpha}\mu \,\mathrm{d}\mu$$

and using (11.21) to eliminate $\partial u/\partial\alpha$, one finds

$$\frac{\partial}{\partial\tau}\left[2\int_0^1 I(\tau, \mu, \mu_0; \alpha)u(\mu, \tau + \alpha)\mu \,\mathrm{d}\mu\right]$$
$$+4\int_0^1 u(\mu', \tau + \alpha)\,\mathrm{d}\mu' \int_0^1 I(\tau, \mu, \mu_0; \alpha)B(0, \mu', \mu; \tau + \alpha)\,\mathrm{d}\mu$$
$$= 2\int_0^1 B(\tau, \mu, \mu_0; \alpha)u(\mu, \tau + \alpha)\mu \,\mathrm{d}\mu. \tag{11.44}$$

Function $B(\tau, \mu, \mu_0; \alpha)$ is determined by the right-hand side of (11.26). Therefore using (11.21) once again, we can rewrite (11.44) in the following way:

$$\frac{\partial}{\partial\tau}\left[2\int_0^1 I(\tau, \mu, \mu_0; \alpha)u(\mu, \tau + \alpha)\mu \,\mathrm{d}\mu\right] = -\mu_0\frac{\partial}{\partial\tau}\left[u(\mu_0, \tau + \alpha)\mathrm{e}^{-\tau/\mu_0}\right].$$

After the integration has been performed, we ultimately find

$$2\int_0^1 I(\tau, \mu', \mu; \alpha)u(\mu', \tau+\alpha)\mu' \,\mathrm{d}\mu' = \mu\left[u(\mu, \alpha) - u(\mu, \tau + \alpha)\mathrm{e}^{-\tau/\mu}\right]. \tag{11.45}$$

This is one of the relations sought. It establishes the relationship between the intensity of radiation in a semi-infinite atmosphere, and the function $u(\mu, \alpha)$, which determines the solution of the Milne problem at the boundary.

Acting similarly and using (11.4), (11.21) and (11.6), one can derive one more relation of the same kind. This one involves only the functions relating to the Milne problem, namely

$$\frac{\partial}{\partial \tau} \int_0^1 I(\tau, \mu; \alpha) u(\mu, \tau + \alpha) \mu \, d\mu = \int_0^1 u(\mu, \tau + \alpha) B(0, \mu; \tau + \alpha) \, d\mu. \quad (11.46)$$

With the help of (11.19), it can be rewritten as follows:

$$\frac{\partial}{\partial \tau} \int_0^1 I(\tau, \mu; \alpha) u(\mu, \tau + \alpha) \mu \, d\mu$$
$$= \frac{\lambda(\tau + \alpha)}{2} \int_0^1 u(\mu, \tau + \alpha) \, d\mu \int_0^1 \chi(\mu, -\mu'; \tau + \alpha) u(\mu', \tau + \alpha) \, d\mu'. \quad (11.47)$$

Further manipulations with (11.47) involve (11.31). Having replaced α with τ, we multiply (11.20) by $2I_\infty(-\tau, -\mu)$ and integrate the product with respect to μ from 0 to 1. Using (11.30) and (11.31) one finds one more relation

$$\lambda(\tau) M \int_0^1 u(\mu, \tau) \, d\mu \int_0^1 \chi(\mu, -\mu'; \tau) u(\mu', \tau) \, d\mu'$$
$$= \frac{d}{d\tau} 2 \int_0^1 u(\mu, \tau) I_\infty(-\tau, -\mu) \mu \, d\mu = \frac{dN(\tau)}{d\tau}, \quad (11.48)$$

where

$$N(\tau) = 2 \int_0^1 u(\mu, \tau) I_\infty(-\tau, -\mu) \mu \, d\mu \quad (11.49)$$

by definition. Hence, one derives the last of the relations sought

$$2M \int_0^1 I(\tau, \mu; \alpha) u(\mu, \tau + \alpha) \mu \, d\mu = N(\tau + \alpha) - N(\alpha), \quad (11.50)$$

which corresponds to (11.47).

Comparing (11.45), (11.48) and (11.50) with (3.42), (3.41) and (3.43), respectively, we can see once again that they resemble each other in appearance.

To conclude this section, we provide conservative limits for (11.48) and (11.50). Subscript "0" will be used, as it always has been, to denote quantities relating to the conservative atmosphere ($\lambda(\tau) \equiv 1$). Instead of (11.48), one finds

$$4 \int_0^1 u_0(\mu, \tau) \, d\mu \int_0^1 \chi(-\mu, \mu'; \tau) u_0(\mu', \tau) \, d\mu' = 3 - x_1(\tau) - 3\gamma_0'(\tau), \quad (11.51)$$

where

$$\gamma_0(\tau) = 2 \int_0^1 u_0(\mu, \tau) \mu^2 \, d\mu, \quad (11.52)$$

and instead of (11.50),

$$8\int_0^1 I_0(\tau,\mu;\alpha)u_0(\mu,\tau+\alpha)\mu\,\mathrm{d}\mu = 3[\tau-\gamma_0(\tau+\alpha)+\gamma_0(\alpha)] - \int_\alpha^{\tau+\alpha} x_1(t)\,\mathrm{d}t. \tag{11.53}$$

These formulas will be proved in Sect. 12.6.

11.6 Integrals of the Transfer Equation

In this chapter we have encountered three problems:

1. the radiation transfer in an infinite atmosphere that has been continued into negative optical depths symmetrically;
2. the Milne problem;
3. the parallel external flux problem.

The solutions to these three problems are provided by three functions, namely $I_\infty(\tau,\mu)$, $I(\tau,\mu;\alpha)$, and $I^m(\tau,\mu,\mu_0;\alpha)$. The three forms of the transfer equations which these functions correspond to, admit integrals that are quadratic with respect to intensities. To derive these integrals we need the following expressions.

The following more general formulas can be written instead of (11.31) and (11.32)

$$I_\infty(-\tau,\mu;\alpha) = M(\alpha)u(\mu,\tau+\alpha) + 2\int_0^1 \rho(\mu,\mu';\tau+\alpha)I_\infty(-\tau,-\mu';\alpha)\mu'\,\mathrm{d}\mu', \tag{11.54}$$

$$I_\infty(\tau,-\mu;\alpha) = 2\int_0^1 \rho(\mu,\mu';\tau+\alpha)I_\infty(\tau,\mu';\alpha)\mu'\,\mathrm{d}\mu', \tag{11.55}$$

where $I_\infty(\tau,\mu;\alpha)$ is determined for all τ by the equation ($\alpha \geq 0$)

$$\mu\frac{\mathrm{d}I_\infty(\tau,\mu;\alpha)}{\mathrm{d}\tau} + I_\infty(\tau,\mu;\alpha) = \frac{\lambda(|\tau|+\alpha)}{2}\int_{-1}^{+1} I_\infty(\tau,\mu';\alpha)\chi(\mu,\mu';|\tau|+\alpha)\,\mathrm{d}\mu', \tag{11.56}$$

and M by the equation

$$M(\alpha) = 2\int_{-1}^{+1} I_\infty(\tau,\mu;\alpha)I_\infty(-\tau,\mu;\alpha)\mu\,\mathrm{d}\mu \tag{11.57}$$

The following normalizing condition is supposed to hold:

$$\frac{\lambda(\alpha)}{2}\int_{-1}^{+1} I_\infty(0,\mu;\alpha)\,\mathrm{d}\mu = 1. \tag{11.58}$$

Function $I_\infty(\tau, \mu; \alpha)$ provides the solution of the problem for two-sided infinity, whose optical properties are symmetrical with respect to the origin of coordinates $\tau = 0$. The cuts are placed at the levels $\tau = \alpha$ and $\tau = -\alpha$ in this atmosphere, and the layer of optical thickness 2α lying between them is discarded.

Besides (11.54) and (11.55), we will also use (11.9) and (11.10). All these four relations include the function $\rho(\mu, \mu'; \alpha)$, which is symmetrical with respect to permutations of μ and μ'. Therefore, acting as we did in deriving (10.18) we find from (11.9) that

$$2\int_{-1}^{+1} I^m(\alpha, \mu', \mu_0; \tau) I^m(\tau, -\mu', \mu; \alpha)\mu' \, \mathrm{d}\mu'$$
$$= I^m(\alpha, -\mu_0, \mu; \tau)\mathrm{e}^{-\tau/\mu_0}\mu_0 - I^m(\tau, -\mu, \mu_0; \alpha)\mathrm{e}^{-\alpha/\mu}\mu. \qquad (11.59)$$

We introduce the following quantity

$$^*I^m(\tau, \mu, \mu_0; \alpha) = I^m(\tau, \mu, \mu_0; \alpha) + \frac{1}{2}\delta(\mu - \mu_0)\mathrm{e}^{-\tau/\mu}, \qquad (11.60)$$

where δ is the Dirac function. This means that we are introducing the total intensity of direct + diffuse radiation for the corresponding azimuthal harmonic. Also, we will denote integrals of the transfer equation by $\mathcal{I}_{ij}(i, j = 1, 2, 3)$ where i and j numerate the number of the problems mentioned at the beginning of this section and whose solutions are involved in the integration. As a result, using (11.60), one can rewrite (11.59) in the following way

$$\mathcal{I}_{33} = 2\int_{-1}^{+1} {}^*I^m(\tau, \mu', \mu_0; \alpha)\, {}^*I^m(\alpha, -\mu', \mu; \tau)\mu' \, \mathrm{d}\mu' = 0. \qquad (11.61)$$

This is one of the integrals sought. Combining (11.54), (11.55), (11.9) and (11.10) and deriving as for (11.59), one can obviously find the other integrals. They can be subdivided into two groups: a) "pure" integrals, which quadratically include intensities of the same problem; b) "mixed" integrals, which involve intensities of different problems. The "pure" integral $\mathcal{I}_{11}$ has been derived earlier and is called the M integral

$$\mathcal{I}_{11} = M = 2\int_{-1}^{+1} I_\infty(-\tau, \mu) I_\infty(\tau, \mu)\mu \, \mathrm{d}\mu. \qquad (11.62)$$

The other integrals are

$$\mathcal{I}_{22} = 2\int_{-1}^{+1} I(\tau, \mu; \alpha) I(\alpha, -\mu; \tau)\mu \, \mathrm{d}\mu = [N(\tau) - N(\alpha)]M^{-1}; \qquad (11.63)$$

$$\mathcal{I}_{12}^{(1)} = 2\int_{-1}^{+1} I(\tau, -\mu'; \alpha) I_\infty(\tau, \mu'; \alpha)\mu' \, \mathrm{d}\mu' = 1, \tag{11.64}$$

$$\mathcal{I}_{12}^{(2)} = 2\int_{-1}^{+1} I(\tau, -\mu'; \alpha) I_\infty(-\tau, -\mu'; \alpha)\mu' \, \mathrm{d}\mu' = N(\alpha), \tag{11.65}$$

$$\mathcal{I}_{13}^{(1)} = 2\int_{-1}^{+1} {}^*I^0(\tau, -\mu', \mu; \alpha) I_\infty(\tau, \mu'; \alpha)\mu' \, \mathrm{d}\mu' = 0, \tag{11.66}$$

$$\mathcal{I}_{13}^{(2)} = 2\int_{-1}^{+1} {}^*I^0(\tau, \mu', \mu; \alpha) I_\infty(-\tau, \mu'; \alpha)\mu' \, \mathrm{d}\mu' = M(\alpha) u(\mu, \alpha)\mu, \tag{11.67}$$

and only one "mixed" integral of the type 23:

$$\mathcal{I}_{23}^{(1)} = 2\int_{-1}^{+1} {}^*I^0(\tau, \mu', \mu; \alpha) I(\alpha, -\mu'; \tau)\mu' \, \mathrm{d}\mu' = u(\mu, \alpha)\mu. \tag{11.68}$$

To derive (11.63) and (11.67) we have used (11.50) and (11.45).

First of all, it can easily be seen that all the integrals but $\mathcal{I}_{22}$ are invariant over the optical depth τ. As to $\mathcal{I}_{22}$, one should note that the quantity $\partial\mathcal{I}_{22}/\partial\alpha$ is invariant, whereas $\mathcal{I}_{22}$ itself is not. For the conservative case, "mixed" integrals of types (11.64–67) turn into ordinary integrals of flux for the Milne problem and the parallel external flux problem. It can be easily proved that for $\mathcal{I}_{22}$ the following equality holds:

$$\begin{aligned} \mathcal{I}_{22} &= 2\int_{-1}^{+1} I_0(\tau, \mu; \alpha) I_0(\alpha, -\mu; \tau)\mu \, \mathrm{d}\mu \\ &= \frac{3}{4}\left[\tau - \alpha + \gamma_0(\alpha) - \gamma_0(\tau) - \frac{1}{3}\int_\alpha^\tau x_1(t)\mathrm{d}t\right]. \end{aligned} \tag{11.69}$$

Asymptotic constants M and N as well as function $u(\mu, \alpha)$ (we will see later that they arise in the study of asymptotic properties of fields in optically thick atmospheres) are integrals for corresponding problems.

11.7 The Concept of an Inverted Semi-Infinite Atmosphere

We have already seen that the concept of the *truncated* atmosphere is very important for the theory we are developing. It is this very idea that allowed us to write a number of equations and relations for an inhomogeneous atmosphere, for instance, the invariance relations (11.1) and (11.2). All these equations and relations are very similar to those known from the theory of radiation transfer in a homogeneous atmosphere.

To establish asymptotic properties of radiation fields in inhomogeneous atmospheres however, we will need the concept of the ***inverted*** semi-infinite inhomogeneous atmosphere, which will be called ***inverted atmosphere*** for short. This atmosphere is constructed in the following way.

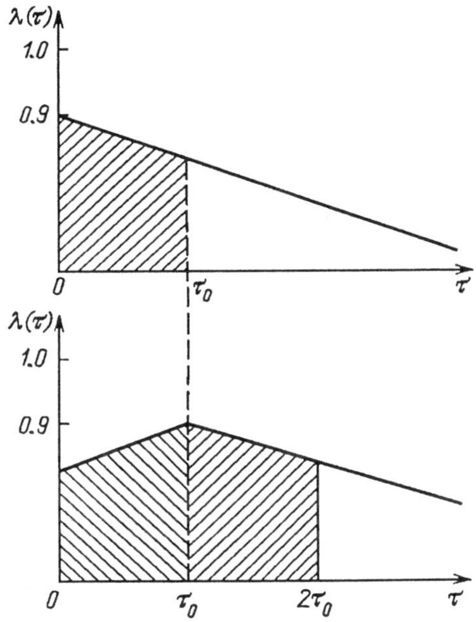

Fig. 11.2. The concept of inverted semi-infinite atmosphere. Initial semi-infinite medium (above) and corresponding inverted atmosphere with the inversion parameter τ_0 (below)

We make a cut at an arbitrary optical depth τ_0 in an "ordinary" (i.e., at $\alpha = 0$) semi-infinite atmosphere. Then we turn the layer of optical thickness τ_0 upside down and place it on the *initial* atmosphere from above. For such an inhomogeneous semi-infinite atmosphere the parameter τ_0 is called the ***inversion parameter***. The optical properties of the inverted atmosphere, with inversion parameter τ_0 can be described in the following way:

$$\lambda(\tau) = \begin{cases} \lambda(\tau_0 - \tau) & \text{for} \quad \tau \in [0, \tau_0], \\ \lambda(\tau - \tau_0) & \text{for} \quad \tau \geq \tau_0; \end{cases} \tag{11.70}$$

$$\chi(\gamma, \tau) = \begin{cases} \chi(\gamma, \tau_0 - \tau) & \text{for} \quad \tau \in [0, \tau_0], \\ \chi(\gamma, \tau - \tau_0) & \text{for} \quad \tau \geq \tau_0. \end{cases}$$

Figure 11.2 shows, in general, the profile of function $\lambda(\tau)$ in an ordinary atmosphere (the upper profile) and in the corresponding inverse atmosphere with the parameter τ_0 (the lower one). It is clearly seen that for $\tau > \tau_0$ we have the initial atmosphere, which has been used to construct the inverted one (by turning the shaded area of the upper part of the figure upside down).

One should mention the following points regarding the concept introduced in this section.

1. As follows from the definition, $\lambda(\tau)$ and $\chi(\gamma, \tau)$ in the inverted atmosphere are continuous with respect to τ if these functions have this property in the ordinary atmosphere.
2. The inverted atmosphere can be treated as a *truncated atmosphere with a negative truncation parameter.* This implies that the truncation parameter α can be positive or negative; positive α means the truncation of atmosphere, and negative means the inversion of atmosphere.
3. Because the truncated atmosphere is an inhomogeneous semi-infinite medium with the optical properties determined by (11.70), all the equations and formulas in the present chapter have analogies in the inverted atmosphere. Moreover, to get the formulas for the inverted atmosphere with the inversion parameter τ_0 we just need to set $\alpha = -\tau_0$ in the corresponding formulas of the truncated atmosphere (if they are known, of course).

Now we are to give examples of such formulas, since we will need them later. We use tilde and subscript τ_0, which provides the value of the inversion parameter, to denote quantities relating to the inverted semi-infinite atmosphere. For instance, $\rho(\mu, \mu_0; -\tau_0) = \tilde{\rho}_{\tau_0}(\mu, \mu_0)$ and $I(\tau, \mu; -\tau_0) = \tilde{I}_{\tau_0}(\tau, \mu)$ are the reflection coefficient (for the zeroth harmonic) and the radiation intensity in the Milne problem, respectively.

Revert to Fig. 1.5 to consider the cut at $\tau = \tau_0$. Then the inverted atmosphere with the inversion parameter τ_0 is on the left of the cut and the truncated atmosphere with the truncation parameter τ_0 is on the right. Reasoning similar to that for deriving (11.39) yields ($\tau \geq 0$):

$$I_\infty(\tau_0 - \tau, -\mu) = M\tilde{I}_{\tau_0}(\tau, -\mu)$$
$$+2\int_0^1 \tilde{I}_{\tau_0}(\tau, \mu, \mu')I_\infty(\tau_0, -\mu')\,\mathrm{d}\mu' + I_\infty(\tau_0, -\mu)\mathrm{e}^{-\tau/\mu}\theta(\mu). \quad (11.71)$$

Similarly, if we place the cut at $\tau = \tau_0$, which means that the inverted atmosphere is on the right of the cut, we find

$$I_\infty(\tau - \tau_0, \mu) = 2\int_0^1 \tilde{I}_{\tau_0}(\tau, \mu, \mu')I_\infty(-\tau_0, \mu')\,\mathrm{d}\mu' + I_\infty(-\tau_0, \mu)\mathrm{e}^{-\tau/\mu}\theta(\mu). \quad (11.72)$$

Assuming $\tau = 0$ in (11.71) and (11.72), one finds

$$I_\infty(\tau_0, \mu) = M\tilde{u}_{\tau_0}(\mu) + 2\int_0^1 \tilde{\rho}_{\tau_0}(\mu, \mu')I_\infty(\tau_0, -\mu')\mu'\,\mathrm{d}\mu', \quad (11.73)$$

$$I_\infty(-\tau_0,\mu) = 2\int_0^1 \tilde{\rho}_{\tau_0}(\mu,\mu')I_\infty(-\tau_0,\mu')\mu'\,\mathrm{d}\mu'. \tag{11.74}$$

In this case the escape function for the inverted atmosphere obeys the normalizing condition

$$2\int_0^1 u(\mu,-\tau_0)I_\infty(-\tau_0,\mu)\mu\,\mathrm{d}\mu = 2\int_0^1 \tilde{u}_{\tau_0}(\mu)I_\infty(-\tau_0,\mu)\mu\,\mathrm{d}\mu = 1. \tag{11.75}$$

Equations (11.72–75) are the analogies for the inverted atmosphere of (11.40), (11.31), (11.32) and (11.34) respectively.

The analogy of (11.43a) for the inverted conservative atmosphere looks as follows:

$$\tilde{I}_{0\tau_0}(\tau,\mu) = \frac{3}{4}\left\{\tau\left[1-\frac{\bar{x}_1(\tau)}{3}\right] + \frac{1}{3}(\tau-\tau_0)\bar{x}_1(\tau_0-\tau)\right.$$
$$\left.+2\int_0^1 \tilde{I}_{0\tau_0}(\tau,\mu,\mu')\mu'\,\mathrm{d}\mu' - \mu\left[1-\mathrm{e}^{-\tau/\mu}\theta(\mu)\right]\right\}. \tag{11.76}$$

11.8 Discussion of the General Approach to the Solution of the Stated Problems

Proceeding from the equations derived in this chapter, one can suggest an algorithm for finding the radiation field in inhomogeneous atmospheres with the truncation parameter $\alpha \geq 0$ (for both the parallel external flux problem and the Milne problem). Let us consider, for the sake of simplicity, a more specific problem. Suppose that an inhomogeneous layer of optical thickness $\alpha_0 \geq 0$ is adjacent to a homogeneous semi-infinite atmosphere for which the solutions of both problems are known (these solutions can, for instance, be found with the help of the methods described in Chap. 3). Hence, all the azimuthal harmonics $I^m(\tau,\mu,\mu_0)$ and the solution of the Milne problem $I(\tau,\mu)$ ($\mu \in [0,1]$) are assumed to be given for $\tau \geq \alpha_0$. In particular, functions $\rho^m(\mu,\mu_0)$ and $u(\mu)$ ($\mu \in [0,1]$) are also considered to be known.

So, in our case the calculation of the radiation field involves several stages, namely:

1. Evaluating the radiation field in an infinite medium, whose optical properties have been continued into negative optical depths symmetrically. In other words, we need to find the function $I_\infty(\tau,\mu)$ under the condition that
$$\frac{\lambda(0)}{2}\int_{-1}^{+1} I_\infty(0,\mu)\,\mathrm{d}\mu = 1 \tag{11.77}$$
for all $\tau \in [0,\alpha_0]$ with (11.62) holding for all τ.

2. Determining $\rho^m(\mu,\mu_0;\alpha)$ and $B^m(0,\mu,\mu_0;\alpha)$ from the system of equations (11.16), (11.18) for all $\alpha \in [0,\alpha_0]$ under the condition $\rho^m(\mu,\mu_0;\alpha_0) = \rho^m(\mu,\mu_0)$ with (11.32) holding for all $\tau = \alpha \in [0,\alpha_0]$ and $m = 0$.
3. Finding $I^m(\tau,\mu,\mu_0;\alpha)$ from (11.27) for $\mu \in [0,1]$ under the boundary conditions $I^m(0,\mu,\mu_0;\alpha) = 0$, $\partial I^m/\partial\alpha\big|_{\alpha=\alpha_0} = 0$.
4. Calculating the functions $I^m(\tau,-\mu,\mu_0;\alpha)$ ($\mu \in [0,1]$) with help of (11.9) with (11.61) holding. At $m = 0$, (11.66) should hold for $\alpha > 0$ and

$$I_\infty(\tau,\mu) = 2\int_0^1 I(\tau,\mu,\mu';0)I_\infty(0,\mu')\,\mathrm{d}\mu' + I_\infty(0,\mu)\mathrm{e}^{-\tau/\mu}\theta(\mu). \quad (11.78)$$

5. Use of the formula

$$I_\infty(-\tau,-\mu) = MI(\tau,\mu;0) + 2\int_0^1 I(\tau,\mu,\mu';0)I_\infty(0,-\mu')\,\mathrm{d}\mu' + I_\infty(0,-\mu)\mathrm{e}^{-\tau/\mu}\theta(\mu) \quad (11.79)$$

to solve the Milne problem. Equations (11.63–65) and (11.67) should hold with $\alpha = 0$.

Of course, the above scheme is an abstract one and needs to be adjusted to the particular problem (optical properties of the media, etc). However there are two points that can be inferred from this scheme.

If α_0 is not too large, the solution of the above-mentioned problems does not involve considerable difficulties. The application of one of the modern methods for solution of the Cauchy problem makes it easy to reach the goal. Similarly, one encounters no problems if the medium absorbs strongly, even in an optically thick surface layer. If the latter is the case, the influx of the photons from below is insignificant compared with the influx of photons from the boundary, and these photons, which come from below, cannot substantially influence the radiation field in the surface layer of atmosphere. Then, the application of an iterative method allows us to solve the problem easily.

So the most significant problems arise when α_0 is comparatively large and especially when the scattering is close to conservative ($1-\lambda(\tau) \ll 1$). In this case we need to create an extended difference network, that demands substantial computation time and memory. We may in that case utilize the asymptotic properties of radiation fields which will be discussed in the next chapter for the case of an inhomogeneous atmosphere. However, before starting this chapter, a review of Sect. 3.8 “Separation of Variables. Angular Relaxation of Photon” is recommended.

12. Asymptotic Properties of Radiation Fields in Inhomogeneous Atmospheres

The fact that the number of variables and parameters is large is known to be one of the major causes of calculating difficulties in the transfer theory. For an inhomogeneous medium, the situation is even more difficult because of additional parameters characterizing the inhomogeneity of the atmosphere. Therefore, there is a practical necessity to study the cases in which the number of parameters can be reduced.

In Chap. 9 we have studied the rigorous separation of variables in the problem of diffuse transmission and reflection of light by a plane layer in an inhomogeneous atmosphere. This separation is due to the reversibility relations (9.18) and the fact that phase function can be expanded in terms of Legendre polynomials. The present chapter deals with the *asymptotic separation of variables*, which is known to be caused by the angular relaxation of photons i.e., by the process in which the photon "forgets" its initial direction of movement. It is this very feature that is responsible for the separation of angular variables (see Sect. 3.8). We will also consider nearly conservative asymptotic formulas, which are due to the fact that the difference is very small between the radiation field for nearly conservative scattering and that for conservative scattering. This will enable us to derive a number of very simple rigorous asymptotic formulas. At the end of the next chapter one can find examples of numerical calculations made with the help of these asymptotic relations as well as their comparison with the numerical calculations made according to rigorous formulas.

12.1 Radiative Transfer in an Infinite Medium

Let us first consider radiative transfer in an infinite medium with sources of infinite power placed at minus infinity. The necessity to consider this problem is caused by the facts that first, a finite medium can in certain conditions be regarded an infinite one, and second, the knowledge of the radiation field in an infinite medium substantially facilitates the study of the field in both the deep layers of a semi-infinite medium, and an atmosphere of considerable optical thickness. The formulation of the problem and the main initial formulas have already been given in Sect. 1.7. One should remind the reader that the functions $\lambda(\tau)$ and $\chi(\gamma,\tau)$ suppose the existence of a non-trivial solution of the Milne problem for each of the adjacent half-spaces.

Let us introduce the notations

$$y_j(\tau) = \frac{1}{2}\int\limits_{-1}^{+1} I_\infty(\tau,\mu)P_j(\mu)\,\mathrm{d}\mu. \tag{12.1}$$

Using (1.56) at $m = 0$, (12.1) and (1.77), we obtain the expression for the source function

$$B_\infty(\tau,\mu) = \lambda(\tau)\sum_{j=0}^{\infty} x_j(\tau)y_j(\tau)P_j(\mu). \tag{12.2}$$

Taking into account that $y_j(\tau)$ are the coefficients of expansion of the function $I_\infty(\tau,\mu)$ in Legendre polynomials, we have

$$I_\infty(\tau,\mu) = \sum_{j=0}^{\infty}(2j+1)y_j(\tau)P_j(\mu). \tag{12.3}$$

Applying the recursion relation for Legendre polynomials, we obtain from (1.76), (12.1) and (12.2) a system of differential equations for the functions $y_j(\tau)$ $(j = 0,1,2,\ldots)$:

$$(j+1)y'_{j+1}(\tau) + jy'_{j-1}(\tau) + [2j+1-\lambda(\tau)x_j(\tau)]y_j(\tau) = 0. \tag{12.4}$$

Successively eliminating the derivatives from (12.4) we arrive at two other important relations:

$$\begin{aligned} y'_0(\tau) = &-[3-\lambda(\tau)x_1(\tau)]y_1(\tau) + \frac{2}{3}[7-\lambda(\tau)x_3(\tau)]y_3(\tau) \\ &-\frac{2\cdot 4}{3\cdot 5}[11-\lambda(\tau)x_5(\tau)]y_5(\tau) + \ldots, \end{aligned} \tag{12.5}$$

$$\begin{aligned} [1-\lambda(\tau)]y_0(\tau) = &\frac{1}{2}[5-\lambda(\tau)x_2(\tau)]y_2(\tau) - \frac{1\cdot 3}{2\cdot 4}[9-\lambda(\tau)x_4(\tau)]y_4(\tau) \\ &+\frac{1\cdot 3\cdot 5}{2\cdot 4\cdot 6}[13-\lambda(\tau)x_6(\tau)]y_6(\tau) - \ldots. \end{aligned} \tag{12.6}$$

Equations (12.2–4) show that the problem of determining the source function and the radiation intensity in an infinite medium reduces to finding the functions $y_j(\tau)$ $(j = 0,1,2,\ldots)$, determined by the system of linear differential equations (12.4). This system implies that in order to find $y_j(\tau)$ $(j = 0,1,2,\ldots)$ it suffices to determine the function $y_0(\tau)$.

Let us consider the boundary conditions for the system of equations (12.4). First of all, (1.77a) and (12.1) result in the following:

$$\lambda(0)y_0(0) = 1. \tag{12.7}$$

If the medium under consideration is such that for any arbitrary $\tau = \tau^*$, $\lambda(\tau) \not\equiv 1$ for all $\tau > \tau^*$, then we have to append the following condition to (12.4) for all j:

$$y_j(\tau) \to 0 \qquad \text{as} \quad \tau \to \infty, \tag{12.8}$$

which is a result of the fact that in an absorbing medium, the radiation intensity attenuates with decreasing the distance from priming sources. Otherwise (i.e., when $\lambda \equiv 1$ for all $\tau > \tau^*$) the boundary condition (12.8) must be altered.

Now we consider particular cases when the radiation intensity in an infinite medium can be determined easily.

12.1.1 Isotropization of Radiation. P_1 Approximation

We assume that the medium is purely scattering, i.e., $\lambda \equiv 1$ for arbitrary τ. Then, just as in the case of a homogeneous medium, taking into account the normalization of the phase function

$$\frac{1}{2}\int_{-1}^{+1} \chi(\mu, \mu'; \tau)\,\mathrm{d}\mu' = 1, \tag{12.9}$$

from (1.77) and (1.78) we immediately find that $I_\infty(\tau, \mu) \equiv 1 = y_0(\tau)$.

Thus, *the radiation field in an inhomogeneous conservative infinite atmosphere becomes completely isotropic.* This fact is of fundamental importance for all further considerations.

With a greater departure of $\lambda(\tau)$ from unity, the radiation becomes more and more anisotropic and, consequently, one should keep more terms in expansion (12.3). Therefore, in full analogy with the method of spherical harmonics (see, e.g., [21], Sect. 10.1), we shall say that the system of equations (12.4) is solved in P_n approximation if we assume that $y'_{n+1} \equiv 0$. We thereby convert from the infinite system of equations (12.4) to a finite one.

Let us derive the corresponding equation in the P_1 approximation. In this case, instead of (12.4) we have (for simplicity, we will not mark the dependence of λ and x_j on τ)

$$y'_1(\tau) + (1 - \lambda)y_0(\tau) = 0, \tag{12.10}$$

$$y'_0(\tau) + (3 - \lambda x_1)y_1(\tau) = 0. \tag{12.11}$$

From this we find the following second-order differential equation with variable coefficients:

$$\left[\frac{y'_0(\tau)}{3 - \lambda x_1}\right]' = (1 - \lambda)y_0(\tau) \tag{12.12}$$

with the boundary conditions

$$\lambda(0)y_0(0) = 1, \quad y_0(\tau) \to 0 \quad \text{as} \quad \tau \to \infty. \tag{12.13}$$

By analogy with a homogeneous medium, (12.12) is called the ***characteristic equation*** in the P_1 approximation.[1] In this approximation, we find from (12.3) and (12.11)

$$I_{\infty}^{(1)}(\tau, \mu) = y_0(\tau) - \frac{3y_0'(\tau)}{3 - x_1}\mu. \tag{12.14}$$

The superscript "1" denotes here the number of approximations of function $I_{\infty}(\tau, \mu)$.

12.1.2 P_2 Approximation

In this approximation we have

$$y_1'(\tau) + (1 - \lambda)y_0(\tau) = 0, \tag{12.15}$$

$$2y_2'(\tau) + y_0'(\tau) + (3 - \lambda x_1)y_1(\tau) = 0, \tag{12.16}$$

$$2y_1'(\tau) + (5 - \lambda x_2)y_2(\tau) = 0. \tag{12.17}$$

From this we find

$$y_1(\tau) = -\frac{1}{3 - \lambda x_1}\left[g(\tau)y_0(\tau)\right]', \tag{12.18}$$

$$y_2(\tau) = \frac{2(1 - \lambda)}{5 - \lambda x_2}y_0(\tau), \tag{12.19}$$

where we designate

$$g(\tau) = 1 + \frac{4(1 - \lambda)}{5 - \lambda x_2}. \tag{12.20}$$

Substituting (12.18) into (12.15) we arrive at the following second-order differential equation for $y_0(\tau)$:

$$\left\{\frac{1}{3 - \lambda x_1}\left[g(\tau)y_0(\tau)\right]'\right\}' = (1 - \lambda)y_0(\tau), \tag{12.21}$$

which is the characteristic equation in P_2 approximation. The boundary conditions are the following:

$$\lambda(0)y_0(0) = 1; \qquad y_0(\infty) = 0. \tag{12.22}$$

Equation (12.21) can be simplified considerably. We convert to the new variable

$$z = \int_0^{\tau} [3 - \lambda(t)x_1(t)]\,\mathrm{d}t. \tag{12.23}$$

[1] Strictly speaking, we have a certain terminological difference here. The characteristic equation for a homogeneous medium is the equation which determines a discrete spectrum of values of *constants* k (see Sect. 2.3). Here we consider the equation for determining the *function* $y_0(\tau)$.

Then, instead of (12.21), we obtain

$$\frac{\mathrm{d}^2 v(z)}{\mathrm{d}z^2} = a(z)v(z), \tag{12.24}$$

where

$$v(z) = g(z)y_0(z), \quad a(z) = \frac{1-\lambda}{3-\lambda x_1} g^{-1}(z), \tag{12.25}$$

while instead of the boundary conditions (12.22), we have

$$v(0) = g(0)/\lambda(0); \quad v(\infty) = 0. \tag{12.26}$$

Thus, if (12.21) (or (12.24) equivalent to it) is solved for all $\tau \geq 0$, in the P_2 approximation, it follows from (12.3), (12.18) and (12.19) that we have the function

$$I_\infty^{(2)}(\tau,\mu) = y_0(\tau) - \frac{3}{3-\lambda x_1}[g(\tau)y_0(\tau)]'\mu + \frac{10(1-\lambda)}{5-x_2} y_0(\tau)P_2(\mu), \tag{12.27}$$

where the superscript to I_∞ still denotes the approximation number.

The fourth term in (12.27), $7y_3(\tau)P_3(\mu)$, can also be assumed approximately known, since it follows from (12.5) that

$$y_3(\tau) = \frac{3}{2}\frac{y_0'(\tau) + (3-\lambda x_1)y_1(\tau)}{7-\lambda x_3}.$$

Hence, for both P_1 and P_2 approximations, the problem of determining the radiation field in an infinite medium can be reduced to the solution of a differential equation of the second order of the type (12.24). This equation can be reduced to the Riccati differential equation and has, in general, to be solved numerically, which in certain cases causes no problems. For a homogeneous atmosphere, (12.12) gives the well-known result

$$y_0(\tau) = \frac{1}{\lambda}\exp[-\tau\sqrt{(1-\lambda)(3-x_1)}]. \tag{12.28}$$

Formally, both (12.12) and (12.24) can be solved for an arbitrary function $\lambda(\tau)$. However, the solution of these equations will be rigorous only in the asymptotic sense, namely when for all τ: $\lambda(\tau) \to 1$ (i.e. for the nearly conservative scattering $[1-\lambda(\tau)] \ll 1$)[2]. It can be shown however, that in the diffuse approximation (one-dimensional inhomogeneous semi-infinite medium), the determination of the radiation field is equivalent to solving (12.12), where x_1 is the portion of the radiation that has been scattered forward in an elementary collision. So, the solutions of both (12.12) and (12.24) can be considered approximately valid for any $\lambda(\tau)$.

[2] The condition $\lambda(\tau) \to 1$ needs elucidating. Function $\lambda(\tau)$ can be considered to depend on a number of free parameters, whose domain of definition is such that the condition $\lambda(\tau) \in [0,1]$ holds for all τ. On the other hand, the condition $\lambda(\tau) \to 1$ requires that the values of free parameters are fixed in such a way that $\lambda(\tau)$ is very close to unity for *all* τ.

12.1.3 M Integral. Relationship Between $y_0(-\tau)$ and $y_0(\tau)$

Substituting (12.14) into (11.37) and setting $\tau = 0$, one finds

$$M_1 = -\frac{8y_0'(0)}{3 - x_1(0)}, \tag{12.29}$$

where the subscript of M denotes the order of approximation.

With the help of the fact that M is invariant with respect to τ, one can derive the following important result. Within the P_2 approximation, function $y_0(-\tau)$ can be expressed in terms of function $y_0(\tau)$ for $\tau \geq 0$. To prove this fact, first of all we find the constant M, defined by (11.37), in the P_2 approximation, from which we can find (cf. (2.31))

$$M = 4\sum_{j=0}^{\infty}(j+1)[y_j(\tau)y_{j+1}(-\tau) + y_j(-\tau)y_{j+1}(\tau)]. \tag{12.30}$$

Hence, in the P_2 approximation we have

$$\frac{1}{4}M_2 = y_0(\tau)y_1(-\tau) + y_0(-\tau)y_1(\tau) + 2[y_1(\tau)y_2(-\tau) + y_1(-\tau)y_2(\tau)]. \tag{12.31}$$

Using (12.18) and (12.19) and taking into account that $M_2 = \text{const}$, instead of (12.31) we obtain a first-order differential equation for the function $y_0(-\tau)$, the solution of which is

$$y_0(-\tau) = \left[1 + \frac{1}{4}M_2 \int_0^{\tau} \frac{3 - \lambda(t)x_1(t)}{g^2(t)y_0^2(t)}\mathrm{d}t\right] y_0(\tau). \tag{12.32}$$

This important equation makes it easy to find the function $y_0(-\tau)$ if the function $y_0(\tau)$ is known. Besides, it will be further seen that this expression helps us to take a limit $\lambda(\tau) \to 1$.

As for the quantity M_2, taking $\tau = 0$ in (12.31) and allowing for (12.19), we obtain

$$M_2 = 8\frac{y_1(0)g(0)}{\lambda(0)}, \tag{12.33}$$

where $y_1(0)$ can be found by solving (12.21) and (12.24) with the use of (12.18).

It is clear that in the homogeneous case, all the above formulas turn into the expressions which have been derived in Chap. 2.

12.2 Deep Layer Regime in a Semi-Infinite Atmosphere

The physical reasoning that will allow us to derive the asymptotic expression sought, will be similar to that used in Sects. 3.8 and 3.9, which led us to the Sobolev asymptotic formula (3.57).

The radiation intensity loses its dependence on azimuth with an increase of optical depth, just as is the case for a homogeneous atmosphere. Also, the *relative* angular structure of the function $I^{\rm as}(\tau,\mu,\mu_0)$ is mainly determined by the optical properties of the medium in the domain $\tau \pm \tau_0$, where $\tau_0 \gg 1$, (i.e. in *a zone of angular relaxation of the photon*). Therefore

$$I^{\rm as}(\tau,\mu,\mu_0) = C(\mu_0) I_\infty(\tau,\mu), \tag{12.34}$$

for all $\tau > \tau_0$, where $C(\mu_0)$ is a normalizing factor. To set $\alpha = 0$ and discard the second term in the right-hand side of (11.45), and after that to find $C(\mu_0)$, one has to substitute (12.34) into the *rigorous* integral relation (11.45). With the help of (11.34) one ultimately finds

$$I^{\rm as}(\tau,\mu,\mu_0) = u(\mu_0) I_\infty(\tau,\mu)\mu_0. \tag{12.35}$$

This is the asymptotic formula for the radiation intensity in deep layers of a semi-infinite atmosphere. This formula generalizes the well-known formula (3.57) to the case of an inhomogeneous medium.

It causes no difficulties to find a similar expression for an inverted semi-infinite atmosphere with the parameter of inversion τ_0. By similar reasoning one finds ($\tau \geq \tau_0$; $\tau_0 \gg 1$)

$$\tilde{I}^{\rm as}_{\tau_0}(\tau,\mu,\mu_0) = \tilde{u}_{\tau_0}(\mu_0) I_\infty(\tau-\tau_0,\mu)\mu_0, \tag{12.36}$$

where the escape function $\tilde{u}_{\tau_0}(\mu_0)$ is normalized according to (11.75).

The substitution of (12.35) and (12.36) into (11.39) and (11.71) respectively yields the asymptotic formulas for the radiation intensity in the Milne problem. Namely, for $\tau \gg 1$ and $\mu \in [1,+1]$ one has ($\tau \geq \tau_0$; $\tau_0 \gg 1$)

$$MI^{\rm as}(\tau,\mu) = I_\infty(-\tau,-\mu) - NI_\infty(\tau,\mu), \tag{12.37}$$

$$M\tilde{I}^{\rm as}_{\tau_0}(\tau,\mu) = I_\infty(\tau_0-\tau,-\mu) - \tilde{N}_{\tau_0} I_\infty(\tau-\tau_0,\mu), \tag{12.38}$$

where N is determined by (11.49) at $\tau = 0$, and

$$\tilde{N}_{\tau_0} = 2\int\limits_0^1 \tilde{u}_{\tau_0}(\mu) I_\infty(\tau_0,-\mu)\mu \,\mathrm{d}\mu. \tag{12.39}$$

The asymptotic formula (3.58) is an analogy of (12.37) and (12.38) for a homogeneous medium. The conservative limit of (12.37) and (12.38) will be found a little later.

12.3 Separation of Angular Variables in the Problem of Light Scattering in an Optically Thick Layer

Let us consider the standard problem of determining the radiation field in a plane inhomogeneous semi-infinite atmosphere illuminated by parallel rays. In such an atmosphere we make an imaginary cut at an optical depth τ_0. Then the radiation intensity in a semi-infinite atmosphere for all $\tau \in [0, \tau_0]$ is simultaneously the radiation intensity in a layer of optical thickness τ_0. The upper boundary of this layer is illuminated by parallel rays while the lower one is illuminated by diffuse radiation propagating at an optical depth τ_0 in a semi-infinite medium. Thus, for any azimuthal harmonic of the radiation intensity we can write, on the basis of the generalized invariance principle ($\tau \in [0, \tau_0]; \mu \in [-1, +1]$)

$$I^m(\tau, \mu, \mu_0) = I^m(\tau, \mu, \mu_0; \tau_0) + I^m(\tau_0, \mu, \mu_0) \mathrm{e}^{\frac{\tau_0 - \tau}{\mu}} \theta(-\mu)$$
$$+2 \int_0^1 I^m(\tau_0, -\mu', \mu_0) \tilde{I}^m(\tau, \mu, \mu'; \tau_0) \, \mathrm{d}\mu'. \tag{12.40}$$

Similarly, for an inverted semi-infinite atmosphere with an inversion parameter τ_0 we have

$$\tilde{I}^m_{\tau_0}(\tau, \mu, \mu_0) = I^m(\tau_0 - \tau, -\mu, \mu_0; \tau_0) + \tilde{I}^m_{\tau_0}(\tau_0, \mu, \mu_0) \mathrm{e}^{\frac{\tau_0 - \tau}{\mu}} \theta(-\mu)$$
$$+2 \int_0^1 \tilde{I}^m_{\tau_0}(\tau_0, -\mu', \mu_0) I^m(\tau_0 - \tau, -\mu, \mu'; \tau_0) \, \mathrm{d}\mu'. \tag{12.41}$$

Here, as before, $I^m(\tau, \mu, \mu_0; \tau_0)$ is the m-th azimuthal harmonic of the radiation intensity in a layer of optical thickness τ_0 which is "cut-off" from the semi-infinite medium; $\tilde{I}^m(\tau, \mu, \mu_0; \tau_0)$ is the same quantity for a layer illuminated from below. For $\tau \gg 0$ and $m = 0$ the use of asymptotic formulas (12.35) and (12.36) brings the system of equations (12.40) and (12.41) towards a system of integral equations with degenerate kernel. Solving this system and making some manipulations, we obtain

$$I(\tau, \mu, \mu_0) = I(\tau, \mu, \mu_0; \tau_0) + \frac{u(\mu_0)\mu_0}{1 - N\tilde{N}_{\tau_0}} \Bigg\{ I_\infty(\tau_0, \mu) \mathrm{e}^{\frac{\tau_0 - \tau}{\mu}} \theta(-\mu)$$
$$+2 \int_0^1 \tilde{I}_{\tau_0}(\tau_0 - \tau, -\mu, \mu') I_\infty(\tau_0, -\mu') \, \mathrm{d}\mu'$$
$$-\tilde{N}_{\tau_0} \left[I_\infty(0, -\mu) \mathrm{e}^{-\tau/\mu} \theta(\mu) + 2 \int_0^1 I(\tau, \mu, \mu') I_\infty(0, -\mu') \, \mathrm{d}\mu' \right] \Bigg\}, \tag{12.42}$$

$$\tilde{I}_{\tau_0}(\tau,\mu,\mu_0) = \tilde{I}(\tau_0-\tau,-\mu,\mu_0;\tau_0) + \frac{\tilde{u}_{\tau_0}(\mu_0)\mu_0}{1-N\tilde{N}_{\tau_0}}\Bigg\{ I_\infty(0,\mu)\mathrm{e}^{\frac{\tau_0-\tau}{\mu}}\theta(-\mu)$$
$$+2\int_0^1 I(\tau_0-\tau,-\mu,\mu')I_\infty(0,-\mu')\,\mathrm{d}\mu'$$
$$-N\left[I_\infty(\tau_0,-\mu)\mathrm{e}^{-\tau/\mu}\theta(\mu)+2\int_0^1 \tilde{I}_{\tau_0}(\tau,\mu,\mu')I_\infty(\tau_0,-\mu')\,\mathrm{d}\mu'\right]\Bigg\}. \tag{12.43}$$

Using (11.71) and (11.72), expressions (12.42) and (12.43) can be simplified to the following form:

$$I(\tau,\mu,\mu_0) = I(\tau,\mu,\mu_0;\tau_0) + \frac{u(\mu_0)\mu_0}{1-N\tilde{N}_{\tau_0}}\Big\{ I_\infty(\tau,\mu)$$
$$-M\tilde{I}_{\tau_0}(\tau_0-\tau,-\mu) - \tilde{N}_{\tau_0}\left[I_\infty(-\tau,-\mu)-MI(\tau,\mu)\right]\Big\}, \tag{12.44}$$

$$\tilde{I}_{\tau_0}(\tau,\mu,\mu_0) = \tilde{I}(\tau_0-\tau,-\mu,\mu_0;\tau_0) + \frac{\tilde{u}_{\tau_0}(\mu_0)\mu_0}{1-N\tilde{N}_{\tau_0}}\Big\{ I_\infty(\tau-\tau_0,\mu)$$
$$-MI(\tau_0-\tau,-\mu) - N\left[I_\infty(\tau_0-\tau,-\mu)-M\tilde{I}_{\tau_0}(\tau,\mu)\right]\Big\}. \tag{12.45}$$

Recall that asymptotic formulas (12.42–45) are valid at $\tau_0 \gg 1$ for all $\tau \in [0,\tau_0]$.

Equations (12.44) and (12.45) acquire an especially simple form far from the both boundaries ($\tau \gg 1$; $\tau_0-\tau \gg 1$):

$$I(\tau,\mu,\mu_0;\tau_0) = \frac{Mu(\mu_0)\mu_0}{1-N\tilde{N}_{\tau_0}}\tilde{I}^{\mathrm{as}}_{\tau_0}(\tau-\tau_0,-\mu), \tag{12.46}$$

$$\tilde{I}(\tau,\mu,\mu_0;\tau_0) = \frac{M\tilde{u}_{\tau_0}(\mu_0)\mu_0}{1-N\tilde{N}_{\tau_0}}I^{\mathrm{as}}(\tau,\mu), \tag{12.47}$$

i.e, to determine the radiation field in an inhomogeneous plane layer of large optical thickness in this case, it suffices to know $u(\mu_0)$ and to solve the problem of radiative transfer in an infinite medium.

And finally, for the boundaries, we obtain from (12.44) and (12.45)

$$\rho(\mu,\mu_0) = \rho^{\mathrm{as}}(\mu,\mu_0;\tau_0) + \tilde{N}_{\tau_0}u(\mu)u(\mu_0)t^*(\tau_0), \tag{12.48}$$

$$\sigma^{\mathrm{as}}(\mu,\mu_0;\tau_0) = u(\mu_0)\tilde{u}_{\tau_0}(\mu)t^*(\tau_0), \tag{12.49}$$

$$\tilde{\rho}(\mu,\mu_0) = \tilde{\rho}^{\mathrm{as}}(\mu,\mu_0;\tau_0) + N\tilde{u}_{\tau_0}(\mu)\tilde{u}_{\tau_0}(\mu_0)t^*(\tau_0), \tag{12.50}$$

where the modified transmission factor is

$$t^*(\tau_0) = \frac{M}{1-N\tilde{N}_{\tau_0}}. \tag{12.51}$$

This differs from the transmission factor $t(\tau_0)$ which is defined for a homogeneous medium by (4.46), where $t^* = M/(1 - N^2 \mathrm{e}^{-2k\tau_0})$.

Using (12.37), (12.38) and (12.51) we can rewrite (12.44) and (12.45) in a more simple form:

$$I(\tau, \mu, \mu_0) = I(\tau, \mu, \mu_0; \tau_0) + [\tilde{I}^{\mathrm{as}}_{\tau_0}(\tau_0 - \tau, -\mu)$$
$$-\tilde{I}_{\tau_0}(\tau_0 - \tau, -\mu) + \tilde{N}_{\tau_0} I(\tau, \mu)] t^*(\tau_0) u(\mu_0) \mu_0, \qquad (12.51\mathrm{a})$$

$$\tilde{I}_{\tau_0}(\tau, \mu, \mu_0) = \tilde{I}(\tau_0 - \tau, -\mu, \mu_0; \tau_0)$$
$$+[I^{\mathrm{as}}_{\tau_0}(\tau_0 - \tau, -\mu) - I(\tau_0 - \tau, -\mu) + N \tilde{I}_{\tau_0}(\tau, \mu)] t^*(\tau_0) u(\mu_0) \mu_0 \qquad (12.51\mathrm{b})$$

We will consider the transition to the conservative limit somewhat later, but now we will turn our attention to the following important circumstance.

If we consider the asymptotic formulas (12.44), (12.45), and (12.48–50) from the point of view of the theory of radiative transfer in homogeneous atmospheres, which means that we regard these formulas as the expressions that enable us to calculate the radiation field in an optically thick layer providing that this problem for a semi-infinite and infinite medium has already been solved, then the following question arises. How do we expand the optical properties of the medium, which are characterized by functions $\lambda(\tau)$ and $\chi(\gamma, \tau)$, for the optical depths $\tau > \tau_0$?

Generally, it can be done in different ways. The only condition to be fulfilled is that if we want, for example, to consider a medium with nearly conservative scattering, then the attached layer should possess the same quality. Therefore, the simplest way to extend the optical properties is to attach a homogenous semi-infinite layer whose optical properties are characterized by the quantities $\lambda = \lambda(\tau_0)$ and $\chi(\gamma, \tau_0)$ to an optically thick inhomogeneous layer. Taking this approach we achieve in particular, a conjugation of the layers which is continuous in optical properties. However, if we have a choice between the semi-infinite media to be attached, it would be most sensible to choose the one that provides the simplest and most convenient solution to the characteristic equation, which actually enables us to find the desired function $I_\infty(\tau, \mu)$.

There is, however, one more aspect of the problem. In general, it is not possible to solve the problem of radiation transfer *in an inhomogeneous medium* analytically. We have to exploit numerical methods for this purpose, namely, we have mentioned in Sect. 10.3 that the problem can be reduced to solving the differential equation of the type (10.20). The smaller the optical thickness of the layer, the easier we can solve this equation by standard finite-difference methods. Considerable calculating problems arise for a semi-infinite atmosphere, especially when the scattering is close to conservative. In this case in order to obtain sufficient accuracy we have to solve (10.20) for a very large interval of optical depth. In other words, in contrast to a homogeneous medium, the numerical solution to the problem of radiative transfer in a layer of finite optical thickness can be found more simply, in general, than for a semi-infinite atmosphere.

However, if the solution to the parallel external flux problem is found for a surface layer of optical thickness $\tau_0 \gg 1$ which is "cut-off" from the semi-infinite inhomogeneous medium under consideration, then from (12.40) and with the help of (12.35) we find at once that for $\tau \in [0, \tau_0]$

$$I(\tau, \mu, \mu_0) = I(\tau, \mu, \mu_0; \tau_0) + \left[I_\infty(\tau_0, \mu) \mathrm{e}^{\frac{\tau_0 - \tau}{\mu}} \theta(-\mu) + 2 \int_0^1 \tilde{I}(\tau, \mu, \mu'; \tau_0) I_\infty(\tau_0, -\mu') \, \mathrm{d}\mu' \right] u(\mu_0)\mu_0, \quad (12.52)$$

while for $\tau > \tau_0$ one must use (12.35). The function $u(\mu)$ can also be expressed through the function $I(0, -\mu, \mu_0; \tau_0) = \rho^{\mathrm{as}}(\mu, \mu_0; \tau_0)\mu_0$. Indeed, substituting (12.48) into (11.31) (with $\tau = 0$), and using (11.49), we obtain

$$I_\infty(0, \mu) = t^*(\tau_0)u(\mu) + 2 \int_0^1 \rho^{\mathrm{as}}(\mu, \mu'; \tau_0) I_\infty(0, -\mu')\mu' \, \mathrm{d}\mu'. \quad (12.53)$$

Similarly, from (11.73) we find the following equation for determining the function $\tilde{u}_{\tau_0}(\mu)$ which is necessary for finding the transmission coefficient:

$$I_\infty(\tau_0, \mu) = t^*(\tau_0)\tilde{u}_{\tau_0}(\mu) + 2 \int_0^1 \tilde{\rho}^{\mathrm{as}}(\mu, \mu'; \tau_0) I_\infty(\tau_0, -\mu')\mu' \, \mathrm{d}\mu'. \quad (12.54)$$

The quantity t^* can be found directly from (12.53) and (12.54) with allowance made for (11.33) and (11.75). The same equations together with (11.49) and (12.39) allow us to determine N and $\tilde{N}_{\tau_0}$ independently.

Thus the numerical solution of the radiative transfer problem involves three stages:

1. the calculation of the radiation field in an optically thick surface layer which is "cut-off" from a semi-infinite medium, and illuminated by parallel rays from below and from above (corresponding equations are provided in Chaps. 9 and 10);
2. the calculation of the radiation field in an infinite medium;
3. the use of (12.52), (12.53) and (12.35).

Van de Hulst and Grossman [170] used a similar method to find the intensity of radiation emerging from a homogeneous semi-infinite atmosphere, where for small optical thicknesses they used the method of doubling layers (see Sect. 4.5). For a homogeneous atmosphere, there exist other methods also allowing us to find $\rho(\mu, \mu_0)$, whereas the use of the asymptotic formula (12.48) seems to be the simplest way to obtain this quantity in the case of inhomogeneous scattering. Let us discuss the first two stages somewhat more elaborately.

For a gradual increase in the optical thickness of a layer one has to find the value of $\tau = \tau_0 = \tau_0^{\text{as}}$ where the asymptotic formulas (12.52), (12.53) and (12.35) can be used. It can be shown easily that the value of τ_0^{as} can be found with the help of a computer. Indeed, the calculations include the calculation of the transmission coefficient $\sigma(\mu, \mu_0, \tau)$, which becomes the matrix $\sigma = (\sigma_{ik})$ as a result of the discretization in μ and μ_0. Equation (12.49) clearly suggests that the asymptotic regime is reached when the equality

$$\sigma_{ii}\sigma_{kk} = \sigma_{ik}\sigma_{ki}. \tag{12.55}$$

holds within the accuracy taken for all i and k at certain $\tau_0 = \tau_0^{\text{as}}$.

The second problem to be solved is to find the radiation field in an inhomogeneous medium. The basic equations applicable to this case have been provided earlier. This problem is solved very easily when the scattering is conservative in the entire medium. There are also no considerable problems when the scattering is nearly conservative. One should, however, draw the reader's attention to the following circumstance.

With the increase of absorption, solution of the characteristic equation (see Sects. 12.1–12.3) becomes more and more difficult. It is not sufficient to use the P_2 approximation to obtain the solution, and, therefore, one has to use the P_N approximation with $N > 2$. However, the asymptotic correction in the formula (12.52) (the second term on the right-hand side) becomes smaller with the increase of absorption in the surface layers of the semi-infinite atmosphere. This is owing to the photons that have reached level $\tau = \tau_0 \gg 1$. In other words, if the true absorption is considerable in the surface layers of a semi-infinite atmosphere, then one can, acting within the accuracy taken, choose such a level $\tau = \tau_0$ that $I(\tau, \mu, \mu_0) = I(\tau, \mu, \mu_0; \tau_0)$ there will be no need to solve the characteristic equation. Hence, the case of nearly conservative scattering, when level τ_0 is high and appropriate asymptotic corrections are made, acquires special importance. We now proceed to a detailed consideration of this situation.

12.4 Reflection Coefficient for a Semi-Infinite Atmosphere with Nearly Conservative Scattering

Quantities for the conservative scattering we denote by subscript "0" like we did earlier. We set $m = 0$ in (1.63), divide both parts of the equation obtained by $\lambda(\tau)$ and then subtract the result from the equation for the function $B_0(\tau, \mu, \mu_0)$. In this case we have

$$B_0(\tau,\mu,\mu_0) - B(\tau,\mu,\mu_0) = \frac{1-\lambda(\tau)}{\lambda(\tau)} B(\tau,\mu,\mu_0)$$
$$+\frac{1}{2}\int_0^1 \chi(\mu,\mu';\tau)\,\mathrm{d}\mu' \int_0^\tau [B_0(\tau',\mu',\mu_0) - B(\tau',\mu',\mu_0)]\mathrm{e}^{-\frac{\tau-\tau'}{\mu'}} \frac{\mathrm{d}\tau'}{\mu'}$$
$$+\frac{1}{2}\int_0^1 \chi(\mu,-\mu';\tau)\,\mathrm{d}\mu' \int_\tau^\infty [B_0(\tau',-\mu',\mu_0) - B(\tau',-\mu',\mu_0)]\mathrm{e}^{-\frac{\tau'-\tau}{\mu'}} \frac{\mathrm{d}\tau'}{\mu'}. \tag{12.56}$$

From a comparison of (12.56) and the conservative case of (1.73) it is seen, that as $\lambda(\tau) \to 1$ the first equation provides the asymptotic solution

$$B_0(\tau,\mu,\mu_0) - B(\tau,\mu,\mu_0) = C(\mu_0) B_0(\tau,\mu), \tag{12.57}$$

where $C(\mu_0)$ is some normalizing factor. We multiply both sides of (12.57) by $\mathrm{e}^{-\tau/\mu}$ and integrate over τ from 0 to ∞. Then allowing for (9.4) (with $\tau_0 = \infty$ and $m = 0$) and (1.74), as well as the fact that the reflection coefficient is a symmetrical function relative to μ and μ_0, we obtain

$$\rho_0(\mu,\mu_0) = \rho(\mu,\mu_0) + C u(\mu) u(\mu_0). \tag{12.58}$$

This formula can also be found from the following simple physical considerations. The quantity $\rho_0(\mu,\mu_0)$ differs from $\rho(\mu,\mu_0)$ owing to the fact that some fraction of the photons was absorbed in the medium. If

$$1 - \lambda(\tau) \ll 1 \tag{12.59}$$

in the entire medium (or at least in its surface layer, i.e., for $\tau \in [0, \tau_0]$, where $\tau_0 \gg 1$), then in an overwhelming majority of cases the photons are absorbed in deep layers. But if they were "generated" again, then in order to pass out through the boundary they would have to diffuse through a layer of large optical thickness, and hence their number would be proportional to the quantity $u_0(\mu)$. Therefore, allowing for symmetry of the reflection coefficient relative to μ and μ_0, we would at once write (12.58). But we preferred to obtain it analytically, and only then explain the physical meaning.

To determine C in (12.58) we employ (11.32) at $\tau = 0$ and the asymptotic formula (12.14). As a result, after simple manipulations we obtain

$$C = -\frac{4y_0'(0)}{3 - x_1(0)}. \tag{12.60}$$

So, in place of (12.58) we finally find

$$\rho(\mu,\mu_0) = \rho_0(\mu,\mu_0) + \frac{4y_0'(0)}{3 - x_1(0)} u_0(\mu) u_0(\mu_0). \tag{12.61}$$

This is the asymptotic equation sought for the reflection coefficient for an inhomogeneous semi-infinite atmosphere with $\lambda(\tau) \to 1$. Let us discuss some consequences emerging from consideration of (12.61)

1. To calculate the reflection coefficient for weak true absorption one must know the reflection coefficient $\rho_0(\mu, \mu_0)$ for conservative scattering, as well as the quantity $y_0'(0)$. The latter is found after the solution of the differential equation (12.12). If the phase function in the atmosphere does not depend on τ then finding the function $\rho_0(\mu, \mu_0)$ presents no particular difficulty, since the methods for its numerical calculation are well developed (see Sect. 3.5). We note also that as soon as the solution of (12.12) is obtained, and hence the function $y_0(\alpha)$ determined for all $\alpha \geq 0$, an expression is thereby found (when the condition (12.59) is satisfied) for the reflection coefficient for a truncated atmosphere. In this case instead of (12.61) we have

$$\rho(\mu, \mu_0; \alpha) = \rho_0(\mu, \mu_0; \alpha) + \frac{4y_0'(\alpha)u_0(\mu, \alpha)u_0(\mu_0, \alpha)}{[3 - x_1(\alpha)]y_0(\alpha)}, \tag{12.62}$$

where $\rho(\tau, \tau_0; \alpha)$ is the reflection coefficient for a truncated semi-infinite atmosphere.
2. We note in advance that the reflection coefficient for an inhomogeneous plane layer of large optical thickness for conservative scattering is given by (12.89) and (12.87). It follows from a comparison of them with (12.61) that if one sets

$$\tau_0 = \frac{1}{3}\int_0^{\tau_0} x_1(t)\,dt - \gamma_0 - \tilde{\gamma}_0 - \frac{3 - x_1(0)}{3y_0'(0)}, \tag{12.63}$$

then $\rho_0^{\text{as}}(\mu, \mu_0; \tau_0) = \rho(\mu, \mu_0)$. In other words, an outside observer can in no way distinguish a semi-infinite inhomogeneous atmosphere with weak true absorption from a conservatively scattering inhomogeneous plane atmosphere of large optical depth.
If the phase function does not vary in the medium, then in place of (12.63) we have

$$\tau_0 = -\frac{1}{y_0'(0)} - \frac{6\gamma_0}{3 - x_1}. \tag{12.63a}$$

In particular, for a homogeneous medium $y_0'(0) = -\sqrt{(1-\lambda)(3-x_1)} = -k$, therefore instead of (12.63a) one finds

$$\tau_0 = \frac{1}{k} - \frac{6\gamma_0}{3 - x_1}. \tag{12.63b}$$

It is clear from physical considerations that (12.63–63b) determine some average level $\tau = \tau_0$ in a nearly conservatively scattering atmosphere in the vicinity of which photons predominantly "perish". Actually, it is just at this depth in a conservatively scattering semi-infinite atmosphere that one should place a fully absorbing layer (or, in other words, simply discard the entire atmosphere for $\tau > \tau_0$) in order to satisfy the equality $\rho_0^{\text{as}}(\mu, \mu_0; \tau_0) = \rho(\mu, \mu_0)$.

If the second term on the right-hand side of (12.63b) is neglected in comparison with the first, then one arrives at Chamberlain's formula [171] for the average depth of photon absorption in a nearly conservatively scattering homogeneous atmosphere. Using Chamberlain's definition of the average depth $\langle\tau\rangle$, one can show that, with the condition (12.59) in an inhomogeneous atmosphere,

$$\langle\tau\rangle = -1/y_0'(0). \tag{12.64}$$

This equation also follows from (12.63a) if the second term on its right side is neglected in comparison with the first.

3. Let us assume that in the semi-infinite inhomogeneous atmosphere under consideration the phase function does not depend on τ. Then, by comparing (12.61) with the corresponding equation (3.80) for a homogeneous atmosphere, one can conclude that in the case of small true absorption under consideration, the reflection coefficient for an inhomogeneous atmosphere will be exactly the same as that for a homogeneous one with some effective λ_e equal to

$$\lambda = \lambda_e = 1 - \frac{[y_0'(0)]^2}{3 - x_1}. \tag{12.65}$$

Consequently, if the phase function in a nearly conservative medium does not vary with depth, then one cannot establish, by any extra-atmospheric measurements of the intensity of diffusely reflected radiation, whether the atmosphere is inhomogeneous or not.

12.5 Escape Function, Albedo of Atmospheres and Other Quantities for Small True Absorption

The expression (12.61) obtained above allows one to find the asymptotic equations for a number of quantities characterizing an inhomogeneous atmosphere for nearly conservative scattering.

1. The escape function $u(\mu)$. We use (11.31) with $\tau = 0$, and represent $u(\mu)$ in the form

$$u(\mu) = u_0(\mu) + u_1(\mu), \tag{12.66}$$

where the function $u_1(\mu) \to 0$ as $\lambda(\tau) \to 1$. After substitution of (12.66), (12.61) and (12.14) into (11.31) with allowance made for (12.29) we obtain

$$u_1(\mu) = u_0(\mu) \frac{3\gamma_0 y_0'(0)}{3 - x_1(0)}. \tag{12.67}$$

So finally,

$$u(\mu) = u_0(\mu)\left[1 + \frac{3\gamma_0 y_0'(0)}{3 - x_1(0)}\right]. \tag{12.68}$$

We note that since the function $u(\mu)$ is the intensity of the escaping radiation in the Milne problem, the structure of the equation $u(\mu) = u_0(\mu)(1 + \text{const})$ for small true absorption has the same simple physical interpretation, as does (12.61) for the reflection coefficient.

The analogy of (12.68) for the case of an inverted semi-infinite atmosphere with the inversion parameter τ_0 has the form

$$\tilde{u}_{1\tau_0}(\mu) = R(\tau_0)\tilde{u}_{0\tau_0}(\mu), \tag{12.69}$$

where

$$R(\tau_0) = y_0^{-1}(-\tau_0)\left\{1 - \frac{3y_0'(-\tau_0)\tilde{\gamma}_0(\tau_0)}{[3 - x_1(\tau_0)]y_0(-\tau_0)}\right\}, \tag{12.70}$$

$$\tilde{\gamma}_0(\tau_0) = 2\int_0^1 \tilde{u}_{0\tau_0}(\mu)\mu^2\,\mathrm{d}\mu. \tag{12.71}$$

In the conservative case, the escape function is related to the reflection coefficient by (11.31a). To derive this equation it suffices to substitute (12.62) and (12.14) into (11.31) and take into account (12.31) and (12.32) (in the P_1 approximation). It should be born in mind that for conservative scattering, in place of the normalizing condition (11.34) we have

$$2\int_0^1 u_0(\mu,\tau)\mu\,\mathrm{d}\mu = 1,$$

with

$$2\int_0^1 \rho_0(\mu,\mu_0,\tau)\mu_0\,\mathrm{d}\mu_0 = 1,$$

because the plane albedo of a semi-infinite atmosphere at $\lambda = 1$ is equal to unity.

2. The plane albedo. This quantity, as we know, is defined by the equation

$$A(\mu) = 2\int_0^1 \rho(\mu,\mu')\mu'\,\mathrm{d}\mu'. \tag{12.72}$$

It can be derived using method used in Sect. 3.11.6 to obtain the quantity in the homogeneous case. Namely, substituting the expression (12.27) for $I_\infty^{(2)}(0,\mu)$ which is found in the P_2 approximation into (11.31) and (11.32) with $\tau = 0$, adding term-by-term the relations obtained and taking into account (12.29), (12.61) and (12.68), we obtain in this approximation

$$A_2(\mu) = 1 + \frac{4y_0'(0)}{3 - x_1(0)} u_0(\mu)$$
$$+ 12\gamma_0 \left[\frac{y_0'(0)}{3 - x_1(0)}\right]^2 u_0(\mu) + \frac{15[1 - \lambda(0)]}{5 - x_2(0)} v_0(\mu), \tag{12.73}$$

where $\gamma_0 = \gamma_0(0)$ is given by (11.52),

$$v_0(\mu) = \mu - 2 \int_0^1 \rho_0(\mu, \mu') \mu'^2 \, d\mu', \tag{12.74}$$

and $y_0'(0)$ can be found by solving (12.21) or (12.24).

3. The spherical albedo. Since this quantity is defined by the equation

$$A_s = 2 \int_0^1 A(\mu) \mu \, d\mu, \tag{12.75}$$

by substituting (12.73) into it, we at once find the simple formula

$$A_{2s} = 1 + \frac{4y_0'(0)}{3 - x_1(0)} \left[1 + \frac{3\gamma_0 y_0'(0)}{3 - x_1(0)}\right]. \tag{12.76}$$

4. The approximate equation for determining $y_0'(0)$. It can be seen that the equations obtained above all include the quantity $y_0'(0)$, for the determination of which one must solve the differential equation (12.12) or (12.21). This can be done analytically only in exceptionally rare cases. It is therefore of interest to obtain an equation which would permit one to estimate $y_0'(0)$, if only approximately, but sufficiently simply.

Let the function $\lambda(\tau)$ be monotonous everywhere (or at least in a sufficiently thick surface layer of the atmosphere). We equate the effective value of λ given by (12.65) to the value of the function $\lambda(\tau)$ at the average optical depth of photon absorption, i.e., we set $\lambda_e = \lambda(\langle\tau\rangle)$. With allowance for (12.64) and (12.65), this approximate equality can be written as

$$1 - \frac{[y_0'(0)]^2}{3 - x_1(0)} = \lambda\left[-\frac{1}{y_0'(0)}\right]. \tag{12.77}$$

This equation, transcendental in the general case, can serve for the approximate determination of the quantity $y_0'(0)$. For a homogeneous atmosphere it gives the well-known rigorous result $y_0'(0) = -\sqrt{(1 - \lambda)(3 - x_1)}$.

5. Asymptotic constants N and C. These are defined by equations

$$N = 2 \int_0^1 u(\mu) I_\infty(0, -\mu) \mu \, d\mu, \tag{12.78}$$

$$C = 2\int_0^1 u(\mu)\mu \, \mathrm{d}\mu. \tag{12.79}$$

Taking into account the normalizing condition (11.33) as well as (12.27), (12.29), and (12.68), we find in P_2 approximation

$$N_2 = 1 - 6\gamma_0 y_1(0) + \frac{1}{2}\left(\frac{3}{4}M_1\gamma_0\right)^2, \tag{12.80}$$

$$C_2 = 1 - 3\gamma_0 y_1(0) + \frac{1}{2}\left(\frac{3}{4}M_1\gamma_0\right)^2 - \frac{15[1-\lambda(0)]}{5-x_2(0)}[\varepsilon_0 - x_2(0)], \tag{12.81}$$

where

$$\varepsilon_0 = 6\int_0^1 u_0(\mu)\mu^3 \, \mathrm{d}\mu, \tag{12.82}$$

and $y_1(0)$ is determined by (12.18). The method of deriving (12.80) and (12.81) is the same as that for (3.82) and (3.83).

Similarly, for the inverted atmosphere, using (11.75), (12.39) and (12.27) it is not difficult to find that

$$\tilde{N}_{2\tau} y_0(-\tau) = y_0(\tau) - \frac{3}{4}M_1\tilde{\gamma}_0(\tau)R(\tau), \tag{12.83}$$

$$\tilde{C}_{2\tau} = \frac{1+\tilde{N}_{2\tau}}{y_0(-\tau)} + \frac{5[1-\lambda(\tau)]}{5-x_2}[\tilde{\varepsilon}_0(\tau) - 1], \tag{12.84}$$

where

$$\tilde{\varepsilon}_0(\tau) = 6\int_0^1 \tilde{u}_{0\tau}(\mu)\mu^3 \, \mathrm{d}\mu. \tag{12.85}$$

6. The modified factor of transmission $t^*(\tau_0)$ is given by (12.51). Equations (12.32), (12.80) and (12.83) allow us to get rid of the indeterminacy in (12.51) when $\lambda(\tau) \to 1$. Namely,

$$t^*(\tau_0) = \frac{4}{3}\left\{\left[\frac{1}{3}\int_0^{\tau_0} \frac{3-x_1(t)}{y_0^2(t)}\mathrm{d}t + \gamma_0 - \frac{3}{8}M_1\gamma_0^2\right]\left[1 + \frac{1}{4}M_1\int_0^{\tau_0}\frac{3-x_1(t)}{y_0^2(t)}\mathrm{d}t\right]^{-1} \right.$$
$$\left. + \tilde{\gamma}_0(\tau_0)y_0^{-1}(-\tau_0)R(\tau_0)\left[1 - \frac{3}{4}M_1\gamma_0\right]\right\}^{-1}. \tag{12.86}$$

The P_1 approximation has been used to write (12.86) only for the sake of simplicity. It is not difficult to get the expression for $t^*(\tau_0)$ also in the P_2 approximation. However, the formula turns out to be a bulky one.

12.6 An Inhomogeneous Atmosphere with Conservative Scattering

If the atmosphere scatters light conservatively, i.e., $\lambda(\tau) \equiv 1$ throughout the medium, then the situation becomes far more simple. The simplification is so essential that it turns out to be possible to find rigorous asymptotic solutions to some specific problems of radiative transfer theory in an inhomogeneous atmosphere. A number of such formulas have already been provided in the book. Some of them are provided without proof. Therefore now we are not only going to provide appropriate results but also to present proofs wherever necessary.

So, setting $\lambda(\tau) \to 1$ in (12.86), one derives the explicit expression for $t_0^*(\tau_0)$

$$t_0^*(\tau_0) = \frac{4}{3}\left\{\tau_0\left[1 - \bar{x}_1(\tau_0)/3\right] + \gamma_0 + \tilde{\gamma}_0(\tau_0)\right\}^{-1}, \tag{12.87}$$

where

$$\bar{x}_1(\tau_0) = \frac{1}{\tau_0}\int_0^{\tau_0} x_1(t)\,\mathrm{d}t. \tag{12.88}$$

As a result, we find the following formulas instead of (12.48–50):

$$\rho_0(\mu,\mu_0) = \rho_0^{\mathrm{as}}(\mu,\mu_0;\tau_0) + u_0(\mu)u_0(\mu_0)t_0^*(\tau_0), \tag{12.89}$$

$$\sigma_0^{\mathrm{as}}(\mu,\mu_0;\tau_0) = u_0(\mu_0)\tilde{u}_{0\tau_0}(\mu)t_0^*(\tau_0), \tag{12.90}$$

$$\tilde{\rho}_0(\mu,\mu_0) = \tilde{\rho}_0^{\mathrm{as}}(\mu,\mu_0;\tau_0) + \tilde{u}_{0\tau_0}(\mu)\tilde{u}_{0\tau_0}(\mu_0)t_0^*(\tau_0). \tag{12.91}$$

It is these very formulas that many of the results of Sect. 8.4 for a multilayer conservative atmosphere come from.

Equation (11.39), which establishes a relationship between the intensity in the Milne problem and that in the parallel external flux problem also takes a simple form. First, for $\lambda(\tau) \equiv 1$ the quantity $I_\infty(\tau,\mu) = 1$. Therefore, proceeding from (11.40) one derives

$$2\int_0^1 I_0(\tau,\mu,\mu')\,\mathrm{d}\mu' = 1 - \mathrm{e}^{-\tau/\mu}\theta(\mu). \tag{12.92}$$

Second. Using (12.14) and (12.32), one finds from (11.39) and (11.40)

$$I_0(\tau,\mu) = \frac{3}{4}\left\{\left[1 - \bar{x}_1(\tau_0)/3\right]\tau - \mu + 2\int_0^1 I_0(\tau,\mu,\mu')\mu'\,\mathrm{d}\mu' + \mu\mathrm{e}^{-\tau/\mu}\theta(\mu)\right\}. \tag{12.93}$$

For $\tau \gg 1$ we find the following formulas instead of (12.37) and (12.38)

$$I_0(\tau,\mu) = \frac{3}{4}\left\{\left[1 - \bar{x}_1(\tau_0)/3\right] - \mu + \gamma_0\right\}, \tag{12.94}$$

$$\tilde{I}^{\text{as}}_{0\tau_0}(\tau,\mu) = \frac{3}{4}\left\{\tau - \frac{1}{3}\int\limits_{\tau_0-\tau}^{\tau_0} x_1(t)\,\mathrm{d}t - \mu + \tilde{\gamma}_0(\tau_0)\right\}, \tag{12.95}$$

and we have

$$I_0(\tau,\mu,\mu_0;\tau_0) = \tilde{I}^{\text{as}}_{0\tau_0}(\tau-\tau_0,-\mu)t_0^*(\tau_0)u_0(\mu_0)\mu_0, \tag{12.96}$$

$$\tilde{I}_0(\tau,\mu,\mu_0;\tau_0) = I_0^{\text{as}}(\tau,\mu)t_0^*(\tau_0)\tilde{u}_{0\tau_0}(\mu_0)\mu_0. \tag{12.97}$$

which correspond to (12.46) and (12.47), respectively.

The conservative analogies for (11.48) and (11.50) can be derived in a similar way. It can be easily shown that for a semi-infinite truncated atmosphere the following asymptotic formula corresponds to (12.68) (to show this it suffices just to set $\tau_0 = -\tau$ in (12.69)):

$$u(\mu,\tau) = \frac{u_0(\mu,\tau)}{y_0(\tau)}\left\{1 + \frac{3\gamma_0(\tau)y_0'(\tau)}{[3-x_1(\tau)]y_0(\tau)}\right\}, \tag{12.98}$$

where $\gamma_0(\tau)$ is determined by (11.52). Using (11.34) and (11.49) and making allowance for (12.14), (12.32), and (12.98), one derives

$$N_1(\tau) = 1 + \frac{M_1}{4}\left[\int\limits_0^{\tau_0}\frac{3-x_1(t)}{y_0^2(t)}\mathrm{d}t - \frac{3\gamma_0(\tau)}{y_0^2(\tau)}\right].$$

The conservative limits of (11.48) and (11.50), i.e. (11.51) and (11.53), can be found very easily with the help of this formula.

Thus, many formulas take a very simple form for the case of a conservatively scattering inhomogeneous atmosphere. For instance, it follows from (12.90) and (12.94–97) that to calculate the radiation field in deep layers of an optically thick medium and the transmission coefficient of this medium we do not need to solve the transfer equation. For this purpose it is sufficient to find the escape functions $u_0(\mu)$ and $\tilde{u}_{0\tau_0}(\mu)$. The quantity $\tilde{\gamma}_0(\tau)$ depends on the form of phase function only very weakly, which enables us to derive the following *explicit* expression for the relative variation of radiation intensity in deep atmospheric layers with height:

$$\frac{I_0(\tau,\mu,\mu_0;\tau_0)}{I_0(\tau_1,\mu,\mu_0;\tau_0)} = \frac{\tilde{I}_{\tau_0}(\tau_0-\tau,-\mu)}{\tilde{I}_{\tau_0}(\tau_0-\tau_1,-\mu)} = \frac{\tau_0 - \tau - \frac{1}{3}\int_\tau^{\tau_0} x_1(t)\,\mathrm{d}t + \mu + \tilde{\gamma}_0(\tau_0-\tau)}{\tau_0 - \tau_1 - \frac{1}{3}\int_{\tau_1}^{\tau_0} x_1(t)\,\mathrm{d}t + \mu + \tilde{\gamma}_0(\tau_0-\tau_1)}.$$

12.7 Conservatively Scattering Atmosphere Above a Reflecting Surface

The case of conservative scattering is realized rather often; for instance, in the atmospheres of Venus and the Earth. Therefore we have to consider the situation when an inhomogeneous atmosphere is adjacent to an isotropically reflecting surface with given albedo A. To solve this problem, which is of considerable importance, we will use the results obtained in Sect. 10.4.

Having performed rather cumbersome manipulations, we derive the following formula for the radiation field in an optically thick layer ($\tau_0 \gg 1$):

$$\bar{I}_0(\tau,\mu,\mu_0;\tau_0) = I_0(\tau,\mu,\mu_0) - \frac{(1-A)\mu_0}{1-A[1-t_0^*(\tau_0)]}[I_0(\tau,\mu) \\ -\tilde{I}_{0\tau_0}(\tau_0-\tau,-\mu) + \tilde{I}_{0\tau_0}^{\mathrm{as}}(\tau_0-\tau,-\mu)]u_0(\mu_0)t_0^*(\tau_0). \tag{12.99}$$

Hence, we find the following expression for the transmission coefficient

$$\bar{\sigma}_0^{\mathrm{as}}(\mu,\mu_0;\tau_0) = \frac{u_0(\mu_0)t_0^*(\tau_0)}{1-A[1-t_0^*(\tau_0)]}[A + (1-A)\tilde{u}_{0\tau_0}(\mu)]. \tag{12.100}$$

Far from both boundaries ($\tau \gg 1$, $\tau_0 - \tau \gg 1$), (12.99) gives

$$\bar{I}_0^{\mathrm{as}}(\tau,\mu,\mu_0;\tau_0) = \left\{1 - \frac{(1-A)I_0^{\mathrm{as}}(\tau,\mu)}{1-A[1-t_0^*(\tau_0)]}t_0^*(\tau_0)\right\}u_0(\mu_0)\mu_0. \tag{12.101}$$

One can derive a number of other formulas, most of which are corollaries of two expressions valid for a conservatively scattering inhomogeneous atmosphere of arbitrary optical thickness. Let us use the definition of the projection of the flux vector onto the axis τ, which is provided by (1.87). To eliminate the constant C we put $\tau = 0$ in the flux integral (1.112) and then we put $\tau = \tau_0$. Thus we derive

$$2\int_0^1 [\rho_0(\mu,\mu';\tau_0) + \sigma_0(\mu,\mu';\tau_0)]\mu'\,\mathrm{d}\mu' = 1 - \mathrm{e}^{-\tau_0/\mu}. \tag{12.102}$$

Similarly, for the layer illuminated from below, we find

$$2\int_0^1 [\tilde{\rho}_0(\mu,\mu';\tau_0) + \tilde{\sigma}_0(\mu,\mu';\tau_0)]\mu'\,\mathrm{d}\mu' = 1 - \mathrm{e}^{-\tau_0/\mu}. \tag{12.103}$$

With the help of the notations of Sect. 10.4, one can rewrite the last two formulas as follows:

$$A_0(\mu_0,\tau_0) + V_0(\mu_0,\tau_0) = 1, \tag{12.104}$$

$$\tilde{A}_0(\mu_0,\tau_0) + \tilde{V}_0(\mu_0,\tau_0) = 1. \tag{12.105}$$

Physical considerations suggest that at any level of a conservative atmosphere, the total amount of descending radiative energy plus the total amount of energy absorbed by the surface is equal to the energy of descending radiation. This allows us to write the following formula:

$$\bar{E}_0^{\uparrow}(\tau,\mu_0) = \bar{E}_0^{\downarrow}(\tau,\mu_0) - \pi\mu_0(1-A)V_0(\mu_0,\tau_0). \tag{12.106}$$

Here we have used the definitions of the illuminance from above and below, which are provided by (1.101) and (1.102) respectively. Using (12.102), (12.103) and (9.18), one derives

$$A_{0s}(\tau_0) = \tilde{A}_{0s}(\tau_0), \tag{12.107}$$

i.e., the spherical albedo of atmosphere does not depend on the side it is illuminated from. The absorption function equals zero in our case, which, together with (10.40), (10.45) and (12.107) gives

$$V_{0s}(\tau_0) = 1 - A_{0s}(\tau_0), \tag{12.108}$$

$$(1-A)\bar{V}_0(\mu_0,\tau_0) = 1 - \bar{A}_0(\mu_0,\tau_0). \tag{12.109}$$

The latter formula can also be derived from (12.106). Equations (12.107) and (12.108) enable us to rewrite (10.46) into the following simple form:

$$\bar{A}_{0s}(\tau_0) = 1 - \frac{(1-A)[1-A_{0s}(\tau_0)]}{1-AA_{0s}(\tau_0)}. \tag{12.110}$$

It is interesting to note that this formula, which has also been derived in the textbook by Sobolev [172], Sect. 19.4, for an isotropically scattering homogeneous atmosphere, turns out to be valid for an inhomogeneous atmosphere with phase function which arbitrarily depends on
tau. Equation (12.110) can also be derived from

$$\bar{A}_0(\mu_0,\tau_0) = 1 - \frac{(1-A)[1-A_0(\mu_0,\tau_0)]}{1-AA_{0s}(\tau_0)}. \tag{12.111}$$

which, in turn, follows from (10.41), (12.104) and (12.108). In particular, it results from (12.111) that the quantity $[1-\bar{A}_0(\mu_0,\tau_0)]/[1-A_0(\mu_0,\tau_0)]$ is independent of μ_0.

Substantial simplifications occur if $\tau_0 \gg 1$. Then

$$A_0^{\text{as}}(\tau_0) = 1 - t_0^*(\tau_0), \tag{12.112}$$

and the following expression corresponds to (12.110):

$$\bar{A}_{0s}^{\text{as}}(\tau_0) = 1 - \frac{(1-A)t_0^*(\tau_0)}{1-A[1-t_0^*(\tau_0)]}. \tag{12.113}$$

For the plane albedo we find

$$A_0^{\mathrm{as}}(\mu_0,\tau_0) = 1 - u_0(\mu_0)t_0^*(\tau_0), \tag{12.114}$$

which gives the following expression in place of (12.104):

$$V_0^{\mathrm{as}}(\mu_0,\tau_0) = u_0(\mu_0)t_0^*(\tau_0). \tag{12.115}$$

Then with the allowance made for (12.109) and (12.111), one finds

$$\bar{A}_0^{\mathrm{as}}(\mu_0,\tau_0) = 1 - \frac{(1-A)t_0^*(\tau_0)u_0(\mu_0)}{1-A[1-t_0^*(\tau_0)]}, \tag{12.116}$$

$$\bar{V}_0^{\mathrm{as}}(\mu_0,\tau_0) = \frac{t_0^*(\tau_0)u_0(\mu_0)}{1-A[1-t_0^*(\tau_0)]}. \tag{12.117}$$

As it is the spherical albedo of the planet that can be determined with the help of observations, it is convenient to rearrange the last two formulas as follows

$$\bar{V}_0^{\mathrm{as}}(\mu_0,\tau_0) = \frac{1-\bar{A}_{0s}^{\mathrm{as}}(\tau_0)}{1-A}u_0(\mu_0), \tag{12.118}$$

$$\bar{A}_0^{\mathrm{as}}(\mu_0,\tau_0) = 1 - [1-\bar{A}_{0s}^{\mathrm{as}}(\tau_0)]u_0(\mu_0). \tag{12.119}$$

One should draw the reader's attention to the following feature. The way the relative variation of the illumination of planetary surface depends on the Sun zenith distance is specified by the expression $\bar{V}(\mu_0,\tau_0)/\bar{V}(0,\tau_0) = u_0(\mu_0)/u_0(0)$. Function $u_0(\mu_0)$, and hence the relative illumination measured at the *planetary surface*, actually depends only on the phase function in the *upper* layers of the atmosphere.

In conclusion, let us consider one more interesting fact. Using (12.94) and (12.101) one can find

$$\frac{d\bar{I}_0^{\mathrm{as}}(\tau,\mu,\mu_0;\tau_0)}{\mathrm{d}\tau} = -\frac{1}{4}\frac{(1-A)t_0^*(\tau_0)[3-x_1(\tau)]}{1-A[1-t_0^*(\tau_0)]}u_0(\mu_0)\mu_0. \tag{12.120}$$

In particular, it follows from here that a noticeable jump should occur in the atmospheric layers when the function $x_1(\tau)$ itself undergoes a sharp change (for instance, at the boundary between cloudy and gaseous layers). This is also the case, for instance, in the Venusian atmosphere (see Sect. 14.2 in this book).

Thus, we can see that a number of asymptotic formulas take very simple forms for a conservatively inhomogeneous atmosphere. Moreover, some of these formulas, for example (12.94), (12.98) and (12.113), provide explicit analytical solutions to certain problems in the theory of radiative transfer in inhomogeneous media.

12.8 Radiation Field in an Atmosphere with Nearly Conservative Scattering

The results of Sects. 12.4–12.7 relate to the radiation fields emerging from an inhomogeneous layer. On the other hand, it is known (see Sect. 3.11) that for a homogeneous semi-infinite atmosphere there exist rigorous asymptotic formulas that — at an *arbitrary* optical depth — express the zeroth azimuthal harmonic of the radiation intensity for nearly conservative scattering in terms of the corresponding quantity for conservative scattering. Therefore, the question arises of whether it is possible to generalize this result to an inhomogeneous atmosphere.

When we peruse what is written in Sect. 3.11.1 after (3.74), about the physical meaning of asymptotic formulas (3.71) and (3.73), it becomes clear that the physical reasoning given there does not depend on whether the atmosphere is homogeneous or not. The only important thing is that the medium can be considered as nearly conservatively scattering, i.e., the condition $[1-\lambda(\tau)] \ll 1$ holds in the entire medium.

This allows us to write the relation for the Milne problem

$$I_1(\tau,\mu) - I_1^{\mathrm{as}}(\tau,\mu) = a[I_0(\tau,\mu) - I_0^{\mathrm{as}}(\tau,\mu)], \tag{12.121}$$

where the factor a is to be determined from the boundary condition $I(0,\mu) = 0$ for $\mu \in (0,1]$. We are to derive it now.

The independence of a from μ follows from the fact that any relatively small number of photons extinguished in the surface layer cannot significantly influence the angular structure of the radiation field. As for its constancy with optical depth, it is easily established as follows:

For $1-\lambda(\tau) \ll 1$, the transfer equation for the Milne problem has the form

$$\mu\frac{\mathrm{d}I(\tau,\mu)}{\mathrm{d}\tau} + I(\tau,\mu) = \frac{1}{2}\int\limits_{-1}^{+1} I(\tau,\mu')\chi(\mu,\mu';\tau)\,\mathrm{d}\mu' = B(\tau,\mu), \tag{12.122}$$

where $\chi(\mu,\mu',\tau)$ is the zeroth azimuthal harmonic of the phase function, while $B(\tau,\mu)$ is the source function. Multiplying both sides of (12.121) by $\frac{1}{2}\chi(\mu,\mu';\tau)$ and integrating over μ' in the limits from -1 to 1, we arrive at the analogous expression where the intensities will be replaced by the corresponding source functions. On the other hand, by differentiating (12.121) with respect to τ and using (12.122), we can arrive at the same equation for the source functions, only under the condition that a does not depend on τ.

Similar considerations apply to the analogy of (3.73) for the parallel external flux problem, namely,

$$\begin{aligned} I_1(\tau,\mu,\mu_0) - I_1^{\mathrm{as}}(\tau,\mu,\mu_0) \\ = I_0(\tau,\mu,\mu_0) - I_0^{\mathrm{as}}(\tau,\mu,\mu_0) + \alpha(\mu_0)[I_0(\tau,\mu) - I_0^{\mathrm{as}}(\tau,\mu)]. \end{aligned} \tag{12.123}$$

To find $\alpha(\mu_0)$, we can use the boundary condition $I(0,\mu,\mu_0) = 0$ for $\mu \in (0,1]$.

Using the afore-mentioned boundary conditions, we can derive from (12.121)

$$a = I_1^{\rm as}(0,\mu)/I_0^{\rm as}(0,\mu), \tag{12.124}$$

and from (12.123) we get

$$\alpha(\mu_0) = [I_1^{\rm as}(0,\mu,\mu_0) - I_0^{\rm as}(0,\mu,\mu_0)]/I_0^{\rm as}(0,\mu). \tag{12.125}$$

Using (12.37), with the aid of (12.14), (12.18), (12.29) and (12.80) we find

$$I_1^{\rm as}(0,\mu) = \frac{3}{4}(\gamma_0 - \mu)\left[1 + \frac{3\gamma_0 y_0'(0)}{3 - x_1}\right]. \tag{12.126}$$

Taking into account (12.94) and (12.124), we can rewrite as follows:

$$a = 1 + \frac{3\gamma_0 y_0'(0)}{3 - x_1(0)}. \tag{12.127}$$

Analogously, with the aid of (12.14), (12.35), (12.68) and (12.94) instead of (12.125) we can get

$$\alpha(\mu_0) = \frac{4y_0'(0)}{3 - x_1} u_0(\mu_0)\mu_0. \tag{12.128}$$

Substituting (12.127) and (12.128) into (12.121) and (12.123) respectively, one can derive the following final result:

$$I_1(\tau,\mu) = I_1^{\rm as}(\tau,\mu) + \left[1 + \frac{3\gamma_0 y_0'(0)}{3 - x_1(0)}\right][I_0(\tau,\mu) - I_0^{\rm as}(\tau,\mu)], \tag{12.129}$$

$$\begin{aligned} I_1(\tau,\mu,\mu_0) &= I_0(\tau,\mu,\mu_0) + I_1^{\rm as}(\tau,\mu,\mu_0) - I_0^{\rm as}(\tau,\mu,\mu_0) \\ &+ \frac{4y_0'(0)}{3 - x_1(0)}[I_0(\tau,\mu) - I_0^{\rm as}(\tau,\mu)]u_0(\mu_0)\mu_0. \end{aligned} \tag{12.130}$$

It is easy to ascertain that the asymptotic expressions (12.129) and (12.130) satisfy not only the corresponding transfer equations, but also the integral relations, which were given in the previous chapter (under the condition $1 - \lambda(\tau) \ll 1$, of course).

A formal derivation of (12.129) and (12.130), which is similar to that given in Chap. 3.11 when the analogous expressions for the homogeneous medium were been found, can also be given. In particular, at the boundary of the atmosphere ($\tau = 0$) from (12.129) and (12.130) follow (12.68) and (12.61), respectively.

The explicit expressions for the functions involved in the right-hand parts of (12.129) and (12.130) have already been provided in preceding sections of the present chapter (see (12.14), (12.35), (12.67) and (12.94)). We only have to note the asymptotic formula

$$M_2 I_1^{\rm as}(\tau,\mu) = I_\infty^{(2)}(-\tau,-\mu) - N_2 I_\infty^{(2)}(\tau,\mu), \tag{12.130a}$$

where the function $I_\infty^{(2)}(\tau,\mu)$ is given by the expression (12.27), and the quantities M_2 and N_2 are given by (12.33) and (12.80) respectively.

12.8.1 Radiation Field in an Inverted Atmosphere and in Optically Thick Layer

It is already known that in order to solve some problems we must know the radiation field not only in a semi-infinite atmosphere but also in the corresponding inverted semi-infinite medium (see Sect. 11.7). If the optical properties of the latter one allow us to treat this medium as nearly conservative, then one can easily derive the corresponding analogies of (12.129) and (12.130) for this case. All the necessary formulas will be given below without proof. For convenience, we repeat some that have been provided earlier.

Proceeding exactly as in the derivation of the asymptotic equations (12.129) and (12.130), for an inverted, semi-infinite atmosphere with an inversion parameter τ_0, we can find that for all $\tau \geq 0$,

$$\tilde{I}_{1\tau_0}(\tau,\mu,\mu_0) = \tilde{I}^{\mathrm{as}}_{1\tau_0}(\tau,\mu,\mu_0) + \tilde{I}_{0\tau_0}(\tau,\mu,\mu_0) - \tilde{I}^{\mathrm{as}}_{0\tau_0}(\tau,\mu,\mu_0)$$
$$-\frac{4}{3}\tilde{\gamma}_0^{-1}(\tau_0)[1 - R(\tau_0)y_0(-\tau_0)][\tilde{I}_{0\tau_0}(\tau,\mu) - \tilde{I}^{\mathrm{as}}_{0\tau_0}(\tau,\mu)]\tilde{u}_{0\tau_0}(\mu_0)\mu_0, \quad (12.131)$$

$$\tilde{I}_{1\tau_0}(\tau,\mu) = \tilde{I}^{\mathrm{as}}_{1\tau_0}(\tau,\mu) + R(\tau_0)[\tilde{I}_{0\tau_0}(\tau,\mu) - \tilde{I}^{\mathrm{as}}_{0\tau_0}(\tau,\mu)], \quad (12.132)$$

$$R(\tau_0) = y_0^{-1}(-\tau_0)\left\{1 - \frac{3y_0'(-\tau_0)\tilde{\gamma}_0(\tau_0)}{[3 - x_1(\tau_0)]y_0(-\tau_0)}\right\}. \quad (12.133)$$

Here

$$\tilde{I}^{\mathrm{as}}_{1\tau_0}(\tau,\mu,\mu_0) = \tilde{u}_{1\tau_0}(\mu_0)\mu_0 I^{(1)}_{\infty}(\tau - \tau_0,\mu), \quad (12.134)$$

where the boundary intensity $\tilde{u}_{1\tau_0}(\mu)$ in the Milne problem is expressed through the same quantity $\tilde{u}_{0\tau_0}(\mu)$ for conservative scattering through the equations

$$\tilde{u}_{1\tau_0}(\mu) = R(\tau_0)\tilde{u}_{0\tau_0}(\mu), \quad (12.135)$$

$$\tilde{\gamma}_0(\tau_0) = 2\int_0^1 \tilde{u}_{0\tau_0}(\mu)\mu^2\,\mathrm{d}\mu. \quad (12.136)$$

Moreover, as follows from (12.134) and (12.135),

$$\tilde{I}^{\mathrm{as}}_{0\tau_0}(\tau,\mu,\mu_0) = \tilde{u}_{0\tau_0}(\mu_0)\mu_0, \quad (12.137)$$

while

$$M_2\tilde{I}^{\mathrm{as}}_{1\tau_0}(\tau,\mu) = I^{(2)}_{\infty}(\tau_0 - \tau,-\mu) - \tilde{N}_2(\tau_0)I^{(2)}_{\infty}(\tau - \tau_0,\mu), \quad (12.138)$$

where by definition

$$\tilde{N}(\tau_0) = 2\int_0^1 \tilde{u}_{\tau_0}(\mu)I_{\infty}(\tau_0,-\mu)\mu\,\mathrm{d}\mu, \quad (12.139)$$

with

$$\tilde{N}_2(\tau_0)y_0(-\tau_0) = y_0(\tau_0) - \frac{3}{4}M_2\tilde{\gamma}_0(\tau_0)R(\tau_0). \qquad (12.140)$$

The function $y_0(\tau)$ is assumed to be found in P_2 approximation, i.e., from (12.21). For conservative scattering in an inverted atmosphere, we have

$$\tilde{I}^{\text{as}}_{0\tau_0}(\tau,\mu) = \frac{3}{4}\left\{\tau\left[1-\frac{\bar{x}_1(\tau)}{3}\right] + \frac{1}{3}(\tau_0-\tau)\bar{x}_1(\tau_0-\tau) - \mu + \tilde{\gamma}_0(\tau_0)\right\}, \qquad (12.141)$$

where $\bar{x}_1(\tau)$ is determined by (1.117), and

$$\tilde{I}_{0\tau_0}(\tau,\mu) = \frac{3}{4}\left\{\tau\left[1-\frac{\bar{x}_1(\tau)}{3}\right] + \frac{1}{3}(\tau_0-\tau)\bar{x}_1(\tau_0-\tau) + 2\int_0^1 \tilde{I}_{0\tau_0}(\tau,\mu,\mu')\mu'\mathrm{d}\mu' - \mu\left[1-\mathrm{e}^{-\tau/\mu}\theta(\mu)\right]\right\}. \qquad (12.142)$$

Thus, in the case of an inverted atmosphere, the equations given above also express the radiation intensity for nearly conservative scattering through the intensity for conservative scattering.

Equation (12.131) allows us to derive a more general result. As is already known (see Chap. 12.3), knowledge of the radiation field in a semi-infinite atmosphere allows one to find the radiation field in a layer of large optical depth, which is "cut off" from this atmosphere. For this, (12.51a) is used. We only have to note that an indeterminacy of 0/0 arises in $t^*(\tau_0)$ as $\lambda(\tau) \to 1$. Therefore the quantities on the right-hand side in (12.51) should be taken in P_2 approximation.

Using the results given in Chap. 10.4, it is also easy to obtain the corresponding equation in the case when an optically thick layer of inhomogeneous atmosphere is adjacent to an isotropically reflecting surface.

12.8.2 Radiation Flux

As before, we designate the projection of the vector of total radiation flux onto the τ axis as

$$H_\tau(\tau,\mu_0) = 2\pi\int_{-1}^{+1} I(\tau,\mu,\mu_0)\mu\,\mathrm{d}\mu + \pi\mu_0\mathrm{e}^{-\tau/\mu_0}. \qquad (12.143)$$

Then from (12.130), with allowance for (12.14), (12.35), (12.68), (12.93) and (12.94), we have the following simple asymptotic equation for an inhomogeneous, semi-infinite atmosphere, valid for all $\tau \geq 0$:

$$\frac{1}{4\pi}H_\tau(\tau,\mu_0) = -\frac{y_0'(\tau)}{3-x_1(\tau)}u_0(\mu_0)\mu_0, \qquad (12.144)$$

Its analogy for a homogeneous medium is provided by (3.92).

In the case of a layer of large optical depth, from (12.143) and (12.51a) with allowance for (12.144), we find

$$\frac{1}{4\pi}H_\tau(\tau,\mu_0;\tau_0) = -\frac{[y_0'(\tau) - \tilde{N}_2(\tau_0)y_0'(-\tau)]}{[1 - N_2\tilde{N}_2(\tau_0)][3 - x_1(\tau)]}u_0(\mu_0)\mu_0. \tag{12.145}$$

For $\tau_0 \to \infty$, as it follows from (12.139) and (12.8), $\tilde{N}(\tau_0) \to 0$. As a result, (12.145) turns into (12.144).

The reader could notice that in the chapters relating to inhomogeneous atmospheres we did not specify the explicit form of the inhomogeneity, i.e., all the results were concerned with practically arbitrary dependence of phase function and probability of photon survival on depth. The specific dependence of $\lambda(\tau)$ (for simplicity, scattering is considered to be isotropic) will be treated in the next chapter. This will allow, in particular, the accuracy of a number of asymptotic expressions obtained above to be estimated.

13. Atmospheres with Exponentially Varying Characteristics

It has already been said that no restrictions have so far been imposed on the way the functions $\lambda(\tau)$ and $\chi(\tau)$ depend on optical depth. Moreover, it turns out that even if we consider particular forms of these functions, which at the same time admit considerable generality, it is possible to derive several relatively simple equations and relations. The significance of these equations and relations is due to the following reasons. First, for real inhomogeneous atmospheres, the dependence of optical properties on depth can, more or less accurately, be described by a concrete function. Second, one can design a simple algorithm for the numerical solution of these equations. Thus, all the above points allow us both to estimate the influence of inhomogeneity of actual media on the radiation field, and provide a testing tool for more complicated methods of solution.

In the present chapter we first consider the problem of diffuse reflection of light from a semi-infinite atmosphere whose dependence of optical properties on optical depth can be represented as a superposition of exponents. We also study the case of nearly conservative scattering, which will allow us to estimate the accuracy of the rigorous asymptotic formulas derived earlier (Sects. 12.4, 12.5). Then we reveal the method for the calculation of the internal radiation field based on the use of the doubling formula. And ultimately, we derive the integral equation that determines the intensity of radiation emerging from a semi-infinite atmosphere.

For simplicity, we will consider only isotropic scattering in this chapter, i.e. we will assume $\chi(\gamma, \tau) = 1$.

13.1 Coefficient of Reflection from a Semi-Infinite Atmosphere

We have assumed that the scattering is isotropic. Therefore all the formulas of Chap. 9 become considerably simpler. In particular, it can be derived from (9.54) that the reflection coefficient for a semi-infinite atmosphere with α being the truncation parameter, is

$$\rho(\mu, \mu_0, \alpha) = \frac{1}{4} \int\limits_{\alpha}^{\infty} \lambda(t)\varphi(\mu_0, t)\varphi(\mu, t) \mathrm{e}^{-(t-\alpha)\left(\frac{1}{\mu}+\frac{1}{\mu_0}\right)} \frac{\mathrm{d}t}{\mu\mu_0}; \tag{13.1}$$

here function $\varphi(\mu, t)$ is determined by the following integral equation:

$$\varphi(\mu, t) = 1 + \frac{1}{2} \int\limits_{t}^{\infty} \lambda(t')\varphi(\mu, t')\,\mathrm{d}t' \int\limits_{0}^{1} \varphi(\mu', t') \mathrm{e}^{-(t'-t)\left(\frac{1}{\mu}+\frac{1}{\mu'}\right)} \frac{\mathrm{d}\mu'}{\mu'}. \tag{13.2}$$

If function $\varphi(\mu, \alpha)$ is known for all $\mu \in [0, 1]$ and $\alpha \geq 0$, then the intensity of descending radiation ($\mu \in [0, 1]$) in the atmosphere is determined by an equation following from (10.13). The intensity of ascending radiation can be found from a corresponding formula following from (10.14).

It is also important here that knowledge of the function $\varphi(\mu, t)$ for all $t \geq 0$ enables us to use (13.1) to find the reflection coefficient for an arbitrary truncated atmosphere, i.e. for an arbitrary $\alpha \geq 0$.

There are several ways to construct the algorithm for solving (13.2), each one depending on a particular form of the function $\lambda(\tau)$. Let us assume now that

$$\lambda(\tau) = \sum_{k=0}^{\infty} \lambda_k \mathrm{e}^{-km\tau}, \qquad (\lambda(\tau) \in (0, 1]), \tag{13.3}$$

where λ_k and m are some parameters.

Then it follows from (13.2) that

$$\varphi(\mu, t) = \sum_{k=0}^{\infty} \varphi^{(k)}(\mu) \mathrm{e}^{-km\tau}, \tag{13.4}$$

where functions $\varphi^{(k)}(\mu)$ for $k \geq 1$ are determined by the equations

$$\varphi^{(k)}(\mu) = \frac{\mu}{2} \int\limits_{0}^{1} \frac{\mathrm{d}\mu'}{\mu + \mu' + km\mu\mu'} \sum_{i=0}^{k} \lambda_i \sum_{j=0}^{k-i} \varphi^{(j)}(\mu)\varphi^{(k-i-j)}(\mu'), \tag{13.5}$$

which result from setting the coefficients at the same exponents equal to each other. Function $\varphi^{(0)}(\mu)$ is determined by the Ambartsumian equation

$$\varphi^{(0)}(\mu) = 1 + \frac{\lambda_0}{2} \varphi^{(0)}(\mu)\mu \int\limits_{0}^{1} \frac{\varphi^{(0)}(\mu')\mathrm{d}\mu'}{\mu + \mu'} \tag{13.6}$$

and may be considered to be known.

In turn, at $\alpha = 0$ the reflection coefficient is

$$\rho(\mu, \mu_0) = \sum_{k=0}^{\infty} \rho^{(k)}(\mu, \mu_0), \tag{13.7}$$

where

$$\rho^{(k)}(\mu, \mu_0) = \frac{1}{4(\mu + \mu_0 + km\mu\mu_0)} \sum_{i=0}^{k} \lambda_i \sum_{j=0}^{k-i} \varphi^{(j)}(\mu)\varphi^{(k-i-j)}(\mu_0). \quad (13.8)$$

In the general case, i.e. for an arbitrary truncated atmosphere, we have from (13.3), (13.4) and (13.1):

$$\rho(\mu, \mu_0; \alpha) = \sum_{k=0}^{\infty} \rho^{(k)}(\mu, \mu_0) \mathrm{e}^{-km\alpha}, \quad (13.9)$$

Thus (13.4) and (13.9) show that there is a separation of angular variables (μ, μ_0), and the spatial one τ (or α), which simplifies the problem substantially. Therefore, to find the reflection coefficient for an arbitrary truncated semi-infinite atmosphere it is sufficient to find the coefficients $\rho^{(k)}(\mu, \mu_0)$ $(k = 0, 1, 2, \dots)$ for a non-truncated atmosphere. For this, it is necessary to know the functions $\varphi^{(k)}(\mu)$, which can be found from (13.5) one by one. Note that at $\lambda_0 = 0$ the problem becomes substantially simpler, since (13.5) turns into a recursion relation for finding successive functions $\varphi^{(k)}(\mu)$ for $\varphi^{(0)}(\mu) = 1$.

The plane albedo of atmosphere $A(\mu)$ can be expressed in terms of functions $\varphi^{(k)}(\mu)$ (see (1.104)). As

$$A(\mu) = 2 \int_0^1 \rho(\mu, \mu')\mu' \, \mathrm{d}\mu', \quad (13.10)$$

then using (13.7), one finds

$$A(\mu) = \sum_{k=0}^{\infty} A_k(\mu), \quad (13.11)$$

where

$$A_k(\mu) = 2 \int_0^1 \rho^{(k)}(\mu, \mu')\mu' \, \mathrm{d}\mu'. \quad (13.12)$$

With the help of (13.5) and (13.8) one derives

$$A_k(\mu) = \frac{1}{1 + km\mu} \left[\frac{1}{2} \sum_{i=0}^{k} \lambda_i \sum_{j=0}^{k-i} \alpha_0^{(j)} \varphi^{(k-i-j)}(\mu) - \varphi^{(k)}(\mu) \right], \quad (13.13)$$

where

$$\alpha_0^{(j)} = \int_0^1 \varphi^{(j)}(\mu) \, \mathrm{d}\mu. \quad (13.14)$$

13.2 Results of Calculations and an Estimation of Accuracy of Asymptotic Formulas for $\lambda(\tau) = \lambda_1 e^{-m\tau}$

Consider the case when

$$\lambda(\tau) = \lambda_1 e^{-m\tau}, \tag{13.15}$$

for all $\tau \geq 0$; $\lambda_1 \in [0,1]$ and $m \geq 0$. One should recall what was said about this kind of $\lambda(\tau)$ in Sect. 1.6. Namely, one has to take into consideration that here we encounter, in fact, the case of an atmosphere of *finite* optical thickness. This problem can be considered to be the Milne problem only if $m \ll 1$.

When $\lambda(\tau)$ is provided by (13.15), functions $\varphi^{(k)}(\mu)$ are, according to (13.5), found with the help of the following recursion formula

$$\varphi^{(k)}(\mu) = \frac{\lambda_1}{2}\mu \int_0^1 \frac{d\mu'}{\mu + \mu' + km\mu\mu'} \sum_{j=0}^{k-1} \varphi^{(j)}(\mu)\varphi^{(k-1-j)}(\mu') \tag{13.16}$$

for $\varphi^{(0)}(\mu) = 1$, and the reflection coefficient is provided by (13.7), where

$$\rho^{(k)}(\mu, \mu_0) = \frac{\lambda_1}{4(\mu + \mu_0 + km\mu\mu_0)} \sum_{j=0}^{k-1} \varphi^{(j)}(\mu)\varphi^{(k-1-j)}(\mu_0), \quad (k \geq 1). \tag{13.17}$$

Obviously, for a given m, it is sufficient to find $\varphi^{(k)}(\mu)$ at $\lambda_1 = 1$, since the following relations hold

$$\varphi^{(k)}(\mu; \lambda_1) = \lambda_1^k \varphi^{(k)}(\mu; 1), \tag{13.18}$$

$$\rho^{(k)}(\mu, \mu_0; \lambda_1) = \lambda_1^k \rho^{(k)}(\mu, \mu_0; 1). \tag{13.19}$$

Equations (13.18) and (13.19) are direct consequences of (13.16) and (13.17).

It is always possible to take $\lambda_1 = e^{-m\alpha}$ with α being the parameter of truncation. This means that the solution of the problem of diffuse reflection of light from a non-truncated atmosphere at $\lambda_1 = 1$, instantly gives us the corresponding solution for the truncated atmosphere.

Calculations show that functions $\varphi^{(k)}(\mu)$ decrease quite fast with an increase of k. This is exemplified by Table 13.1, where functions $\varphi^{(k)}(\mu)$ are provided at $m = 1$ and $\lambda_1 = 1$. Also, Tables XXXII–XXXIV contain values of functions $\varphi(\mu, \lambda_1)$, $\rho(\mu, 1; \lambda_1)$, and $A(\mu, \lambda_1)$ calculated for different λ_1 and m. This information may turn out to be useful in both interpreting results of observations and testing programs designed for more complicated dependencies of optical properties on depth.

Table 13.1 Values of $\varphi^{(k)}(\mu)$ for $m = 1$ and $\lambda_1 = 1$

μ	k					
	1	2	3	4	5	6
0	0	0	0	0	0	0
0.1	0.113	0.029	0.010	0.004	0.002	0.001
0.2	0.132	0.049	0.018	0.007	0.003	0.001
0.3	0.193	0.064	0.024	0.010	0.004	0.002
0.4	0.215	0.074	0.029	0.012	0.005	0.002
0.5	0.231	0.083	0.032	0.013	0.006	0.002
0.6	0.244	0.089	0.036	0.015	0.006	0.003
0.7	0.254	0.095	0.038	0.016	0.007	0.003
0.8	0.262	0.099	0.040	0.017	0.007	0.003
0.9	0.269	0.103	0.042	0.017	0.007	0.003
1.0	0.275	0.106	0.043	0.018	0.008	0.003

As m decreases, the decrease of $\varphi^{(k)}(\mu)$ with the increase of k occurs more slowly. So, for sufficiently small m it becomes almost impossible to find the solution. The physical reason for this can be explained with the help of (13.7) and (13.19). Indeed, these formulas yield

$$\rho(\mu, \mu_0; \lambda_1) = \sum_{n=1}^{\infty} \lambda_1^n \rho^{(n)}(\mu, \mu_0; 1). \tag{13.20}$$

Hence, for sufficiently small m, $\rho^{(n)}(\mu, \mu_0; 1)$ is determined mostly by photons that have undergone n scatterings. The decrease of m is known to be equivalent to the increase of optical thickness of atmosphere. Therefore at $\lambda_1 = 1$ and at sufficiently small m, there exists a great number of photons making considerable contribution to the intensity of radiation emerging from the atmosphere, even at large n. At $\lambda_1 = 1$, a photon may stop diffusing only by leaving the media. For sufficiently small m there is hardly more than one boundary at which the photon may do this.

Hence, one encounters substantial problems in applying the above method to calculating $\rho(\mu, \mu_0)$ for $m \ll 1$ and $\lambda_1 = 1$. However, (13.15) suggests that for this case the radiative transfer in an optically thick surface layer (m^{-1} in order of magnitude) of atmosphere, occurs for nearly conservative scattering. Therefore one can use the corresponding asymptotic formulas. The comparison between the calculations performed according to these formulas and those made with the help of computational formulas of high accuracy will allow us to estimate the accuracy of the rigorous asymptotic relations found.

First of all, one has to find the solution of the corresponding characteristic equation. Proceeding from (12.12) and (13.15) and taking a P_1 approximation, one finds the following differential equation

$$y_0''(\tau) = 3\left(1 - e^{-m\tau}\right) y_0(\tau) \tag{13.21}$$

under the boundary conditions

$$y_0(0) = 1, \qquad y_0(\tau) \to 0 \quad \text{for} \quad \tau \to \infty. \tag{13.22}$$

The following expression can be shown to be the solution of the above equation:

$$y_0(\tau) = \frac{J_\nu(\nu e^{-\sqrt{3}\tau/\nu})}{J_\nu(\nu)}, \tag{13.23}$$

where

$$\nu = 2\sqrt{3}/m \gg 1, \tag{13.24}$$

and $J_\nu(z)$ is a Bessel function of the first kind. As we are primarily interested in optical depths such that $\tau/\nu \ll 1$, the following expressions can be easily derived by neglecting all terms in (13.23) except the first two (see, for instance, [173] Chap. 9):

$$y_0(\tau) = 1 - 3g_0\tau\left(\frac{m}{3}\right)^{1/3}, \tag{13.25}$$

$$y_0'(\tau) = -3g_0\left(\frac{m}{3}\right)^{1/3}\left(1 + \frac{m}{2}\tau\right), \tag{13.26}$$

where $g_0 = \Gamma(2/3)/\Gamma(1/3) = 0.505468\ldots$. In particular, we have

$$y_0'(0) = -3g_0\left(\frac{m}{3}\right)^{1/3}. \tag{13.27}$$

It is of certain interest to compare the rigorous result provided by (13.27) with the approximate one of (12.77). Proceeding from (12.77) and taking into account that $m \ll 1$, one finds $y_0'(0) = -(3)^{2/3}(m/3)^{1/3} = -2.08(m/3)^{1/3}$. At the same time, (13.27) yields $y_0'(0) = -1.52(m/3)^{1/3}$. Hence, for the case considered, the approximate method exploited to find $y_0'(0)$ yields a *rigorous* dependency of this quantity on m; the numerical coefficient turns out, however, to be a little bit greater.

Consider now the asymptotic formulas providing the characteristics of intensity of the radiation emerging from the atmosphere for nearly conservative isotropic scattering. If $\lambda(\tau) \equiv 1$ throughout the atmosphere, then the atmosphere is homogeneous and the reflection coefficient

$$\rho_0(\mu, \mu_0) = \frac{1}{4}\frac{\varphi_0(\mu)\varphi_0(\mu_0)}{\mu + \mu_0}, \tag{13.28}$$

which follows from (3.36) at $m = 0$. Here, the Ambartsumian function $\varphi_0(\mu)$ is determined by (13.6), where one should put $\lambda_0 = 1$. It can be obtained easily from (13.6) that at $\lambda_0 = 1$ this equation can also be written as follows (see, for instance, RT, Sect. 38):

$$\frac{\mu}{\varphi_0(\mu)} = \frac{1}{\sqrt{3}} - \frac{1}{2}\int_0^1 \frac{\varphi_0(\mu')\mu'^2\,\mathrm{d}\mu'}{\mu+\mu'}. \tag{13.29}$$

Equations (3.14), (13.28) and (13.29) suggest that for isotropic conservative scattering the escape function takes a form:

$$u_0(\mu) = \frac{\sqrt{3}}{4}\varphi_0(\mu). \tag{13.30}$$

For non-conservative scattering, $\varphi(\mu)$ and $u(\mu)$ are connected by the relation (7.17).

As a result, asymptotic formulas (12.61), (12.73) and (12.76) take the following simple form:

$$\rho(\mu,\mu_0) = \frac{1}{4}\varphi_0(\mu)\varphi_0(\mu_0)\left[(\mu+\mu_0)^{-1} + y_0'(0)\right], \tag{13.31}$$

$$A_2(\mu) = 1 + \frac{y_0'(0)}{\sqrt{3}}[1+\gamma_0 y_0'(0)]\varphi_0(\mu), \tag{13.32}$$

$$A_{2s} = 1 + \frac{4y_0'(0)}{3}[1+\gamma_0 y_0'(0)], \tag{13.33}$$

where $y_0'(0)$ is provided by (13.27); $\gamma_0 = q(\infty) = 0.710446\ldots$ and is called the ***Hopf constant***. It can be found very often in the problems of transfer of monochromatic radiation for isotropic scattering. The Ambartsumian function is tabulated very elaborately and corresponding tables are normally provided in handbooks on the theory of radiation transfer. In particular, if we do not need high accuracy, the function $\varphi_0(\mu)$ can be found with the help of Table VI in the Appendix and (13.30). For non-conservative scattering this can be done by exploiting (7.17) and corresponding tables in the Appendix.

Note that as $\varphi(\mu) = 1 + 2\mu\int_0^1 \rho(\mu,\mu')\,\mathrm{d}\mu'$ by definition, (13.31) results in the following asymptotic formula:

$$\varphi(\mu) = [1 + y_0'(0)\mu]\,\varphi_0(\mu). \tag{13.34}$$

To determine how well the asymptotic formulas that have been found work, it is most convenient to make use of the expression for spherical albedo in a semi-infinite layer for $\lambda(\tau)$ given by (13.15) at $\lambda_1 = 1$. Here the point is that for this case there are detailed tables of A_s for different m given in the work by Garsia and Siewert [174]. There, all figures are provided to six significant digits.

Table 13.2

m	10^{-6}	10^{-3}	10^{-2}
A_s	0.986085	0.869837	0.742082
A_{2s}	0.986087	0.870405	0.747483

Table 13.3

m	10^{-6}		10^{-3}	
μ	R	R_1	R	R_1
0.05	1.45551	1.45541	1.33392	1.32325
0.1	1.51670	1.51659	1.38317	1.37155
0.2	1.60153	1.60139	1.44609	1.43275
0.3	1.66083	1.66067	1.48469	1.46969
0.4	1.70558	1.70541	1.50941	1.49271
0.5	1.74078	1.74059	1.52504	1.50655
0.6	1.76925	1.76903	1.53431	1.51391
0.7	1.79272	1.79249	1.53892	1.51646
0.8	1.81237	1.81212	1.54000	1.51531
0.9	1.82901	1.82876	1.53837	1.51125
1.0	1.84325	1.84297	1.53462	1.50486

In the first line of Table 13.2 one can find exact values of A_s according to [174], and the second contains the results of simple calculations by (13.33). The work [174] includes quantities $R(\mu, \mu_0) = 2\rho(\mu, \mu_0)\mu_0$ at $\mu_0 = 0.9$. Here the reflection coefficient ρ is calculated with the help of (13.31). The results of calculations (R_1) are given in Table 13.3, along with the exact values (R). One sees that (13.31) cannot provide as accurate values as (13.33) can. The latter takes into account terms of the order of magnitude $m^{2/3}$, whereas (13.31) uses terms of the order of magnitude $m^{1/3}$. One can make the calculations by (13.33) more accurate by finding $y_0'(0)$ in a P_2 approximation.

Besides the above example, the case when

$$\lambda(\tau) = \lambda_0 + \lambda_1 e^{-m\tau} \tag{13.35}$$

is also of certain interest for practical applications. If $\lambda_1 < 0$, then $\lambda(\tau)$ increases with τ. This case is, however, a little bit more complicated than the last one since (13.5) does not turn into a recursion relations for $\lambda_0 \neq 0$.

For isotropic nearly conservative scattering, (12.130), which provides the radiation intensity at all optical depths, also takes a very simple form. Using (12.14), (12.35), (12.67), (12.93) and (12.94), we find the following formula instead of (12.130)

$$I(\tau,\mu,\mu_0) = I_0(\tau,\mu,\mu_0) - \left\{ y_0'(\tau)\mu + [1 - y_0(\tau)][1 + q(\infty)y_0'(0)] \right.$$

$$\left. - y_0'(0)\left[2\int_0^1 I_0(\tau,\mu,\mu')\mu'\,\mathrm{d}\mu' + \mu\mathrm{e}^{-\tau/\mu}\theta(\mu)\right]\right\}\frac{\sqrt{3}}{4}\varphi_0(\mu)\mu_0. \quad (13.36)$$

In this case, $y_0(\tau)$ and $y_0'(\tau)$ are given by (13.25) and (13.26), respectively. Function $I_0(\tau,\mu,\mu_0)$ relates to the case of a homogeneous atmosphere. Therefore, it can be found either by numerical methods (see Sect. 3.10) or with the help of rigorous analytical expression.

So, we can see that for isotropic nearly conservative scattering, the calculation of the radiation field is reduced to the solution of (12.21) and (12.12). Also, this problem causes no difficulties if the phase function does not depend on depth. The only situation when problems may arise, is when both λ and $\chi(\gamma)$ depend on depth, because in this case one has to solve the parallel external flux problem for a conservatively scattering inhomogeneous atmosphere. Any general analytical results are unlikely to be derived without assumptions about the way the phase function depends on depth being made.

13.3 Algorithm for Calculating Internal Radiation Field

Until now, we have been considering the solution of the problem of diffuse reflection when the dependence of λ on τ is provided by (13.3). However, it can be shown very easily that one can design a method for finding the radiation intensity $I(\tau,\mu,\mu_0)$ for all $\tau > 0$. This algorithm is very similar to that for calculating the radiation field in a semi-infinite atmosphere. The latter method, which exploits the doubling method, has been described in detail in Sect. 3.10.

So, by putting $\tau = \tau_1$ in (11.1) one derives ($\mu \in [-1,+1]$):

$$I(2\tau,\mu,\mu_0;\alpha) = I(\tau,\mu,\mu_0;\tau+\alpha)\mathrm{e}^{-\tau/\mu_0} + I(\tau,\mu,\mu_0;\alpha)\mathrm{e}^{-\tau/\mu}\theta(\mu)$$

$$+2\int_0^1 I(\tau,\mu',\mu_0;\alpha)I(\tau,\mu,\mu';\tau+\alpha)\,\mathrm{d}\mu'. \quad (13.37)$$

This formula is an analogy for the doubling formula (3.63) for an inhomogeneous medium.

If $\lambda(\tau)$ is given by (13.3), one can try to find the radiation intensity in the form

$$I(\tau,\mu,\mu_0;\alpha) = \sum_{k=0}^{\infty} I_k(\tau,\mu,\mu_0)\mathrm{e}^{-km\alpha}. \quad (13.38)$$

The substitution of (13.38) into (13.37) yields the following recursion relation

$$I_k(2\tau,\mu,\mu_0) = I_k(\tau,\mu,\mu_0)\mathrm{e}^{-(1/\mu_0+km\tau)} + I_k(\tau,\mu,\mu_0)\mathrm{e}^{-\tau/\mu}\theta(\mu)$$
$$+2\sum_{i=0}^{k}\mathrm{e}^{-(k-i)m\tau}\int_0^1 I_i(\tau,\mu',\mu_0)I_{k-i}(\tau,\mu,\mu')\,\mathrm{d}\mu'. \quad (13.39)$$

The relation (13.39) is the basis for the method for calculating the internal radiation field.

Like the case of a homogeneous medium (see Sect. 3.10), at first it is sufficient to find the components of descending radiation for a small initial value $\tau = \Delta\tau$. Namely, using the relation for $I(\tau,\mu,\mu_0;\alpha)$ for $\mu \in [0,1]$, which is analogous to (10.16) at $z = \infty$, and also (10.14) and (13.38), one easily derives

$$I_k(\tau,\mu,\mu_0) = \frac{\lambda_k}{4}\frac{\mathrm{e}^{-\Delta\tau/\mu} - \mathrm{e}^{-\Delta\tau(1/\mu_0+km)}}{\mu-\mu_0+km\mu\mu_0}\mu_0$$
$$+\frac{1}{2}\mathrm{e}^{-\frac{\Delta\tau}{\mu_0}}\mu_0\sum_{i=0}^{k}\lambda_{k-i}\int_0^1\rho^{(i)}(\mu',\mu_0)\frac{\mathrm{e}^{-m(k-i)\Delta\tau} - \mathrm{e}^{-\Delta\tau\left(\frac{1}{\mu}+\frac{1}{\mu'}\right)}}{\mu+\mu'-m\mu\mu'(k-i)}\mu'\,\mathrm{d}\mu'. \quad (13.40)$$

In particular, if $\lambda(\tau)$ is determined by (13.15), then $I_0 \equiv 0$ and

$$I_1(\tau,\mu,\mu_0) = \frac{\lambda_1}{4}\frac{\mathrm{e}^{-\Delta\tau/\mu} - \mathrm{e}^{-\Delta\tau(1/\mu_0+m)}}{\mu-\mu_0+m\mu\mu_0}\mu_0, \quad (13.40a)$$

$$I_k(\tau,\mu,\mu_0) = \frac{\lambda_1}{2}\mathrm{e}^{-\frac{\Delta\tau}{\mu_0}}\mu_0\int_0^1\rho^{(k-1)}(\mu',\mu_0)\frac{\mathrm{e}^{-m\Delta\tau} - \mathrm{e}^{-\Delta\tau\left(\frac{1}{\mu}+\frac{1}{\mu'}\right)}}{\mu+\mu'-m\mu\mu'}\mu'\,\mathrm{d}\mu',$$
$$(k \geq 2). \quad (13.40b)$$

Then we use the doubling formula (13.39), which allows us to find the intensity at depths $2\Delta\tau$, $4\Delta\tau$, ... To find the intensity of ascending radiation, we use the following relation

$$I_k(\tau,-\mu,\mu_0) = \rho^{(k)}(\mu,\mu_0)\mathrm{e}^{-(1/\mu_0+km)\tau}$$
$$+2\sum\nolimits_{i=0}^{k}\mathrm{e}^{-(k-i)m\tau}\int_0^1 I_i(\tau,\mu',\mu_0)\rho^{(k-i)}(\mu,\mu')\mu'\,\mathrm{d}\mu', \quad (13.41)$$

which follows from (10.14), (13.9) and (13.38). Other ways of using (11.1), which are similar to those described in Sect. 3.10, are also possible.

Hence, if the single scattering albedo is determined by expression (13.3), one can solve the parallel external flux problem in a simple and efficient way and, thus, find the intensity of diffuse radiation spreading in arbitrary directions at each point of the inhomogeneous layer.

13.4 Linear Integral Equation for Intensity of Radiation Emerging from Isotropically Scattering Semi-Infinite Atmosphere

The following formula is a natural generalization of (13.3)

$$\lambda(\tau) = \int_0^\infty L(m)\mathrm{e}^{-m\tau}\mathrm{d}m, \tag{13.42}$$

where $L(m)$ is an arbitrary function of m. It turns out that this generalization enables us to obtain equations of a new kind, which directly determine the intensity of diffuse radiation emerging from the atmosphere.

For simplicity, we will consider a semi-infinite atmosphere with isotropic scattering. Internal sources are distributed in the atmosphere, their power depending on only one spatial coordinate, τ, and being determined by a certain function $g(\tau)$. Then the transfer equation can be written as follows:

$$\mu\frac{\mathrm{d}I(\tau,\mu)}{\mathrm{d}\tau} + I(\tau,\mu) = B(\tau), \tag{13.43}$$

where the source function

$$B(\tau) = \frac{\lambda(\tau)}{2}\int_{-1}^{+1} I(\tau,\mu')\,\mathrm{d}\mu' + g(\tau). \tag{13.44}$$

As usual, we assume at the boundary of the atmosphere that

$$I(0,\mu) = 0 \qquad \text{for} \qquad \mu \in (0,1]. \tag{13.45}$$

With the source function $B(\tau)$ found, the intensity of radiation emerging from the atmosphere is determined by the formula

$$I(0,-\mu) = \int_0^\infty B(\tau)\mathrm{e}^{-\tau/\mu}\frac{\mathrm{d}\tau}{\mu}. \tag{13.46}$$

To derive the equation for the boundary intensity $I(0,-\mu)$, we multiply (13.44) by $\mathrm{e}^{-\tau/\mu}$ and integrate with respect to τ from 0 to ∞. Using (13.42) and (13.46), one finds

$$I(0,-\mu) = \frac{1}{2}\int_{-1}^{+1}\mathrm{d}\mu'\int_0^\infty L(m)\,\mathrm{d}m\int_0^\infty I(\tau,-\mu')\mathrm{e}^{-\tau/\nu}\frac{\mathrm{d}\tau}{\mu} + I^*(0,-\mu), \tag{13.47}$$

where

$$\nu = \frac{\mu}{1+m\mu}, \tag{13.48}$$

$$I^*(0,-\mu) = \int_0^\infty g(\tau)\mathrm{e}^{-\tau/\mu}\frac{\mathrm{d}\tau}{\mu}. \tag{13.49}$$

On the other hand, substituting μ with -μ' in (13.43), multiplying it by $\mathrm{e}^{-\tau/\nu}$ and integrating with respect to τ from 0 to ∞, we obtain

$$\int_0^\infty I(\tau,-\mu')\mathrm{e}^{-\tau/\nu}\,\mathrm{d}\tau = \frac{\nu I(0,-\nu) - \mu' I(0,-\mu')}{\nu - \mu'}\nu. \tag{13.50}$$

The insertion of (13.50) into (13.47) yields

$$I(0,-\mu) = \frac{1}{2}\int_{-1}^{+1}\mathrm{d}\mu'\int_0^\infty \frac{\nu I(0,-\nu) - \mu' I(0,-\mu')}{\nu - \mu'}\frac{L(m)}{1+m\mu}\mathrm{d}m + I^*(0,-\mu). \tag{13.51}$$

As the boundary condition (13.45) has to hold, (13.51) takes the following ultimate form:

$$\begin{aligned} I(0,-\mu) = {} & \frac{1}{2}\int_0^{+1}\mathrm{d}\mu'\int_0^\infty \frac{\nu I(0,-\nu) - \mu' I(0,-\mu')}{\nu - \mu'}\frac{L(m)}{1+m\mu}\mathrm{d}m \\ & + \frac{1}{2}\int_0^\infty I(0,-\nu)\nu \ln\frac{1+\nu}{\nu}\frac{L(m)}{1+m\mu}\mathrm{d}m + I^*(0,-\mu). \end{aligned} \tag{13.52}$$

Equation (13.52) is the linear equation that we sought. Its solution determines the intensity of radiation emerging from the atmosphere. A notable peculiarity of (13.52) is that it is valid for a large variety of functions for which the Laplace transform (13.42) can be reversed, i.e., for which one can find the function $L(m)$.

For a homogeneous atmosphere, equations like (13.52) have long been known (see LSPA, Chap. 7), even for the case of the anisotropic scattering inclusive. It is of substantial interest to generalize this case to the case of anisotropic scattering in inhomogeneous atmospheres.

14. Astrophysical, Geophysical, and Other Possible Applications of the Theory

We now in fact enter the closing chapter. It's aim is to show how the previously elaborated methods can be applied to various problems in astrophysics, geophysics, and other areas of science. My concern of "other areas" looks superficial, for the following reason. Working in astrophysics and, partly, geophysics, I have gained, I suppose, sufficient experience in resolving the problems arising therein. For this reason, I am cautious of entering other fields. However, one may expect that these methods can be utilized somewhere else.

There is yet another point to which I would like to draw the readers' attention. The results presented in this book may not always be used in a direct and straightforward way. In other words, some work is required to adapt a method to a specific problem. On the other hand I hope that the reader who has mastered the basic elements of the scalar theory of radiation transfer, will manage with other specific problems.

Winding up this Preface, I would like to note that the fragmentary structure of this chapter reflects the applied aspects of the author's work.

14.1 Effect of Inhomogeneity of a Cloudless Earth Atmosphere on the Radiation Field

The optics of the Earth's atmosphere is one of the primary areas where the results and methods of the theory of radiative transfer are applied. Much work has already been done there. Many aspects are presented in LSPA, Sect. 10.3 and MLS, Chap. 19, Minin's book [14], and in other monographs (see [28, 29], [31]–[37] and others).

To a rather crude approximation, in visible region of the spectrum one may treat two basic types of atmosphere pattern. We call them the model of clouded atmosphere, and the model of cloudless atmosphere. The clouded atmosphere is considered as consisting of an optically thick homogeneous cloud layer surrounded by optically thin gas-aerosol layers from above and below. The cloudless atmosphere is, according to the model, an inhomogeneous layer of optical thickness $\tau \lesssim 1$, whose optical properties depend on height above the Earth's surface. In both cases, the layers are horizontally-homogeneous, and the atmosphere is assumed to be plane. The problem of radiative transfer in a horizontally inhomogeneous atmosphere is really an independent and extremely difficult one. In fact, we are now at the very beginning of the path that may lead to its resolution (see, e.g., [47]).

It is clear that for the model of clouded atmosphere, the field of diffuse radiation is mainly determined by the optical properties of the cloud layer (the alignments of the paths of optical probing are supposed not to be too sloping). The corresponding asymptotic formulas are then applicable, and the first approximation can be rather easily obtained with the help of the formulas given in the second part of this book. The examples of using asymptotic formulas to estimate the optical thickness of the Earth's clouds and other optical properties can, for example, be found in [175, 176].

In the case of the cloudless atmosphere the situation appears to be slightly different. A comparatively small optical thickness of atmosphere prevents the use of those asymptotic formulas. On the other hand, strong stratification of aerosol by height suggests use of the model of inhomogeneous atmosphere. The latter makes the calculations more complicated. Of course, modern numerical methods elaborated to solve transfer equations enable us to obtain solutions for an atmosphere of arbitrary complexity. However, because the amount of experimental data is usually very large, it is desirable to have a rather simple and efficient algorithm for preliminary processing of the data.

The simplest case is that one may assume the optical thickness of the atmosphere to be $\tau_0 \ll 1$ in some interval of wavelengths. Using the single scattering approximation, from (1.41) and the boundary conditions (1.44), one easily finds ($\mu \in [0,1]$),

$$I^{(1)}(\tau,\mu,\varphi) = \frac{1}{4}\mathrm{e}^{-\tau/\mu}\int_0^{\tau} \lambda(\tau')\chi(\gamma,\tau')\mathrm{e}^{-\tau'\left(\frac{1}{\mu_0}-\frac{1}{\mu}\right)}\frac{\mathrm{d}\tau'}{\mu}, \tag{14.1}$$

$$I^{(1)}(\tau,-\mu,\varphi) = \frac{1}{4}\mathrm{e}^{\tau/\mu}\int_{\tau}^{\tau_0} \lambda(\tau')\chi(\gamma,\tau')\mathrm{e}^{-\tau'\left(\frac{1}{\mu_0}+\frac{1}{\mu}\right)}\frac{\mathrm{d}\tau'}{\mu}. \tag{14.2}$$

We then have for reflection and transmission coefficients

$$\rho^{(1)}(\mu,\mu_0,\varphi) = \frac{1}{4}\int_0^{\tau_0} \lambda(\tau')\chi(\gamma,\tau')\mathrm{e}^{-\tau'\left(\frac{1}{\mu_0}+\frac{1}{\mu}\right)}\frac{\mathrm{d}\tau'}{\mu\mu_0}, \tag{14.3}$$

$$\sigma^{(1)}(\mu,\mu_0,\varphi) = \frac{1}{4}\mathrm{e}^{-\tau_0/\mu}\int_0^{\tau_0} \lambda(\tau')\chi(\gamma,\tau')\mathrm{e}^{-\tau'\left(\frac{1}{\mu_0}-\frac{1}{\mu}\right)}\frac{\mathrm{d}\tau'}{\mu\mu_0}. \tag{14.4}$$

Because the medium is inhomogeneous, we should consider the same layer illuminated by parallel rays from below. Then we derive these formulas

$$\tilde{I}^{(1)}(\tau,\mu,\varphi) = \frac{1}{4}\mathrm{e}^{-\left(\frac{\tau_0}{\mu_0}+\frac{\tau}{\mu}\right)}\int_0^{\tau} \lambda(\tau')\chi(\gamma,\tau')\mathrm{e}^{\tau'\left(\frac{1}{\mu_0}+\frac{1}{\mu}\right)}\frac{\mathrm{d}\tau'}{\mu}, \tag{14.5}$$

$$\tilde{I}^{(1)}(\tau,-\mu,\varphi) = \frac{1}{4}\mathrm{e}^{-\left(\frac{\tau_0}{\mu_0}-\frac{\tau}{\mu}\right)}\int_{\tau}^{\tau_0} \lambda(\tau')\chi(\gamma,\tau')\mathrm{e}^{-\tau'\left(\frac{1}{\mu_0}-\frac{1}{\mu}\right)}\frac{\mathrm{d}\tau'}{\mu}, \tag{14.6}$$

and the corresponding ones for the reflection and transmission coefficients,

$$\tilde{\rho}^{(1)}(\mu, \mu_0, \varphi) = \frac{1}{4} \int_0^{\tau_0} \lambda(\tau')\chi(\gamma, \tau') \mathrm{e}^{-(\tau_0 - \tau')\left(\frac{1}{\mu_0} + \frac{1}{\mu}\right)} \frac{\mathrm{d}\tau'}{\mu\mu_0}, \tag{14.7}$$

$$\tilde{\sigma}^{(1)}(\mu, \mu_0, \varphi) = \frac{1}{4} \mathrm{e}^{-\tau_0/\mu_0} \int_0^{\tau_0} \lambda(\tau')\chi(\gamma, \tau') \mathrm{e}^{-\tau'\left(\frac{1}{\mu_0} - \frac{1}{\mu}\right)} \frac{\mathrm{d}\tau'}{\mu\mu_0}. \tag{14.8}$$

Superscript “1” is used in (14.1–8) to denote that the formulas have been obtained within the single scattering approximation and $\cos\gamma$ is calculated according to (1.43), where for the use of (14.2), (14.3), (14.5) and (14.7), μ must be replaced by $-\mu$.

The above formulas allow us also to obtain the analytic expression for the intensity of diffuse radiation in the case when the atmosphere is adjacent to an isotropically reflecting surface. Equation (10.30) is then employed. To obtain the reflection and transmission coefficients, we use (10.33) and (10.34) respectively, $\alpha = 0$ and $z = \tau_0$ are substituted.

If the optical thickness of the atmosphere is low ($\tau_0 \lesssim 1$), it seems quite natural to offer the following approximate method for calculating the radiation field. For a *homogeneous* layer of the atmosphere we will approximately calculate the contribution to the field of diffuse radiation of photons scattered two or more times; the optical properties of the layer are fixed by certain *average* values of the functions and $\lambda(\tau)$ and $\chi(\gamma, \tau)$ in the interval $\tau \in [0, \tau_0]$. The contribution of single-scattered photons is evaluated exactly by means of (14.1) and (14.2) or (14.5) and (14.6). Let us denote

$$\bar{\lambda}\bar{\chi}(\gamma) = \int_0^{z_0} g(z)\lambda(z)\chi(\gamma, z)\,\mathrm{d}z, \tag{14.9}$$

$$\bar{\lambda} = \int_0^{z_0} g(z)\lambda(z)\,\mathrm{d}z, \tag{14.10}$$

where $g(z)$ is a weight function satisfying the normalizing condition

$$\int_0^{z_0} g(z)\,\mathrm{d}z = 1. \tag{14.11}$$

Then, the intensity of radiation in the atmosphere is proposed to be calculated via the approximate formulas ($\mu \in [0, 1]$),

$$I(\tau, \mu, \varphi) = I_h(\tau, \mu, \varphi) + I^{(1)}(\tau, \mu, \varphi) - \frac{1}{4} \frac{\bar{\lambda}\bar{\chi}(\gamma)}{\mu - \mu_0} \mu_0 \left(\mathrm{e}^{-\tau/\mu} - \mathrm{e}^{-\tau/\mu_0}\right), \tag{14.12}$$

$$I(\tau, -\mu, \varphi) = I_h(\tau, -\mu, \varphi) + I^{(1)}(\tau, -\mu, \varphi) - \frac{1}{4} \frac{\bar{\lambda}\bar{\chi}(\gamma)}{\mu + \mu_0} \mu_0 \left(\mathrm{e}^{-\tau/\mu_0} - \mathrm{e}^{-\tau_0\left(\frac{1}{\mu} + \frac{1}{\mu_0}\right) + \frac{\tau}{\mu}}\right), \tag{14.13}$$

where the intensities $I_h(\tau, \pm\mu, \varphi)$ are evaluated for a homogeneous layer of optical thickness τ_0 with the properties determined by (14.9) and (14.10).

Corresponding expressions for an atmosphere illuminated from below can be obtained similarly. For example, the radiation intensity $I_h(\tau, \pm\mu, \varphi)$ can be easily calculated by the method outlined in Sect. 4.6. In particular, using (14.12) and (14.13) results in the following expression for the reflection and transmission coefficients:

$$\rho(\mu, \mu_0, \varphi) = \rho_h(\mu, \mu_0, \varphi) + \rho^{(1)}(\mu, \mu_0, \varphi) - \frac{1}{4}\frac{\bar{\lambda}\bar{\chi}(\gamma)}{\mu + \mu_0}\left[1 - \mathrm{e}^{-\tau_0\left(\frac{1}{\mu}+\frac{1}{\mu_0}\right)}\right], \tag{14.14}$$

$$\sigma(\mu, \mu_0, \varphi) = \sigma_h(\mu, \mu_0, \varphi) + \sigma^{(1)}(\mu, \mu_0, \varphi) - \frac{1}{4}\frac{\bar{\lambda}\bar{\chi}(\gamma)}{\mu - \mu_0}\left(\mathrm{e}^{-\tau_0/\mu} - \mathrm{e}^{-\tau_0/\mu_0}\right). \tag{14.15}$$

Functions ρ_h and σ_h for a homogeneous layer can be easily found numerically. In this case, the doubling method (see Sect. 4.5) can be employed.

It is worth noticing that the key idea used in deriving the approximate equations (14.12) and (14.13) is rather close to that offered by Sobolev to obtain the approximate formulas for the radiation field in an atmosphere with an arbitrary phase function (see LSPA, Sect. 8.4). Sobolev was the first to propose the exact treatment of the first-order scattering for an arbitrary phase function, and the approximate treatment of scattering of higher orders for the phase function $\chi(\gamma) = 1 + x_1 \cos\gamma$.

It is obvious that for an arbitrarily inhomogeneous atmosphere, the smaller the optical thickness of atmosphere, the more accurate (14.12–15) become. On the other hand, the lower the degree of inhomogeneity of atmosphere, the wider the range of optical thicknesses where these formulas become more accurate. If the atmosphere is adjacent to an isotropically reflecting surface, the corresponding generalization can easily be done with the use of (10.30), (10.33) and (10.34). Moreover, if the reflection law is different from Lambert's, the generalization can be also made easily (see [14], Sect. 3.4).

The effect of inhomogeneity of the Earth's cloudless atmosphere on the radiation field was studied by Gavrilova [177, 178]. The basic feature of her treatment distinguishing it from what is done by most of other authors is to deal with *realistic* functions $\lambda(\tau)$ and $\chi(\gamma, \tau)$ that exist in the atmosphere under more or less standard conditions.

The method of obtaining the transmission and reflection coefficients was based on the system of integral equations (9.30) and (9.31) determining the generalized Ambartsumian's functions $\tilde{\varphi}_k^m(\mu, \tau)$ and $\tilde{\psi}_k^m(\mu, \tau)$. This system was resolved numerically. At the initial stage, she studied the isotropic scattering, which would help her to develop further the efficient numerical algorithm and to find out the basic features brought about by the inhomogeneity of atmosphere. Functions $\tilde{\varphi}(\mu, \tau_0)$ and $\tilde{\psi}(\mu, \tau_0)$ were calculated for six profiles $\lambda(z)$ in the following wavelength range, $\bar{\lambda} = 0.4 - 0.8$ μm. The corresponding average values of particles albedo λ used in the model of homogeneous atmosphere were found with the weight function $g(z) = \alpha(z)/\tau_0$.

The effect of atmospheric inhomogeneity on the calculated values of transmission and reflection coefficients turned out to be most significant for sloping paths ($\mu = \mu_0 \simeq 0.2$). Error then reaches 40%. However, accounting for inhomogeneity corrections in the first-order scattering reduced this error by four times. The transmission coefficient turned out to be far less sensitive to the choice of model. The system of equations (9.30) and (9.31), to determine the functions $\tilde{\varphi}_k^m(\mu,\tau)$ and $\psi_k^m(\mu,\tau)$, was later used by Gavrilova [178] to calculate the surface illumination $\bar{E}(\mu_0,\tau_0)$, plane albedo $\bar{A}(\mu_0,\tau_0)$ and absorption function $\bar{V}(\mu_0,\tau_0)$. It is important here to notice that to reach the required accuracy it sufficed to use no more than the first four terms in the expansion series of phase function in terms of Legendre polynomials. This is a consequence of the directional average of radiation. It is also important to notice that use of the model of homogeneous atmosphere was well justified. For instance, the error in evaluating $\bar{E}(\mu_0,\tau_0)$ via this model was less than 2.5% (for further details, see [178] and [14], Sect. 3.5).

One thus concludes that our formulas can in many cases be successfully used both to calculate the radiation field in a cloudless Earth atmosphere and to solve corresponding inverse problems as well. At least, this approach can be used at the preliminary stage of processing the experimental data. These formulas can also be applied to the Martian atmosphere without, certainly, dust storms.

14.2 Vertical Structure of the Venusian Atmosphere According to Data Obtained by Probes

The ultimate goal of astrophysical observation of any object is to study its physical properties. Radiation is included in much of the information from the upper layers of the atmosphere. That is why, with distant observations, we are unable in fact to study the vertical structure of planetary atmospheres in detail.

Use of probes for studying the vertical structure of the planetary atmosphere that surrounds Venus, has crucially changed the situation. The amount of optical measurement done at different heights above this surface has enabled us to study the atmosphere of Venus in great detail. Fairly complete lists of works on this topic are given elsewhere, [179] and [180].

The venusian atmosphere appears to be strongly inhomogeneous with height. For this reason, the theory of radiative transfer in inhomogeneous atmospheres with anisotropic scattering, which can be found in this book, can be used to process the data of optical measurements collected by probes. This section aims to show this in some particular cases.

Let us draw the reader's attention to the following important fact. Venus is the only planet of the solar system, excepting the Earth of course, for which the vertical structure of the atmosphere has been thoroughly studied.

This enables us to use Venus as a test object to verify the applicability of the interpretations of those distant observations of giant planets that are to determine the vertical structure of planetary atmospheres.

14.2.1 Vertical Distribution of the Absorption Coefficient in an Atmosphere

It was found in Sect. 1.10 that if the gradient of the total flux of radiation with respect to height $\mathrm{d}H_\tau(h,\mu_0)/\mathrm{d}h$ and its average intensity $\bar{I}(h,\mu_0)$ are known, we can resolve one of the inverse problems of the optics of atmosphere that consist of determining the volume absorption coefficient $\kappa(h)$ as a function of height h. This can be done by means of the following exact relationship (see (1.110)):

$$\frac{\mathrm{d}H_\tau(h,\mu_0)}{\mathrm{d}h}/\bar{I}(h,\mu_0) = \kappa(h). \tag{14.16}$$

Equation (14.16) cannot be used directly to estimate $\kappa(h)$ with the help of results of optical measurements carried out by probes, for the reason that none of the modules measured $\bar{I}(h,\mu_0)$ directly. However, some works give us the measured values of illuminations $E^{\downarrow}(h,\mu_0)$ and $E^{\uparrow}(h,\mu_0)$, which enable us to calculate the flux, since

$$H_\tau(h,\mu_0) = E^{\downarrow}(h,\mu_0) - E^{\uparrow}(h,\mu_0). \tag{14.17}$$

These data are sufficient to find the mean intensity for deep layers of the atmosphere

$$\bar{I}(h,\mu_0) = 2\pi\int_{-1}^{+1} I(h,\mu,\mu_0)\,\mathrm{d}\mu. \tag{14.18}$$

This can be done, however, only in the case if the Venusian atmosphere can be regarded as a nearly conservatively scattering one in the respective regions of the spectrum.

Indeed, the atmosphere considered, namely an inhomogeneous atmosphere of finite optical thickness τ_0 adjacent to a radiation reflecting surface, may be treated as a semi-infinite inhomogeneous medium. Optical properties of this medium are defined in such a way that the reflection coefficient of this semi-infinite layer of the "atmosphere" is the same as that of the surface with $\tau > \tau_0$. Then, the asymptotic formula (12.35) is valid for $\tau \gg 1$. If $[1-\lambda(\tau)] \ll 1$, then the (12.27) gives us the intensity $I_\infty(\tau,\mu)$ in P_2 approximation, at least, when we consider the layers that are situated far enough from both upper and lower boundaries of the atmosphere. From (1.90), (1.91), (12.35), (12.27) and (14.18) we find then

$$\bar{I}(h,\mu_0) = 2[E^{\downarrow}(h,\mu_0) + E^{\uparrow}(h,\mu_0)] + O[1-\lambda(\tau)]. \tag{14.19}$$

So we arrive at the following asymptotic formula for estimating the volume absorption coefficient:

Table 14.1

$\tilde{\lambda}$, $[10^{-6}\text{m}]$	$A_s^{(I)}$	x_1	x_2	$\lambda^{(I)}$	$A_s^{(T)}$	$\lambda^{(T)}$
0.315	0.45	2.29	3.43	0.97256	0.49	0.96886
0.359	0.52	2.26	3.37	0.98013	0.56	0.98156
0.393	0.56	2.23	3.33	0.98315	0.61	0.98681
0.415	0.66	2.22	3.30	0.99162	0.71	0.99368
0.457	0.77	2.18	3.23	0.99647	0.80	0.99738
0.501	0.79	2.15	3.16	0.99699	0.82	0.99800
0.626	0.94	2.09	3.01	0.99979	0.97	0.99994
0.730	0.93	2.09	2.96	0.99970	0.95	0.99984
0.860	0.89	2.13	2.96	0.99926	0.92	0.99957
1.060	0.89	2.21	3.00	0.99932	0.88	0.99917

$$\kappa(h) = \frac{\mathrm{d}H_\tau(h,\mu_0)/\mathrm{d}h}{2[E^{\downarrow}(h,\mu_0) + E^{\uparrow}(h,\mu_0)]}. \tag{14.20}$$

Let us revert to discussing the problem of whether the Venusian atmosphere can be considered as a conservatively scattering one. It can be resolved without relying on the probe data, due to the fact that the Venusian brightness can be measured at almost all angles of the phase α (the angle whose vertex is at the center of the Sun and whose sides intersect the planet and the Earth). This results in determining the spherical albedo for different regions of spectrum.

The second column of Table 14.1 contains the monochromatic values of the spherical albedo of Venus $A_s^{(I)}$ calculated by Irvine [181] integrating the phase dependencies of planetary brightness in ten different spectral regions.

Let us deal with the simplest model of the Venusian atmosphere. Precisely, let us suggest that it is a semi-infinite homogeneous layer. Assuming further that the Henyey-Greenstein phase function with $0 \le g \le 0.9$ is valid to first order and comparing the calculated values of A_s given in Table XIX with the values $A_s^{(I)}$ obtained on the basis of observations, one comes to the conclusion that $1-\lambda \le 0.02$ for the wavelength range under study. Then, if the values of the parameters x_1 and x_2 are known, one can use (3.99) to evaluate λ. The values of the parameters can be found with the help of the data of terrestrial polarization observations of Venus (for details see [182, 183]); they are also provided in Table 14.1. And the final comment: it follows from Table XXI that the quantities γ_0 and ε_0 only weakly depend on the form of the phase function. Therefore, taking the values of x_1 into account, γ_0 and ε_0 are given for $g = 0.75$. The fifth column of Table 14.1 contains values of λ calculated via (3.99).

To find A_s, Irvine [181] exploits two methods. However each of them introduces inaccuracies that can hardly be estimated. Furthermore, within the V system, Irvine uses a value of Sun's magnitude of $-26^m.81$, whereas

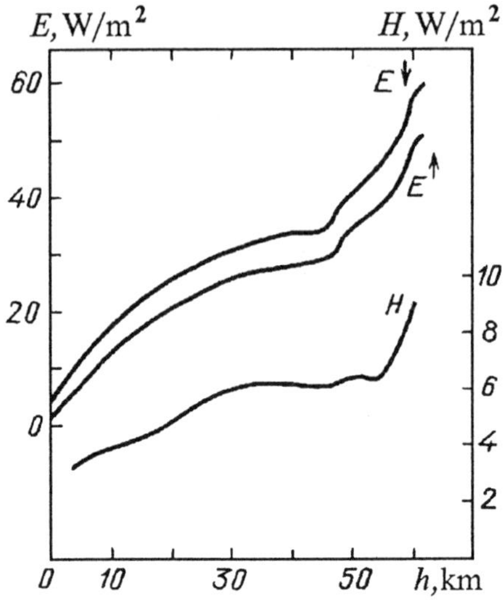

Fig. 14.1. Data of Pioneer–Venus measurements of illuminations of plane area from above $E^{\downarrow}$ and from below $E^{\uparrow}$, and radiation flux H for various heights above the Venusian surface. Effective wavelength is 0.63 μm. For the lower curve, the ordinate scale is on the right

a more precise value is $-26^m.74$ (Johnson [184]). Using the latter value and theoretical results of the phase curves for the Venusian atmosphere [183], Travis [185] determines the numerical values of the spherical albedo A_s and λ. They are given in Table 14.1 and denoted by the superscript T. These data indicate that the reflectivity of the visible surface of Venus is even higher.

Assuming the values $A_s^{(I)}$ to be as minimal as possible, one finds that $(1-\lambda)$ obeys the inequality $0 \leq (1-\lambda) \leq (1-\lambda^{(I)})$. This inequality is valid owing to the assumption about the semi-infiniteness of the atmosphere results, on average (note, the actual atmosphere is inhomogeneous), in the upper bound of $1-\lambda$. Therefore, the Venusian atmosphere can *on the average*, be regarded as a conservatively scattering one over a rather wide range of wavelengths $0.6 \lesssim \tilde{\lambda} \lesssim 1$ μm in continuum.

Let us return to (14.20) and try to apply it for interpreting Pioneer–Venus data of the absolute measurements of $E^{\downarrow}(h,\mu_0)$ and $E^{\uparrow}(h,\mu_0)$ at different heights [186]. The illumination is measured as well, and the flux $H_\tau(h,\mu_0)$ is estimated for the effective wavelength $\tilde{\lambda} = 0.63$ μm and in the height interval $4-63$ km. These results are presented in Fig. 14.1 taken from [186].

The asymptotic formula (14.20) is applicable since the measurements mentioned above deal with deep layers. One should stress the following: this equation can be derived using the Schwarzschild-Schuster approximation. It is known that the more isotropic the radiation in both upper and lower hemispheres is, the better this approximation works. Hence, (14.20) is approximately valid up to the surface.

The figure shows that the upper atmosphere (above 35 km) can be subdivided into four almost homogeneous layers, two of which scatter light conservatively. Table 14.2 demonstrates the volume absorption coefficient calculated for the other two layers and for heights less than 35 km.

Table 14.2

Height, [km]	κ, [km^{-1}]	α, [km^{-1}]	λ
57.5–61.3	0.0039	1.4	0.9972
53–57.5	0	–	1
51–53	0.0014	1.0	0.9986
35–51	0	–	1
30	0.0005	0.139	0.9964
25	0.0015	0.212	0.9929
20	0.0012	0.287	0.9958
15	0.0011	0.344	0.9968
10	0.0009	0.511	0.9982
5	0.0012	0.677	0.9982

Using the measurements carried out by the particle size spectrometer installed in Pioneer–Venus [186], we calculate the extinction coefficient $\alpha(h)$. Afterwards, we estimate the single scattering albedo $\lambda(h)$ (see Table 14.2). $\alpha(h)$ is calculated according to (1.8) for pure CO_2 at heights in the interval $5-35$ km and the number $N(h)$ of molecules per unit of volume is evaluated with the help of the physical model of the Venusian atmosphere offered by Kuz'min and Marov [187]. Doing this, we increase the numerical value of $\alpha(h)$ by 1.31 times (see Sect. 14.2.2). One sees that the values of $\lambda(h)$ given in Table 14.2 are in good agreement with the average value of λ calculated by means of terrestrial observations for corresponding wavelengths (Table 14.1). The agreement looks even better after assuming that the cloud layer of Venus is conservatively scattering for heights above 61.3 km. The work [188], which has been completed later, presented a more elaborate processing of the data [186], which was intended to determine κ, $(3-x_1)\alpha$ and λ. The obtained results are mostly in good agreement with the data provided in Table 14.2.

One should notice the following point. If an atmosphere is nearly conservatively scattering, and the data of measurements of $E^{\downarrow}(h,\mu_0)$ and $E^{\uparrow}(h,\mu_0)$ are known, then one can easily show that the radiation field can be completely reconstructed for deep layers. Equivalently the radiation field can be calculated for all $\mu \in [-1,+1]$ at any given μ_0. Indeed, it follows from (14.17), (1.90), (1.91), (12.35) and (12.27) that

$$I(h,\mu,\mu_0) = \frac{1}{2\pi}\left[E^{\downarrow}(h,\mu_0) + E^{\uparrow}(h,\mu_0) + \frac{3}{2}H_\tau(h,\mu_0)\mu\right]. \tag{14.21}$$

Evidently this simple formula is a consequence of the fact that if the true absorption is not intensive, the radiation in deep atmospheric layer becomes closer to isotropic.

If instead of illumination a probe measures intensities of radiation at certain discrete directions, (14.21) appears to be useful. For instance, Moroz et al. provide the values of $H_\tau(h,\mu_0)$ [189], Figs. 7, 8 over a wide range of heights, namely, from 1 to 60 km. These values were measured by Venera-11 and Venera-12 for four spectral intervals from 0.49 to 1.3 μm. At the same time, Figs. 2 and 3 of the above work show the values of intensities of radiation measured from zenith $I(h, 1, \mu_0)$ (i.e., for $\mu = 1$). Using (14.17) and (14.21) we find

$$E^{\downarrow}(h, \mu_0) + E^{\uparrow}(h, \mu_0) = 2\pi I(h, 1, \mu_0) - \frac{3}{2} H_\tau(h, \mu_0). \tag{14.22}$$

Hence, in this case we arrive at the following expression for the estimation of the volume absorption coefficient instead of (14.20):

$$\kappa(h) = \frac{\mathrm{d}H_\tau(h, \mu_0)/\mathrm{d}h}{4\pi I(h, 1, \mu_0) - 3H_\tau(h, \mu_0)}. \tag{14.23}$$

to evaluate the volume absorption coefficient. Making use of both (14.23) and the data provided in [189], one may easily estimate $\kappa(h)$ for the Venusian atmosphere.

Let us note the following. Dealing with the deep layers of a nearly conservatively scattering atmosphere, (14.21) is valid for all $\mu \in [-1, +1]$ and for the value of μ_0^m corresponding to actual measurements. As follows from (12.35), in order to pass to an arbitrary $\mu_0 \in [0, 1]$, we have to know the function $u_0(\mu_0)$. This function is a universal one in the sense that it is almost independent of the way in which the optical properties (phase function) change at heights below $h \simeq 61$ km. This function solely depends on the properties of the upper layers of the Venusian atmosphere. As inferred from the data of polarization observations [183], these layers are very stable. If the particle size distribution function obeys the normal logarithmic law (1.14), we can choose [182, 183] $r_0 = 0.74$ μm, $\sigma^2 = 0.12$, and the real part of the refraction index $n = 1.435$ for these layers. Table 14.3 shows the results of

Table 14.3 Values of the function $u_0(\mu_0)$

μ_0	λ [μm]				
	0.6	0.7	0.8	0.9	1.0
0	0.316	0.321	0.323	0.324	0.324
0.023	0.372	0.374	0.374	0.373	0.371
0.077	0.455	0.455	0.453	0.451	0.449
0.157	0.550	0.550	0.548	0.546	0.544
0.259	0.651	0.651	0.651	0.650	0.649
0.375	0.757	0.757	0.758	0.758	0.758
0.500	0.865	0.865	0.866	0.866	0.867
0.625	0.970	0.970	0.970	0.971	0.971
0.741	1.066	1.066	1.066	1.067	1.067
0.843	1.149	1.149	1.149	1.149	1.149
0.923	1.214	1.213	1.213	1.213	1.213
0.977	1.257	1.256	1.256	1.255	1.255
1.000	1.276	1.275	1.274	1.274	1.274

calculations of the function $u_0(\mu_0)$ with the above parameters. One sees that the function $u_0(\mu_0)$ hardly depends on wavelength. Tables of this function enable us to pass to an arbitrary value $\mu_0 \in [0,1]$ because it can be rewritten as follows

$$I(h,\mu,\mu_0) = I(h,\mu,\mu_0^m)\frac{u_0(\mu_0)\mu_0}{u_0(\mu_0^m)\mu_0^m}. \tag{14.24}$$

Therefore, if the illuminations in the Venusian atmosphere $E^{\downarrow}(h,\mu_0)$ and $E^{\uparrow}(h,\mu_0)$ have been measured correctly, the radiation field in deep layers can be completely reconstructed for a wide range of wavelengths (evidently, from 0.5 to 1.2 μm).

Of course, the above approach developed for estimating how the absorption coefficient varies with height is applicable to any planetary atmosphere for which the corresponding measurements are available. By the way, Gavrilova [74] applies this approach to find $\kappa(h)$ via the measurements of spectral fluxes in a cloudless Earth atmosphere. These measurements were carried out onboard aircraft. The model problem was then used and the data of measurements processed.

14.2.2 Optical Parameters of Atmospheres in Different Spectral Regions

We have already started to discuss this problem in the previous section. Here we outline two methods to find the structure of the Venusian atmosphere using the data of optical measurements of a probe. In other words, these methods allow us to determine the optical thicknesses of layers, the particles albedo, and phase function for each layer.

The first method makes a use of strict asymptotic formulas of the theory of radiative transfer in multilayer atmospheres consisting of optically thick layers. The main formulas dealing with this case are derived in Chap. 8. Let us show how they can be applied using the example of the optical measurements made by Venera-10 [190, 191].

The results of these measurements are shown in Fig. 14.2 with points and circles which correspond to four of the five effective wavelengths $\tilde{\lambda}_{\text{eff}}$ =0.52, 0.59, 0.65, 0.72 and 0.85 μm. The above effective wavelengths are used to measure the illumination from above $E^{\downarrow}(h,\mu)$ at $\mu_0 = 0.891$ for $1.5 \lesssim h \lesssim 63$ km (for $\tilde{\lambda}_{\text{eff}} = 0.85$ μm the measurements are done at heights $h \lesssim 10$ km).

As seen from Fig. 14.2, a sharp change in the dependence of illumination on height occurs nearly at the height of 48 km. Let us assume that the Venusian atmosphere scatters light conservatively at least for $\tilde{\lambda}_{\text{eff}} = 0.72$ μm. It follows then from (12.120) that the above-mentioned height coincides with that of the boundary dividing cloud from subcloud layers, the scattering properties of which are radically different. To the first order one assumes that the scattering in the subcloud atmosphere is of purely Rayleigh type, and that a homogeneous cloud layer is above the subcloud atmosphere. Then

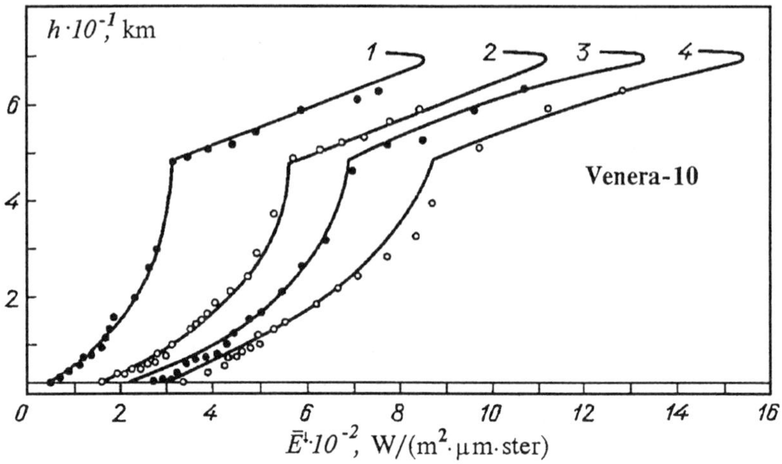

Fig. 14.2. The data of Venera-10 measurements of illuminations from above as a function of height above the surface of planet (dots and circles), and the results of calculations (lines). Curves 1,2,3,4 correspond to $\tilde{\lambda}_{\text{eff}} = 0.72, 0.65, 0.59$ and $0.52\ \mu$m. Curves 2, 3 and 4 are each displaced from one another to the right by one graduation

one should justify these assumptions by comparing the results of calculations with measurements.

Before turning to calculations, we define a model of the atmosphere. Let us, taking into consideration the points stated above, choose the following simplest two-layer model of Venusian atmosphere (indices "1" and "2" are used to denote quantities describing the upper and lower layers, respectively).

1. The atmosphere consists of two layers of optical thickness τ_1 and τ_2, so that $\tau_1, \tau_2 \gg 1$.
2. Both layers weakly absorb, i.e., $1 - \lambda_1 \ll 1$, $1 - \lambda_2 \ll 1$.
3. The upper layer consists of spherical particles whose size is distributed via the normal logarithmic law (1.14). The refraction index does not vary within the spectrum and is equal to 1.435; the geometric mean of radii r_0 and their dispersion σ^2 are 0.74 μm and 0.12, respectively (see [182, 183]). Hence, the phase function $\chi_1(\gamma)$ is known.
4. The atmosphere is adjacent to an isotropically reflecting surface, whose albedo A does not vary within the spectrum and is equal to 0.1. (According to Avduevskii at al. [190, 191], the measured values of albedo are within the interval $0.02 \le A \le 0.2$). One should here take into account that illumination from above weakly depends on albedo in the case when the albedo is within the mentioned interval. For Venus, estimations of the wavelength dependence of albedo are given in [192].

To estimate the parameters of atmosphere, one can use the formulas given in Sect. 8.7. For this, one finds the parameters τ_1, τ_2, λ_1 and the phase function $\chi_2(\gamma)$ in the lower layer. Using Fig. 6 of Ref. [191] one obtains the values of $\bar{E}^{\downarrow}(h, \mu_0)$ measured at different heights, in particular at height 48 km, which is the height of the surface dividing the layers, and at the minimal height 1.5 km. The ratio $\bar{E}_{12}^{\downarrow}/\bar{E}_2^{\downarrow}$ is calculated with the use of this data for all wavelengths. Here $\bar{E}_{12}^{\downarrow}$ and $\bar{E}_2^{\downarrow}$ are the illuminations of the boundaries of gaseous and cloud layers, respectively. On the other hand, assuming the lower layer to be a purely scattering one ($\lambda_2 = 1$), the formulas given in Sect. 8.7 yield

$$4\frac{E_{12}^{\downarrow}(\tau_1, \mu_0)}{E_2^{\downarrow}(\tau_2, \mu_0)} = (1 - A)[(3 - x_{12})\tau_2 + 2 + 3\gamma_2 + 4A(1 - A)^{-1}]. \qquad (14.25)$$

It follows from (14.25) that the quantity considered is independent of the optical properties of the upper layer. This situation remains valid even for more general ratio of intensities if the scattering is nearly conservative in both layers. The conservatism of scattering can be easily verified with the help of the formulas of Sect. 8.7. This conclusion is of substantial practical importance for the solution of inverse problems of atmospheric optics because it allows us to reduce significantly the number of parameters.

Equation (14.25) is used to derive the dependence of $(3 - x_{12})\tau_2$ on the wavelength, which appears to be close to that of Rayleigh. Namely, the application of the least square method to these results yields

$$\tau_2 = 1.82 \cdot \tilde{\lambda}^{-4}, \qquad (14.26)$$

where the wavelength is measured in microns. Hence, we can assume that $x_{12} = 0$ for the lower layer.

Following on to the physical model of the Venusian atmosphere used by Kuz'min and Marov [187] and using (1.8), we have $\tau_2 = 1.39\tilde{\lambda}^{-4}$ at a height of $h = 48$ km. Thus, the value of τ_2 calculated according to (1.8) is considerably less than that of τ found via measurements and the application of (14.25). This discrepancy leads us to assume the presence of a considerable amount of fine aerosol (the radius of particles ≤ 0.03 μm) in the lower layer. Then, the dependence on wavelength remains of the Rayleigh type for the range of wavelengths considered, however the coefficient of $\tilde{\lambda}^{-4}$ increases. The Venusian atmosphere is very dry, which causes fine particles to stay in the atmosphere for an arbitrarily long period.

In order to determine the parameters of the upper layer, we take the following steps. Using the third assumption and the formulas of the Mie theory, we find the phase function $\chi_1(\gamma)$ and polydisperse cross-section of a particle Q. Variation of Q within the spectrum is shown in Fig. 14.3 (the ordinate scale is on the right). Then, making allowance for the above results, we assume that in cloud layer the single scattering albedo for $\tilde{\lambda} = 0.72$ μm is $\lambda = 1$. Using the asymptotic formula, it follows from (8.77),

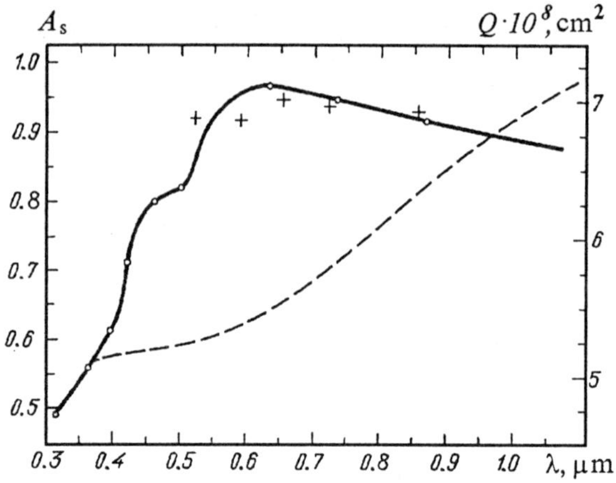

Fig. 14.3. Spectral dependence of spherical albedo of Venus according to terrestrial observations (circles). Crosses mark the results of calculations. Dashed line shows spectral dependence of polydisperse cross-section Q (the ordinate scale is on the right)

$$\bar{E}_{12}^{\downarrow}(\tau_1, \mu_0) = \pi\mu_0 u_1^{(0)}(\mu_0) \frac{(3 - x_{12})\tau_2 + 6(\gamma_2 + a) + 2}{(3 - x_{11})\tau_1 + (3 - x_{12})\tau_2 + 3(\gamma_1 + \gamma_2 + 2a)}, \tag{14.27}$$

where $a = 2A/3(1 - A)$ and the value of $\bar{E}_{12}^{\downarrow}$ measured for $\tilde{\lambda}_{\text{eff}} = 0.72\ \mu$m, taken from [190], Fig. 3, that we obtain $\tau_1 = 51$. The values of τ_1 corresponding to other wavelengths can be easily determined with the use of the variation of the cross-section within the spectrum (Fig. 14.3); here we choose $\gamma_1 = 0.714$, $\gamma_2 = 0.711$. Function $u_1^{(0)}(\mu_0)$ is tabulated in Table 14.3.

For the other wavelengths, the value of the single scattering albedo in cloud layer can be found by setting equal the calculated and measured values of $\bar{E}_{12}^{\downarrow}(\tau_1, \mu_0)$ and/or $\bar{E}_2^{\downarrow}(\tau_2, \mu_0)$. The comparatively simple expressions for the above quantities can be easily derived from the formulas given in Sect. 8.7 if the scattering is close to conservative. Hence, there is no need to make cumbersome calculations to find out the optical parameters of the atmosphere for the case in question.

Table 14.4 shows τ_1, τ_2, λ_1 and λ_2 found by aforementioned procedure. It is worth noticing that earlier we assumed $\lambda_2 = 1$ for the whole spectrum interval. However, at $\tilde{\lambda}_{\text{eff}} = 0.52\ \mu$m, we introduced slight absorption in the lower layer ($\lambda_2 = 0.999$) to get a better agreement between the calculated and measured dependence of illumination on height.

The radiation field (intensity, illumination) corresponding to the given values of optical parameters is calculated for a number of μ and μ_0 at different heights above the surface of Venus. Intensities $I_1^{\infty}(\tau, \mu, \mu_0)$ and $I_2^{\infty}(\tau, \mu, \mu_0)$ are found by the method described in Sect. 3.10. It is interesting to compare these results of our calculations with the measurements provided by Venera-

Table 14.4

$\tilde{\lambda}$ [mu]	τ_1	λ_1	τ_2	λ_2
0.52	46.4	0.9996	24.9	0.999
0.59	47.4	0.9996	15.0	1.0
0.65	48.9	1.0	10.2	1.0
0.72	51.0	1.0	6.77	1.0
0.85	55.2	1.0	3.49	1.0

10. For this purpose, in Fig. 14.2 we use points and circles to show the data of the illumination from above taken from [190], Fig. 3. Solid lines are reserved for our results. The numbers 1, 2, 3 and 4 correspond to the wavelengths $\tilde{\lambda}_{\text{eff}} = 0.72; 0.65; 0.59; 0.52$ μm. One sees that even when the largest discrepancy takes place at $\tilde{\lambda}_{\text{eff}} = 0.52$ μm, the difference is less than 10%.

The height of the upper bound of the cloud layer was estimated by imposing the condition of the best agreement between the results of calculations and the data of measurement. It appears to be 70 km. Why? In order to understand this, we take the optical thickness of the upper layer and the height of the lower bound to be fixed at $\simeq 50$ and $\simeq 48$ km, respectively. Calculations give us the dependence of the illumination on the optical depth in the layer, and the results of the measurements provide us with the dependence on geometrical depth. The best agreement between calculations and measurements while varying the geometrical thickness Δh of the cloud layer ($\Delta h \geq 15$ km) is obtained at $\Delta h \simeq 22$ km, i.e., $h \simeq 70$ km. The upper bound is most likely fuzzy. Therefore, this height should be treated as an effective one.

With the thickness of the cloud layer ($\simeq 22$ km), its optical thickness, and the polydisperse cross-section of a single particle (Fig. 14.3) all known, it is fairly easy to find the number of particles in a unit volume. It is approximately 400 cm^{-3}.

We pay further attention to the fact shown by Fig. 14.2 that the illumination increases with increasing τ, reaches its peak at a certain optical depth $\tau = \tau'$, and then decreases. This interesting fact can be observed for a certain interval $(0 < \mu_1 \leq \mu_0 \leq 1)$ and is really the direct consequence of the asymptotic equations (3.87) and (3.60) in the case of semi-infinite homogeneous atmosphere. If $\tau \gg 1$ these formulas yield

$$E^{\downarrow}(\tau, \mu_0) = u_0(\mu_0)\mathrm{e}^{-k\tau}, \tag{14.28}$$

$$E_0^{\downarrow}(\tau, \mu_0) = u_0(\mu_0). \tag{14.29}$$

Indeed, Tables IV, VI, VIII, X, XII, XIV and XVI show that for $\mu_0 \in [\mu_1, 1]$,

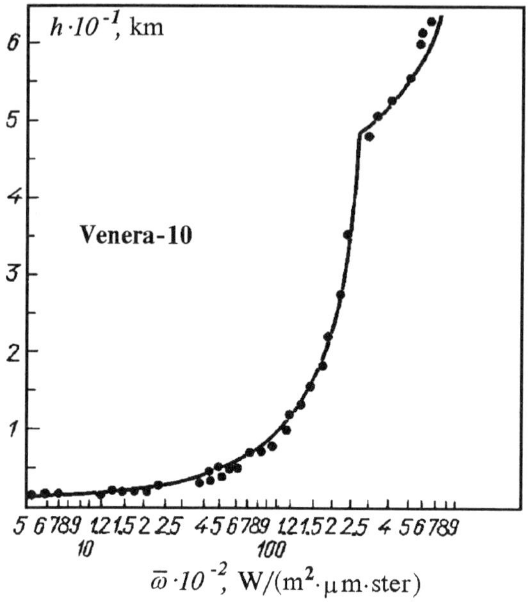

Fig. 14.4. As Fig. 14.2 for $\tilde{\lambda}_{\text{eff}} = 0.72\ \mu$m. Dots represent the measurements at 23° to the nadir

$u_0(\mu_0) \geq 1$; here $\mu_1 \simeq 0.66$ (the dependence of the latter quantity on elongation of the phase function is very weak). Hence, for $\mu_0 \in (\mu_1, 1]$, the illumination in deep layers of a semi-finite conservatively scattering medium always exceeds that at the boundary. If the medium is nearly conservatively scattering, it follows from (14.28) that for any arbitrarily small k, there always exists τ (which depends on μ_0), so that for each $\mu_0 \in [\mu_1, 1]$ we have $E^{\downarrow}(\tau, \mu_0) = 1$. Then, the fact that $E^{\downarrow}(\tau, \mu_0)$ is a continuous function of τ results in the fact that there is such a depth $\tau = \tau'$, at which the function reaches its maximum value.

One can give a simple physical interpretation of this phenomenon. If angles of incidence of light to the upper boundary of the atmosphere are small and the single scattering albedo is close or equal to unity, then when the depth begins to increase, the number of scattered photons increases dramatically. This results in the photons being gathered in the layer adjacent to the boundary, and so the illumination in this layer becomes greater than that at the boundary (i.e., atmosphere works like a lens). Increasing the depth further, the number of scattered photons begins to decrease (which is a consequence of the influence of the surface and, if $\lambda \neq 1$, the true absorption). This implies a decrease in the illumination. If μ_0 is small and/or the absorption is large enough, photons in the layer adjacent to the surface do not have enough time to be gathered (they only have enough time either to cross the boundary or to be absorbed). Therefore, as the depth increases, the illumination monotonically decreases.

The corresponding formulas and calculations indicate that for the optically thick atmosphere, an extremum of radiation intensity may also occur under certain conditions. Minin (for details see [14] Sect. 3.4) shows that for an optically thin atmosphere ($\tau_0 \lesssim 1$), the relative illumination may exceed unity. These effects are also found in the calculations by Kattawar and Plass [108].

Let us now come back to the interpretation of probe measurements. Besides the measurements with wide directional diagram, Venera-10 provides us also with measurements with a narrow diagram ([191], Fig. 3); where the radiation energy is from the zenith at an angle of 23 degrees with the nadir. Figure 14.4 compares the calculations and the results of nadir measurements ([191], Fig. 5) for a wavelength of $\tilde{\lambda} = 0.72$ μm. One realizes that the calculations are in good agreement with measurements for all heights.

One may verify the validity of the values of parameters in the Venusian atmosphere by comparing the calculations of spectral values of spherical albedo of the planet (Fig. 14.3, crosses) with the terrestrial observations (solid line according to Table 14.1, the ordinate scale is on the left). The agreement is quite reasonable.

The second method of determining the optical parameters of an atmosphere by means of a probe's optical observations should be used whenever it is clear beforehand that the simple model discussed above is not applicable.

Fig. 14.2 demonstrates that for $\tilde{\lambda} = 0.52$ μm and heights 30–48 km, the agreement between calculated and measured values of illumination is appreciably worse than that for other wavelengths. It can be assumed that there is a haze in the atmosphere at the above heights and that the influence of this haze in the long wave band is rather small. This assumption is confirmed by the measurements of Venera-11, 12, 13 and 14 [193, 194] and Pioneer-Venus [195]. This is why the Venusian atmosphere should be treated as consisting of a minimum of three layers, one of which at least is not optically thick. This implies that comparatively simple formulas are inapplicable to estimate the parameters of these layers. A unique alternative is to make a comparison of the probe measurements with the whole body of calculations made for various sets of parameters describing each of the layers. The best set of parameters is determined by the principle of the smallest residual. It is precisely this principle that Dlugach [196] applies to analyze the data of the measurements of Venera-11, 13 and 14.

The following three-layer model of the Venusian atmosphere is used in [196].

1. A homogeneous cloud layer between the heights 49(50) and 70 km; its properties have been discussed above.
2. A layer consisting of gas and fine particles of radius $r \simeq 0.1$ μm at the height 32–49 km.

3. A layer adjacent to the surface of the planet ($h < 32$ km) which scatters radiation according to the Rayleigh law. The albedo of the surface is taken to be 0.1 in the band considered.

We are dealing with the data of the ascending intensity of radiation obtained by Venera-11 [197, 198] and Venera-13 and 14 [199] for four effective wavelengths. They are presented in Table 14.5.

Values of the optical parameters of the model of the atmosphere are chosen by means of rigorous calculating the radiation field in a multilayer medium. The algorithm for this calculation is described in Chap. 6 (Sects. 6.4, 6.6). For each homogeneous layer considered separately, the internal field is calculated via the method outlined in Sect. 4.6.

For each set of measurements, we start to select parameters using a wavelength of $\tilde{\lambda} = 1.01$ μm. The optical thickness of the whole cloud layer is estimated by imposing the condition of best coincidence of the calculated

Table 14.5

Venera-11

Heights [km]	1.01 [μm] τ	λ	0.735 [μm] τ	λ	0.635 [μm] τ	λ	0.450 [μm] τ	λ
70-60	16	1.0	13	0.9997	13	1.0	12	1.0
60-57	7	1.0	6	0.9997	5	1.0	5	1.0
57-51	12	1.0	10	0.9997	10	1.0	9	1.0
51-49	8	1.0	7	0.9997	6	1.0	6	1.0
49-32	0.6	1.0	1.5	1.0	2.1	1.0	5.5	0.996
32-0	1.5	1.0	5.3	0.999	9.6	0.997	38.0	0.996

Venera-13

Heights [km]	1.01 [μm] τ	λ	0.950 [μm] τ	λ	0.620 [μm] τ	λ	0.485 [μm] τ	λ
70-60	7	1.0	7	1.0	5.5	1.0	5.5	0.9996
60-57	8	1.0	8	1.0	6.5	1.0	6	0.9996
57-52	20	1.0	19	1.0	16	1.0	15.5	0.9996
52-49	25	1.0	24	1.0	20	1.0	19	0.9996
49-32	1.2	1.0	1.5	1.0	4.2	1.0	10.3	1.0
32-0	1.7	1.0	2.2	0.8	11.6	0.9996	32.0	0.997

Venera-14

Heights [km]	1.01 [μm] τ	λ	0.950 [μm] τ	λ	0.620 [μm] τ	λ	0.485 [μm] τ	λ
70-60	17	1.0	16	1.0	13.5	1.0	13	1.0
60-58	3	1.0	3	1.0	3	1.0	2	1.0
58-52	12	1.0	11	1.0	9.5	1.0	9	1.0
52-50	8	1.0	8	1.0	6	1.0	6	1.0
50-32	1.7	1.0	2.1	1.0	5.9	1.0	14.8	1.0
32-0	1.7	1.0	2.2	0.5	11.6	1.0	32.0	0.998

and measured profiles of radiation intensity. Considering the heights of 60–49(50) km, one should distinguish three optical layers of different density. For the above wavelength, the scattering is taken to be conservative at all levels. The optical thicknesses of layers for other wavelengths are obtained using the spectral dependence of cross-section. It appears that for the lower layer, which is of Rayleigh type, the dependence of its optical thickness on wavelength should be described by the expression $\tau_R = 1.56\tilde{\lambda}^{-4}$ (Venera-11), $\tau_{\mathrm{R}} = 1.77\tilde{\lambda}^{-4}$ (Venera-13 and 14). On the other hand, this dependence should be described by the formula $\tau_{\mathrm{R}} = 1.29\tilde{\lambda}^{-4}$ for the layer of gaseous atmosphere situated below the height $h = 32$ km (note that $\tilde{\lambda}$ is always measured in microns). Hence, in general, these experiments confirm also the assumption that there is a considerable quantity of fine aerosol ($r \lesssim 0.03$ μm) in the lower strata of the Venusian atmosphere. Table 14.5, taken from [196], contains the obtained values of the parameters of atmospheric layers (for further details see [196]). Figures confirming that the calculations and experiments are in good agreement can be also found in [196].

Analysis of the data of Table 14.4 and Table 14.5 reveals significant variation of the optical characteristics of the Venusian atmosphere, particularly those of optical thickness and concentration of absorbing substances in the layers. It is likely that this variation of optical parameters could be caused by temporal and spatial changes in atmospheric properties.

One should emphasize that the data given in Tables 14.4 and 14.5 may include considerable errors. This may result not only from error of measurement, but also from deviations of local optical thickness of cloud layer, which may occur above the probe during the hour-long landing process.

The point is that the data confirming the aforementioned deviations are obtained during the television experiment conducted on the surface of the planet by probes Venera-13 and 14. Namely, there were deviations of brightness of surface elements [200, 201]. The largest deviations turned out to be about 20–25% during observations of 40–60 min. It is clear that the fact that the duration of these measurements and the time of landing of the probe are comparable may induce errors in determining the parameters that are obtained assuming a stationary model of atmosphere.

These errors are estimated by Dlugach [202]. She demonstrates that a nonstationary atmosphere could result in the fact that the optical thickness of the layer at heights 49(50)–60 km may differ considerably (several times) from that given in Table 14.5. This, in turn, makes a correct determination of the structure of atmosphere using only the data of measurements of lighting regime very problematic.

14.3 Absorption Line Formation in an Inhomogeneous Planetary Atmosphere. Basic Concepts and General Formulas

Following LSPA (Sect. 9.3) we will recall the main points of the theory of absorption line formation in the spectra of planets with multiple light scattering. These lines belong to different absorption bands which have, in most cases, a very complicated structure.

Entering the atmosphere of a planet, the sunlight is either scattered by aerosol particles and molecules or undergoes true absorption. True absorption in a continuous spectrum is caused by aerosol particles, while absorption in line frequencies is due not only to particles, but principally to the corresponding molecules responsible for the appearance of these lines. Therefore, the following quantities must be involved in the theory of absorption line formation in planetary spectra (see Sect. 1.2): the volume coefficient of true absorption $\kappa_\nu(z)$ for the frequency ν of the corresponding line at a geometrical depth z in the atmosphere, the volume coefficient of true absorption of the aerosols $\kappa(z)$, the volume coefficient of the molecular and aerosol scattering $\sigma(z)$ and the phase function $\chi(\gamma, z)$. Within an absorption line, $\kappa_\nu(z)$ strongly depends on frequency, whereas the other three quantities can be assumed to be approximately independent of ν in the same frequency range. In principle, the quantities $\chi(\gamma, z)$, $\sigma(z)$ and $\kappa(z)$ can be found from the results of spectrophotometric and polarimetric studies of planets in the continuous spectrum. However, they can be found from spectroscopic observations in appropriate absorption lines too. For simplicity, we will at this stage consider an isolated absorption line whose profile is described by the function $\kappa_\nu(z)$.

Let $I_{\nu_1}(\mu, \mu_0, \varphi; \tau_0^{\nu_1})$ be the intensity of the radiation diffusely reflected by a planetary atmosphere in the continuous spectrum for a frequency ν_1 near the line under consideration (see Fig. 14.5). As we know, this intensity is determined by $\chi_{\nu_1}(\gamma, z)$, the single scattering albedo $\lambda_{\nu_1}(\mathrm{z})$, the optical thickness $\tau_0^{\nu_1}$ of the atmosphere and the albedo A_{ν_1} of the surface, with

$$\tau_0^{\nu_1} = \int_0^{z_0} [\sigma_{\nu_1}(z) + \kappa_{\nu_1}(z)]\, \mathrm{d}z, \tag{14.30}$$

$$\lambda_{\nu_1}(z) = \frac{\sigma_{\nu_1}(z)}{\sigma_{\nu_1}(z) + \kappa_{\nu_1}(z)}, \tag{14.31}$$

where z_0 is the geometrical thickness of the atmosphere.

We denote the corresponding intensity of the radiation in line frequencies by $I_\nu(\mu, \mu_0, \varphi; \tau_0^\nu)$. For this intensity

$$\tau_0^\nu = \int_0^{z_0} [\sigma_{\nu_1}(z) + \kappa_{\nu_1}(z) + \kappa_\nu(z)]\, \mathrm{d}z \tag{14.32}$$

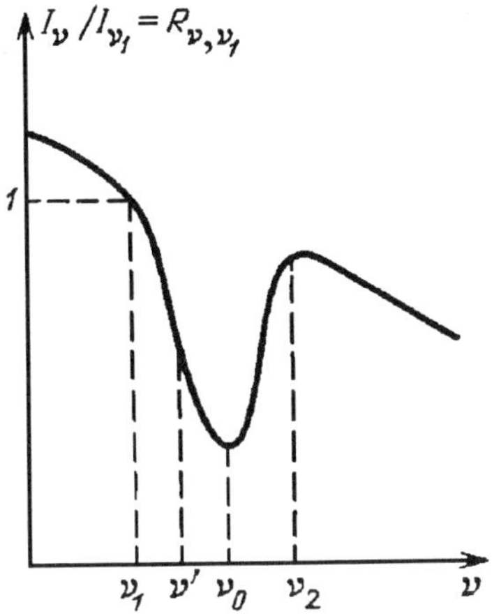

Fig. 14.5.

and

$$\lambda_\nu(z) = \frac{\sigma_{\nu_1}(z)}{\sigma_{\nu_1}(z) + \kappa_{\nu_1}(z) + \kappa_\nu(z)}. \tag{14.33}$$

The quantity

$$R_{\nu,\nu_1}(\mu, \mu_0, \varphi) = \frac{I_\nu(\mu, \mu_0, \varphi; \tau_0^\nu)}{I_{\nu_1}(\mu, \mu_0, \varphi; \tau_0^{\nu_1})} = \frac{\rho_\nu(\mu, \mu_0, \varphi; \tau_0^\nu)}{\rho_{\nu_1}(\mu, \mu_0, \varphi; \tau_0^{\nu_1})} \tag{14.34}$$

is usually used to characterize the absorption line profile (see Fig. 14.5). In (14.34) we switched from intensities to reflection coefficients ρ_ν by using (1.95). The ***equivalent width*** of the absorption line

$$W_{\nu_1,\nu_2}(\mu, \mu_0, \varphi) = \int_{\nu_1}^{\nu_2} [1 - R_{\nu',\nu_1}(\mu, \mu_0, \varphi)]\, \mathrm{d}\nu' \tag{14.35}$$

is used as a spectrum-integrated characteristic. Here the integration over frequency is performed within the absorption line limits (see Fig. 14.5). The theory usually uses the following formula for the equivalent line width:

$$W(\mu, \mu_0, \varphi) = \int_0^\infty [1 - R_{\nu'}(\mu, \mu_0, \varphi)]\, \mathrm{d}\nu'. \tag{14.36}$$

In a similar way, we can consider an observer who is on the surface of a planet (for example, the Earth). Then, instead of (14.34) and (14.36) we have

$$R^{(t)}_{\nu,\nu_1}(\mu, \mu_0, \varphi) = \frac{\sigma_\nu(\mu, \mu_0, \varphi; \tau_0^\nu)}{\sigma_{\nu_1}(\mu, \mu_0, \varphi; \tau_0^{\nu_1})}, \tag{14.37}$$

$$W^{(t)}(\mu, \mu_0, \varphi) = \int_0^\infty \left[1 - R^{(t)}_{\nu'}(\mu, \mu_0, \varphi)\right] \mathrm{d}\nu', \tag{14.38}$$

where the superscript t indicates the fact that the above quantities now pertain to the light diffusely transmitted by the atmosphere, the transmission coefficient of which is denoted by σ_ν. Let us, however, revert to the case when a planet is studied by an external observer.

The formulas given above determine the profile and equivalent width of the line for an arbitrary point on the disk of the planet. In spectroscopic observations, there is usually a slit cutting out a part of the disk. In this case, according to LSPA, Sect 9.2, the illumination $E_e(\alpha)$ of the Earth caused by a part of the planetary disk observed in the continuous spectrum for the phase angle α is given by the formula:

$$E_e(\alpha) = F\frac{R_p^2}{\Delta^2}H(\alpha), \tag{14.39}$$

where

$$H(\alpha) = \iint \rho(\mu, \mu_0, \varphi)\mu\mu_0 \cos\psi \, d\omega \, d\psi. \tag{14.40}$$

Here, πF is the illumination of an element normal to sunlight at the upper boundary of the planetary atmosphere, R_p is the radius of the planet and Δ is the distance from the planet to the Earth. In (14.40) the integration extends over the observable area of the planetary disk and μ and μ_0 are related to the planetocentric latitude ψ and longitude ω by means of the following relations:

$$\mu_0 = \cos\psi\cos(\alpha - \omega), \tag{14.41}$$

$$\mu = \cos\psi\cos\omega. \tag{14.42}$$

For example, if the spectrograph slit is parallel to the intensity equator of the planet and covers the entire area of the disk between latitudes ψ_0 and ψ_1, we have, instead of (14.40), the following expression:

$$H(\alpha) = \int\limits_{\alpha-\pi/2}^{\pi/2} \int\limits_{\psi_0}^{\psi_1} \rho(\mu, \mu_0, \varphi)\mu\mu_0 \cos\psi \, d\omega \, d\psi. \tag{14.43}$$

Similar relations hold for the frequencies within the absorption line, which can be shown by substituting $\rho_\nu(\mu, \mu_0, \varphi)$ for $\rho(\mu, \mu_0, \varphi)$ in (14.40) and (14.43) (for simplicity, from now on we do not indicate the dependence of these quantities on ν_1 and $\tau_0^{\nu_1}$). The corresponding value of the function $H(\alpha)$ within the absorption line is denoted by $H_\nu(\alpha)$. Then its profile can be characterized by the quantity

$$R_\nu(\alpha) = \frac{H_\nu(\alpha)}{H(\alpha)}. \tag{14.44}$$

With the help of (14.44), (14.40) and (14.34) we obtain

$$[1 - R_\nu(\alpha)]H(\alpha) = \iint [1 - R_\nu(\mu, \mu_0, \varphi)]\rho(\mu, \mu_0, \varphi)\mu\mu_0 \cos\psi \, d\omega \, d\psi, \tag{14.45}$$

and, taking into account (14.34), we come to the following expression:

$$[1 - R_\nu(\alpha)]H(\alpha) = \iint [\rho(\mu, \mu_0, \varphi) - \rho_\nu(\mu, \mu_0, \varphi)\mu\mu_0 \cos\psi \, \mathrm{d}\omega \, \mathrm{d}\psi. \quad (14.46)$$

As a result, by analogy with (14.35), for the corresponding equivalent line width we have

$$W(\alpha) = \int_{\nu_1}^{\nu_2} [1 - R_{\nu'}(\alpha)] \mathrm{d}\nu' \quad (14.47)$$

(indices ν_1 and ν_2 are omitted). Substituting (14.45) into (14.47) and taking (14.35) into account we find

$$W(\alpha)H(\alpha) = \iint W(\mu, \mu_0, \varphi)\rho(\mu, \mu_0, \varphi)\mu\mu_0 \cos\psi \, \mathrm{d}\omega \, \mathrm{d}\psi. \quad (14.48)$$

Now we can apply the general formulas derived above to find R_ν and $R_\nu^{(t)}$ and the equivalent widths W and $W^{(t)}$ when the optical thickness of the atmosphere in the continuum is large ($\tau_0 \gg 1$) and the true absorption is negligible in both the continuum and the absorption line.

14.4 Absorption Line in an Optically Thick Nearly Conservatively Scattering Atmosphere

Let the atmosphere be nearly conservatively scattering in both the continuum and the frequencies of the absorption line, i.e.

$$1 - \lambda \ll 1 \quad \text{and} \quad 1 - \lambda_\nu \ll 1. \quad (14.49)$$

In this case, as we will see later, the formulas derived earlier can be simplified substantially.

For an atmosphere of considerable optic thickness the expression (12.48) can be rewritten within the frequencies of the absorption line as follows:

$$\rho_\nu(\mu, \mu_0, \varphi; \tau_0^\nu) = \rho_\nu(\mu, \mu_0, \varphi) - \frac{M_\nu \tilde{N}_\nu(\tau_0^\nu)}{1 - N_\nu \tilde{N}_\nu(\tau_0^\nu)} u_\nu(\mu) u_\nu(\mu_0), \quad (14.50)$$

where $\rho_\nu(\mu, \mu_0, \varphi)$ is the reflection coefficient for a semi-infinite atmosphere and the subscript ν denotes dependence on the frequency. In the same way, for the continuum we have

$$\rho(\mu, \mu_0, \varphi; \tau_0) = \rho(\mu, \mu_0, \varphi) - \frac{M\tilde{N}(\tau_0)}{1 - N\tilde{N}(\tau_0)} u(\mu) u(\mu_0), \quad (14.51)$$

where it is implied that all the quantities involved in (14.51) pertain to the frequency ν_1 of the continuum and the designation $\tilde{N}_{\tau_0} = \tilde{N}(\tau_0)$ has been introduced.

If the conditions of (14.49) are satisfied, we can, according to (12.61) and (12.29), write

$$\rho_\nu(\mu, \mu_0, \varphi) = \rho_0(\mu, \mu_0, \varphi) - \frac{1}{2} M_\nu u_0(\mu) u_0(\mu_0). \tag{14.52}$$

Taking into consideration (12.68) and (12.29), we have

$$u_\nu(\mu) = \left(1 - \frac{3}{8}\gamma_0 M_\nu\right) u_0(\mu). \tag{14.53}$$

To obtain formulas for the continuous spectrum it is sufficient to omit the subscript ν in the above formulas. By substituting (14.52) and (14.53) into (14.50) and (14.51) we obtain the relations, which, after being substituted into (14.34) and omitting the subscript ν_1, give the following formula:

$$1 - R_\nu(\mu, \mu_0, \varphi) = \frac{1}{2}\left[M_\nu \frac{1 + N_\nu \tilde{N}_\nu(\tau_0)}{1 - N_\nu \tilde{N}_\nu(\tau_0)} - M \frac{1 + N\tilde{N}(\tau_0)}{1 - N\tilde{N}(\tau_0)}\right] \frac{u_0(\mu) u_0(\mu_0)}{\rho(\mu, \mu_0, \varphi)}. \tag{14.54}$$

Here it has been taken into account that (12.80), (12.29) and (14.53) result in

$$u_\nu(\mu) u_\nu(\mu_0) \simeq N_\nu u_0(\mu) u_0(\mu_0) \tag{14.55}$$

as well as that if (14.49) holds, then as follows from (14.30), we can put $\tau_0^\nu = \tau_0^{\nu_1} = \tau_0$ for all the frequencies within the line.

The formula (14.54) is what we sought. It determines the profile of a weak absorption line at an arbitrary point on the planetary disk as a product of two factors, one of which depends only on the frequency and another only on the position on the disk. Thus, the profile of a weak absorption line in relative units is the same over the entire disk (if, of course, the atmosphere is horizontally homogeneous). This important inference was drawn by Sobolev (see LSPA, Sect. 9.3) when he considered weak absorption line formation in a semi-infinite homogeneous atmosphere. As we can see, this conclusion remains valid for an inhomogeneous atmosphere of greater thickness.

Note that $\rho(\mu, \mu_0, \varphi)$, which appears in (14.54), can be regarded as known from observations. We also know (see, for instance, LSPA, Sect. 8.1) that the function $u_0(\mu)$ depends on the optical properties of the atmosphere only very weakly. So, having specified the function $u_0(\mu)$ for a certain phase function whose form is established on the basis of a priori considerations, we can regard the value

$$\frac{[1 - R_\nu(\mu, \mu_0, \varphi)]\rho(\mu, \mu_0, \varphi)}{u_0(\mu) u_0(\mu_0)} \tag{14.56}$$

as known. Thus, taking a particular optical model of the atmosphere (by specifying $\lambda_\nu(\tau)$ and $\chi(\gamma, \tau)$), we can find the parameters of the atmosphere by comparing model calculations of the quantities in the square brackets in the right-hand side of (14.54) with the value of (14.56) known from observations.

As for the absorption line observed from the surface of a planet with an optically thick atmosphere, it follows from (14.37) and (14.38) that, taking into account (12.49), (12.68) and (12.69), $R_\nu^{(t)}$ and $W^{(t)}$ *do not depend at all* on the celestial sphere position and the zenith distance to the Sun. In other words, the outline of a weak absorption line in an inhomogeneous nearly conservatively scattering atmosphere is determined only by its optical properties. Namely, using (14.37), (12.49), (12.51), (12.68) and (12.69), and omitting subscript ν_1 we have

$$R_\nu^{(t)}(\mu,\mu_0,\varphi) \equiv R_\nu^{(t)}(\tau_0) = \frac{t_\nu^{(1)}(\tau_0)}{t^{(1)}(\tau_0)}, \tag{14.57}$$

where the transmission factor $t^{(1)}(\tau_0)$, which is found in the P_1 approximation (i.e., under the condition of nearly conservative scattering), is given by the equation

$$t^{(1)}(\tau_0) = \frac{M\left[1+\dfrac{3\gamma_0 y_0'(0)}{3-x_1(0)}\right]R(\tau_0)}{1-N\tilde{N}(\tau_0)}, \tag{14.58}$$

and $R(\tau_0)$ can be found with the help of the expression (12.70). As for the equivalent widths, we can find from (14.38) and (14.57) that

$$W^{(t)}(\mu,\mu_0,\varphi) \equiv W^{(t)}(\tau_0) = \frac{1}{t^{(1)}(\tau_0)}\int_0^\infty [t^{(1)}(\tau_0) - t_\nu^{(1)}(\tau_0)]\,\mathrm{d}\nu. \tag{14.59}$$

Now, let us consider the case when observations are carried out with the help of a slit cutting out part of the planetary disk. Substituting (14.54) into (14.55) we obtain

$$1 - R_\nu(\alpha) = \left[M_\nu\frac{1+N_\nu\tilde{N}_\nu(\tau_0)}{1-N_\nu\tilde{N}_\nu(\tau_0)} - M\frac{1+N\tilde{N}(\tau_0)}{1-N\tilde{N}(\tau_0)}\right]\frac{h(\alpha)}{H(\alpha)}, \tag{14.60}$$

where

$$h(\alpha) = \frac{1}{2}\iint u_0(\mu)u_0(\mu_0)\mu\mu_0\cos\psi\,\mathrm{d}\omega\,\mathrm{d}\psi, \tag{14.61}$$

and $H(\alpha)$ is determined by (14.40).

In its turn, the substitution of (14.60) into (14.47) gives the following expression for the equivalent width of an absorption line:

$$W(\alpha,\tau_0) = \Psi(\alpha)Q(\tau_0), \tag{14.62}$$

where

$$\Psi(\alpha) = \frac{h(\alpha)}{H(\alpha)}, \tag{14.63}$$

$$Q(\tau_0) = \int_{\nu_1}^{\nu_2}\left[M_\nu\frac{1+N_\nu\tilde{N}_\nu(\tau_0)}{1-N_\nu\tilde{N}_\nu(\tau_0)} - M\frac{1+N\tilde{N}(\tau_0)}{1-N\tilde{N}(\tau_0)}\right]\mathrm{d}\nu. \tag{14.64}$$

Now it is worthwhile to analyze the behavior of the functions $\Psi(\alpha)$ and $Q(\tau_0)$ separately. Let us begin with the first of these functions.

14.4.1 Dependence of the Observed Spectrum on the Width and Orientation of Spectrograph Slit

From the above discussion it follows that this dependence is determined by the function $\Psi(\alpha)$. To all appearances the width of a slit may exert insignificant influence on the *relative* change of $\Psi(\alpha)$. According to the estimates [203], which were made for different values of the width of a slit aligned with the intensity equator, the deviations of the relative change of $\Psi(\alpha)$ did not exceed 3% of that obtained for the entire planetary disk (infinitely wide slit, see also Table XXXVII). In this case, in (14.40) and (14.61) the integration should to be performed from $\alpha - \pi/2$ to $\pi/2$ over ω and from 0 to $\pi/2$ over ψ. The results of calculations of the function $\bar{\Psi}(\alpha)$ for the entire planetary disk taken from [204] are given in Table 14.6 (the overbar means here that the quantity is averaged over the whole planetary disk). Here the first four columns correspond to the Henyey-Greenstein phase function and the A and B phase functions are given by the formula (1.17) for the following parameters: A) $g_1 = 0.8$; $g_2 = -0.75$; $c = 0.95$; B) $g_1 = 0.8$; $g_2 = -0.5$; $c = 0.95$. The A and B phase functions have, in contrast to that of (1.15), a local maximum at the angle $\gamma = 180°$. The phase function A approximates the actual phase function of Venus, found by means of polarimetric observations [183] at a wavelength $\tilde{\lambda} \simeq 0.8\ \mu m$.

As can be seen from Table 14.6, within the interval $0 \leq \alpha \leq 120°$ the function $\bar{\Psi}(\alpha)$ depends on the shape of the phase function comparatively weakly. Nevertheless, an intriguing fact takes place. Namely, when $g \geq 0$, the one-parameter Henyey-Greenstein phase function of (1.15), which decreases monotonously with γ, corresponds to the function $\Psi(\alpha)$, which also decreases monotonously. However, the fact that the function $\chi(\gamma)$ has a local maximum at $\gamma = 180°$ (i.e., the phase function increases in the backward direction considerably) causes, as can be seen in the instance of the A and B phase

Table 14.6 Values of the function $\bar{\Psi}(\alpha)$

α	g					
	0	0.6	0.7	0.85	A	B
0°	0.559	0.600	0.604	0.611	0.427	0.544
20°	0.559	0.598	0.602	0.608	0.523	0.557
40°	0.550	0.580	0.585	0.589	0.574	0.568
60°	0.520	0.542	0.545	0.549	0.565	0.551
80°	0.472	0.475	0.478	0.479	0.509	0.496
100°	0.405	0.375	0.373	0.372	0.405	0.397
120°	0.320	0.242	0.237	0.230	0.257	0.254
140°	0.221	0.106	0.097	0.087	0.099	0.100
160°	0.113	0.023	0.017	0.011	0.014	0.013
170°	0.056	0.006	0.004	0.002	0.002	0.002
180°	0	0	0	0	0	0

functions, a shift of the absolute maximum of the function $\bar{\Psi}(\alpha)$. For small angles, a weaker rise of the phase function in the backward direction causes a steeper increase in the function $\bar{\Psi}(\alpha)$. This effect was found in [205, 203] and called the "inverse phase effect".

Recall that as follows from (14.62), a quantity Q can be evaluated, if the quantity $W(\alpha)$ (or $R_\nu(\alpha)$) is known and we calculate the function $\Psi(\alpha)$ from some considerations. If a planetary atmosphere is horizontally-homogeneous, then the function $Q = \text{const}$. at any point (any area) of the planetary disk. Hence, knowing Q for different areas of the disk, one has qualitative information about the horizontal structure of the atmosphere, i.e. about the change in its optical properties (optical thickness, see (14.66) and (14.68) below). This is naturally a subject of much interest.

Using the above reasoning, we give corresponding formulas and results for estimate of the function Ψ also for the case that the spectrum is observed through an infinitely narrow slit which is parallel (Ψ_y) or perpendicular (Ψ_x) to the intensity equator.

For definiteness, the model of a semi-infinite, homogeneous, conservatively scattering atmosphere will be considered. This atmosphere is comprised of polydisperse system of a spherical particles with distribution (1.10), where effective radius $r_{\text{eff}} = 1.05\mu\text{m}$, and $b = 0.07$. The real part of the refraction index is set to be 1.44. These parameters correspond to those of the atmospheric aerosol of Venus with wavelength close to $\lambda = 0.8\mu\text{m}$ (see [183]). The parameter x_1 of the phase function appears to be equal 2.01.

This choice of the orientation of the slit and wavelength is caused by the fact that in the late 60s and early 70s, a series of observations of weak CO_2 absorption bands in the near IR spectral region (0.782–1.063 μm) were performed. Observations were conducted by means of a spectrograph with the slit mostly oriented parallel or perpendicular to the intensity equator. The width of the slit was significantly less than the visible diameter of the planet. These results have been published in a series of works by L. Young et al. Most of these works are mentioned in L. Young's review [206].

If an infinitely narrow slit is parallel to the intensity equator of a planet, then, using (14.63), (14.61) and (14.41) for the function Ψ we find

$$\Psi_y(\alpha, \psi) = \frac{1}{2} \frac{\int\limits_{\alpha-\pi/2}^{\pi/2} u_0(\mu) u_0(\mu_0) \cos(\alpha - \omega) \cos\omega \, d\omega}{\int\limits_{\alpha-\pi/2}^{\pi/2} \rho_0(\mu, \mu_0, \varphi) \cos(\alpha - \omega) \cos\omega \, d\omega} .$$

The situation becomes more complicated when the slit is perpendicular to the equator. In this case we turn to the Cartesian coordinate system by projecting a planet onto the plane perpendicular to the line of sight; the x-axis is the projection of the intensity equator and the y-axis is the projection of a certain meridian onto this plane. Then, with allowance for the fact that $y = \sin\psi$,

$x = \cos\psi \sin\omega$ for an arbitrary point on the planetary disk, and taking into consideration (14.41) and (14.42), one can find after some manipulations

$$\Psi_x(\alpha,\omega) = \frac{1}{2}\frac{\int\limits_a^b [\sqrt{\cos^2\omega - y^2}\cos\alpha + \sin\omega\sin\alpha] u_0(\mu)u_0(\mu_0)\,dy}{\int\limits_a^b [\sqrt{\cos^2\omega - y^2}\cos\alpha + \sin\omega\sin\alpha] \rho_0(\mu,\mu_0,\varphi)\,dy},$$

where a and b take the following values:
1) for $0 \le \alpha \le \pi/2$

$$a = 0, \quad b = \sqrt{1 - \tfrac{\sin^2\omega}{\cos^2\alpha}} \quad \text{at} \quad \alpha - \pi/2 \le \omega \le 0,$$
$$a = 0, \quad b = \cos\omega \quad \text{at} \quad 0 \le \omega \le \pi/2;$$

2) for $\pi/2 \le \alpha \le \pi$

$$a = \sqrt{1 - \tfrac{\sin^2\omega}{\cos^2\alpha}}, \quad b = \cos\omega \quad \text{at} \quad 0 \le \omega \le \alpha - \pi/2,$$
$$a = 0, \quad b = \cos\omega \quad \text{at} \quad \alpha - \pi/2 \le \omega \le \pi/2.$$

Here ω denotes the planetocentric longitude of the position of the slit $[\omega \in (-\pi/2, +\pi/2)]$. The results of numerical calculations of the functions $\Psi y(\alpha,\psi)$ and $\Psi x(\alpha,\omega)$ for this model of atmosphere are presented in Tables XXXVII and XXXVIII in the Appendix. The last column of Table XXXVII contains the results of calculations of the function $\bar{\Psi}(\alpha)$ for infinitely wide slit. One can see that the profile of this function approximately corresponds to that of $\Psi_y(\alpha,\psi)$ for $\psi = 30°$.

Further details can be found in our work [207], where some applications of the obtained results to a qualitative study of the global optical properties of the cloud layer of Venus are offered. It is found, in particular, that the longitude-averaged optical thickness of the Venusian atmosphere sharply decreases from the intensity equator to the poles.

14.4.2 Profile and Equivalent Line Width for Different Models of Atmospheres

Let us, as usual, assume that the profile of the absorption coefficient is of the Lorentzian type

$$\kappa_\nu(z) = \frac{\kappa_0(z)}{1 + \left(\dfrac{\nu - \nu_0}{\Delta\nu(z)}\right)^2} \tag{14.65}$$

where $\kappa_0(z)$ is the volume absorption coefficient at the central frequency ν_0 of the line and $\Delta\nu(z)$ is its half-width. First let us consider the simplest possible case.

14.4.2.1 A Homogeneous Atmosphere Adjacent to an Isotropic Reflecting Surface. In this case it is more convenient to derive a special form of the general formula (14.64), which can be done by using the above technique and the asymptotic expressions given in 5.5.3. Assuming $\nu_1 = 0$ and $\nu_2 = \infty$, after simple transformations, we have

$$\bar{Q}(\tau_0) = \frac{8}{\sqrt{3-x_1}} \int_0^\infty (\sqrt{1-\lambda_\nu} \coth \bar{l}\sqrt{1-\lambda_\nu} - \sqrt{1-\lambda} \coth \bar{l}\sqrt{1-\lambda})\, \mathrm{d}\nu, \tag{14.66}$$

where

$$\bar{l} = \sqrt{3-x_1} \left[\tau_0 + \frac{6\gamma_0}{3-x_1} + \frac{4A}{(1-A)(3-x_1)} \right]. \tag{14.67}$$

The overbar, as usual, means that the case when the atmosphere is adjacent to the surface is being considered. As for the line profile, it is described by the following formula:

$$1 - \bar{R}_\nu(\alpha, \tau_0) = \frac{8\Psi(\alpha)}{\sqrt{3-x_1}} (\sqrt{1-\lambda_\nu} \coth \bar{l}\sqrt{1-\lambda_\nu} - \sqrt{1-\lambda} \coth \bar{l}\sqrt{1-\lambda}), \tag{14.68}$$

and λ_ν and λ must satisfy the inequalities (14.49). Moreover, using (14.31), (14.33) and (14.65) we find

$$1 - \lambda_\nu = 1 - \lambda + \frac{\lambda^2 b}{1 + \lambda b + \xi^2} \simeq 1 - \lambda + \frac{b}{1+\xi^2}, \tag{14.69}$$

where

$$b = \frac{\kappa_0}{\sigma}, \tag{14.70}$$

$$\xi = \frac{\nu - \nu_0}{\Delta\nu}. \tag{14.71}$$

Using the expansion

$$\coth x = \frac{1}{x} + 2 \sum_{n=1}^{\infty} \frac{x}{x^2 + (\pi n)^2}, \tag{14.72}$$

(14.66) can be rewritten as follows:

$$\bar{Q}(\tau_0) = \frac{16\bar{l}}{\sqrt{3-x_1}} \sum_{n=1}^{\infty} \frac{(\pi n)^2}{\bar{l}^2(1-\lambda) + (\pi n)^2} \int_0^\infty \frac{(\lambda - \lambda_\nu)\, \mathrm{d}\nu}{\bar{l}^2(1-\lambda_\nu) + (\pi n)^2}. \tag{14.73}$$

Inserting (14.69) and taking into consideration the inequalities of (14.49), after integration we obtain

$$\bar{Q}(\tau_0) = \frac{16\pi b \bar{l} \Delta\nu}{\sqrt{3-x_1}} \sum_{n=1}^{\infty} \frac{(\pi n)^2}{\left[(b+1-\lambda)\bar{l}^2 + (\pi n)^2\right]^{1/2} \left[(1-\lambda)\bar{l}^2 + (\pi n)^2\right]^{3/2}}. \tag{14.74}$$

If there is no absorption in the continuum (i.e. $\lambda \equiv 1$), we have

$$\bar{Q}_0(\tau_0) = \frac{16b\bar{l}\Delta\nu}{\sqrt{3-x_1}} \sum_{n=1}^{\infty} \frac{1}{n\left[b\bar{l}^2 + (\pi n)^2\right]^{1/2}} \tag{14.75}$$

instead of (14.74). So, when $b \to 0$ ("superweak" line) we have

$$\bar{Q}_0(\tau_0) \sim \frac{8\pi b\bar{l}\Delta\nu}{3\sqrt{3-x_1}}. \tag{14.76}$$

Equation (14.75) can be used to obtain an approximate formula expressing $\bar{Q}_0(\tau_0)$ in terms of elementary functions. To do that, we substitute the summation with an integration calculated with the aid of the trapezoidal rule. As a result we have

$$\bar{Q}_0(\tau_0) \approx \frac{16b\bar{l}\Delta\nu}{\sqrt{3-x_1}} \left[\frac{1}{2\sqrt{b\bar{l}^2+\pi^2}} + \frac{1}{\bar{l}\sqrt{b}} \ln\left(\frac{\bar{l}\sqrt{b} + \sqrt{b\bar{l}^2+\pi^2}}{\pi}\right)\right]. \tag{14.77}$$

The accuracy of this expression can be estimated, for example, with the help of the following fact. When $b \to 0$, (14.77) gives us (14.76) with an accuracy up to a constant factor; the constant factor turns out to be equal to $3/\pi$ instead of the exact value $\pi/3$. Thus, the deviation does not exceed 10% of the exact value. For $b\bar{l}^2 \to \infty$, using (14.77), we find

$$\bar{Q}_0(\tau_0) \sim \frac{8\sqrt{b}\Delta\nu}{\sqrt{3-x_1}} \left(1 + 2\ln\frac{2\bar{l}\sqrt{b}}{\pi}\right). \tag{14.78}$$

Let us introduce the function

$$f_r(y,z) = y \sum_{n=1}^{\infty} \frac{(\pi n)^2}{(y^2+\pi^2 n^2)^{1/2}(z^2y^2+\pi^2n^2)^{3/2}},$$

where

$$y^2 = (1-\lambda+b)\bar{l}^2, \qquad z^2 = \frac{1-\lambda}{1-\lambda+b}.$$

It is useful to note that for $z \neq 0$

$$f_r(\infty,z) = \frac{K(\sqrt{1-z^2}) - E(\sqrt{1-z^2})}{\pi(1-z^2)}, \qquad \lim_{y\to 0}\frac{f_r(y,z)}{y} = \frac{1}{6},$$

$$f_r(y,1) = \frac{\sinh 2y - 2y}{8\sinh^2 y}$$

and if $y \to \infty$

$$f_r(y,0) \simeq \frac{1}{2\pi}(1 + 2\ln\frac{2y}{\pi}).$$

Here $K(x)$ and $E(x)$ are the full elliptic integrals of the first and second kinds respectively. Formula (14.74) can be rewritten as follows:

$$\bar{Q}(\tau_0) = \frac{16\pi b \Delta\nu}{\sqrt{(1-\lambda+b)(3-x_1)}} f_r(y,z).$$

For $(1-\lambda) \to 0$, from (14.74) it can be easily found that

$$\bar{Q}(\tau_0) = \bar{Q}_0(\tau_0)\left[1 + O\left(\frac{1-\lambda}{b}\right)\right]. \tag{14.79}$$

So, if the absorption in the continuum is substantially weaker than that at the center of the line (i.e., $(1-\lambda) \ll b$), then the equivalent width is practically independent of the absorption properties of the atmosphere in the continuum.

An interesting result can be obtained in the opposite case, when $b \ll (1-\lambda)$ (of course, the inequalities of (14.49) are assumed to hold simultaneously). In this case we have

$$\bar{Q}(\tau_0) = \frac{16\pi b\bar{l}\Delta\nu}{\sqrt{3-x_1}} \sum_{n=1}^{\infty} \frac{(\pi n)^2}{\left[(1-\lambda)\bar{l}^2 + (\pi n)^2\right]^2}$$

instead of (14.74). The sum appearing in the right-hand side of the above formula can be expressed in elementary functions, so, finally $\bar{Q}(\tau_0)$ takes the following form:

$$\bar{Q}(\tau_0) = \frac{4\pi b\Delta\nu}{\sqrt{(3-x_1)(1-\lambda)}}\left(1 - \frac{2\bar{l}\sqrt{1-\lambda}}{\sinh 2\bar{l}\sqrt{1-\lambda}}\right)\coth \bar{l}\sqrt{1-\lambda}\,. \tag{14.80}$$

Now, we consider the case when a weak line is observed from the surface of a planet. Using (5.61) we find

$$\begin{aligned} 1 - \bar{R}_\nu^{(t)} = \frac{\sinh \bar{l}\sqrt{1-\lambda}}{\sqrt{1-\lambda}}\Big(&\sqrt{1-\lambda}\sinh^{-1}\bar{l}\sqrt{1-\lambda} \\ &- \sqrt{1-\lambda_\nu}\sinh^{-1}\bar{l}\sqrt{1-\lambda_\nu}\Big) \end{aligned} \tag{14.81}$$

instead of (14.37), where $\bar{l}$ and λ_ν are en given by (14.67) and (14.69) respectively.

Taking into account that

$$\sinh^{-1} x = \frac{1}{x} + 2x \sum_{n=1}^{\infty} \frac{(-1)^n}{x^2 + (\pi n)^2}, \tag{14.82}$$

after substituting (14.81) and (14.82) into (14.38) and carrying out the integration we obtain the following expression for the equivalent line width:

$$\begin{aligned} \bar{W}^{(t)}(\tau_0) = \frac{2\pi b\bar{l}\Delta\nu \sinh(\bar{l}\sqrt{1-\lambda})}{\sqrt{1-\lambda}} \\ \cdot \sum_{n=1}^{\infty} \frac{(-1)^{n+1}(\pi n)^2}{\left[(b+1-\lambda)\bar{l}^2 + (\pi n)^2\right]^{1/2}\left[(1-\lambda)\bar{l}^2 + (\pi n)^2\right]^{3/2}}. \end{aligned} \tag{14.83}$$

In the case when there is no absorption in the continuum (i.e. $\lambda \equiv 1$) we get

$$\bar{W}_0^{(t)}(\tau_0) = 2\pi b\bar{l}^2 \Delta\nu \sum_{n=1}^{\infty} \frac{(-1)^{n+1}}{(\pi n)\left[b\bar{l}^2 + (\pi n)^2\right]^{1/2}} \tag{14.84}$$

instead of (14.83). In particular, when $(1-\lambda) \gg b$, as follows from (14.83), we have a simple analogy of (14.80) for the transmitted radiation

$$\bar{W}^{(t)}(\tau_0) = \frac{\pi}{2}\frac{b\Delta\nu}{1-\lambda}\left(\bar{l}\sqrt{1-\lambda}\coth\bar{l}\sqrt{1-\lambda} - 1\right). \tag{14.85}$$

Let us introduce the function

$$f_t(y,z) = y\sum_{n=1}^{\infty}\frac{(-1)^{n+1}(\pi n)^2}{(y^2+\pi^2 n^2)^{1/2}(z^2y^2+\pi^2n^2)^{3/2}}.$$

Then, instead of (14.83), one can write

$$\bar{W}^{(t)}(\tau_0) = \frac{2\pi b\Delta\nu\sinh\bar{l}\sqrt{1-\lambda}}{\sqrt{(1-\lambda)(1-\lambda+b)}}f_t(y,z).$$

It is helpful to remember that

$$\lim_{y\to 0}\frac{f_t(y,z^2)}{y} = \frac{1}{12}, \quad f_t(y,1) = \frac{y\cosh y - \sinh y}{4\sinh^2 y},$$

and that as $y \to \infty$

$$f_t(y,0) \approx \frac{\ln 2}{\pi}.$$

The functions $f_r(y,z)$ and $f_t(y,z)$ are essentially *modified curves of growth* for diffusely reflected and transmitted radiation, since they determine the corresponding equivalent widths as functions of the number of absorbing molecules, which is characterized by b. These functions are tabulated in Table 14.7.

Thus, if the profile of the volume absorption coefficient is of Lorentzian type (as 14.65), the contour of a weak absorption line forming via multiple scattering in a optically thick atmosphere is given by (14.68), and the equivalent width is given by (14.62) and (14.74) for the light reflected by the entire planet and by (14.81) and (14.83) for the diffusely transmitted radiation, where λ_ν is described by (14.69–71).

It is interesting to consider some particular cases. Let us assume, for simplicity, that there is no absorption in the continuum (i.e., $\lambda = 1$) and $\bar{l} = \tau_0\sqrt{3-x_1}$ (note that $\tau_0 \gg 1$). Let the line be also so weak that $b\bar{l}^2 \ll 1$. For such a "superweak" line, by using the above formulas we easily find that

$$1 - R_{0\nu}(\alpha,\tau_0) \simeq \frac{8}{3}\Psi_0(\alpha)\frac{\kappa_\nu}{\sigma}\tau_0 = \frac{8\Psi_0(\alpha)}{3}\tau_\alpha^\nu \tag{14.86}$$

Table 14.7 Modified curves of growth $f_r(y,z) \times 10$ and $f_t(y,z) \times 10$

y	z^2															
	0		0.05		0.1		0.2		0.4		0.6		0.8		1.0	
	f_r	f_t	f_r	f_t	f_r	f_t	f_r	f_t	f_r	f_t	f_r	f_t	f_r	f_t	f_r	f_t
0	0	0	0	0	0	0	0	0	0	0	0	0	0	0	0	0
0.5	0.83	0.41	0.82	0.41	0.82	0.41	0.82	0.41	0.82	0.40	0.81	0.40	0.81	0.40	0.81	0.39
1.0	1.61	0.79	1.61	0.78	1.60	0.77	1.58	0.76	1.55	0.74	1.53	0.71	1.50	0.69	1.47	0.67
1.5	2.34	1.11	2.31	1.08	2.29	1.07	2.24	1.03	2.15	0.95	2.07	0.89	2.00	0.83	1.93	0.77
2.0	2.98	1.37	2.93	1.32	2.88	1.28	2.78	1.19	2.60	1.05	2.45	0.93	2.33	0.83	2.21	0.74
2.5	3.56	1.57	3.46	1.48	3.36	1.41	3.19	1.27	2.92	1.04	2.69	0.87	2.51	0.74	2.36	0.63
3.0	4.06	1.72	3.90	1.59	3.76	1.47	3.51	1.27	3.12	0.97	2.84	0.76	2.62	0.61	2.43	0.50
3.5	4.50	1.83	4.27	1.64	4.08	1.48	3.74	1.22	3.26	0.86	2.92	0.63	2.67	0.48	2.47	0.38
4.0	4.90	1.91	4.59	1.66	4.33	1.45	3.91	1.13	3.34	0.74	2.96	0.51	2.69	0.37	2.49	0.28
4.5	5.26	1.98	4.85	1.65	4.53	1.39	4.03	1.03	3.39	0.61	2.99	0.40	2.70	0.27	2.50	0.19
5.0	5.58	2.02	5.07	1.62	4.68	1.32	4.11	0.92	3.42	0.50	3.00	0.30	2.71	0.20	2.50	0.13
6.0	6.14	2.08	5.41	1.51	4.90	1.14	4.21	0.70	3.44	0.32	3.01	0.17	2.71	0.10	2.50	0.06
7.0	6.62	2.12	5.65	1.38	5.03	0.95	4.25	0.51	3.45	0.19	3.01	0.09	2.71	0.05	2.50	0.03
8.0	7.04	2.14	5.82	1.23	5.11	0.78	4.28	0.37	3.45	0.11	3.01	0.05	2.71	0.02	2.50	0.01
9.0	7.41	2.16	5.94	1.09	5.15	0.63	4.28	0.26	3.45	0.07	3.01	0.02	2.71	0.01	2.50	0.00
10.0	7.74	2.17	6.02	0.95	5.18	0.49	4.29	0.18	3.45	0.04	3.01	0.01	2.72	0.00	2.50	0.00

for the reflected light and

$$1 - R_{0\nu}^{(t)}(\tau_0) \simeq \frac{(3-x_1)}{6} \frac{\kappa_\nu}{\sigma} \tau_0^2 \tag{14.87}$$

for the transmitted light, where κ_ν is determined by (14.65), σ is the volume scattering coefficient and τ_α^ν is the optical thickness of gas at the frequency ν.

Similarly, for the equivalent width we have

$$W_0(\tau_0) \simeq \frac{8\pi \Delta\nu}{3} \Psi_0(\alpha) \frac{\kappa_0}{\sigma} \tau_0 = \frac{8\pi \Delta\nu \tau_\alpha^{\nu_0}}{3} \Psi_0(\alpha), \tag{14.88}$$

and

$$W_0^{(t)}(\tau_0) \simeq \frac{\pi(3-x_1)\Delta\nu}{6} \frac{\kappa_0}{\sigma} \tau_0^2, \tag{14.89}$$

where $\tau_0^{\nu_0}$ is the optical thickness of the absorbing gas at the central frequency of the line.

Here one should explain that the notion of "weak line" does not imply that such a line is of small central depth, or, in other words, can hardly be observed in the continuous spectrum. It follows, for instance, from (14.86) and (14.87) that under certain conditions even a "superweak line" may be of arbitrary depth. In other words, the notion of "weak line" means here that for the atmosphere where this line is formed, the condition (14.49) should hold. In particular, it follows from (14.87) that in the case under consideration the second of the above conditions takes the form $b \leq 6/(3-x_1)\tau_0^2$. Then (14.89) gives the estimation $W_0^{(t)}(\tau_0) \leq \pi \Delta\nu$, which is also valid for $W_0(\tau_0)$.

There are some interesting corollaries of (14.86–89). Firstly, it is found that the contour of the "superweak" line does not differ from the original Lorentzian profile of (14.65). In other words, multiple scattering does not deform the profile of the "superweak" line. The same thing happens if we consider profile of the line forming in an optically thin layer. This result, which is paradoxical at first glance, is a consequence of the fact that at the frequencies of the "superweak" line, the probability of a photon colliding with an absorbing particle and consequently vanishing, is very low, so the entire multiple scattering process actually takes place in nonabsorbing media. All these things make it possible to extend the results obtained above for an isolated line to a molecular band, if its model is given. Since there arises no difference between the procedure which is being discussed now and that which was thoroughly developed in the spectroscopy for thin layers of gas (i.e., without the consideration of multiple light scattering) we refer the reader to the Goody's monograph [32] (its second edition [208] in particular) for further details.

Secondly, in the case of light reflection, both the profile and equivalent width of the line are proportional to τ_0, while these quantities are proportional to τ_0^2 in the case of the transmitted radiation. The cause of such a relation can be established on the basis of general considerations of a photon random walk, from which it follows that when we consider the photons forming the reflected light, the average time of residence in the medium is proportional to τ_0, while that of the photons transmitted by the layer is proportional to τ_0^2. Roughly speaking, the situation is as follows: the reflected photons walk for the most part in the surface layer of the medium, while the transmitted photons are driven to diffuse through an optically thick layer and are much more likely to vanish. As a result, the equivalent width is considerably larger. These points are discussed in greater detail in [127] and Chamberlain's monograph [209], Sect. 4.2.

Finally, it should be stressed that for fixed values of λ and λ_ν the formulas defining the absorption line profile and the equivalent width are much more exact for transmitted radiation than for reflected radiation. The point is that the original asymptotic formula for the transmission coefficient (5.61), is accurate up to the terms of order of smallness $1-\lambda$ inclusive, while for the reflected radiation the accuracy of corresponding formulas is determined by a small quantity of the order $(1-\lambda)^{1/2}$.

14.4.2.2 An Inhomogeneous Isothermal Atmosphere. Let us return to the consideration of an inhomogeneous atmosphere. It follows from (14.54) and (14.60) that the basic problem here lies with the determination of the difference $r_\nu^*(\tau_0) = r_\nu^{(1)}(\tau_0) - r^{(1)}(\tau_0)$, where

$$r_\nu^{(1)}(\tau_0) = \frac{1}{2} M_\nu \frac{1 + N_\nu \tilde{N}_\nu(\tau_0)}{1 - N_\nu \tilde{N}_\nu(\tau_0)} \tag{14.90}$$

is the reflection factor found in the P_1 approximation for a layer of thickness τ_0 (the scattering is supposed to be nearly conservative). The quantities M, N and $\tilde{N}_\nu(\tau_0)$ are, in turn, expressed in terms of the solution $y_0(\tau)$ of the characteristic equation (the corresponding formulas are given in Sects. 12.1.3, 12.5, 12.1.1).

Suppose that the atmosphere is homogeneous in the continuum, i.e., that σ, κ and x_1 do not vary with depth. We will assume that the volume absorption coefficient κ_ν is the only quantity that is depth-dependent. Then, converting to the geometric depth z and taking (14.33) and (14.49) into consideration, the system of characteristic equations (12.10) and (12.11) can, in the P_1 approximation, be rewritten as follows:

$$y_1'(z) + [\kappa + \kappa_\nu(z)]y_0(z) = 0, \tag{14.91}$$

$$y_0'(z) + (3 - x_1)\sigma y_1(z) = 0. \tag{14.92}$$

As a result, we arrive at the following characteristic equation:

$$y_0''(z) = (3 - x_1)\sigma\,[\kappa + \kappa_\nu(z)]\,y_0(z), \tag{14.93}$$

which is subject to the boundary conditions

$$y_0(0) = 1, \qquad y_0(\infty) = 0. \tag{14.94}$$

As before, we use the Lorentzian profile of (14.65). The quantity $\kappa_0(z)$ can be represented in the form

$$\kappa_0(z) = \frac{S(j,T)n(z)}{\pi \Delta\nu(P,T)}, \tag{14.95}$$

where

$$\Delta\nu(P,T) = \Delta\nu_0 \left(\frac{T_0}{T(z)}\right)^{1/2} \frac{P(z)}{P(0)}. \tag{14.96}$$

$S(j,T)$ is the integral section of absorption in a line with initial state j at temperature T, $\Delta\nu_0$ is the half-width of the line at the temperature $T_0 = T(0)$ and pressure $P_0 = P(0)$ (i.e. at the boundary of the atmosphere, where $z = 0$) and $n(z)$ is the number of absorbing gas molecules per unit volume.

Let us consider the case of an isothermal atmosphere, i.e., when $T = \text{const}$. Let

$$P(z) = P_0 e^{z/H}, \tag{14.97}$$

where H is the height of the homogeneous atmosphere. Denoting $n_0 = n(0)$, instead of (14.65) we have

$$\kappa_\nu(z) = \frac{\kappa_0}{1 + \xi_\nu^2 e^{-2z/H}}, \tag{14.98}$$

where

$$\kappa_0 = \frac{n_0 S}{\pi \Delta\nu_0}, \tag{14.99}$$

$$\xi_\nu = \frac{\nu - \nu_0}{\Delta\nu_0}. \tag{14.100}$$

Substituting (14.98) into (14.93) and introducing a new variable

$$x = \xi_\nu^2 \mathrm{e}^{-2z/H}, \tag{14.101}$$

we arrive at the equation

$$x\frac{\mathrm{d}^2 y_0(x)}{\mathrm{d}x^2} + \frac{\mathrm{d}y_0(x)}{\mathrm{d}x} = \frac{(3-x_1)H^2\sigma[\kappa_0 + \kappa(1+x)]}{4x(1+x)} y_0(x). \tag{14.102}$$

The solution of this equation can be directly expressed in terms of the hypergeometric function $F(a,b,c;z)$. Applying the boundary conditions of (14.94) we have

$$y_0(z) = \frac{F(a,b,c;-\xi_\nu^2 \mathrm{e}^{-2z/H})}{F(a,b,c;-\xi_\nu^2)} \mathrm{e}^{-k_{\nu_0}\sigma z}, \tag{14.103}$$

where

$$k_{\nu_0}^2 = \frac{(3-x_1)(\kappa+\kappa_0)}{\sigma} \simeq (1-\lambda_{\nu_0})(3-x_1), \tag{14.104}$$

$$a = \frac{1}{2}(k_{\nu_0} + k_0)\sigma H; \quad b = \frac{1}{2}(k_{\nu_0} - k_0)\sigma H; \quad c = 1 + k_{\nu_0}\sigma H, \tag{14.105}$$

$$k_0^2 = \frac{(3-x_1)\kappa}{\sigma} \simeq (1-\lambda)(3-x_1). \tag{14.106}$$

From (14.103) we find

$$y_0'(z) = -\frac{2}{H}\frac{V(a,b,c;-\xi_\nu^2 \mathrm{e}^{-2z/H})}{F(a,b,c;-\xi_\nu^2)} \mathrm{e}^{-k_{\nu_0}\sigma z}, \tag{14.107}$$

where

$$V(a,b,c;x) = bF(a,b+1,c;x) - \left(b - \frac{c-1}{2}\right) F(a,b,c;x). \tag{14.108}$$

If there is no absorption in the continuum ($\lambda \equiv 1$), the corresponding calculations with the above formulas become much simpler because we calculate the function

$$F(a_0, a_0, 1+2a_0; -x) = F_1(a_0; -x)$$
$$= \frac{\Gamma(2a_0)}{a_0\Gamma^2(a_0)}\left\{1 - a_0\int_0^1 \left[1 - \left(\frac{1-t}{1+xt}\right)^{a_0}\right]\frac{\mathrm{d}t}{t^{1-a_0}}\right\}, \tag{14.109}$$

which depends on only one parameter $a_0 = \frac{1}{2}k_{\nu_0}\sigma H$ instead of the function $F(a,b,c;-x)$ which depends on two parameters.

Thus, we have obtained an analytical representation of the functions $y_0(z)$ and $y_0'(z)$ for all $z \geq 0$. In order to find the corresponding expressions for these functions at $z < 0$ it is sufficient to use (12.32), which in the P_1 approximation has the following form:

$$y_0(-z) = \left[1 + \frac{M_1\sigma}{4}\int_0^z \frac{3 - x_1(t)}{y_0^2(t)}\mathrm{d}t\right] y_0(z), \tag{14.109a}$$

where M_1 is defined by (12.29). As a result we have an analytical representation for the functions N and $\tilde{N}$. Thus, as follows from (14.103), (14.54) and (14.64), the profile of the line and its equivalent width in inhomogeneous isothermic atmosphere can be expressed by quadratures.

14.4.2.3 Polytropic Model of the Atmosphere. Let us consider a more general case when the gaseous component of the atmosphere satisfies the polytrope equation

$$PT^{-\alpha} = \text{const}, \tag{14.110}$$

where $\alpha = m/(m-1)$, and m is the polytrope exponent. Using (14.110), we can include in (14.96) the dependence of the line half-width on both pressure and temperature. Its relation to the depth z can be stated by the linear law

$$T(z) = T_0(1 + tz), \tag{14.111}$$

where t is a parameter. From (14.110) and (14.111) we find

$$P(z) = P_0(1 + tz)^\alpha. \tag{14.112}$$

As a result, neglecting the dependence of the integral section S on the temperature we can write

$$\kappa_0(z) = \kappa_0(1 + tz)^{-1/2}, \tag{14.113}$$

instead of (14.95); here κ_0 is defined by (14.99). Hence, (14.65) takes the following form:

$$\kappa_\nu(z) = \frac{\kappa_0(1 + tz)^{-1/2}}{1 + \xi_\nu^2(1 + tz)^{1-2\alpha}}, \tag{14.114}$$

where ξ_ν is given by (14.100).

Venus is a planet for which the polytrope model works well. According to the model presented in [180], Sect. 2.05, at $\alpha = 6.86$ and $t = 3.07 \cdot 10^{-2} km^{-1}$, (14.111) and (14.112) describe the variation of the temperature and pressure within the depth interval $z = 0$–70 km with errors not exceeding 16% and 10%, respectively.

For simplicity, let us consider the case of the central line frequency $\nu - \nu_0$ and assume that there is no absorption in the continuous spectrum (i.e., $\kappa = 0$). Then, instead of (14.91) and (14.92), we find

$$y_0''(z) = \frac{(3 - x_1)\sigma\kappa_0}{(1 + tz)^{1/2}} y_0(z). \tag{14.115}$$

This equation can be reduced to the Bessel equation, whose solution with the boundary conditions of (14.94) is given by

$$y_0(z) = (1+tz)^{1/2} \frac{K_{2/3}\left[\beta(1+tz)^{3/4}\right]}{K_{2/3}(\beta)}, \tag{14.116}$$

where

$$\beta = \frac{4\sqrt{(3-x_1)\sigma\kappa_0}}{3t}, \tag{14.117}$$

and $K_n(z)$ is the modified Bessel function (the Macdonald function). Using (14.91) and (14.92) we find

$$y_1(z) = \frac{4}{3}\frac{\kappa_0}{\beta t}(1+tz)^{1/4} \frac{K_{1/3}\left[\beta(1+tz)^{3/4}\right]}{K_{2/3}(\beta)}. \tag{14.118}$$

The corresponding formulas for negative z can be obtained with the help of (14.109a).

Thus, we have all the formulas necessary to calculate the intensity of radiation at the center of a weak absorption line; these relations allow for the fact that temperature varies with depth according to the law of (14.111) and there is no dependence of the intensity on pressure. At the same time, as follows from (14.114), the behavior of the intensity in the wings of the line depends on pressure, (especially if $2\alpha \gg 1$).

Taking this circumstance into consideration, let us return to (14.103). As we can see, the function $y_0(z)$ can be expressed as a product of two factors. The first factor, which is equal to $\exp(-k_{\nu_0}\sigma z)$, determines the behavior of $y_0(z)$ at $\nu = \nu_0$ (i.e at the central frequency); another factor describes the line profile for other frequencies. Therefore, in our case (i.e., when $(1-\lambda) \ll \kappa_0/\sigma$) the approximate formula

$$y_0(z) = \frac{F_1(a_0; -\xi_\nu^2 e^{-z/H})(1+tz)^{1/2} K_{2/3}\left[\beta(1+tz)^{3/4}\right]}{F_1(a_0; -\xi_\nu^2) K_{2/3}(\beta)} \tag{14.119}$$

can be suggested to calculate the function $y_0(z)$; here the function $F_1(a_0, -x)$ is determined by (14.109). In other words, the factor in (14.103) that determines the intensity of radiation at the center of the line (in this case the atmosphere is homogeneous), has been replaced by a factor that accounts for the inhomogeneity of the atmosphere due to the vertical variation of the temperature.

Thus, the calculations made with the help of (14.119) allow for the dependence of both the line profile and its equivalent width on the vertical variation of pressure (parameter H) and temperature (parameter t).

Above we have considered an atmosphere which is homogeneous in the continuous spectrum. It is characterized by volume scattering coefficient $\sigma =$ const . and absorption coefficient $\kappa =$ const. However, when the scattering in a medium is predominantly gaseous (for example, in the subcloud layer of the Venusian atmosphere), allowance must be made for the fact that the scattering coefficient will increase with depth. Thus, we consider the following model of the atmosphere.

Suppose that there is no absorption in the continuous spectrum ($\kappa \equiv 0$) and that the volume scattering coefficient varies in accordance with a polytropic law, i.e.

$$\sigma(z) = \sigma_0(1+tz)^{\alpha-1}. \tag{14.119a}$$

Since the atmosphere is assumed to be purely gaseous, we put $x_1 = 0$ in (14.92). Let us consider the central frequency of the line. In this case the absorption coefficient is determined by (14.113). Substituting (14.119a) and (14.113) into (14.91) and (14.92) and then introducing the new variable

$$x = 1 + tz,$$

we obtain the equation

$$xy_1''(x) + \frac{1}{2}y_1'(x) - \frac{3\kappa_0\sigma}{t^2}x^{\alpha-1/2}y_1(x) = 0, \tag{14.119b}$$

under the boundary conditions

$$y_1(1) = -\kappa_0/t, \qquad y_1'(\infty) = 0.$$

The solution of (14.119b) under these boundary conditions can also be expressed in terms of the Macdonald function. It has a form

$$y_1(z) = \sqrt{\frac{\kappa_0}{3\sigma_0}}(1+tz)^{1/4}\frac{K_l[s(1+tz)^{1/4l}]}{K_{l-1}(s)}. \tag{14.119c}$$

Hence

$$y_1'(z) = -\kappa_0(1+tz)^{1/4l-3/4}\frac{K_{l-1}[s(1+tz)^{1/4l}]}{K_{l-1}(s)}, \tag{14.119d}$$

where

$$l = (2\alpha+1)^{-1}, \qquad s = \frac{4\sqrt{3\kappa_0\sigma_0}}{(2\alpha+1)t}.$$

As we see, the intensity at the center of the line now depends strongly on pressure in the case being considered.

Equations (14.116), (14.118), (14.119c) and (14.119d) can be used to find the intensity of the radiation emerging from the atmosphere at the central frequency of a line. This line is formed in a two-layer medium whose upper layer is a homogeneous aerosol medium in the continuous spectrum, while its lower layer is a pure gaseous medium whose properties vary with depth in accordance with the formula (14.119a) (simplest model of Venusian atmosphere).

14.4.3 Arbitrary Model of an Atmosphere. Reduction to the Cauchy Problem

Now let us turn to the general case in which the optical properties of the atmosphere are described by *arbitrarily* specified functions of optical depth. We continue to consider a weakly absorbing medium, so as before, we assume that the conditions of (14.49) are satisfied. In this case the main objective is to solve the following *boundary-value problem* (see Sect. 12.1.1)

$$\left.\begin{aligned} y_1'(\tau) &= -p(\tau)y_0(\tau),\\ y_0'(\tau) &= -q(\tau)y_1(\tau),\\ y_0(0) &= 1; \quad y_0(\infty) = 0, \end{aligned}\right\} \tag{14.120}$$

where

$$p(\tau) = 1 - \lambda(\tau); \qquad q(\tau) = 3 - \lambda(\tau)x_1(\tau). \tag{14.121}$$

If this problem is solved, the reflection and transmission factors for an optically thick layer can easily be calculated.

Thus, the boundary-value problem formulated above is of auxiliary nature. The prime objective is to find both the reflection factors $r_\nu^{(1)}(\tau_0)$ and the transmission factors $t_\nu^{(1)}(\tau_0)$ in the P_1 approximation, which are defined by formulas (14.90) and (14.58) respectively. In this way the profile of the weak absorption line is found. Therefore, a question naturally arises: could this factor have been determined directly bypassing the solution of the boundary-value problem of (14.120), i.e., without finding the functions $y_0(\tau)$ and $y_1(\tau)$.

This can be done as follows: define the functions $r_\nu^{(1)}(\tau_0)$ and $t_\nu^{(1)}(\tau_0)$ with the help of the equations

$$r^{(1)}(\tau_0) = \frac{4y_1(0)}{y_0(0) + 3\gamma_0 y_1(0)} + \frac{M\tilde{N}(\tau_0)}{[1 - N\tilde{N}(\tau_0)][y_0(0) + 3\gamma_0 y_1(0)]^2}, \tag{14.121a}$$

$$t^{(1)}(\tau_0) = \frac{M}{[1 - N\tilde{N}(\tau_0)][y_0(0) + 3\gamma_0 y_1(0)][y_0(-\tau_0) + 3\tilde{\gamma}_0(\tau_0)y_1(-\tau_0)]}. \tag{14.122}$$

It is not difficult to verify that (14.121a) and (14.122) can be reduced to (14.90) and (14.58) in the P_1 approximation at $[1 - \lambda(\tau)] \ll 1$.

We introduce four auxiliary functions $V_0(\tau)$, $V_1(\tau)$, $W_0(\tau)$ and $W_1(\tau)$ with the help of the following equalities:

$$\left.\begin{aligned} 2y_0(0)W_0(\tau) &= y_0(-\tau) + y_0(\tau),\\ 2y_0(0)W_1(\tau) &= y_1(-\tau) - y_1(\tau),\\ 2y_1(0)[V_0(\tau) - 3\gamma_0 W_0(\tau)] &= y_0(-\tau) - y_0(\tau),\\ 2y_1(0)[V_1(\tau) - 3\gamma_0 W_1(\tau)] &= y_1(-\tau) + y_1(\tau). \end{aligned}\right\} \tag{14.123}$$

Then (14.121a) and (14.122) can be rewritten as follows:

$$r^{(1)}(\tau) = t^{(1)}(\tau)[W_0(\tau) + 3\tilde{\gamma}_0(\tau_0)W_1(\tau)], \tag{14.124}$$

$$t^{(1)}(\tau) = 4[V_0(\tau) + 3\tilde{\gamma}_0(\tau_0)V_1(\tau)]^{-1}. \tag{14.125}$$

Allowing for the fact that the functions $y_0(\tau)$ and $y_1(\tau)$ in the P_1 approximation are solutions of the boundary value problem of (14.120), the following ***Cauchy problem*** on the interval $[0, \tau_0]$ can be formulated to find $V_0(\tau)$, $V_1(\tau)$, $W_0(\tau)$, $W_1(\tau)$:

$$\left.\begin{aligned} &V_1'(\tau) = p(\tau)V_0(\tau),\\ &V_0'(\tau) = q(\tau)V_1(\tau),\\ &V_0(0) = 3\gamma_0; \quad V_1(0) = 1, \end{aligned}\right\} \tag{14.126}$$

and

$$\left.\begin{aligned} &W_1'(\tau) = p(\tau)W_0(\tau),\\ &W_0'(\tau) = q(\tau)W_1(\tau),\\ &W_0(0) = 1; \quad W_1(0) = 0. \end{aligned}\right\} \tag{14.127}$$

These four functions are not independent ones and, as follows from (14.126) and (14.127), they satisfy the equality

$$V_1(\tau)W_0(\tau) - V_0(\tau)W_1(\tau) = 1, \tag{14.128}$$

which may serve as an accuracy test when solving the equations.

In particular, for a homogeneous atmosphere, (14.126) and (14.127) yield

$$W_0(\tau) = \cosh k_0\tau, \qquad W_1(\tau) = \frac{k_0}{3 - x_1}\sinh k_0\tau;$$

$$V_0(\tau) = (3 - x_1)\left[\frac{\sinh k_0\tau}{k_0} + \frac{3\gamma_0}{3 - x_1}\cosh k_0\tau\right],$$

$$V_1(\tau) = \cosh k_0\tau + \frac{3\gamma_0 k_0}{3 - x_1}\sinh k_0\tau,$$

where $k_0 = \sqrt{(1-\lambda)(3 - x_1)}$. Substituting these expressions into (14.124) and (14.125), we obtain the reflection and transmission factors that appeared in (4.66a) and (4.66).

The main advantage of the Cauchy problem over the boundary-value problem is that the Cauchy problem can usually be solved numerically. In the case of the boundary-value problem, that can not be stated confidently, even if the uniqueness and existence of the solution are known in advance.

The above method of reducing the boundary-value problem to the Cauchy problem is adopted from Fomin's work [210], where the reader can find further details.

14.5 Effect of Inhomogeneity of the Stellar Photospheres on the Continuous Spectrum

It is characteristic of stellar photospheres that not only the emission and absorption of radiation but also the scattering of radiation occurs there. The radiation in photospheres of hot stars is scattered by free electrons and the radiation in photospheres of cold stars by molecules.

A number of methods have been developed (see, for instance, Michalas book [211], Chap. 6) to solve the problem of the formation of continuous spectra in stars with scattering photospheres. Here we are to consider this problem by exploiting the fact that there is a relationship between this problem and that of the diffuse reflection of light by a photosphere.

A characteristic feature of scattering photospheres is that they are all inhomogeneous media in which the probability of the photon's survival in scattering depends on the optical depth. This feature will be considered first.

As well as being applicable to stellar photospheres, the theory may also be applied to extended atmospheres of Wolf-Rayet stars, where the radiation scattering by free electrons is of considerable importance. Estimations show (see, for instance [212, 213]) that the electron scattering should also be taken into account to consider the radiative transfer in shells of supernovas. The mechanism of electron scattering was also used (see [214, 215] and other works) to explain the energy distribution in the spectra of X-rays sources.

14.5.1 Basic Equations and Relations

Let a photosphere consist of plane parallel layers and $I_\nu(r,\theta)$ be the intensity of radiation of frequency ν; the radiation is propagating at a distance r from the center of the star and at an angle θ to the radius vector. If the scattering is isotropic, the intensity $I_\nu(r,\theta)$ is known to be determined by the following transfer equation:

$$\cos\theta\frac{\mathrm{d}I_\nu}{\mathrm{d}r} + (\kappa_\nu + \sigma_\nu)I_\nu = \sigma_\nu \int I_\nu \frac{\mathrm{d}\omega}{4\pi} + \kappa_\nu B_\nu^*(T), \tag{14.129}$$

where $\kappa_\nu(r)$ and $\sigma_\nu(r)$ are the volume coefficients of true absorption and scattering, respectively, and $B_\nu^*(T)$ is Plank's radiation intensity at temperature T. The dependence of T on r is assumed to be known from the model of the photosphere, which makes allowance for non-transparency caused by both the true absorption and scattering. In (14.129), integration is performed with respect to all spatial angles.

If we introduce optical depth

$$\tau_\nu = \int_r^\infty (\kappa_\nu + \sigma_\nu)\mathrm{d}r \tag{14.130}$$

for frequency ν in the photosphere and the probability $\lambda_\nu = \sigma_\nu/(\sigma_\nu+\kappa_\nu)$ that a photon survives in the scattering then we can rewrite (14.129) as follows:

$$\mu \frac{\mathrm{d}I_\nu}{\mathrm{d}\tau_\nu} + I_\nu = \lambda_\nu \int I_\nu \frac{\mathrm{d}\omega}{4\pi} + (1 - \lambda_\nu) B_\nu^*(T), \tag{14.131}$$

with $\mu = -\cos\theta$. At different frequencies the quantity λ_ν is different, therefore, the effect of scattering on the continuous spectrum of a star is of complicated character. This effect has already been studied for the simplest case of a homogeneous atmosphere (i.e., when λ_ν does not depend on depth); Slyusarev [216], for example, considered this problem within the Eddington approximation. Here we are to give rigorous consideration to this problem making allowance for the dependence of λ_ν on depth.

Note that usually λ_ν decreases with the depth of the photosphere. For instance, if the scattering is due to free electrons, the scattering coefficient is proportional to the square of density, whereas the density increases with depth. This situation will be considered later. Now we are to derive basic relations that link the problem of finding the intensity of the radiation leaving photosphere to that of diffuse reflection.

14.5.2 Reduction to the Standard Problem

The solution of (14.129) is equivalent to that of the equation for the source function

$$B_\nu(\tau_\nu) = \frac{\lambda_\nu(\tau_\nu)}{2} \int_0^\infty E_1(|\tau_\nu - t_\nu|) B_\nu(t_\nu)\,\mathrm{d}t_\nu + [1 - \lambda_\nu(\tau_\nu)] B_\nu^*(T), \tag{14.132}$$

where $E_1(\tau) = \int_0^1 \mathrm{e}^{-\tau/\mu} \frac{\mathrm{d}\mu}{\mu}$,

$$B_\nu(\tau_\nu) = \lambda_\nu(\tau_\nu) \int I_\nu \frac{\mathrm{d}\omega}{4\pi} + [1 - \lambda_\nu(\tau_\nu)] B_\nu^*(T). \tag{14.133}$$

If the function $B_\nu(\tau_\nu)$ known, the radiation intensity $I_\nu(-\mu)$ at the boundary of the photosphere is determined for all $\mu \in [0, 1]$ by the formula

$$I_\nu(-\mu) = \int_0^\infty B_\nu(\tau) \mathrm{e}^{-\tau/\mu} \frac{\mathrm{d}\tau}{\mu}. \tag{14.134}$$

Then the flux H_ν of the radiation emerging from the photosphere is determined by the expression

$$H_\nu = 2\pi \int_0^1 I_\nu(-\mu) \mu\,\mathrm{d}\mu. \tag{14.135}$$

Here we are interested only in the distribution of energy in the continuous spectrum of the star. Therefore we are to find only $I_\nu(-\mu)$ and H_ν.

To find the above quantities, let us consider the standard equation for the source function $\tilde{B}(\tau, \mu)$ of the parallel external flux problem:

$$\tilde{B}_\nu(\tau_\nu, \mu_0) = \frac{\lambda_\nu(\tau_\nu)}{2} \int_0^\infty \tilde{B}_\nu(t_\nu, \mu_0) E_1(|\tau_\nu - t_\nu|)\, \mathrm{d}t_\nu + \frac{\lambda_\nu(\tau_\nu)}{4} F \mathrm{e}^{-\tau_\nu/\mu_0}, \tag{14.136}$$

where $\pi F \mu_0$ is the illumination of the upper atmospheric boundary by parallel rays incident at angle $\cos^{-1}\mu_0$. The reflection coefficient $\rho_\nu(\mu, \mu_0)$ is determined then by the expression

$$\rho_\nu(\mu, \mu_0) = \frac{1}{F} \int_0^\infty \tilde{B}_\nu(\tau, \mu_0) \mathrm{e}^{-\tau/\mu} \frac{\mathrm{d}\tau}{\mu\mu_0}. \tag{14.137}$$

It is important for us here that the intensity $I_\nu(-\mu)$, which is provided by (14.134), can be expressed in terms of the function $\tilde{B}_\nu(\tau_\nu, \mu_0)$. So in other words, the initial problem is reduced to the standard parallel external flux problem. To derive the above-mentioned expression one has to multiply (14.136) by $B_\nu(\tau_\nu)/\lambda_\nu(\tau_\nu)$ and then integrate it with respect to τ_ν from 0 to ∞. Using (14.132) and (14.134) we have

$$I_\nu(-\mu) = \frac{4}{F} \int_0^\infty \frac{1 - \lambda_\nu(\tau)}{\lambda_\nu(\tau)} B_\nu^*(T) \tilde{B}_\nu(\tau, \mu) \frac{\mathrm{d}\tau}{\mu}. \tag{14.138}$$

Thus, to find the radiation intensity $I_\nu(-\mu)$ for an *arbitrary* dependence of T on τ, it is sufficient to know only function $\tilde{B}_\nu(\tau, \mu)$, which may be considered to be a standard one.

14.5.3 Various Methods of Solution

Let us assume, for simplicity, that

$$B_\nu^*(T) = B_\nu^*(T_0)(1 + \beta_\nu \tau), \tag{14.139}$$

where T_0 is surface temperature of star and β_ν a certain parameter.

Multiplying (14.136) by $(1+\beta_\nu\tau)/\lambda_\nu(\tau)$, and integrating then with respect to τ from 0 to ∞ we get an expression that, after (14.137) and (14.138) being used and simple manipulations performed, yields

$$I_\nu(-\mu) = B_\nu^*(T_0)[1 - A_\nu(\mu) + \beta_\nu D_\nu(\mu)], \tag{14.140}$$

where $A_\nu(\mu)$ is the plane albedo of the atmosphere, and

$$D_\nu(\mu) = \mu + 2 \int_0^1 \rho_\nu(\mu, \mu') \mu'^2 \,\mathrm{d}\mu'. \tag{14.141}$$

Substituting (14.140) into (14.135), we arrive at the following formula for the flux of radiation emerging from the photosphere:

$$H_\nu = \pi B_\nu^*(T_0)(1 - A_s^{(\nu)} + \beta_\nu D_\nu), \tag{14.142}$$

where $A_s^{(\nu)}$ is the spherical albedo of the photosphere and

$$D_\nu = \frac{2}{3} + 2\int_0^1 A_\nu(\mu)\mu^2\,\mathrm{d}\mu. \tag{14.143}$$

It is important here that if the Planck's function is represented as in (14.139), then it is not necessary to find the source function $\tilde{B}_\nu(\tau,\mu)$ for all τ to solve the problem. Formula (14.140) suggests that it is sufficient just to solve the problem of diffuse reflection, i.e., to find the reflection coefficient $\rho_\nu(\mu,\mu_0)$. In other words, what we really need is to find the function $\varphi_\nu(\mu,\tau)$, in terms of which this coefficient can be expressed (see (13.1) and (13.2)). In some cases this way to solve the problem appears to be much simpler than a direct solution of (14.136) with subsequent use of the formula (14.138). An example will be considered later exemplifying the situation we have just discussed.

At first, however, we are to obtain two relations which directly express the functions $A_\nu(\mu)$ and $D_\nu(\mu)$ in terms of $\varphi_\nu(\mu,\tau)$. Let us multiply (13.2) by $\mathrm{e}^{-t/\mu}$ and integrate it with respect to t from 0 to ∞. As a result we have

$$A_\nu(\mu) = 1 - \int_0^\infty \varphi_\nu(\mu,t)\left[1 - \frac{\lambda_\nu(t)}{2}\alpha_0^{(\nu)}(t)\right]\mathrm{e}^{-t/\mu}\frac{\mathrm{d}t}{\mu}, \tag{14.144}$$

where $(n = 0, 1, 2, \ldots)$

$$\alpha_n^{(\nu)}(\tau) = \int_0^1 \varphi_\nu(\mu,\tau)\mu^n\,\mathrm{d}\mu. \tag{14.145}$$

By multiplying (13.2) by $t\mathrm{e}^{-t/\mu}$ and integrating with respect to t from 0 to ∞, we find

$$D_\nu(\mu) = \int_0^\infty \varphi_\nu(\mu,t)\left[t - \frac{\lambda_\nu(t)}{2}\left(t\alpha_0^{(\nu)}(t) - \alpha_1^{(\nu)}(t)\right)\right]\mathrm{e}^{-t/\mu}\frac{\mathrm{d}t}{\mu}. \tag{14.146}$$

Equations (14.144) and (14.146) are the relations sought.

For an arbitrary dependence of λ on τ, (13.2) can easily be solved numerically. Very simple algorithms can be derived to calculate the function $\varphi(\mu,\tau)$ for certain particular forms of this dependence (which are at the same time rather general). One of them has been described in detail in Sect. 13.1 for the case when the function $\lambda(\tau)$ can be represented as a superposition of exponents. The numerical values of functions $A(\mu)$ and $D(\mu)$ as well as A_s and D are given in Tables 14.8–14.11 for $\lambda(\tau) = \lambda_1\mathrm{e}^{-m\tau}$, $m \geq 0$.

Table 14.8 Plane albedo $A(\mu)$ at $\lambda_1 = 1$

μ	m							
	0	0.1	0.2	0.5	1	2	5	10
0	1	0.793	0.757	0.706	0.667	0.628	0.584	0.557
0.1	1	0.734	0.682	0.602	0.531	0.451	0.333	0.242
0.2	1	0.686	0.624	0.527	0.443	0.353	0.234	0.155
0.3	1	0.644	0.574	0.468	0.380	0.290	0.181	0.114
0.4	1	0.606	0.532	0.421	0.333	0.237	0.147	0.090
0.5	1	0.573	0.494	0.382	0.296	0.214	0.124	0.075
0.6	1	0.542	0.462	0.350	0.267	0.190	0.107	0.064
0.7	1	0.514	0.433	0.323	0.242	0.170	0.095	0.056
0.8	1	0.489	0.408	0.300	0.222	0.154	0.044	0.049
0.9	1	0.466	0.385	0.279	0.205	0.141	0.075	0.044
1.0	1	0.445	0.365	0.261	0.191	0.130	0.070	0.040

Table 14.9 Values of $D(\mu)$ at $\lambda_1 = 1$

μ	m							
	0	0.1	0.2	0.5	1	2	5	10
0	0.577	0.432	0.408	0.374	0.348	0.324	0.297	0.281
0.1	0.720	0.534	0.500	0.447	0.402	0.353	0.283	0.231
0.2	0.837	0.618	0.576	0.313	0.459	0.403	0.331	0.285
0.3	0.948	0.699	0.652	0.582	0.526	0.469	0.402	0.363
0.4	1.056	0.781	0.730	0.657	0.599	0.545	0.483	0.450
0.5	1.162	0.862	0.809	0.735	0.678	0.626	0.571	0.541
0.6	1.267	0.945	0.891	0.816	0.761	0.712	0.661	0.635
0.7	1.371	1.030	0.974	0.900	0.847	0.801	0.754	0.731
0.8	1.474	1.115	1.059	0.986	0.936	0.892	0.848	0.827
0.9	1.577	1.201	1.146	1.074	1.025	0.984	0.944	0.925
1.0	1.679	1.288	1.234	1.164	1.127	1.077	1.040	1.022

Table 14.10 Spherical albedo A_s

λ_1	m							
	0	0.1	0.2	0.5	1	2	5	10
0	0	0	0	0	0	0	0	0
0.50	0.147	0.139	0.133	0.117	0.100	0.079	0.050	0.032
0.60	0.195	0.183	0.172	0.150	0.126	0.098	0.062	0.039
0.70	0.257	0.236	0.220	0.198	0.155	0.119	0.073	0.046
0.80	0.342	0.304	0.279	0.232	0.188	0.141	0.086	0.054
0.90	0.438	0.396	0.354	0.283	0.224	0.166	0.099	0.061
0.95	0.697	0.457	0.400	0.313	0.244	0.179	0.105	0.065
1.00	1.000	0.532	0.454	0.346	0.266	0.192	0.112	0.069

Table 14.11 Values of D

λ_1	m							
	0	0.1	0.2	0.5	1	2	5	10
0.	0.667	0.667	0.667	0.667	0.667	0.667	0.667	0.667
0.50	0.758	0.748	0.747	0.738	0.727	0.713	0.695	0.684
0.60	0.788	0.774	0.771	0.758	0.742	0.724	0.702	0.688
0.70	0.828	0.807	0.801	0.781	0.760	0.737	0.708	0.692
0.80	0.883	0.849	0.833	0.808	0.779	0.750	0.715	0.696
0.90	0.973	0.907	0.885	0.840	0.802	0.764	0.723	0.700
0.95	1.050	0.946	0.914	0.859	0.814	0.772	0.726	0.702
1.00	1.333	0.994	0.949	0.879	0.827	0.780	0.730	0.705

Another way to solve the problem can be proposed if $\lambda_\nu(\tau)$ can be represented in the following form:

$$\lambda_\nu(\tau_\nu) = \int_0^\infty L_\nu(m) \mathrm{e}^{-m\tau_\nu}\, dm, \tag{14.147}$$

where $L_\nu(m)$ is an arbitrary function of m. This method exploits the linear integral equation (13.52) for the radiation intensity $I_\nu(0,-\mu)$, where one has to put

$$I_\nu^*(0,-\mu) = \int_0^\infty [1 - \lambda(\tau_\nu)] B_\nu^*(T) \mathrm{e}^{-\tau_\nu/\mu} \frac{\mathrm{d}\tau_\nu}{\mu}. \tag{14.148}$$

One of the methods for solving (13.52) numerically is the following: the function $I_\nu(0,-\mu)$ can be represented by the expression

$$I_\nu(0,-\mu) = \sum_{n=0}^{N} c_{\nu n} \mu^{n/2}, \tag{14.149}$$

where $c_{\nu n}$ are the coefficients to be found. After substituting (14.149) into (13.52) we perform the integration with respect to μ'. Then, successively setting $\mu = 0, 1/N, 2/N, \ldots, 1$ and integrating with respect to m, one arrives at the system of linear algebraic equations that determines $c_{\nu n}$.

14.5.4 Isothermic Photosphere with a Density Decreasing by the Barometric Law

The formulas derived are valid for arbitrary dependencies of λ_ν and $B_\nu^*(T)$ on τ_ν. Now we will consider the situation when the density ρ decreases in the photosphere by the barometric law, which has already been used in Sect. 14.4.2.2 to study the formation of a line in a planetary atmosphere. Namely, let us assume

$$\rho = \rho_0 \mathrm{e}^{-\frac{r-r_0}{H}}, \tag{14.150}$$

where H is the height of a homogeneous atmosphere. Let us also assume that the photosphere is isothermic.

Obviously, the scattering coefficient $\sigma \sim \rho$. At the same time, it is often assumed (see, for example, [172], Sect. 1.5) that the coefficient of true absorption $\kappa_\nu \sim \rho^2$. Then,

$$\sigma(r) = \sigma_0 \mathrm{e}^{-\frac{r-r_0}{H}}, \tag{14.151}$$

$$\kappa_\nu(r) = \kappa_\nu^{(0)} \mathrm{e}^{-2\frac{r-r_0}{H}}. \tag{14.152}$$

Hence,

$$\frac{\kappa_\nu(r)}{\sigma^2(r)} = \frac{\kappa_\nu^{(0)}}{\sigma_0^2} = m_\nu H = \text{const}, \tag{14.153}$$

where

$$m_\nu = \frac{\kappa_\nu^{(0)}}{\sigma_0^2 H}. \tag{14.154}$$

On the other hand, one finds by substituting (14.151) and (14.152) into (14.130) that

$$\tau_\nu(r) = \left[\sigma(r) + \frac{1}{2}\kappa_\nu(r)\right] H. \tag{14.155}$$

So we have

$$\lambda_\nu^2(r) = \frac{\sigma^2(r)}{\sigma^2(r) + 2\kappa_\nu(r)\left[\sigma(r) + \frac{\kappa_\nu(r)}{2}\right]}. \tag{14.156}$$

Using (14.154), (14.155) and (14.156) we obtain the expression sought

$$\lambda_\nu(\tau_\nu) = \frac{1}{\sqrt{1 + 2m_\nu\tau_\nu}}, \tag{14.157}$$

which determines the relation between λ_ν and optical depth.

Figure 14.6, which is taken from [217], lets us get a general idea of how (14.157) describes the situation in actual stellar photospheres. This figure shows the calculations of $\lambda_\nu(\tau_\nu)$ for the wavelength 5000 Å (solid line) in the photosphere of star with $T_{\text{eff}} = 15000$ K and $\lg g = 2.5$. The calculations are based on models of the atmosphere of stars with normal chemical composition, which were elaborately determined in [218, 219]. The dashed line corresponds to the profile calculated according to (14.157) at $m_\nu = 65$.

It can be seen easily that the actual profile of $\lambda_\nu(\tau_\nu)$ differs from that given by (14.157). The actual profile is not as steep as the calculated one, which is a consequence of the assumption that the photosphere is isothermic (for details see [217]). Nevertheless, it may be supposed that the dependence (14.157) can sometimes be used at least as the first approximation.

One can easily verify that to derive (14.157) from (14.147) it is sufficient to put

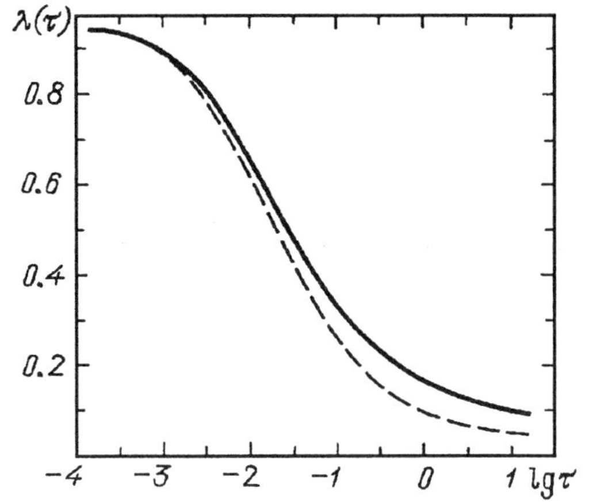

Fig. 14.6. Calculations of $\lambda_\nu(\tau_\nu)$ for the wavelength 5000 Å in the photosphere of a star with $T_{\rm eff} = 15000$ K and $\lg g = 2.5$ (solid line). Dashed line presents the results of calculations according to (14.157) at $m_\nu = 65$

$$L_\nu(x) = \frac{1}{\sqrt{2\pi m_\nu x}} \mathrm{e}^{-\frac{x}{2m_\nu}}. \tag{14.158}$$

Substituting (14.158) into (13.52) and (14.148), we arrive at the following equation to find the distribution of brightness over the stellar disk at a fixed frequency:

$$I_\nu(0,-\mu) = \frac{1}{\sqrt{\pi}} \int_0^1 \mathrm{d}\eta \int_0^\infty \frac{\zeta I_\nu(0,-\zeta) - \eta I_\nu(0,-\eta)}{\zeta-\eta} \frac{\mathrm{e}^{-z^2}\mathrm{d}z}{1+2m_\nu \mu z^2}$$
$$+\frac{1}{\sqrt{\pi}} \int_0^\infty I_\nu(0,-\zeta)\zeta \ln\frac{1+\zeta}{\zeta} \frac{\mathrm{e}^{-z^2}\mathrm{d}z}{1+2m_\nu\mu z^2} + I_\nu^*(0,-\mu), \tag{14.159}$$

where

$$\zeta = \frac{\mu}{1+2m_\nu z^2 \mu}, \tag{14.160}$$

$$I_\nu^*(0,-\mu) = B_\nu^*(T_0)\left[1 - \frac{2}{\sqrt{\pi}} \int_0^\infty \mathrm{e}^{-z^2} \frac{\mathrm{d}z}{1+2m_\nu\mu z^2}\right]. \tag{14.161}$$

Equation (14.159) was solved numerically for the case when the function $I_\nu(0,-\mu)$ is presented by the expression (14.149). Table 14.12 gives the numerical values of $I_\nu(0,-\mu)/B_\nu^*(T_0)m_\nu^{1/3}$ provided by the above-mentioned solution at various values of the parameter m_ν, which is determined by (14.154). The last line of Table 14.12 contains numerical values of $H_\nu m_\nu^{-1/3}/\pi B_\nu^*(T_0)$ for various m_ν.

Table 14.12 Values of $I_\nu(0,-\mu)/B_\nu^*(T_0)m^{1/3}$

μ	m							
	0	0.01	0.025	0.05	0.125	0.25	0.50	1.00
0	0.607	0.511	0.489	0.464	0.424	0.389	0.352	0.313
0.1	0.757	0.643	0.619	0.589	0.548	0.513	0.476	0.439
0.2	0.881	0.752	0.725	0.692	0.646	0.608	0.564	0.519
0.3	0.997	0.854	0.821	0.787	0.734	0.686	0.635	0.577
0.4	1.111	0.952	0.913	0.874	0.810	0.754	0.690	0.622
0.5	1.222	1.044	0.999	0.953	0.878	0.811	0.737	0.658
0.6	1.332	1.133	1.081	1.026	0.938	0.862	0.777	0.687
0.7	1.441	1.216	1.156	1.094	0.994	0.906	0.811	0.712
0.8	1.550	1.300	1.231	1.156	1.044	0.945	0.840	0.733
0.9	1.658	1.379	1.300	1.216	1.088	0.979	0.866	0.751
1.0	1.765	1.453	1.365	1.270	1.120	1.011	0.888	0.767
$\dfrac{H_\nu m_\nu^{-1/3}}{\pi B_\nu^*(T_0)}$	1.402	1.179	1.118	1.056	0.960	0.875	0.784	0.690

The quantity $H_\nu/\pi B_\nu^*(\tau_0)$ is the ratio of the flux of radiation emerging from the isothermic photosphere in the presence of electron scattering to that in the absence of electron scattering (i.e., when the flux is determined by Plank's formula). Table 14.12 indicates that the above ratio decreases with a decrease of m_ν. This may be ascribed to the following: when m_ν decreases, the path of the photon in the photosphere becomes longer and, consequently, the probability of true absorption also becomes higher. The quantity m_ν is different for different regions of the spectrum, which results in the fact that electron scattering may considerably distort the spectrum of the star.

For actual photospheres the quantity $B_\nu^*(T_0)$ is not a constant. It depends on optical depth τ_ν, the frequency dependence of which is, in turn, determined by both true absorption and electron scattering. At the same time, λ_ν is not determined by the simple formula (14.157). Therefore, the effect of electron scattering on the spectrum will be more complicated than that for an isothermic photosphere. The problem under consideration can be solved in even more general cases by means of either (13.2) or (14.159).

14.5.5 Asymptotic Formulas

If electron scattering considerably dominates true absorption in the upper layers of the photosphere, then one can obtain rigorous asymptotic formulas for $I_\nu(0,-\mu)$ and H_ν. If the photosphere is isothermic and $m_\nu \ll 1$, these formulas are valid when $\lambda(\tau_\nu)$ is determined by (14.157) or (13.15). Both formulas suggest that at $m_\nu \ll 1$, the quantity $\lambda_\nu(\tau_\nu)$ remains close to unity up to depths of order $1/m_\nu$.

On the other hand, it follows from (14.140) and (14.142) that the intensity and the flux of the radiation emerging from an isothermic photosphere can be expressed in terms of plane $A_\nu(\mu)$ and spherical $A_s^{(\nu)}$ albedos as follows:

$$\frac{I_\nu(-\mu)}{B_\nu^*(T_0)} = 1 - A_\nu(\mu), \tag{14.162}$$

$$\frac{H_\nu}{\pi B_\nu^*(T_0)} = 1 - A_s^{(\nu)}. \tag{14.163}$$

Therefore, if $m_\nu \ll 1$, we can use the rigorous asymptotic formulas (13.32) and (13.33). As a result, instead of (14.162) and (14.163) we have

$$\frac{I_\nu(-\mu)}{B_\nu^*(T_0)} = 0.8756 \left(\frac{m_\nu}{3}\right)^{1/3} \left[1 - 1.077 \left(\frac{m_\nu}{3}\right)^{1/3}\right] \varphi_0(\mu), \tag{14.164}$$

$$\frac{H_\nu}{\pi B_\nu^*(T_0)} = 0.6740 \left(\frac{m_\nu}{3}\right)^{1/3} \left[1 - 1.077 \left(\frac{m_\nu}{3}\right)^{1/3}\right]. \tag{14.165}$$

These formulas are derived for the case when λ_ν falls with τ_ν exponentially. One can, however, assume that they correctly describe the respective dependencies for the case when λ_ν decreases according to (14.157) because both laws asymptotically coincide for $m_\nu \ll 1$. And what is more, (14.164) and (14.165) are valid for the photosphere of any star for which the dependence of λ_ν on τ_ν in surface atmospheric layers can be adequately described by the linear function $\lambda_\nu(\tau_n u) = 1 - m_\nu \tau_\nu$ for $m_\nu \ll 1$.

Now we are to consider some consequences of (14.164) and (14.165). It is seen from (14.164) that the *relative* distribution of brightness over the disk does not depend on frequency. It is however connected with the following: for $m_\nu \ll 1$ the annihilation of photons occurs mostly in very deep photospheric layers, whereas the angular distribution of the outgoing radiation is determined only by the optical properties of the surface layers of the photosphere, which hardly absorb radiation.

As for the radiation flux, taking (14.154) into consideration and discarding the second term, we rewrite (14.165) as follows:

$$\frac{H_\nu}{\pi B_\nu^*(T_0)} = 0.6740 \left(\frac{\kappa_\nu^{(0)}}{3\sigma^2 H}\right)^{1/3}. \tag{14.166}$$

At the same time, if the photosphere is homogeneous and close to a conservatively scattering one (i.e. when $\kappa_\nu^{(0)} \ll \sigma$), it follows from (14.163) and (3.99) that

$$\frac{H_\nu}{\pi B_\nu^*(T_0)} = \frac{4}{\sqrt{3}}\sqrt{1-\lambda} \simeq \frac{4}{\sqrt{3}} \left(\frac{\kappa_\nu^{(0)}}{\sigma}\right)^{1/2}. \tag{14.167}$$

Hence, if the stellar photosphere is inhomogeneous and close to conservatively scattering, then the dependence of the flux on frequency is "softer" than that for the respective homogeneous photosphere: $H_\nu \sim [\kappa_\nu^{(0)}]^{1/3}$ for an inhomogeneous photosphere , whereas $H_\nu \sim [\kappa_\nu^{(0)}]^{1/2}$ for a homogeneous one.

It is worth noticing that (166) (within a constant factor of order one) was derived in [220] and used to study X-ray sources.

14.6 Other Fields of Possible Application of the Theory

Other fields of application of the theory have already been mentioned in the introduction. To sum up: the results that have been obtained are applicable in any field of science where we have either to solve an equation that can be reduced to a scalar equation of the transfer of monochromatic radiation in plane medium, or to give interpretation to the data of corresponding observations. Now we are to give a brief enumeration of the above fields. For the sake of the integrity, we will sometimes have to repeat things discussed in the introduction.

1. Other Fields of Astrophysics. In addition to what has been said about the application of the theory to stellar photospheres and planetary atmospheres (see also LSPA Chaps. 9, 10 and MLS Chaps. 17, 18), a number of results can be used to study the radiation diffusion in gas and dust nebula, dust envelopes of stars, etc. They are especially applicable to situations when the optical thicknesses of the objects to be studied are large and we can, therefore, use the formulas of the asymptotic theory.

2. The Optics of the Earth Atmosphere. Some aspects of this branch of geophysics were touched upon in Sect. 14.1. A big variety of problems which arise when studying the atmosphere from both the surface and outer space are described in a number of monographs (see, for instance, [28, 29],[31]–[37], [14]). The books of Kuo-Nan Liou [36], Goody [32] and Goody and Young [208] can serve as a good introduction to the problems considered in atmospheric optics. One should stress that recently, considerable attention has been paid to the analysis of the change in atmospheric properties caused by man-made factors (for recent results concerning the above problem, see, for example, the works of Kondratiev et al. [221, 222]).

A corresponding review can be found in MLS, Chap. 19. The bibliography, which covers the period up to 1977 and which can be found at the end of the above chapter, provides us with the list of works that describe almost all the problems arising in the field that the theory of multiple light scattering can be applied to. A review of the numerical methods that can be applied to solve this problem, done by Stamnes, can be found in [223].

3. Marine Optics is another important part of geophysics. The sea, like an atmosphere, is a turbid medium where atmospheric light scattering occurs on both various suspensions and air bubbles. In general, the distribution of these "scatterers" is inhomogeneous, therefore one can use the methods and results of the theory of radiative transfer. A certain specific feature is brought about by the presence of the interface separating atmosphere and sea. Its influence can be taken into account by the introduction of a suitable boundary condition. Generally speaking, one should consider the problem as follows: the transfer of radiation in inhomogeneous medium consisting of sea and atmosphere. The problems of dealing with the above approach are described in the books of Erlov [30, 224], Ivanov [225] and Shifrin [226].

4. Marine Acoustics is one more part of geophysics to which the theory of radiative transfer can be applied. The point is that to conduct studies in marine acoustics it is often necessary to pick up signals amongst the noise generated by the boundary layer sea – atmosphere (see, for instance, Brekhovskih [227]). The methods developed in the scalar theory of radiative transfer turn out to be applicable to the noise field. Travelling through water, sound undergoes multiple scattering on inhomogeneities (for example, on shoals of fish). The above scattering is described by the transfer equation in standard form, which determines the range of problems, both direct and inverse, that can be treated with the help of this approach (for details, see, for instance, [228]–[230]).

5. The Theory of Neutron Transfer in the approximation of constant sections, or in other words, one-speed (or one-group) theory, is basically the same as the theory of monochromatic radiative transfer under anisotropic scattering. Therefore, the results obtained by physicists in this field have been widely used by astrophysicists and vice versa (for details, see [20]–[23]).

6. Heat Transfer. Even a quick look through the books [25]–[27] allows us to get an idea of how big the variety of problems is that can be reduced to the solution of the transfer equation in the standard form. It is the problem of how to determine the radiation field in a medium that acquires special significance here. The exposition of the new approach to this problem is given in Sects. 3.12 and 4.10. Necessary substitutions made, this approach can also be applied to solve problems of heat transfer.

Here we finish enumerating the fields in which the results and methods of radiative transfer theory in inhomogeneous media can be used. It should be clear, however, that these methods are of considerable practical importance for remote diagnostics of materials which scatter radiation. Grass and forest canopies, colloid solutions and other substances are examples of such materials, and these methods are used not only in science but also in industry.

For many cases, however, it is the method of remote probing that turns out to be able to provide more information. This method is based on polarimetric techniques of measurements, which analyze all four Stokes parameters of the scattered radiation, not just intensity. Therefore, it is sensible to extend some methods and results of this book to the vector variant of the theory of radiative transfer. Sometimes it can be done, at least formally, without any obstacles. The work [231] is an example of such extension. It generalizes the results of scalar theory of light scattering in multilayer media, which are provided in Chap. 6, to the transfer of polarized radiation. Another example is Loskutov and Sobolev's article [232], where the results provided in Sect. 13.4 are generalized to Rayleigh polarization in a medium.

One should also add the following. The whole closing chapter "Miscellaneous Applications" of van de Hulst's MLS deals with various practical applications of radiative transfer. The bibliography that can be found there extends up to 1978 and should be regarded as an additional source of information.

Bibliographical Comments and Additions to Part III

The first analytical study of light scattering in inhomogeneous atmospheres relates to the simplest case of isotropic light scattering in a semi-infinite atmosphere. Nearly at the same time and entirely independently, equations (9.54–57) for the isotropic scattering were derived by Sobolev [233] with the help of his method of "truncation" of atmosphere, and by Bellman and Kalaba by the method of invariant imbedding [234].

Preisendorfer noticed [235] that to describe the diffuse transmission and reflection of light from an inhomogeneous plain layer, one needs not two, but four functions. He also derived functional relationships involving these functions. The work by Ueno [236] deals with the generalization of the results [233], [234] to the case of the layer of finite optical thickness and spherical phase function (see also LSPA, Sect. 3.5). The angular separation of variables was made in this paper for the problem of diffuse reflection and transmission through a plane inhomogeneous layer. This work also contains the derivation of the integro-differential equations for the reflection and transmission coefficients. Another form of these equations was obtained in the article by Sobolev [237]. However, the considerable anisotropy of the scattering and its dependence on height is characteristic of planetary atmospheres. The generalization taking both above features into consideration was made in the work by the author [238]; the results of this work are described in Sect. 9.1.

The paper by Busbridge [239] studies the relationship between the methods used by Sobolev [233] and Ueno [236]. This paper contains the proof of the existence of the solution of (9.55) for $0 \leq \lambda(\tau) \leq 1$. Later, various aspects of isotropic scattering in an inhomogeneous atmosphere were studied by Ueno [240]–[242]. Sobolev [237] derived (9.50–52) for the case of isotropic scattering. The method used by Sobolev is different from that used in this book for anisotropic scattering. More general results, which constitute Sect. 9.2, are reported here for the first time.

Proceeding to Chap. 10, one should note that (10.8), which determines the radiation field in an inhomogeneous atmosphere, was derived by Ueno et al. in [243]. Equation (10.3) was provided by the author in [51] for the case of a homogeneous atmosphere; at $\tau_0 = \infty$ it turns into the equation that was derived by Ivanov [82]. The work [244] deals with the method that enables us to reduce the calculation of the radiation field in an isotropic homogeneous atmosphere to the calculation of the intensity of only the descending radiation. Homogeneous analogies of the integral relations (10.18) and (10.19) were first derived in the work [104] by the author. Most of the results of Sect. 10.2 were obtained in the paper [83] by the author.

Note that the method described in Sect. 10.3 resembles that used by Kagivada et al. [10], Chap. 5 to solve (10.6) and (10.8) with isotropic scattering for particular functions $\lambda(\tau)$ (equation (10.8) is solved at $\tau = \tau_0$ and $\mu < 0$; i.e. equation for the reflection coefficient $\rho(\mu, \mu_0, \tau_0)$. A similar problem was solved in the paper [245] by Viik, where an algorithm was described applicable to the cases of an inhomogeneous layer adjacent to Lambertian and looking-glass surfaces. Examples of concrete calculations and an extensive bibliography can also be found there. Among the works that were made later, one should mention those by Kawabata and Hirata [246], where the method of invariant imbedding was developed, and also the work by Wilson and Wan [247], where the method of moments was used.

There were other methods developed for the solution of the problem of the radiative transfer in a plane inhomogeneous layer, for example the matrix method of Hunt and Grant [248] which is based on the Chandrasekhar principles of invariance (for details, also see the review by Hunt [249]), the method of successive approximations [250], the method of spherical harmonics [251], the method of discrete ordinates [252] and others. However, the concrete calculations provided in these works mostly concern homogeneous media, so it is not clear how applicable these methods are to inhomogeneous problems.

In [10] and [245], the numerical methods for the solution of the transfer equation are realized for functions $\lambda(\tau)$ taken in certain abstract forms. In this respect, the works by Gavrilova [177], [178] are more interesting since there the *actual* functions $\lambda(\tau)$ for the Earth atmosphere were used. The system of integral equations (9.30) and (9.31) was solved numerically to study the impact of the inhomogeneity on the accuracy of calculations of various quantities determining the radiation field. The approximation of isotropic scattering was first considered in [177] and the scattering with actual phase function in [178] (for details, see Sect. 14.1). One should add here that earlier, Tanaka [253]–[255] developed a matrix algorithm for solving the problem of diffuse reflection and transmission for the actual models of the Earth atmosphere. The results of calculations for the scalar case and for the case with polarization can be found in [254], [255] respectively.

General aspects of the numerical solution of the transfer equations are described in the monograph by Marchuk and Lebedev [46]. Among the recent textbooks on numerical methods for solving the transfer equation, the books [256] and [257] may be listed.

Equations (10.33) and (10.34) for the case of an inhomogeneous isotropic atmosphere were first derived in the work [154] by the author. Similar formulas were derived by Ueno [258] for a plane layer that scatters light isotropically. The equations determining the radiation intensity in an inhomogeneous layer adjacent to a reflecting surface were obtained in Sect. 10.4 with the help of the method used by Ivanov in [133] for a homogeneous layer.

The articles [83] and [259] by the author constitute the main part of Chap. 11. The relation (11.43a) in terms of source functions was first derived by Ivanov [82]. The Milne problem for isotropic scattering was also considered by Sobolev in [260], where, in particular, the analogies of (11.10) and (11.13) were derived.

The papers by Stokes and DeMarkus [261] as well as the work by Chow et al. [262] suggest that the application of variational methods to the problems of transfer theory in an inhomogeneous medium is very promising.

Proceeding to Chap. 12, one should note that the main results of Sects. 12.1 and 12.2 were obtained in [70, 263] and the content of Sect. 12.3 corresponds to that of the author's article [259]. Most of the results of Sects. 12.4, 12.5 can be found in the author's work [264] and partly in [259]. The formulas of Sects. 12.6, 12.7 were obtained ibidem (see also the author's works [265, 266]). Section 12.8 corresponds to the article [263].

The content of Sect. 13.1 and partly that of Sect. 13.2 is based on the article by Sobolev and Loskutov [267]. A rigorous estimation of the accuracy of asymptotic formulas that was given in Sect. 13.2 can also be found in the work [263] by Fomin and the author. The numerical data provided in Tables XXXII–XXXIV of the Appendix are taken from the article [268] by Loskutov and Sobolev, where an approximate method for determining the Ambartsumian's functions for an inhomogeneous atmosphere was described. The tables containing the values of these functions are also given in [269]. The algorithm for calculating the internal field described in Sect. 13.3 has never been described before. The content of Sect. 13.4 relies on the work [270] by Sobolev. Earlier [271] he derived an appropriate equation for the more complicated case of complete frequency redistribution of the radiation in the line.

One should note that the problem of radiative transfer in an inhomogeneous medium with the albedo of particles exponentially depending on the depth turned out to be a test problem for inhomogeneous atmospheres, just as the problem of isotropic scattering did for homogeneous media. The work [272] by Martin was among the first ones taking this approach. This problem was also studied in the article [273] by Mullikin and Siewert as well as in other works, most of which were mentioned in the review [274] by Daniel et al; the latter paper covers the period up to the end of 1983.

In Chap. 14, which closes the book, references and bibliographic notes are made in the text. One should add here that the inverse problems of the theory of multiple light scattering in planetary atmospheres, in particular the Venusian atmosphere, were elaborately considered in the works [275]–[282] by Ustinov.

The content of Sect. 14.2 is mostly based on the works [73, 162] and partly on [98].

Sect. 14.4 corresponds to the works [283, 284] by Fomin and the author. Some of the results of Sect. 14.4 were obtained by Fomin and the author earlier in [285, 286]. Sect. 14.5 corresponds to the works [287] and [270] by Sobolev, which are also the sources of the tables in this section.

Appendix
Tables of Some Functions and Constants

I. Diffusion exponent k

g	λ								
	1	0.999	0.995	0.990	0.980	0.950	0.900	0.800	0.700
R	0	0.0547	0.1222	0.1724	0.2428	0.3787	0.5232	0.7048	0.8200
0	0	0.0548	0.1222	0.1725	0.2430	0.3795	0.5254	0.7104	0.8286
0.50	0	0.0387	0.0866	0.1224	0.1731	0.2733	0.3856	0.5418	0.6578
0.75	0	0.0274	0.0614	0.0871	0.1239	0.1990	0.2881	0.4234	0.5345
0.80	0	0.0245	0.0550	0.0781	0.1114	0.1804	0.2639	0.3944	0.5042
0.85	0	0.0212	0.0478	0.0680	0.0974	0.1595	0.2371	0.3624	0.4707
0.90	0	0.0174	0.0392	0.0560	0.0810	0.1354	0.2064	0.3258	0.4325

II. Asymptotical constant M

g	λ								
	1	0.999	0.995	0.990	0.980	0.950	0.900	0.800	0.700
R	0	0.147	0.332	0.478	0.699	1.227	2.087	4.457	8.869
0	0	0.147	0.332	0.477	0.697	1.219	2.057	4.329	8.508
0.5	0	0.207	0.469	0.674	0.983	1.712	2.860	5.811	10.72
0.75	0	0.293	0.664	0.954	1.394	2.434	4.078	8.269	15.11
0.80	0	0.328	0.742	1.068	1.561	2.731	4.585	9.315	17.01
0.85	0	0.378	0.858	1.235	1.808	3.173	5.346	10.90	19.91
0.90	0	0.464	1.053	1.517	2.227	3.931	6.666	13.67	25.03

III. Brightness distribution over the disk for the Rayleigh phase function

μ	λ								
	1	0.999	0.995	0.990	0.980	0.950	0.900	0.800	0.700
1.0	1.000	1.000	1.000	1.000	1.000	1.000	1.000	1.000	1.000
0.9	0.893	0.903	0.915	0.922	0.932	0.949	0.963	0.977	0.985
0.8	0.792	0.810	0.831	0.845	0.864	0.895	0.923	0.952	0.967
0.7	0.697	0.721	0.749	0.768	0.794	0.838	0.879	0.923	0.946
0.6	0.608	0.636	0.669	0.692	0.723	0.778	0.831	0.889	0.921
0.5	0.524	0.554	0.590	0.616	0.651	0.715	0.778	0.851	0.891
0.4	0.446	0.476	0.514	0.541	0.579	0.649	0.720	0.806	0.856
0.3	0.373	0.402	0.439	0.467	0.505	0.579	0.656	0.754	0.814
0.2	0.305	0.332	0.367	0.393	0.430	0.504	0.586	0.693	0.763
0.1	0.240	0.264	0.295	0.319	0.353	0.423	0.504	0.617	0.696
0.0	0.170	0.189	0.213	0.232	0.260	0.320	0.393	0.502	0.591

IV. Escape function $u(\mu)$ for the Rayleigh phase function

μ	λ								
	1	0.999	0.995	0.990	0.980	0.950	0.900	0.800	0.700
1.0	1.265	1.217	1.162	1.123	1.070	0.975	0.879	0.754	0.661
0.9	1.186	1.141	1.086	1.046	0.991	0.884	0.765	0.589	0.440
0.8	1.108	1.065	1.011	0.972	0.915	0.802	0.671	0.477	0.322
0.7	1.029	0.988	0.937	0.898	0.841	0.726	0.591	0.396	0.251
0.6	0.949	0.911	0.863	0.825	0.770	0.656	0.523	0.335	0.203
0.5	0.869	0.834	0.788	0.753	0.700	0.590	0.462	0.286	0.168
0.4	0.788	0.756	0.714	0.680	0.631	0.528	0.408	0.246	0.141
0.3	0.705	0.676	0.638	0.608	0.562	0.467	0.358	0.212	0.120
0.2	0.620	0.595	0.561	0.534	0.493	0.409	0.311	0.182	0.102
0.1	0.530	0.508	0.479	0.456	0.421	0.348	0.265	0.155	0.087
0.0	0.419	0.402	0.379	0.361	0.334	0.277	0.211	0.125	0.071

V. Brightness distribution over the disk for isotropic scattering

μ	λ								
	1	0.999	0.995	0.990	0.980	0.950	0.900	0.800	0.700
1.0	1.000	1.000	1.000	1.000	1.000	1.000	1.000	1.000	1.000
0.9	0.882	0.892	0.901	0.908	0.917	0.933	0.947	0.963	0.973
0.8	0.771	0.787	0.805	0.818	0.835	0.864	0.892	0.923	0.943
0.7	0.666	0.688	0.712	0.730	0.753	0.794	0.833	0.880	0.910
0.6	0.569	0.594	0.623	0.643	0.671	0.722	0.772	0.833	0.873
0.5	0.479	0.505	0.536	0.559	0.590	0.648	0.707	0.782	0.832
0.4	0.396	0.422	0.453	0.477	0.509	0.572	0.638	0.725	0.786
0.3	0.319	0.344	0.374	0.397	0.430	0.494	0.565	0.662	0.732
0.2	0.249	0.271	0.299	0.320	0.351	0.414	0.487	0.591	0.670
0.1	0.184	0.202	0.227	0.245	0.273	0.331	0.401	0.508	0.593
0.0	0.118	0.133	0.149	0.164	0.185	0.232	0.292	0.391	0.479

VI. Escape function $u(\mu)$ for isotropic scattering

μ	λ								
	1	0.999	0.995	0.990	0.980	0.950	0.900	0.800	0.700
1.0	1.259	1.212	1.156	1.117	1.065	0.969	0.873	0.752	0.667
0.9	1.182	1.137	1.083	1.043	0.988	0.882	0.765	0.593	0.445
0.8	1.105	1.062	1.010	0.970	0.914	0.803	0.675	0.485	0.330
0.7	1.028	0.988	0.937	0.898	0.843	0.730	0.598	0.406	0.261
0.6	0.950	0.912	0.864	0.827	0.773	0.662	0.532	0.346	0.213
0.5	0.872	0.837	0.792	0.756	0.705	0.598	0.473	0.299	0.179
0.4	0.792	0.760	0.719	0.686	0.637	0.536	0.419	0.259	0.152
0.3	0.711	0.682	0.644	0.614	0.570	0.477	0.370	0.225	0.131
0.2	0.628	0.603	0.569	0.542	0.502	0.419	0.323	0.195	0.112
0.1	0.540	0.518	0.489	0.466	0.432	0.360	0.277	0.167	0.096
0.0	0.433	0.416	0.392	0.374	0.347	0.290	0.224	0.136	0.079

VII. Brightness distribution over the disk for the Henyey-Greenstein phase function with $g = 0.5$

μ	λ								
	1	0.999	0.995	0.990	0.980	0.950	0.900	0.800	0.700
1.0	1.000	1.000	1.000	1.000	1.000	1.000	1.000	1.000	1.000
0.9	0.859	0.868	0.879	0.887	0.897	0.915	0.932	0.952	0.965
0.8	0.727	0.743	0.762	0.776	0.794	0.827	0.860	0.899	0.925
0.7	0.604	0.625	0.650	0.668	0.692	0.738	0.784	0.841	0.880
0.6	0.492	0.515	0.542	0.563	0.591	0.645	0.701	0.775	0.828
0.5	0.388	0.412	0.440	0.461	0.491	0.550	0.614	0.702	0.767
0.4	0.296	0.317	0.344	0.364	0.394	0.454	0.522	0.621	0.697
0.3	0.212	0.231	0.255	0.273	0.300	0.357	0.426	0.530	0.616
0.2	0.140	0.154	0.173	0.188	0.211	0.261	0.324	0.428	0.519
0.1	0.078	0.088	0.101	0.122	0.128	0.166	0.219	0.313	0.402
0.0	0.025	0.029	0.035	0.040	0.048	0.068	0.099	0.164	0.237

VIII. Escape function $u(\mu)$ for the Henyey-Greenstein phase function with $g = 0.5$

μ	λ								
	1	0.999	0.995	0.990	0.980	0.950	0.900	0.800	0.700
1.0	1.265	1.199	1.124	1.072	1.005	0.889	0.776	0.635	0.534
0.9	1.188	1.124	1.052	1.001	0.933	0.811	0.687	0.526	0.407
0.8	1.109	1.050	0.980	0.930	0.863	0.739	0.610	0.441	0.319
0.7	1.030	0.974	0.908	0.860	0.794	0.671	0.541	0.372	0.255
0.6	0.950	0.898	0.836	0.790	0.727	0.607	0.479	0.316	0.208
0.5	0.869	0.822	0.763	0.720	0.660	0.545	0.424	0.270	0.171
0.4	0.786	0.743	0.689	0.650	0.594	0.486	0.372	0.230	0.141
0.3	0.701	0.662	0.614	0.578	0.527	0.428	0.324	0.196	0.117
0.2	0.612	0.578	0.535	0.503	0.458	0.370	0.278	0.165	0.097
0.1	0.516	0.487	0.451	0.424	0.386	0.311	0.232	0.136	0.079
0.0	0.395	0.373	0.346	0.325	0.295	0.238	0.178	0.105	0.061

IX. Brightness distribution over the disk for the Henyey-Greenstein phase function with $g = 0.75$

μ	λ								
	1	0.999	0.995	0.990	0.980	0.950	0.900	0.800	0.700
1.0	1.000	1.000	1.000	1.000	1.000	1.000	1.000	1.000	1.000
0.9	0.850	0.862	0.876	0.885	0.897	0.919	0.939	0.963	0.976
0.8	0.711	0.731	0.755	0.771	0.793	0.833	0.873	0.919	0.948
0.7	0.583	0.608	0.638	0.660	0.689	0.744	0.801	0.871	0.916
0.6	0.466	0.493	0.526	0.550	0.584	0.650	0.721	0.813	0.874
0.5	0.359	0.385	0.418	0.444	0.480	0.551	0.632	0.744	0.822
0.4	0.264	0.288	0.318	0.342	0.377	0.450	0.537	0.665	0.760
0.3	0.180	0.200	0.226	0.246	0.278	0.346	0.433	0.571	0.680
0.2	0.108	0.122	0.142	0.158	0.182	0.240	0.318	0.454	0.572
0.1	0.051	0.059	0.071	0.081	0.097	0.137	0.198	0.316	0.432
0.0	0.009	0.011	0.014	0.017	0.023	0.037	0.064	0.129	0.209

X. Escape function $u(\mu)$ for the Henyey-Greenstein phase function with $g = 0.75$

μ	λ								
	1	0.999	0.995	0.990	0.980	0.950	0.900	0.800	0.700
1.0	1.270	1.177	1.079	1.016	0.937	0.812	0.702	0.572	0.479
0.9	1.192	1.104	1.009	0.945	0.866	0.733	0.610	0.460	0.354
0.8	1.113	1.030	0.939	0.877	0.797	0.660	0.531	0.374	0.267
0.7	1.033	0.956	0.868	0.808	0.730	0.593	0.462	0.305	0.204
0.6	0.952	0.880	0.798	0.740	0.664	0.530	0.401	0.250	0.158
0.5	0.869	0.803	0.726	0.672	0.600	0.471	0.347	0.206	0.124
0.4	0.783	0.723	0.652	0.602	0.535	0.415	0.299	0.170	0.098
0.3	0.692	0.639	0.576	0.531	0.470	0.360	0.255	0.140	0.078
0.2	0.595	0.549	0.494	0.455	0.401	0.305	0.213	0.113	0.062
0.1	0.484	0.447	0.402	0.370	0.326	0.246	0.170	0.089	0.048
0.0	0.333	0.307	0.276	0.254	0.224	0.169	0.117	0.062	0.033

XI. Brightness distribution over the disk for the Henyey-Greenstein phase function with $g = 0.8$

μ	λ								
	1	0.999	0.995	0.990	0.980	0.950	0.900	0.800	0.700
1.0	1.000	1.000	1.000	1.000	1.000	1.000	1.000	1.000	1.000
0.9	0.849	0.862	0.876	0.886	0.899	0.922	0.944	0.968	0.982
0.8	0.709	0.730	0.755	0.773	0.797	0.840	0.882	0.931	0.959
0.7	0.579	0.606	0.638	0.662	0.693	0.752	0.813	0.886	0.930
0.6	0.461	0.490	0.526	0.552	0.589	0.661	0.737	0.835	0.896
0.5	0.354	0.382	0.418	0.445	0.484	0.562	0.651	0.771	0.850
0.4	0.258	0.284	0.317	0.343	0.381	0.461	0.558	0.697	0.794
0.3	0.174	0.195	0.223	0.245	0.279	0.354	0.451	0.602	0.717
0.2	0.103	0.118	0.139	0.156	0.182	0.246	0.335	0.488	0.615
0.1	0.046	0.053	0.065	0.075	0.091	0.133	0.199	0.329	0.453
0.0	0.007	0.009	0.011	0.014	0.019	0.033	0.060	0.128	0.213

XII. Escape function $u(\mu)$ for the Henyey-Greenstein phase function with $g = 0.75$

μ	λ								
	1	0.999	0.995	0.990	0.980	0.950	0.900	0.800	0.700
1.0	1.271	1.168	1.062	0.995	0.914	0.788	0.681	0.556	0.467
0.9	1.193	1.095	0.992	0.925	0.842	0.707	0.585	0.440	0.337
0.8	1.114	1.022	0.922	0.856	0.772	0.633	0.504	0.351	0.248
0.7	1.034	0.948	0.853	0.788	0.706	0.565	0.434	0.282	0.186
0.6	0.952	0.872	0.783	0.721	0.641	0.503	0.373	0.227	0.141
0.5	0.869	0.795	0.712	0.653	0.577	0.444	0.320	0.185	0.108
0.4	0.782	0.716	0.639	0.585	0.514	0.389	0.274	0.150	0.084
0.3	0.690	0.631	0.563	0.514	0.450	0.336	0.231	0.121	0.065
0.2	0.591	0.540	0.481	0.438	0.382	0.282	0.191	0.097	0.051
0.1	0.476	0.435	0.387	0.353	0.307	0.226	0.151	0.075	0.039
0.0	0.312	0.286	0.254	0.232	0.202	0.149	0.100	0.050	0.026

XIII. Brightness distribution over the disk for the Henyey-Greenstein phase function with $g = 0.85$

μ	λ								
	1	0.999	0.995	0.990	0.980	0.950	0.900	0.800	0.700
1.0	1.000	1.000	1.000	1.000	1.000	1.000	1.000	1.000	1.000
0.9	0.847	0.854	0.877	0.888	0.903	0.929	0.951	0.977	0.987
0.8	0.705	0.717	0.758	0.777	0.804	0.851	0.897	0.948	0.970
0.7	0.575	0.590	0.641	0.666	0.702	0.768	0.834	0.911	0.948
0.6	0.456	0.475	0.528	0.557	0.598	0.679	0.764	0.868	0.922
0.5	0.348	0.368	0.420	0.450	0.493	0.582	0.682	0.812	0.884
0.4	0.252	0.272	0.318	0.346	0.390	0.482	0.592	0.745	0.838
0.3	0.169	0.187	0.222	0.247	0.286	0.373	0.486	0.656	0.770
0.2	0.098	0.113	0.137	0.156	0.186	0.261	0.366	0.542	0.675
0.1	0.041	0.045	0.062	0.073	0.091	0.139	0.217	0.370	0.506
0.0	0.005	0.006	0.009	0.011	0.016	0.030	0.058	0.132	0.222

XIV. Escape function $u(\mu)$ for the Henyey-Greenstein phase function with $g = 0.85$

μ	λ								
	1	0.999	0.995	0.990	0.980	0.950	0.900	0.800	0.700
1.0	1.272	1.154	1.038	0.966	0.882	0.758	0.656	0.539	0.453
0.9	1.193	1.082	0.968	0.896	0.809	0.673	0.554	0.414	0.315
0.8	1.114	1.010	0.899	0.828	0.739	0.597	0.469	0.321	0.223
0.7	1.034	0.936	0.830	0.760	0.672	0.528	0.397	0.251	0.161
0.6	0.953	0.861	0.761	0.694	0.608	0.465	0.337	0.198	0.118
0.5	0.869	0.785	0.691	0.628	0.546	0.408	0.285	0.156	0.088
0.4	0.781	0.705	0.619	0.560	0.484	0.355	0.240	0.124	0.066
0.3	0.687	0.621	0.544	0.491	0.422	0.304	0.200	0.099	0.050
0.2	0.587	0.529	0.463	0.417	0.356	0.254	0.164	0.078	0.038
0.1	0.467	0.421	0.368	0.331	0.282	0.199	0.127	0.059	0.028
0.0	0.287	0.259	0.227	0.204	0.174	0.123	0.079	0.037	0.019

XV. Brightness distribution over the disk for the Henyey-Greenstein phase function with $g = 0.9$

μ	λ								
	1	0.999	0.995	0.990	0.980	0.950	0.900	0.800	0.700
1.0	1.000	1.000	1.000	1.000	1.000	1.000	1.000	1.000	1.000
0.9	0.846	0.863	0.881	0.894	0.912	0.942	0.963	0.983	0.997
0.8	0.703	0.731	0.764	0.786	0.818	0.872	0.921	0.966	0.988
0.7	0.571	0.606	0.648	0.678	0.720	0.797	0.871	0.940	0.975
0.6	0.451	0.489	0.536	0.570	0.620	0.715	0.811	0.907	0.958
0.5	0.343	0.380	0.427	0.463	0.516	0.623	0.739	0.866	0.934
0.4	0.247	0.280	0.324	0.358	0.411	0.524	0.655	0.814	0.901
0.3	0.164	0.190	0.227	0.257	0.305	0.414	0.553	0.739	0.849
0.2	0.094	0.112	0.139	0.162	0.200	0.296	0.431	0.632	0.769
0.1	0.039	0.049	0.064	0.078	0.102	0.169	0.279	0.470	0.627
0.0	0.003	0.004	0.006	0.008	0.013	0.027	0.059	0.141	0.239

XVI. Escape function $u(\mu)$ for the Henyey-Greenstein phase function with $g = 0.9$

μ	λ								
	1	0.999	0.995	0.990	0.980	0.950	0.900	0.800	0.700
1.0	1.272	1.133	1.000	0.923	0.837	0.718	0.625	0.518	0.438
0.9	1.194	1.061	0.931	0.852	0.761	0.626	0.511	0.377	0.283
0.8	1.115	0.989	0.862	0.783	0.689	0.545	0.420	0.278	0.187
0.7	1.035	0.917	0.795	0.717	0.622	0.473	0.345	0.207	0.127
0.6	0.953	0.843	0.727	0.651	0.559	0.411	0.285	0.156	0.088
0.5	0.869	0.768	0.659	0.587	0.498	0.355	0.235	0.119	0.062
0.4	0.781	0.689	0.589	0.522	0.438	0.304	0.193	0.091	0.045
0.3	0.687	0.606	0.516	0.456	0.380	0.257	0.157	0.070	0.033
0.2	0.583	0.514	0.437	0.384	0.318	0.212	0.126	0.053	0.024
0.1	0.458	0.403	0.342	0.301	0.248	0.163	0.096	0.039	0.017
0.0	0.254	0.224	0.190	0.167	0.138	0.092	0.054	0.023	0.010

XVII. Asymptotical constant C

g	λ								
	1	0.999	0.995	0.990	0.980	0.950	0.900	0.800	0.700
R	1	0.961	0.912	0.875	0.822	0.715	0.593	0.420	0.292
0	1	0.961	0.913	0.876	0.824	0.718	0.599	0.429	0.300
0.5	1	0.947	0.883	0.836	0.774	0.657	0.537	0.382	0.276
0.75	1	0.926	0.842	0.784	0.711	0.582	0.461	0.319	0.227
0.80	1	0.918	0.827	0.765	0.688	0.555	0.435	0.298	0.211
0.85	1	0.906	0.805	0.739	0.656	0.522	0.402	0.271	0.190
0.90	1	0.887	0.774	0.697	0.609	0.473	0.356	0.235	0.162

XVIII. Reflection coefficient $\rho(1, 1, \pi)$ of the center of disk

g	λ								
	1.000	0.999	0.995	0.990	0.980	0.950	0.900	0.800	0.700
R	1.102	0.993	0.877	0.801	0.707	0.557	0.430	0.299	0.222
0	1.057	0.949	0.833	0.757	0.663	0.512	0.385	0.255	0.183
0.50	1.095	0.942	0.784	0.684	0.565	0.386	0.252	0.135	0.082
0.75	1.119	0.907	0.701	0.579	0.442	0.260	0.144	0.063	0.034
0.80	1.124	0.889	0.667	0.539	0.399	0.221	0.116	0.048	0.025
0.85	1.129	0.862	0.619	0.484	0.343	0.175	0.085	0.033	0.017
0.90	1.134	0.816	0.544	0.404	0.265	0.120	0.053	0.020	0.010

XIX. Spherical albedo A_s

g	λ								
	1	0.999	0.995	0.990	0.980	0.950	0.900	0.800	0.700
R	1	0.930	0.850	0.795	0.722	0.597	0.479	0.343	0.258
0	1	0.930	0.850	0.795	0.722	0.597	0.478	0.342	0.257
0.50	1	0.902	0.795	0.723	0.633	0.486	0.360	0.232	0.161
0.75	1	0.865	0.724	0.635	0.528	0.369	0.249	0.144	0.093
0.80	1	0.850	0.698	0.603	0.492	0.332	0.217	0.121	0.077
0.85	1	0.829	0.661	0.559	0.443	0.285	0.179	0.096	0.060
0.90	1	0.795	0.603	0.493	0.375	0.224	0.133	0.067	0.041

XX. Geometrical albedo A_g

g	λ								
	1.000	0.999	0.995	0.990	0.980	0.950	0.900	0.800	0.700
R	0.751	0.697	0.637	0.595	0.542	0.451	0.365	0.268	0.205
0	0.689	0.636	0.575	0.534	0.482	0.392	0.309	0.218	0.162
0.50	0.649	0.574	0.493	0.440	0.374	0.271	0.187	0.108	0.069
0.75	0.642	0.537	0.431	0.366	0.290	0.182	0.108	0.052	0.030
0.80	0.640	0.524	0.410	0.341	0.262	0.157	0.088	0.040	0.023
0.85	0.639	0.507	0.381	0.308	0.228	0.126	0.067	0.029	0.016
0.90	0.638	0.481	0.338	0.260	0.180	0.090	0.043	0.018	0.010

XXI. Function $v_0(\mu)$

μ	g						
	R	0	0.5	0.75	0.80	0.85	0.90
1.0	0.477	0.486	0.476	0.470	0.469	0.467	0.466
0.9	0.293	0.299	0.291	0.285	0.284	0.282	0.281
0.8	0.128	0.132	0.126	0.121	0.120	0.119	0.118
0.7	−0.015	−0.014	−0.018	−0.021	−0.022	−0.023	−0.024
0.6	−0.138	−0.140	−0.140	−0.142	−0.142	−0.143	−0.143
0.5	−0.240	−0.245	−0.241	−0.239	−0.239	−0.239	−0.239
0.4	−0.321	−0.328	−0.319	−0.313	−0.312	−0.311	−0.310
0.3	−0.379	−0.389	−0.374	−0.361	−0.358	−0.356	−0.353
0.2	−0.415	−0.428	−0.404	−0.380	−0.375	−0.370	−0.364
0.1	−0.423	−0.440	−0.405	−0.364	−0.354	−0.343	−0.332
0.0	−0.390	−0.410	−0.359	−0.285	−0.263	−0.238	−0.206
δ_0	1.423	1.421	1.424	1.427	1.427	1.428	1.429
ε_0	1.660	1.657	1.662	1.667	1.668	1.669	1.670

XXII. Asymptotic constant N

g	λ								
	1.000	0.999	0.995	0.990	0.980	0.950	0.900	0.800	0.700
R	1.000	0.925	0.840	0.780	0.703	0.566	0.436	0.283	0.186
0	1.000	0.925	0.840	0.781	0.703	0.567	0.436	0.283	0.185
0.5	1.000	0.896	0.781	0.704	0.608	0.449	0.313	0.177	0.105
0.75	1.000	0.855	0.705	0.609	0.494	0.324	0.196	0.091	0.047
0.80	1.000	0.840	0.676	0.574	0.455	0.284	0.163	0.070	0.035
0.85	1.000	0.817	0.636	0.527	0.403	0.234	0.125	0.049	0.023
0.90	1.000	0.781	0.575	0.456	0.329	0.171	0.081	0.028	0.012

XXIII. Functions $u_0(\mu, \tau_0)$ and $w_0(\mu, \tau_0)$ for the Rayleigh phase function

μ	τ_0							
	0.1		0.2		0.3		0.5	
	u_0	w_0	u_0	w_0	u_0	w_0	u_0	w_0
0	0.228	0.199	0.232	0.176	0.241	0.161	0.262	0.145
0.1	0.199	0.149	0.258	0.177	0.290	0.185	0.326	0.179
0.2	0.226	0.166	0.285	0.184	0.323	0.192	0.372	0.193
0.3	0.280	0.215	0.326	0.213	0.363	0.212	0.417	0.210
0.4	0.344	0.277	0.382	0.262	0.415	0.251	0.467	0.236
0.5	0.411	0.343	0.444	0.319	0.474	0.301	0.523	0.273
0.6	0.481	0.412	0.510	0.382	0.537	0.357	0.582	0.318
0.7	0.553	0.483	0.579	0.448	0.603	0.418	0.645	0.370
0.8	0.625	0.554	0.649	0.515	0.671	0.481	0.710	0.425
0.9	0.698	0.627	0.720	0.585	0.740	0.547	0.777	0.484
1.0	0.771	0.700	0.791	0.655	0.810	0.615	0.845	0.546

(continued)

μ	τ_0							
	0.75		1.0		3.0		5.0	
	u_0	w_0	u_0	w_0	u_0	w_0	u_0	w_0
0	0.283	0.131	0.298	0.120	0.356	0.068	0.377	0.047
0.1	0.353	0.164	0.372	0.149	0.444	0.085	0.471	0.059
0.2	0.409	0.184	0.435	0.171	0.521	0.099	0.552	0.069
0.3	0.462	0.201	0.492	0.189	0.592	0.113	0.627	0.078
0.4	0.514	0.220	0.547	0.206	0.661	0.126	0.700	0.087
0.5	0.569	0.248	0.604	0.229	0.729	0.138	0.773	0.096
0.6	0.627	0.283	0.663	0.257	0.796	0.151	0.844	0.105
0.7	0.688	0.325	0.723	0.290	0.863	0.163	0.915	0.114
0.8	0.751	0.371	0.786	0.330	0.929	0.177	0.985	0.122
0.9	0.816	0.422	0.850	0.373	0.996	0.193	1.055	0.131
1.0	0.883	0.476	0.915	0.421	1.062	0.210	1.125	0.141

XXIV. Functions $u_0(\mu, \tau_0)$ and $w_0(\mu, \tau_0)$ for the Henyey-Greenstein phase function with $g = 0.25$

μ	τ_0							
	0.1		0.2		0.3		0.5	
	u_0	w_0	u_0	w_0	u_0	w_0	u_0	w_0
0	0.229	0.207	0.229	0.186	0.236	0.173	0.254	0.160
0.1	0.196	0.157	0.252	0.189	0.282	0.199	0.315	0.196
0.2	0.221	0.175	0.275	0.198	0.312	0.209	0.357	0.213
0.3	0.273	0.223	0.314	0.227	0.349	0.230	0.399	0.232
0.4	0.336	0.286	0.370	0.277	0.399	0.270	0.446	0.260
0.5	0.403	0.351	0.431	0.334	0.456	0.320	0.499	0.298
0.6	0.473	0.420	0.496	0.397	0.518	0.377	0.557	0.344
0.7	0.544	0.491	0.564	0.462	0.583	0.437	0.618	0.396
0.8	0.616	0.562	0.633	0.530	0.650	0.501	0.681	0.453
0.9	0.689	0.635	0.704	0.599	0.718	0.567	0.747	0.512
1.0	0.762	0.708	0.775	0.669	0.788	0.635	0.814	0.574

XXIV. (continued)

μ	τ_0							
	0.75		1.0		3.0		5.0	
	u_0	w_0	u_0	w_0	u_0	w_0	u_0	w_0
0	0.273	0.147	0.287	0.136	0.346	0.083	0.370	0.059
0.1	0.339	0.183	0.357	0.169	0.431	0.103	0.461	0.073
0.2	0.392	0.206	0.416	0.194	0.503	0.120	0.538	0.086
0.3	0.441	0.226	0.469	0.215	0.571	0.136	0.611	0.097
0.4	0.489	0.248	0.521	0.235	0.637	0.152	0.681	0.109
0.5	0.542	0.277	0.575	0.260	0.702	0.167	0.751	0.120
0.6	0.597	0.314	0.630	0.290	0.766	0.182	0.819	0.130
0.7	0.656	0.357	0.688	0.326	0.829	0.197	0.887	0.141
0.8	0.717	0.404	0.748	0.366	0.892	0.213	0.954	0.152
0.9	0.780	0.456	0.809	0.411	0.955	0.231	1.021	0.163
1.0	0.845	0.511	0.873	0.460	1.019	0.251	1.088	0.174

XXV. Functions $u_0(\mu, \tau_0)$ and $w_0(\mu, \tau_0)$ for the Henyey-Greenstein phase function with $g = 0.5$

μ	τ_0							
	0.1		0.2		0.3		0.5	
	u_0	w_0	u_0	w_0	u_0	w_0	u_0	w_0
0	0.201	0.187	0.202	0.173	0.207	0.165	0.222	0.157
0.1	0.177	0.150	0.227	0.183	0.254	0.196	0.283	0.199
0.2	0.206	0.175	0.253	0.199	0.284	0.212	0.325	0.221
0.3	0.261	0.227	0.294	0.233	0.329	0.239	0.366	0.245
0.4	0.326	0.291	0.351	0.286	0.374	0.283	0.413	0.278
0.5	0.394	0.359	0.414	0.346	0.433	0.337	0.467	0.322
0.6	0.465	0.429	0.480	0.412	0.496	0.397	0.525	0.372
0.7	0.537	0.500	0.549	0.479	0.562	0.460	0.588	0.428
0.8	0.609	0.573	0.620	0.549	0.631	0.527	0.652	0.489
0.9	0.683	0.646	0.691	0.619	0.700	0.595	0.719	0.552
1.0	0.756	0.719	0.764	0.691	0.771	0.665	0.788	0.617

XXV. (continued)

μ	τ_0							
	0.75		1.0		3.0		5.0	
	u_0	w_0	u_0	w_0	u_0	w_0	u_0	w_0
0	0.237	0.150	0.250	0.142	0.303	0.097	0.328	0.073
0.1	0.304	0.191	0.321	0.182	0.390	0.125	0.422	0.094
0.2	0.357	0.219	0.379	0.212	0.463	0.148	0.501	0.111
0.3	0.404	0.244	0.430	0.238	0.530	0.169	0.574	0.127
0.4	0.451	0.272	0.481	0.263	0.594	0.190	0.644	0.142
0.5	0.503	0.307	0.533	0.294	0.657	0.209	0.711	0.157
0.6	0.558	0.348	0.587	0.329	0.718	0.228	0.778	0.172
0.7	0.617	0.396	0.644	0.370	0.779	0.248	0.843	0.187
0.8	0.679	0.449	0.704	0.416	0.839	0.269	0.908	0.201
0.9	0.743	0.505	0.766	0.466	0.899	0.292	0.972	0.216
1.0	0.809	0.565	0.829	0.520	0.960	0.317	1.036	0.231

XXVI. Functions $u_0(\mu, \tau_0)$ and $w_0(\mu, \tau_0)$ for the Henyey-Greenstein phase function with $g = 0.75$

μ	τ_0							
	0.1		0.2		0.3		0.5	
	u_0	w_0	u_0	w_0	u_0	w_0	u_0	w_0
0	0.151	0.144	0.153	0.139	0.157	0.137	0.168	0.136
0.1	0.145	0.131	0.184	0.161	0.207	0.176	0.233	0.187
0.2	0.185	0.168	0.218	0.189	0.242	0.202	0.276	0.217
0.3	0.245	0.227	0.266	0.233	0.285	0.239	0.318	0.249
0.4	0.313	0.295	0.327	0.293	0.342	0.292	0.368	0.292
0.5	0.384	0.366	0.394	0.358	0.404	0.353	0.425	0.344
0.6	0.456	0.438	0.464	0.428	0.472	0.419	0.488	0.404
0.7	0.530	0.511	0.535	0.499	0.541	0.488	0.554	0.468
0.8	0.604	0.585	0.608	0.571	0.613	0.558	0.623	0.535
0.9	0.678	0.659	0.681	0.645	0.685	0.630	0.693	0.604
1.0	0.752	0.734	0.755	0.718	0.758	0.703	0.765	0.675

XXVI. (continued)

μ	τ_0							
	0.75		1.0		3.0		5.0	
	u_0	w_0	u_0	w_0	u_0	w_0	u_0	w_0
0	0.179	0.135	0.188	0.133	0.229	0.110	0.250	0.090
0.1	0.252	0.189	0.265	0.187	0.324	0.155	0.355	0.128
0.2	0.303	0.224	0.323	0.225	0.398	0.190	0.437	0.158
0.3	0.349	0.256	0.372	0.257	0.463	0.221	0.508	0.183
0.4	0.396	0.291	0.420	0.290	0.523	0.249	0.574	0.207
0.5	0.450	0.336	0.472	0.329	0.580	0.276	0.637	0.230
0.6	0.508	0.388	0.527	0.375	0.636	0.303	0.698	0.252
0.7	0.571	0.446	0.587	0.428	0.692	0.332	0.750	0.274
0.8	0.636	0.508	0.650	0.485	0.749	0.364	0.816	0.296
0.9	0.704	0.574	0.716	0.546	0.806	0.400	0.875	0.320
1.0	0.774	0.641	0.783	0.611	0.865	0.439	0.934	0.345

XXVII. Function $p_0(\tau_0)$

g	τ_0							
	0.1	0.2	0.3	0.5	0.75	1.0	2.0	3.0	5.0
R	0.534	0.564	0.589	0.632	0.674	0.707	0.792	0.839	0.889
0.25	0.526	0.549	0.570	0.606	0.643	0.674	0.757	0.806	0.862
0.50	0.518	0.534	0.549	0.576	0.606	0.631	0.707	0.757	0.819
0.75	0.509	0.518	0.526	0.541	0.559	0.576	0.630	0.673	0.733

XXVIII. Function $\frac{5-x_2}{15} v_0(\mu, \tau_0)$ for the Rayleigh phase function

μ	τ_0								
	0.1	0.2	0.3	0.5	0.75	1.0	2.0	3.0	5.0
0	−1.251	−1.200	−1.194	−1.225	−1.260	−1.281	−1.309	−1.314	−1.316
0.1	−0.695	−1.011	−1.161	−1.282	−1.336	−1.363	−1.401	−1.409	−1.411
0.2	−0.397	−0.704	−0.899	−1.117	−1.244	−1.304	−1.370	−1.380	−1.383
0.3	−0.235	−0.449	−0.624	−0.868	−1.040	−1.132	−1.243	−1.260	−1.265
0.4	−0.141	−0.279	−0.403	−0.602	−0.772	−0.880	−1.036	−1.062	−1.069
0.5	−0.065	−0.141	−0.218	−0.355	−0.490	−0.586	−0.752	−0.789	−0.800
0.6	−0.005	−0.027	−0.056	−0.121	−0.199	−0.264	−0.404	−0.445	−0.459
0.7	0.049	0.077	0.092	0.099	0.087	0.067	−0.004	−0.035	−0.049
0.8	0.097	0.170	0.228	0.310	0.371	0.406	0.440	0.436	0.430
0.9	0.141	0.258	0.356	0.512	0.650	0.746	0.917	0.961	0.975
1.0	0.183	0.341	0.479	0.708	0.926	1.087	1.422	1.533	1.584

XXIX. Function $\frac{5-x_2}{15} v_0(\mu, \tau_0)$ for the Henyey-Greenstein phase function with $g = 0.25$

μ	τ_0								
	0.1	0.2	0.3	0.5	0.75	1.0	2.0	3.0	5.0
0	−1.244	−1.189	−1.181	−1.208	−1.240	−1.259	−1.283	−1.287	−1.289
0.1	−0.690	−1.001	−1.146	−1.261	−1.311	−1.335	−1.368	−1.375	−1.377
0.2	−0.394	−0.697	−0.888	−1.098	−1.219	−1.276	−1.335	−1.345	−1.348
0.3	−0.233	−0.444	−0.616	−0.853	−1.019	−1.107	−1.211	−1.226	−1.230
0.4	−0.146	−0.276	−0.398	−0.592	−0.756	−0.860	−1.008	−1.033	−1.039
0.5	−0.064	−0.140	−0.214	−0.349	−0.479	−0.572	−0.731	−0.766	−0.777
0.6	−0.005	−0.026	−0.055	−0.119	−0.194	−0.257	−0.392	−0.432	−0.445
0.7	0.049	0.076	0.091	0.098	0.086	0.066	−0.003	−0.033	−0.046
0.8	0.096	0.169	0.225	0.304	0.364	0.397	0.429	0.425	0.418
0.9	0.140	0.255	0.352	0.503	0.637	0.729	0.892	0.934	0.947
1.0	0.182	0.337	0.472	0.695	0.905	1.061	1.382	1.489	1.538

XXX. Function $\frac{5-x_2}{15} v_0(\mu, \tau_0)$ for the Henyey-Greenstein phase function with $g = 0.5$

μ	τ_0								
	0.1	0.2	0.3	0.5	0.75	1.0	2.0	3.0	5.0
0	−1.294	−1.259	−1.264	−1.312	−1.363	−1.395	−1.442	−1.454	−1.458
0.1	−0.719	−1.068	−1.244	−1.400	−1.482	−1.527	−1.595	−1.612	−1.618
0.2	−0.410	−0.743	−0.963	−1.223	−1.388	−1.474	−1.584	−1.607	−1.616
0.3	−0.242	−0.471	−0.664	−0.947	−1.159	−1.281	−1.449	−1.483	−1.494
0.4	−0.144	−0.291	−0.426	−0.652	−0.854	−0.991	−1.210	−1.258	−1.274
0.5	−0.066	−0.145	−0.227	−0.380	−0.536	−0.654	−0.879	−0.939	−0.961
0.6	−0.004	−0.025	−0.055	−0.124	−0.212	−0.289	−0.470	−0.532	−0.557
0.7	0.051	0.082	0.101	0.113	0.104	0.084	−0.001	−0.044	−0.067
0.8	0.100	0.179	0.243	0.338	0.414	0.460	0.515	0.514	0.507
0.9	0.145	0.268	0.376	0.551	0.715	0.833	1.066	1.134	1.161
1.0	0.187	0.352	0.501	0.755	1.007	1.203	1.642	1.806	1.892

XXXI. Function $\frac{5-x_2}{15} v_0(\mu, \tau_0)$ for the Henyey-Greenstein phase function with $g = 0.75$

μ	τ_0								
	0.1	0.2	0.3	0.5	0.75	1.0	2.0	3.0	5.0
0	-1.445	−1.460	−1.499	−1.599	−1.704	−1.778	−1.920	−1.970	−2.000
0.1	−0.801	−1.258	−1.527	−1.818	−2.009	−2.128	−2.352	−2.432	−2.480
0.2	−0.447	−0.858	−1.160	−1.570	−1.881	−2.076	−2.416	−2.528	−2.593
0.3	−0.255	−0.521	−0.768	−1.171	−1.526	−1.768	−2.219	−2.369	−2.455
0.4	−0.147	−0.308	−0.469	−0.765	−1.073	−1.312	−1.828	−2.013	−2.119
0.5	−0.063	−0.143	−0.232	−0.416	−0.632	−0.819	−1.287	−1.486	−1.609
0.6	0.001	−0.013	−0.038	−0.106	−0.208	−0.312	−0.642	−0.816	−0.937
0.7	0.057	0.100	0.131	0.167	0.180	0.171	0.063	−0.031	−0.117
0.8	0.106	0.199	0.280	0.415	0.541	0.633	0.803	0.840	0.839
0.9	0.151	0.289	0.416	0.642	0.878	1.072	1.555	1.773	1.916
1.0	0.192	0.373	0.543	0.855	1.195	1.489	2.306	2.745	3.098

XXXII. Function $\varphi(\mu, \lambda_1)$, $\varphi(0, \lambda_1) \equiv 1.0000$

$m = 0.005$

λ_1	μ									
	0.1	0.2	0.3	0.4	0.5	0.6	0.7	0.8	0.9	1.0
1.000	1.2273	1.4036	1.5630	1.7113	1.8508	1.9830	2.1086	2.2282	2.3425	2.4517
0.999	1.2263	1.4016	1.5598	1.7068	1.8450	1.9757	2.0998	2.2180	2.3307	2.4384
0.995	1.2228	1.3939	1.5477	1.6898	1.8229	1.9484	2.0671	2.1798	2.2869	2.3890
0.990	1.2186	1.3852	1.5338	1.6706	1.7980	1.9176	2.0304	2.1370	2.2381	2.3341
0.975	1.2080	1.3628	1.4989	1.6224	1.7361	1.8417	1.9402	2.0326	2.1194	2.2012
0.950	1.1936	1.3335	1.4540	1.5613	1.6586	1.7477	1.8298	1.9059	1.9767	2.0427
0.900	1.1715	1.2898	1.3885	1.4742	1.5503	1.6185	1.6803	1.7366	1.7882	1.8357
0.850	1.1537	1.2560	1.3394	1.4105	1.4724	1.5272	1.5762	1.6204	1.6605	1.6970
0.800	1.1385	1.2280	1.2995	1.3595	1.4111	1.4563	1.4962	1.5319	1.5641	1.5932
0.700	1.1129	1.1882	1.2358	1.2798	1.3168	1.3487	1.3785	1.4010	1.4228	1.4423
0.600	1.0913	1.1450	1.1855	1.2181	1.2452	1.2681	1.2879	1.3052	1.3205	1.3341
0.500	1.0723	1.1133	1.1437	1.1678	1.1874	1.2039	1.2180	1.2303	1.2410	1.2505

$m = 0.01$

λ_1	μ									
	0.1	0.2	0.3	0.4	0.5	0.6	0.7	0.8	0.9	1.0
1.000	1.2227	1.3928	1.5448	1.6843	1.8141	1.9355	2.0496	2.1571	2.2586	2.3547
0.999	1.2220	1.3912	1.5422	1.6808	1.8095	1.9299	2.0429	2.1494	2.2499	2.3449
0.995	1.2189	1.3849	1.5323	1.6671	1.7919	1.9084	2.0174	2.1199	2.2164	2.3075
0.990	1.2154	1.3775	1.5207	1.6511	1.7715	1.8834	1.9879	2.0859	2.1779	2.2646
0.975	1.2058	1.3577	1.4901	1.6093	1.7183	1.8188	1.9118	1.9984	2.0793	2.1550
0.950	1.1923	1.3304	1.4486	1.5534	1.6479	1.7340	1.8129	1.8856	1.9529	2.0155
0.900	1.1708	1.2882	1.3858	1.4704	1.5450	1.6118	1.6721	1.7268	1.7769	1.8228
0.850	1.1533	1.2551	1.3378	1.4081	1.4693	1.5233	1.5714	1.6147	1.6539	1.6896
0.800	1.1383	1.2274	1.2984	1.3579	1.4090	1.4536	1.4931	1.5282	1.5998	1.5884
0.700	1.1128	1.1819	1.2353	1.2790	1.3158	1.3474	1.3749	1.3991	1.4207	1.4400
0.600	1.0912	1.1448	1.1852	1.2176	1.2445	1.2674	1.2870	1.3042	1.3193	1.3328
0.500	1.0723	1.1132	1.1435	1.1674	1.1870	1.2034	1.2175	1.2296	1.2403	1.2497

$m = 0.02$

λ_1	μ									
	0.1	0.2	0.3	0.4	0.5	0.6	0.7	0.8	0.9	1.0
1.000	1.2172	1.3799	1.5227	1.6518	1.7700	1.8789	1.9798	2.0736	2.1611	2.2428
0.999	1.2167	1.3787	1.5209	1.6493	1.7668	1.8750	1.9753	2.0685	2.1553	2.2364
0.995	1.2141	1.3734	1.5127	1.6381	1.7527	1.8580	1.9554	2.0457	2.1298	2.2082
0.990	1.2110	1.3671	1.5031	1.6252	1.7363	1.8383	1.9324	2.0195	2.1005	2.1759
0.975	1.2026	1.3500	1.4770	1.5900	1.6921	1.7852	1.8706	1.9493	2.0220	2.0895
0.950	1.1902	1.3253	1.4399	1.5405	1.6305	1.7117	1.7856	1.8531	1.9152	1.9725
0.900	1.1697	1.2855	1.3811	1.4634	1.5356	1.5999	1.6575	1.7097	1.7571	1.8004
0.850	1.1526	1.2534	1.3349	1.4038	1.4634	1.5159	1.5625	1.6042	1.6418	1.6760
0.800	1.1378	1.2262	1.2964	1.3549	1.4050	1.4487	1.4871	1.5212	1.5518	1.5794
0.700	1.1125	1.1812	1.2342	1.2774	1.3137	1.3448	1.3718	1.3955	1.4166	1.4354
0.600	1.0910	1.1444	1.1845	1.2167	1.2433	1.2659	1.2852	1.3021	1.3170	1.3302
0.500	1.0722	1.1130	1.1431	1.1668	1.1862	1.2025	1.2164	1.2284	1.2389	1.2482

$m = 0.05$

λ_1	μ									
	0.1	0.2	0.3	0.4	0.5	0.6	0.7	0.8	0.9	1.0
1.000	1.2083	1.3587	1.4866	1.5990	1.6991	1.7890	1.8704	1.9445	2.0121	2.0741
0.999	1.2078	1.3576	1.4851	1.5970	1.6966	1.7861	1.8671	1.9407	2.0080	2.0696
0.995	1.2058	1.3537	1.4792	1.5891	1.6869	1.7747	1.8540	1.9260	1.9918	2.0520
0.990	1.2033	1.3488	1.4719	1.5796	1.6752	1.7608	1.8381	1.9083	1.9722	2.0308
0.975	1.1963	1.3350	1.4515	1.5527	1.6421	1.7220	1.7937	1.8587	1.9178	1.9717
0.950	1.1856	1.3142	1.4209	1.5128	1.5934	1.6648	1.7287	1.7863	1.8385	1.8860
0.900	1.1669	1.2787	1.3695	1.4464	1.5130	1.5714	1.6232	1.6695	1.7111	1.7488
0.850	1.1508	1.2488	1.3271	1.3924	1.4484	1.4970	1.5398	1.5777	1.6117	1.6423
0.800	1.1365	1.2230	1.2908	1.3469	1.3944	1.4354	1.4712	1.5028	1.5309	1.5561
0.700	1.1118	1.1794	1.2311	1.2729	1.3078	1.3375	1.3631	1.3856	1.4054	1.4230
0.600	1.0906	1.1433	1.1827	1.2140	1.2398	1.2615	1.2802	1.2963	1.3105	1.3231
0.500	1.0719	1.1193	1.1419	1.1651	1.1840	1.1998	1.2133	1.2249	1.2350	1.2439

$m = 0.1$

λ_1	μ									
	0.1	0.2	0.3	0.4	0.5	0.6	0.7	0.8	0.9	1.0
1.000	1.2000	1.3388	1.4532	1.5507	1.6353	1.7095	1.7752	1.8338	1.8865	1.9340
0.999	1.1996	1.3380	1.4520	1.5492	1.6334	1.7073	1.7728	1.8317	1.8836	1.9309
0.995	1.1979	1.3348	1.4473	1.5431	1.6261	1.6989	1.7633	1.8207	1.8722	1.9187
0.990	1.1958	1.3308	1.4416	1.5357	1.6172	1.6885	1.7517	1.8079	1.8583	1.9038
0.975	1.1898	1.3193	1.4250	1.5144	1.5916	1.6590	1.7185	1.7714	1.8187	1.8614
0.950	1.1803	1.3015	1.3995	1.4818	1.5525	1.6140	1.6680	1.7160	1.7588	1.7973
0.900	1.1633	1.2700	1.3547	1.4251	1.4850	1.5367	1.5818	1.6217	1.6571	1.6888
0.850	1.1483	1.2426	1.3164	1.3771	1.4283	1.4722	1.5103	1.5438	1.5734	1.5999
0.800	1.1346	1.2183	1.2829	1.3355	1.3795	1.4170	1.4495	1.4778	1.5029	1.5251
0.700	1.1107	1.1766	1.2264	1.2662	1.2990	1.3268	1.3506	1.3712	1.3893	1.4053
0.600	1.0899	1.1415	1.1797	1.2098	1.2344	1.2550	1.2725	1.2876	1.3008	1.3124
0.500	1.0715	1.1111	1.1400	1.1625	1.1806	1.1957	1.2085	1.2194	1.2289	1.2373

$m = 0.2$

λ_1	μ									
	0.1	0.2	0.3	0.4	0.5	0.6	0.7	0.8	0.9	1.0
1.000	1.1901	1.3151	1.4137	1.4947	1.5627	1.6207	1.6709	1.7147	1.7533	1.7875
0.999	1.1897	1.3145	1.4128	1.4935	1.5613	1.6192	1.6692	1.7128	1.7513	1.7854
0.995	1.1883	1.3119	1.4092	1.4890	1.5560	1.6131	1.6625	1.7055	1.7435	1.7772
0.990	1.1866	1.3086	1.4046	1.4833	1.5493	1.6056	1.6542	1.6966	1.7339	1.7670
0.975	1.1814	1.2992	1.3915	1.4669	1.5301	1.5838	1.6301	1.6705	1.7061	1.7376
0.950	1.1732	1.2843	1.3708	1.4412	1.4999	1.5497	1.5926	1.6299	1.6627	1.6918
0.900	1.1580	1.2571	1.3333	1.3947	1.4457	1.4887	1.5256	1.5577	1.5857	1.6105
0.850	1.1442	1.2327	1.2999	1.3538	1.3982	1.4356	1.4675	1.4951	1.5192	1.5405
0.800	1.1315	1.2106	1.2701	1.3173	1.3561	1.3886	1.4163	1.4402	1.4610	1.4793
0.700	1.1087	1.1717	1.2182	1.2547	1.2844	1.3090	1.3299	1.3478	1.3634	1.3770
0.600	1.0887	1.1383	1.1744	1.2024	1.2249	1.2435	1.2592	1.2726	1.2842	1.2944
0.500	1.0706	1.1090	1.1365	1.1576	1.1744	1.1883	1.1999	1.2098	1.2183	1.2258

$m = 0.5$

λ_1	μ									
	0.1	0.2	0.3	0.4	0.5	0.6	0.7	0.8	0.9	1.0
1.000	1.1739	1.2767	1.3516	1.4094	1.4554	1.4931	1.5245	1.5511	1.5739	1.5937
0.999	1.1736	1.2762	1.3510	1.4086	1.4545	1.4921	1.5234	1.5500	1.5728	1.5925
0.995	1.1725	1.2743	1.3484	1.4055	1.4511	1.4883	1.5193	1.5456	1.5682	1.5878
0.990	1.1711	1.2719	1.3453	1.4018	1.4468	1.4836	1.5143	1.5402	1.5625	1.5619
0.975	1.1670	1.2649	1.3360	1.3906	1.4341	1.4697	1.4993	1.5244	1.5459	1.5646
0.950	1.1603	1.2536	1.3210	1.3727	1.4139	1.4474	1.4754	1.4990	1.5193	1.5369
0.900	1.1477	1.2322	1.2929	1.3392	1.3760	1.4060	1.4309	1.4519	1.4899	1.4855
0.850	1.1358	1.2123	1.2669	1.3085	1.3413	1.3680	1.3902	1.4089	1.4249	1.4387
0.800	1.1246	1.1938	1.2429	1.2801	1.3094	1.3332	1.3529	1.3895	1.3837	1.3960
0.700	1.1040	1.1602	1.1996	1.2292	1.2524	1.2712	1.2867	1.2998	1.3109	1.3205
0.600	1.0854	1.1304	1.1615	1.1848	1.2029	1.2176	1.2296	1.2398	1.2484	1.2558
0.500	1.0684	1.1036	1.1276	1.1455	1.1594	1.1706	1.1798	1.1875	1.1940	1.1996

$m = 1.0$

λ_1	μ									
	0.1	0.2	0.3	0.4	0.5	0.6	0.7	0.8	0.9	1.0
1.000	1.1586	1.2418	1.2977	1.3383	1.3692	1.3937	1.4134	1.4298	1.4436	1.4553
0.999	1.1584	1.2414	1.2972	1.3377	1.3686	1.3930	1.4128	1.4291	1.4428	1.4545
0.995	1.1575	1.2396	1.2953	1.3356	1.3662	1.3904	1.4100	1.4262	1.4398	1.4515
0.990	1.1563	1.2381	1.2930	1.3329	1.3633	1.3872	1.4066	1.4227	1.4362	1.4477
0.975	1.1529	1.2326	1.2861	1.3249	1.3544	1.3777	1.3966	1.4122	1.4253	1.4365
0.950	1.1473	1.2237	1.2748	1.3119	1.3401	1.3623	1.3803	1.3952	1.4077	1.4183
0.900	1.1366	1.2066	1.2532	1.2870	1.3127	1.3329	1.3493	1.3628	1.3741	1.3838
0.850	1.1263	1.1903	1.2329	1.2636	1.2869	1.3053	1.3201	1.3323	1.3426	1.3514
0.800	1.1165	1.1748	1.2136	1.2415	1.2626	1.2792	1.2927	1.3037	1.3131	1.3209
0.700	1.0981	1.1463	1.1780	1.2007	1.2179	1.2314	1.2423	1.2513	1.2588	1.2652
0.600	1.0811	1.1202	1.1457	1.1640	1.1777	1.1885	1.1972	1.2044	1.2104	1.2155
0.500	1.0653	1.0962	1.1163	1.1306	1.1413	1.1497	1.1565	1.1620	1.1667	1.1707

XXXIII. Reflection coefficient $\rho(\mu, 1)$

$m = 0.01$

μ	λ_1						
	1.0	0.95	0.9	0.8	0.7	0.6	0.5
0	0.5887	0.4787	0.4101	0.3177	0.2520	0.1999	0.1562
0.1	0.6509	0.5171	0.4354	0.3281	0.2545	0.1980	0.1521
0.2	0.6763	0.5272	0.4381	0.3237	0.2474	0.1902	0.1446
0.3	0.6889	0.5283	0.4340	0.3156	0.2384	0.1816	0.1369
0.4	0.6942	0.5245	0.4266	0.3060	0.2289	0.1730	0.1297
0.5	0.6948	0.5179	0.4176	0.2959	0.2196	0.1649	0.1230
0.6	0.6920	0.5096	0.4076	0.2859	0.2106	0.1573	0.1168
0.7	0.6870	0.5003	0.3973	0.2760	0.2021	0.1503	0.1112
0.8	0.6803	0.4903	0.3869	0.2665	0.1940	0.1437	0.1060
0.9	0.6725	0.4801	0.3766	0.2575	0.1865	0.1376	0.1012
1.0	0.6638	0.4698	0.3665	0.2489	0.1795	0.1320	0.0968

$m = 0.1$

μ	λ_1						
	1.0	0.95	0.9	0.8	0.7	0.6	0.5
0	0.4835	0.4269	0.3800	0.3050	0.2459	0.1969	0.1547
0.1	0.5140	0.4478	0.3938	0.3093	0.2447	0.1925	0.1488
0.2	0.5132	0.4432	0.3868	0.2998	0.2344	0.1826	0.1400
0.3	0.5032	0.4317	0.3745	0.2874	0.2229	0.1724	0.1313
0.4	0.4891	0.4174	0.3604	0.2744	0.2115	0.1627	0.1233
0.5	0.4732	0.4021	0.3459	0.2617	0.2006	0.1537	0.1161
0.6	0.4567	0.3867	0.3317	0.2496	0.1905	0.1454	0.1095
0.7	0.4404	0.3717	0.3179	0.2382	0.1812	0.1379	0.1036
0.8	0.4244	0.3573	0.3049	0.2276	0.1726	0.1310	0.0982
0.9	0.4091	0.3436	0.2927	0.2178	0.1647	0.1248	0.0934
1.0	0.3945	0.3307	0.2812	0.2087	0.1575	0.1191	0.0890

$m = 1$

μ	λ_1						
	1.0	0.95	0.9	0.8	0.7	0.6	0.5
0	0.3638	0.3369	0.3114	0.2642	0.2214	0.1823	0.1463
0.1	0.3336	0.3072	0.2824	0.2371	0.1967	0.1604	0.1275
0.2	0.2944	0.2707	0.2484	0.2079	0.1719	0.1397	0.1107
0.3	0.2613	0.2400	0.2201	0.1839	0.1518	0.1232	0.0975
0.4	0.2341	0.2149	0.1970	0.1644	0.1356	0.1100	0.0870
0.5	0.2118	0.1943	0.1781	0.1486	0.1225	0.0993	0.0785
0.6	0.1931	0.1772	0.1623	0.1354	0.1116	0.0904	0.0714
0.7	0.1774	0.1628	0.1491	0.1243	0.1024	0.0830	0.0655
0.8	0.1641	0.1505	0.1378	0.1149	0.0947	0.0767	0.0605
0.9	0.1525	0.1399	0.1281	0.1068	0.0880	0.0712	0.0563
1.0	0.1425	0.1307	0.1197	0.0997	0.0821	0.0665	0.0525

XXXIV. Albedo of atmosphere $A(\mu)$

$m = 0.01$

μ	λ_1						
	1.0	0.95	0.9	0.8	0.7	0.6	0.5
0	0.8877	0.7644	0.6792	0.5526	0.4540	0.3705	0.2969
0.1	0.8589	0.7164	0.6220	0.4880	0.3890	0.3090	0.2412
0.2	0.8352	0.6815	0.5827	0.4471	0.3506	0.2747	0.2120
0.3	0.8131	0.6511	0.5498	0.4144	0.3207	0.2488	0.1904
0.4	0.7921	0.6239	0.5210	0.3869	0.2963	0.2280	0.1783
0.5	0.7719	0.5990	0.4955	0.3632	0.2757	0.2107	0.1593
0.6	0.7525	0.5762	0.4725	0.3425	0.2580	0.1960	0.1475
0.7	0.7338	0.5551	0.4517	0.3341	0.2425	0.1834	0.1375
0.8	0.7158	0.5355	0.4327	0.3077	0.2289	0.1723	0.1287
0.9	0.6986	0.5172	0.4153	0.2930	0.2168	0.1626	0.1211
1.0	0.6820	0.5001	0.3993	0.2796	0.2059	0.1539	0.1143

$m = 0.1$

μ	λ_1						
	1.0	0.95	0.9	0.8	0.7	0.6	0.5
0	0.7935	0.7172	0.6514	0.5407	0.4481	0.3675	0.2953
0.1	0.7340	0.6514	0.5819	0.4689	0.3785	0.3027	0.2374
0.2	0.6864	0.6027	0.5334	0.4230	0.3369	0.2664	0.2068
0.3	0.6442	0.5611	0.4931	0.3865	0.3049	0.2392	0.1844
0.4	0.6065	0.5249	0.4588	0.3563	0.2790	0.2174	0.1667
0.5	0.5726	0.4930	0.4289	0.3306	0.2573	0.1996	0.1524
0.6	0.5420	0.4646	0.4027	0.3085	0.2389	0.1845	0.1404
0.7	0.5143	0.4393	0.3795	0.2892	0.2230	0.1717	0.1302
0.8	0.4892	0.4164	0.3588	0.2722	0.2092	0.1605	0.1215
0.9	0.4663	0.3958	0.3402	0.2572	0.1970	0.1508	0.1139
1.0	0.4453	0.3771	0.3235	0.2437	0.1861	0.1422	0.1072

$m = 1$

μ	λ_1						
	1.0	0.95	0.9	0.8	0.7	0.6	0.5
0	0.6667	0.6213	0.5779	0.4966	0.4215	0.3516	0.2861
0.1	0.5310	0.4909	0.4529	0.3830	0.3199	0.2624	0.2099
0.2	0.4433	0.4087	0.3761	0.3165	0.2631	0.2149	0.1711
0.3	0.3803	0.3501	0.3218	0.2700	0.2239	0.1824	0.1449
0.4	0.3330	0.3062	0.2812	0.2356	0.1950	0.1586	0.1258
0.5	0.2961	0.2722	0.2498	0.2090	0.1728	0.1404	0.1113
0.6	0.2666	0.2449	0.2246	0.1878	0.1552	0.1260	0.0998
0.7	0.2424	0.2226	0.2042	0.1705	0.1408	0.1143	0.0904
0.8	0.2222	0.2040	0.1871	0.1562	0.1289	0.1046	0.0827
0.9	0.2052	0.1883	0.1726	0.1441	0.1189	0.0964	0.0762
1.0	0.1905	0.1749	0.1603	0.1337	0.1103	0.0894	0.0707

XXXV. Function $i(\mu)$ for the Henyey-Greenstein phase function

$g = 0.25$

μ	λ							
	0.999	0.995	0.990	0.980	0.950	0.900	0.800	0.700
−1.0	0.940	0.873	0.827	0.769	0.671	0.582	0.484	0.424
−0.9	0.946	0.884	0.842	0.788	0.696	0.610	0.514	0.455
−0.8	0.951	0.896	0.858	0.808	0.721	0.640	0.547	0.488
−0.7	0.957	0.908	0.874	0.829	0.749	0.672	0.583	0.526
−0.6	0.963	0.920	0.890	0.850	0.778	0.707	0.623	0.568
−0.5	0.969	0.933	0.907	0.872	0.809	0.745	0.668	0.615
−0.4	0.975	0.946	0.924	0.895	0.842	0.787	0.717	0.669
−0.3	0.981	0.959	0.942	0.920	0.877	0.831	0.772	0.730
−0.2	0.987	0.972	0.961	0.945	0.914	0.880	0.835	0.800
−0.1	0.994	0.986	0.980	0.971	0.954	0.934	0.905	0.882
0	1.000	1.000	0.999	0.999	0.997	0.993	0.985	0.976
0.1	1.006	1.014	1.020	1.027	1.042	1.058	1.077	1.088
0.2	1.013	1.029	1.041	1.058	1.092	1.130	1.183	1.221
0.3	1.019	1.044	1.062	1.089	1.145	1.211	1.308	1.382
0.4	1.026	1.059	1.085	1.122	1.203	1.301	1.454	1.580
0.5	1.032	1.075	1.108	1.157	1.265	1.403	1.630	1.829
0.6	1.039	1.091	1.132	1.194	1.333	1.518	1.843	2.150
0.7	1.046	1.107	1.157	1.232	1.407	1.650	2.108	2.578
0.8	1.053	1.124	1.182	1.273	1.488	1.802	2.443	3.174
0.9	1.060	1.141	1.209	1.316	1.577	1.979	2.881	4.056
1.0	1.067	1.159	1.236	1.361	1.675	2.188	3.475	5.489

$g = 0.5$

μ	λ							
	0.999	0.995	0.990	0.980	0.950	0.900	0.800	0.700
−1.0	0.926	0.844	0.787	0.715	0.594	0.484	0.366	0.296
−0.9	0.933	0.858	0.806	0.738	0.623	0.516	0.397	0.325
−0.8	0.940	0.872	0.824	0.762	0.653	0.550	0.432	0.359
−0.7	0.947	0.886	0.844	0.787	0.685	0.587	0.471	0.397
−0.6	0.955	0.901	0.863	0.812	0.720	0.627	0.514	0.440
−0.5	0.962	0.917	0.884	0.839	0.757	0.671	0.564	0.490
−0.4	0.969	0.932	0.905	0.867	0.796	0.720	0.619	0.548
−0.3	0.977	0.948	0.927	0.897	0.838	0.773	0.682	0.615
−0.2	0.984	0.965	0.949	0.928	0.883	0.831	0.754	0.693
−0.1	0.992	0.981	0.973	0.960	0.931	0.895	0.836	0.785
0	1.000	0.998	0.997	0.993	0.983	0.966	0.929	0.890
0.1	1.007	1.016	1.021	1.029	1.039	1.045	1.041	1.026
0.2	1.015	1.034	1.047	1.065	1.099	1.133	1.169	1.184
0.3	1.023	1.052	1.074	1.104	1.164	1.231	1.319	1.377
0.4	1.031	1.071	1.101	1.145	1.235	1.340	1.496	1.617
0.5	1.039	1.090	1.129	1.187	1.311	1.464	1.708	1.918
0.6	1.048	1.110	1.159	1.232	1.393	1.603	1.964	2.303
0.7	1.056	1.130	1.189	1.279	1.483	1.762	2.276	2.805
0.8	1.064	1.151	1.221	1.329	1.581	1.942	2.663	3.480
0.9	1.073	1.172	1.253	1.381	1.668	2.150	3.150	4.416
1.0	1.081	1.193	1.287	1.436	1.805	2.390	3.775	5.774

$g = 0.75$

μ	λ							
	0.999	0.995	0.990	0.980	0.950	0.900	0.800	0.700
−1.0	0.896	0.781	0.704	0.608	0.453	0.325	0.206	0.148
−0.9	0.906	0.800	0.728	0.636	0.485	0.357	0.232	0.168
−0.8	0.915	0.819	0.753	0.666	0.521	0.392	0.262	0.193
−0.7	0.925	0.839	0.778	0.698	0.559	0.431	0.296	0.221
−0.6	0.935	0.859	0.805	0.731	0.600	0.475	0.336	0.256
−0.5	0.946	0.880	0.832	0.767	0.645	0.524	0.383	0.297
−0.4	0.956	0.902	0.861	0.804	0.694	0.579	0.437	0.347
−0.3	0.967	0.924	0.891	0.843	0.747	0.641	0.502	0.407
−0.2	0.977	0.946	0.922	0.884	0.806	0.712	0.578	0.481
−0.1	0.988	0.970	0.954	0.928	0.869	0.791	0.668	0.571
0	0.999	0.994	0.987	0.975	0.937	0.877	0.762	0.651
0.1	1.010	1.018	1.022	1.024	1.014	0.983	0.906	0.825
0.2	1.021	1.044	1.058	1.076	1.097	1.099	1.062	1.003
0.3	1.032	1.070	1.096	1.131	1.188	1.232	1.252	1.231
0.4	1.044	1.097	1.135	1.189	1.287	1.384	1.484	1.525
0.5	1.055	1.124	1.176	1.250	1.397	1.558	1.770	1.910
0.6	1.067	1.153	1.219	1.316	1.517	1.759	2.124	2.420
0.7	1.079	1.182	1.263	1.385	1.650	1.991	2.567	3.107
0.8	1.091	1.212	1.310	1.459	1.797	2.260	3.126	4.048
0.9	1.103	1.243	1.358	1.537	1.959	2.573	3.839	5.364
1.0	1.115	1.275	1.408	1.620	2.138	2.939	4.758	7.250

$g = 0.9$

μ	λ							
	0.999	0.995	0.990	0.980	0.950	0.900	0.800	0.700
−1.0	0.838	0.668	0.561	0.437	0.266	0.157	0.081	0.053
−0.9	0.853	0.694	0.592	0.470	0.296	0.179	0.094	0.062
−0.8	0.867	0.721	0.624	0.506	0.330	0.205	0.110	0.072
−0.7	0.882	0.749	0.658	0.544	0.368	0.235	0.130	0.086
−0.6	0.898	0.778	0.694	0.586	0.410	0.271	0.153	0.102
−0.5	0.913	0.808	0.732	0.631	0.459	0.314	0.183	0.123
−0.4	0.929	0.840	0.772	0.680	0.514	0.364	0.219	0.149
−0.3	0.945	0.872	0.815	0.733	0.576	0.424	0.265	0.184
−0.2	0.962	0.906	0.860	0.790	0.646	0.495	0.323	0.228
−0.1	0.979	0.942	0.908	0.853	0.727	0.580	0.397	0.288
0	0.996	0.979	0.958	0.919	0.811	0.657	0.410	0.210
0.1	1.013	1.018	1.013	0.994	0.922	0.804	0.613	0.475
0.2	1.031	1.058	1.069	1.073	1.040	0.951	0.773	0.624
0.3	1.049	1.100	1.130	1.160	1.176	1.130	0.984	0.834
0.4	1.067	1.143	1.194	1.254	1.330	1.347	1.263	1.133
0.5	1.086	1.189	1.262	1.356	1.508	1.612	1.639	1.567
0.6	1.105	1.236	1.334	1.468	1.711	1.936	2.148	2.211
0.7	1.124	1.286	1.410	1.589	1.945	2.334	2.845	3.186
0.8	1.144	1.337	1.491	1.721	2.213	2.825	3.811	4.695
0.9	1.164	1.391	1.577	1.864	2.522	3.431	5.165	7.084
1.0	1.185	1.447	1.668	2.021	2.879	4.185	7.084	10.97

XXXVI. Quantity $\zeta = 2\int_0^1 u^2(\mu)\mu^2\,\mathrm{d}\mu$

g	λ								
	1.000	0.999	0.995	0.990	0.980	0.950	0.900	0.800	0.700
R	0.775	0.716	0.647	0.598	0.531	0.412	0.297	0.166	0.092
0.0	0.772	0.714	0.645	0.596	0.531	0.413	0.299	0.169	0.095
0.5	0.776	0.695	0.606	0.547	0.471	0.348	0.241	0.133	0.077
0.75	0.780	0.669	0.555	0.485	0.402	0.279	0.185	0.099	0.056
0.80	0.781	0.658	0.536	0.463	0.378	0.257	0.168	0.089	0.051
0.85	0.782	0.642	0.510	0.433	0.347	0.230	0.148	0.077	0.044
0.90	0.783	0.616	0.469	0.388	0.303	0.194	0.122	0.063	0.036

XXXVII. Functions $8\Psi_y(\alpha,\psi)$ and $8\bar{\Psi}(\alpha)$

α	ψ								
	0°	10°	20°	30°	40°	50°	60°	70°	80°	$8\bar{\Psi}(\alpha)$
0°	4.69	4.62	4.40	4.05	3.58	2.98	2.29	1.53	0.753	4.14
10°	4.81	4.74	4.53	4.18	3.71	3.11	2.42	1.64	0.840	4.27
20°	4.62	4.55	4.33	3.99	3.51	2.92	2.23	1.47	0.703	4.07
30°	4.93	4.86	4.65	4.31	3.84	3.25	2.56	1.79	0.950	4.40
40°	5.07	5.00	4.80	4.46	4.01	3.44	2.77	2.00	1.15	4.56
50°	5.08	5.01	4.81	4.49	4.05	3.50	2.85	2.11	1.28	4.58
60°	4.98	4.92	4.73	4.41	3.99	3.45	2.83	2.12	1.32	4.51
70°	4.79	4.73	4.54	4.24	3.83	3.32	2.72	2.04	1.27	4.33
80°	4.50	4.44	4.26	3.97	3.58	3.09	2.52	1.88	1.15	4.05
90°	4.10	4.05	3.88	3.61	3.24	2.78	2.24	1.64	0.968	3.68
100°	3.60	3.55	3.39	3.14	2.80	2.38	1.89	1.35	0.760	3.20
110°	3.00	2.92	2.81	2.59	2.28	1.92	1.49	1.03	0.550	2.63
120°	2.31	2.28	2.16	1.97	1.72	1.42	1.08	0.722	0.364	2.00
130°	1.62	1.59	1.50	1.36	1.17	0.947	0.704	0.456	0.220	1.37
140°	0.991	0.971	0.913	0.820	0.699	0.559	0.408	0.259	0.121	0.827
150°	0.525	0.514	0.482	0.431	0.366	0.291	0.211	0.133	0.062	0.433
160°	0.200	0.196	0.184	0.165	0.140	0.111	0.081	0.052	0.024	0.165
170°	0.023	0.023	0.022	0.019	0.016	0.013	0.010	0.006	0.003	0.019

XXXVIII. Function $8\Psi_x(\alpha, \omega)$

ω	α									
	0°	10°	20°	30°	40°	50°	60°	70°	80°	
0°	4.69	4.77	4.47	4.63	4.61	4.45	4.22	3.94	3.64	
−10°	4.62	4.63	4.27	4.39	4.35	4.21	4.03	3.90	1.96	10°
−20°	4.40	4.35	3.94	4.04	4.00	3.91	3.84	1.81	2.81	20°
−30°	4.05	3.95	3.50	3.58	3.57	3.55	1.54	2.41	3.05	30°
−40°	3.58	3.43	2.94	3.03	3.07	1.19	1.94	2.61	3.22	40°
−50°	2.98	2.80	2.30	2.40	0.804	1.42	2.08	2.75	3.35	50°
−60°	2.29	2.08	1.59	0.466	0.914	1.51	2.18	2.86	3.46	60°
−70°	1.53	1.31	0.191	0.503	0.957	1.57	2.27	2.96	3.54	70°
−80°	0.753	0.023	0.197	0.525	0.999	1.64	2.37	3.05	3.60	80°
		170°	160°	150°	140°	130°	120°	110°	100°	
	α									ω

XXXVIII. (continued)

ω	α								
	10°	20°	30°	40°	50°	60°	70°	80°	90°
10°	4.77	4.54	4.76	4.78	4.65	4.42	4.10	3.71	3.20
20°	4.63	4.47	4.77	4.86	4.78	4.58	4.28	3.89	3.41
30°	4.35	4.28	4.66	4.83	4.83	4.68	4.42	4.06	3.60
40°	3.95	3.96	4.43	4.70	4.78	4.71	4.51	4.18	3.75
50°	3.43	3.51	4.09	4.45	4.63	4.65	4.53	4.26	3.86
60°	2.79	2.95	3.63	4.10	4.39	4.51	4.48	4.28	3.94
70°	2.07	2.31	3.07	3.66	4.06	4.29	4.36	4.25	3.98
80°	1.31	1.59	2.42	3.12	3.64	4.00	4.17	4.17	3.98

References[1]

1. Chvolson, O.D. Grundzüge einer matematischen Teorie der inneren Diffusion des Lichtes. *Izv. Peterburg. Academii Nauk*, 1890, **33**, 221–265.
2. Hopf, F. Mathematical problems of radiative equilibrium. *Cambridge Mathem. Tract. Cambridge, Univ. Press*, 1934, N 31.
3. Ambartsumian, V.A. Scientific works. v. I. *Erevan, Acad. Sci. Arm.SSR*, 1960 (Russian).
4. Chandrasekhar, S. Radiative transfer. *Oxford, Clarendon Press*, 1950.
5. Sobolev, V.V. A treatise on radiative transfer. *Toronto, Van Nostrand*, 1963.
6. Sobolev, V.V. Light scattering in planetary atmospheres. *Oxford, Pergamon Press*, 1975.
7. Hulst, H.C. van de. Multiple light scattering. Tables, formulas and applications. v. I and II. *New York, Acad. Press*, 1980.
8. Bellman, R.E., Kalaba, R.E., Prestrud, M.C. Invariant imbedding and radiative transfer in slabs of finite thickness. *New York, American Elsevier*, 1963.
9. Casti, J., Kalaba, R.E. Imbedding methods in applied mathematics. *Massachusetts, Addison-Wesley*, 1973.
10. Kagivada, H.H., Kalaba, R., Ueno, S. Multiple scattering processes. Inverse and direct. *Massachusetts, Addison-Wesley*, 1975.
11. Kourganoff, V. Basic methods in transfer problems. *Oxford, Univ. Press*, 1952.
12. Busbridge, I.W. The mathematics of radiative transfer. *Cambridge, Univ. Press*, 1960.
13. Ivanov, V.V. Transfer of radiation in spectral lines. *Washington, U.S. Government Printing Office*, NBS SP N 385, 1973.
14. Minin, I.N. Theory of radiative transfer in atmospheres of planets. *Moscow, Nauka*, 1988 (Russian).
15. Hansen, J.E., Travis, L.D. Light scattering in planetary atmospheres *Space Sci. Rev.*, 1974, **16**, 527–610.
16. De Rooij, W.A. Reflection and transmission of polarized light by planetary atmospheres. (Thesis). *Drukkerij Elinkwijk B.V. Utrecht*, 1985.

[1] This book contains many references to journals that are published in Russian. A number of these journals are translated into English. For readers' convenience we provide here a list of such journals with their English titles.
Astron. vestnik = Solar System Research
Astron. zhurn. = Soviet Astronomy
Astrofizika = Astrophysics
Dokl. AN SSR = Soviet Physics Dokl.
Fiz. atmosf. i okeana = Izvestiya, Atmospheric and Oceanic Physics
Kinemat. i fiz. nebesn. tel = Kinematics and Physics Selestial Bodies
Kosm. Issled. = Cosmical Research
Pisma v astron. zhurn. = Astronomy Letters

17. De Haan, J. Effects of aerosols on the brightness and polarization of cloudless planetary atmospheres (Thesis). *Amsterdam, Vrije Universiteit*, 1987.
18. Wauben, W.M.F. Multiple scattering of polarized radiation in planetary atmospheres. (Thesis). *Amsterdam, Vrije Universiteit*, 1992.
19. Dolginov, A.Z., Gnedin, Yu.N., Silantiev, N.A. Propagation and polarization of radiation in cosmic medium. *New York, Gordon & Breach*, 1994.
20. Case, K.M., de Hoffman, F., Placzec, G. Introduction to the theory of neutron diffusion. *Washington, U.S. Government Printing Office*, 1953.
21. Devison, B. Neutron transport theory. *London, Oxford Univ. Press*, 1957.
22. Shimizu, A., Aoki, K. Application of invariant imbedding to reactor physics. *New York, London, Acad. Press*, 1972.
23. Smelov, V.V., Lectures on neutron transport theory. *Moscow, Atomizdat*, 1960 (Russian).
24. Fano, U., Spencer, L.V., Berger, M.J. Penetration and diffusion of X rays. Handbuch der Physik. Band XXXVIII/2. *Berlin*, 1959.
25. Siegel, R., Howell, J.R. Thermal radiation heat transfer. *New York, Mc Graw Hill*, 1972.
26. Özisik, M.N. Radiative transfer and interactions with conduction and conversion. *New York, A Wiley-Interscience Publ. John Wiley & Sons*, 1973.
27. Rubtsov, N.A. Thermal radiation heat transfer in complex media. *Novosibirsk, Nauka*, 1984 (Russian).
28. Kondratiev, K.Ya. Long-wave radiation transfer in atmosphere. *Moscow, Leningrad, GITTL*, 1950 (Russian).
29. Chamberlain, J.W. Physics of the aurora and airglow. *New York, Academic Press*, 1961.
30. Erlov, N.G. Optical oceanography. *Amsterdam, Elsevier*, 1968.
31. Feigelson, E.M. Radiative processes in stratified clouds. *Moscow, Nauka*, 1964 (Russian).
32. Goody, R.M. Atmospheric radiation. Theoretical basis. *London, Oxford Univ. Press*, 1966.
33. Malkevich, M.S. Satellite study of atmosphere. *Moscow, Nauka*, 1973 (Russian).
34. Zuev, V.E., Kabanov, M.V. Transfer of optical signals in Earth atmosphere. *Moscow, Sov. Radio*, 1977 (Russian).
35. Feigelson, E.M., Krasnokutskaya, L.D. Solar radiation fluxes in clouds. *Leningrad, Gidrometeoizdat*, 1978.
36. Kuo-Nan Liou. An introduction to atmosphere radiation. *New York, Academic Press*, 1980.
37. Smoktiy, O.I. Radiation fields modeling in problems of cosmic spectrophotometry. *Leningrad, Nauka*, 1986 (Russian).
38. Cercignani, C. Theory and application of the Boltzmann equation. *Edinburgh, Scottish Acad. Press*, 1980.
39. Vladimirov, V.S. Mathematical problems of one-speed theory of particles transfer. *Tr. MI AN SSSR*, 1961, **61**, 3–158 (Russian).
40. Maslennikov, M.V. The Milne problem with anisotropic scattering. Ibid., 1968, **47**, 3–132 (Russian).
41. Maslennikov, M.V. Axiomatic model of particles transfer phenomena. *Moscow, Nauka*, 1989 (Russian).
42. Ershov, Yu.I., Shikhov, S.B. Mathematical foundations of transfer theory. I. Foundations of the theory. *Moscow, Energoatomizdat*, 1985 (Russian).
43. Ershov, Yu.I., Shikhov, S.B. Mathematical foundations of transfer theory. II. Applications to reactor physics. *Moscow, Energoatomizdat*, 1985 (Russian).

44. Germogenova, T.A. Local properties of solution to transfer equation. *Moscow, Nauka*, 1986 (Russian).
45. Case, K.M., Zweifel, P.F. Linear transport theory. Massachusetts, *Addison-Wesley*, 1967.
46. Marchuk, G.I., Lebedev, V.I. Numerical methods in neutron transport theory. *Moscow, Atomizdat*, 1981 (Russian).
47. Sushkevich, T.A., Strelkov, S.A., Ioltukhovskii, A.A. Method of characteristics in problems of atmospheric optics. *Moscow, Nauka*, 1990 (Russian).
48. Dlugach, Zh.M., Yanovitskij, E.G. On the behaviour of the azimuth-dependent terms of the radiation intensity in deep layers of semi-infinite medium. *Fiz. atmosf. i okeana*, 1977, **13**(7), 699–709 (Russian).
49. Dlugach, J.M., Yanovitskij, E.G. The optical properties of Venus and Jovian planets. II. Methods and results of calculations of the intensity of radiation diffusely reflected from semi-infinite homogeneous atmospheres. *Icarus*, 1974, **22**(1), 66–81.
50. Dlugach, Zh.M. The computation of the radiation field in a semi-infinite homogeneous atmosphere. *Astron. zhurn.*, 1976, **53**(6), 1295–1305 (Russian).
51. Yanovitskij, E.G. Invariance principles and integral relations for radiation fields in plane atmosphere. *Astron. zhurn.*, 1979, **56**(4), 833–844 (Russian).
52. Hansen, J.E. Radiative transfer by doubling very thin layers. *Astrophys. J.*, 1969, **155**(2), 565–573.
53. Konovalov, N.V. On scope of the asymptotic formulas for calculating monochromatic radiation in optically thick inhomogeneous layer. *Fiz. atmosf. i okeana*, 1975 **11**(12), 1263–1271 (Russian).
54. Dlugach, Zh.M., Yanovitskij, E.G. An estimation of accuracy for depth asymptotic formulas in some radiative transfer problems. *Astrofizika*, 1991, **34**(3), 403–418 (Russian).
55. Yanovitskij, E.G. On the estimation of accuracy for depth asymptotic formulae in some radiative transfer problems. *Kinemat. i fiz. nebesn. tel*, 1994, **10**(5), 21–28 (Russian).
56. Nagirner, D.I. Theory of nonstationary transfer of radiation. *Astrofizika*, 1974, **10**(3), 445–469 (Russian).
57. Shifrin, K.S. Light scattering in turbid medium. *Moscow, GITTL*, 1951 (Russian).
58. Hulst, H.C. van de. Light scattering by small particles. *New York, John Wiley & Sons, Inc. London, Chapman & Hall*, 1957.
59. Bohren, C.F., Huffman, D.R. Absorption and scattering of light by small particles. *New York, A Wiley-Interscience Publ. John Wiley & Sons*, 1983.
60. Born, M., Wolf E. Principles of optics, Forth Edition, *Oxford, Pergamon Press*, 1968.
61. Newton, R.G. Scattering theory of waves and particles. *New York, Mc Grow Hill*, 1966.
62. Deirmendjian, D.E. Electromagnetic scattering on spherical polydispersion. *New York, American Elsevier*, 1969.
63. Loskutov, V.M. Light scattering on spherical particles with refraction index 1.38. *Tr. AO LGU*, 1971, **28**, 14–24 (Russian).
64. Wiscombe, W.J. Improved Mie scattering algorithms. *Appl. Opt.*, 1980, **19**(9), 1505–1509.
65. Bugaenko, O.I. Generalized spherical functions in the Mie problem. *Fiz. atmosf. i okeana*, 1976, **12**(6), 603–611 (Russian).
66. Domke, H. Fourier expansion of the phase matrix for Mie scattering *Z. Meteorol.*, 1975, **25**(6), 357–361.

67. De Rooij, W.A., Van der Stap, C.C.A.H. Expansion of Mie scattering matrices in generalized spherical functions. *Astron. and Astrophys.*, 1984, **131**(2), 237–248.
68. Mishchenko, M.I. Polarization of light reflected by vertically inhomogeneous planetary atmosphere. *Preprint, Kiev, Inst. for theoretical physics, ITF-87-27R*, 1987 (Russian).
69. Dlugach, Zh.M. Expansion of the phase function in Legendre polynomials. *Kinemat. i fiz. nebesn. tel*, 1985, **1**(6), 16–23 (Russian).
70. Yanovitskij, E.G. Anisotropic light scattering in an inhomogeneous atmosphere. The radiation field in deep layers of a semi-infinite atmosphere. *Astron. zhurn.*, 1978, **55**(4), 713–721 (Russian).
71. Rubashevskii, A.A., Yanovitskij, E.G. Calculations of phase dependence of integral brightness of planets. *Kiev, Naukova Dumka*, 1969 (Russian).
72. Fymat, A.L., Lenoble, J. Absorption profile of planetary atmosphere: a proposal for a scattering independent determination. *Appl. Opt.*, 1972, **11**(10), 2249–2259.
73. Yanovitskij, E.G. Height variations of absorption coefficient in Venusian atmosphere. *Kosm. Issled.*, 1984, **22**(2), 273–278 (Russian).
74. Gavrilova, L.A. Determination of the true light absorption volume coefficient from results of aircraft measurements of spectral fluxes in a cloudless atmosphere. *Fiz. atmosf. i okeana*, 1984, **20**(4), 269–273 (Russian).
75. Ivanov, V.V., Volkov, E.V. Radiative transfer: invariance principles for the Green function. *Tr. Astron. observ. LGU*, 1979, **35**, 3–30 (Russian).
76. Ambartsumian, V.A. On diffuse reflection of light by turbid medium. *Dokl. AN SSR*, 1943, **38**(8), 257–261 (Russian).
77. Ambartsumian, V.A. Homogeneous case of the problem of scattering and absorbing medium with finite optical thickness. *Izv. AN ArmSSR Estestv. nauki*, 1944, N 1–2, 31–36 (Russian).
78. Ambartsumian, V.A. On diffuse reflection and transmission of light by anisotropic one-dimensional scattering medium of finite optical thickness. *Dokl. AN ArmSSR*, 1947, **7**, 199–202 (Russian).
79. Pikichian, O.V. Invariance principle in problems of adding of layers with arbitrary properties. *Soobshch. Byurokan. observ.*, 1984, **55**, 5–16 (Russian).
80. Engibarian, N.B., Mnatsakanian, M.A. On linear problems of transfer. *Dokl. AN SSSR*, 1974, **217**(3), 533–535 (Russian).
81. Mnatsakanian, M.A. On reducing the problem of transfer in finite layer to that for half-space. Ibid., 1975, **225**(5), 1049–1052 (Russian).
82. Ivanov, V.V. Invariance principles and internal radiation fields in semi-infinite atmospheres. *Astron. zhurn.*, 1975, **52**(2), 217–226 (Russian).
83. Yanovitskij, E.G. Anisotropic light scattering in an inhomogeneous atmosphere. Invariance principles and integrals of the radiative transfer equation. *Astron. zhurn.*, 1981, **58**(1), 119–129 (Russian).
84. Rogovtsov, N.N. Transfer theory and general invariance principle. *Dokl. AN BSSR*, 1981, **25**(5), 420–423 (Russian).
85. Rogovtsov, N.N. Restoring the internal radiation field by its characteristics at scattering medium boundaries. *Fiz. atmosf. i okeana*, 1980, **16**(3), 244–253 (Russian).
86. Ivanov, V.V. Radiative transfer theory. *Tr. Astron. observ. LGU*, 1981, **37**(59), 70–90 (Russian).
87. Laletin, N.I. Method of surface pseudosources for solution of neutron transfer equation. In: Computational methods in transfer theory. *Moscow, Atomisdat*, 1969, 228–245 (Russian).

88. Rogovtsov, N.N. Correlation between the radiation fields inside the scattering object of arbitrary configuration and at the object boundary on the basis of the method of partially invariant determination. *Zh. Prikl. Spektr.*, 1981, **34**(2), 335–342 (Russian).
89. Rogovtsov, N.N. On some applications of general invariance principle to the radiative transfer theory. Ibid., 1981, **35**(6), 1044–1050 (Russian).
90. Pikichian, O.V. General relation of invariance for investigation of the transfer problems in media with geometrical and physical characteristics of arbitrary complexity. *Dokl. AN SSSR*, 1982, **262**(4), 860–863 (Russian).
91. Pikichian, O.V. Determining the radiation field in arbitrarily chosen macroscopic parts of component medium using known solutions to transfer problems. Ibid., 1982, **262**(3), 601–606 (Russian).
92. Pikichian, O.V. Finding the solution of the problem of transfer in isolated part of volume using the known solution for total volume. Ibid., 1983, **273**(4), 861–865 (Russian).
93. Principle of invariance and its application. Proceedings of all-union symposium. *Izd. AN Arm.SSR*, 1989 (Russian).
94. Loskutov, V.M. Light regime in deep layers of turbid medium with very extended phase function. *Vestnik LGU, Ser. mat., mekh., astron.*, 1969, N 13, 143–149 (Russian).
95. Germogenova, T.A. On the discrete spectrum of characteristic equation of transfer theory. *Zhurn. vychisl. mat. i mat. fiz.*, 1974, **14**(6), 1526–1543 (Russian).
96. Germogenova, T.A., Shulaya, D.A. On the characteristic equation of transfer theory. *Dokl. AN SSSR*, 1976, **231**(4), 841–844 (Russian).
97. Konovalov, N.V. Asymptotical characteristics of monochromatic radiation fields in problems with inhomogeneous plain layer of large optical dimensions. II. Calculations of the basic parameters and functions. *Preprint, Institute of applied mathematics, Moscow*, 1974, N 14 (Russian).
98. Yanovitskij, E.G. The spherical albedo of a planetary atmosphere. *Astron. zhurn.*, 1972, **49**(4), 844–849 (Russian).
99. Shimizu, A. Calculation of the penetration of gamma rays through slabs by the method of invariant imbedding. *Nucl. Sci. and Eng.*, 1968, **32**(2), 184–194.
100. Sobolev, V.V. Diffusion of radiation in the medium of large optical thickness for an isotropic scattering. *Dokl. AN SSSR*, 1968 **179**(1), 41–44 (Russian).
101. Hulst, H.C. van de. Radiative transfer in thick atmospheres with an arbitrary scattering function. *Bull. Astron. Inst. Neth.*, 1968, **20**(1), 77–86.
102. Takahashi, H. The thermal spectrum of neutrons in heterogeneous reactor. *Nucl. Sci. and Eng.*, 1959, **5**(3), 338–348.
103. De Rooij, W.A., Domke, H. On the nonuniqueness of solutions for nonlinear integral equations in radiative transfer theory. *J. Quant. Spectrosc. and Radiat. Transfer*, 1984, **31**(4), 285–299.
104. Yanovitskij, E.G. Radiation field in semi-infinite atmosphere with anisotropic scattering. *Astron. zhurn.*, 1976, **53**(5), 1063 1074 (Russian).
105. Rybicki, G.B. Integrals of the transfer equation: I. Quadratic integrals for monochromatic, isotropic scattering. *Astrophys. J*, 1977, **213**(1), 165–176.
106. Ivanov, V.V. Nonlinear relations in linear radiative transfer problems for plane atmospheres. *Astron. zhurn.*, 1978, **55**(5), 1072–1083 (Russian).
107. Ivanov, V.V. Asymptotical properties of radiation fields in semi-infinite atmospheres. *Astrofizika*, 1974, **10**(2), 193–204 (Russian).
108. Kattawar G. W., Plass G. N. Interior radiances in optically deep absorbing media-III. Scattering from haze L. *J. Quant. Spectrosc. and Radiat. Transfer*, 1975, **15**(1), 61–84.

109. Yanovitskij, E.G. Radiation field in optically thick atmosphere with nearly conservative scattering., *Fiz. atmosf. i okeana*, 1980, **16**(8), 807–815 (Russian).
110. Sobolev, V.V. Anisotropic light scattering in an inhomogeneous atmosphere. I. *Astron. zhurn.*, 1968, **45**(2), 254–263 (Russian).
111. Mel'nikova, N.N, Minin, I.N. On the theory of transfer of monochromatic radiation in cloudy layers. *Fiz. atmosf. i okeana*, 1977, **13**(3), 254–263 (Russian).
112. Sobolev, V.V. Albedo of planetary atmosphere. *Dokl. AN SSSR*, 1969, **184**(2), 318–320 (Russian).
113. Hulst, H.C. van de. The spherical albedo of a planet covered with homogeneous cloud layer. *Astron. and Astrophys.*, 1974, **35**(2), 209–214.
114. Yanovitskij, E.G. A new form of transfer equation in anisotropically scattering atmosphere. *Kinemat. i fiz. nebesn. tel*, 1986, **2**(6), 3–13 (Russian).
115. Minin, I.N. Diffusion of radiation in semi-infinite medium for anisotropic scattering. *Vestnik LGU, Ser. mat., meh., astron.*, 1961, N1, 133–143 (Russian).
116. Peebles, G.H., Plesset, M.S. Transmission of gamma-rays through large thickness of heaxy materials. *Phys. Rev.*, 1951, **81**(3), 430–439.
117. Hulst, H.C. van de. A new look at multiple scattering. *NASA Inst. for Space Studies. New York, Rept.*, 1963.
118. Hulst, H.C. van de. Asymptotic fitting, a method for solving anisotropic transfer problems in thick layers. *J. Comput. Phys*, 1968, **3**(2), 291–306.
119. Hansen, J.E. Multiple scattering of polarized light in planetary atmospheres. Part I. The doubling method. *J. Atmos. Sci.*, 1971, **28**(1), 120–125.
120. Hovenier, J.M. Multiple scattering of polarized light in planetary atmospheres. *Astron. and Astrophys.*, 1971, **13**(1), 7–29.
121. Wiscombe, W.J. On initialization, error and flux conservation in the doubling method. *J. Quant. Spectrosc. and Radiat. Transfer*, 1976, **16**(8), 637–658.
122. Wiscombe, W.J. Doubling initialization revisited. Ibid., 1977, **18**(2), 245–248.
123. Viik, T., Ryym, R., Heinlo, A. The method of approximation of resolvent function in radiative transfer theory. *Tallinn, Valgus*, 1985 (Russian).
124. Ivanov, V.V. Radiative transfer in multilayer optically thick atmosphere. Part I. *Uch. zapiski LGU, N 385, Tr. astron. observ. LGU*, 1976, **32**, 3–23 (Russian).
125. Germogenova, T.A. On character of the solution of transfer equation for plane layer. *Zhurn. vychisl.. mat. i mat. fiz.*, 1961 **1**(6), 1001–1019 (Russian).
126. Danielson, R.E., Moore, D.R., Hulst, H.C. van de. The transfer of visible radiation through clouds. *J. Atmos. Sci.*, 1969, **26**(5), 1078–1087.
127. Ivanov, V.V., Gutshabash, S.D. Propagation of brightness wave in an optically thick atmosphere. *Fiz. atmosf. i okeana*, 1974, **10**(8), 851–863 (Russian).
128. Dlugach, J.M., Yanovitskij, E.G. Surface illuminance and albedo of a planetary atmosphere with nearly conservative scattering. Ibid., 1983, **19**(8), 813–823 (Russian).
129. Minin, I.N. Light absorption in the atmosphere and on the planetary surface with multiple scattering and reflection. Ibid., 1974, **10**(3), 228–235 (Russian).
130. Yanovitskij, E.G. Radiation flux in a weakly absorbing atmosphere of arbitrary optical thickness. *Fiz. atmosf. i okeana*, 1991, **27**(11), 1241–1246 (Russian).
131. Nakajiama, T., King, M.D. Asymptotic theory for optically thick layers: application to the discrete ordinates method. *Appl. Opt.*, 1992, **31**(36), 7669–7683.
132. Yanovitskij, E.G., Dlugach, Zh.M., The radiation field in a multilayer plane atmosphere with arbitrary internal sources. *Kinemat. i fiz. nebesn. tel*, 1992, **8**(5), 12–30 (Russian).

133. Ivanov, V.V. Radiation field in an optically thick planetary atmosphere overlaying a reflecting surface. *Astron. zhurn.*, 1976, **53**(3), 589–595 (Russian).
134. Sobolev, V.V. On brightness coefficients of plane layer of turbid medium. *Dokl. AN SSSR*, 1948, **61**(5), 61–64 (Russian).
135. Hulst, H.C. van de. Scattering in a planetary atmosphere. *Astrophys. J.*, 1948, **107**(1), 220–225.
136. Fomin, N.N. Nonstationary field of radiation in an optically thick atmosphere. *Fiz. atmosf. i okeana*, 1985, **21**(1), 42–49 (Russian).
137. Germogenova, T.A., Konovalov, N.V. Asymptotical characteristics of the solution of transfer equation in problem with inhomogeneous layer. *Zhurn. vychisl. mat. i mat. fiz.*, 1974, **14**(4), 928–946 (Russian).
138. Gutshabash, S.D. Light scattering in two-layer atmosphere. *Vestnik LGU*, 1957, N 1, 158–164 (Russian).
139. Mullikin, T.W. Multiple scattering in homogeneous plane-parallel atmospheres. In: Proc. Interdisciplinary Conference on Electromagnetic Scattering. *Univ. of Massachusetts*, 1965, 697–719.
140. Danialian, E.Kh., Mnatsakanian, M.A. On the problem of light diffusion in a layer of finite optical thickness. *Soobshch. Byurokan. observ.*, 1975, **46**, 101–113 (Russian).
141. Yanovitskij, E.G. Radiation field in a plane atmosphere in the case of anisotropic scattering. Separation of the angular variables. *Astrofizika*, 1980, **16**(2), 363–374 (Russian).
142. Kolesov, A.K. Light reflection from semi-infinite atmosphere. *Tr. astron. observ. LGU*, 1976, **32**, 39–51 (Russian).
143. Kolesov, A.K. Brightness coefficients for two-layer atmosphere for anisotropic scattering. I. *Astrofizika*, 1976, **12**(1), 83–94 (Russian).
144. Kolesov, A.K. Brightness coefficients for two-layer atmosphere for anisotropic scattering. II. Ibid., 1976, **12**(3), 485–494 (Russian).
145. Kolesov, A.K. Light scattering in multilayer atmosphere. *Tr. astron. observ. LGU*, 1978, **34**, 29–45 (Russian).
146. Takashima, T. Emergent radiation from an inhomogeneous, plane-parallel planetary atmosphere calculated by using the "adding" method. *J. Quant. Spectrosc. and Radiat. Transfer*, 1973, **13**(12), 1229–1232.
147. Takashima, T. Matrix method evaluating an internal radiation field in a plane-parallel atmosphere. *Astrophys. and Space Sci*, 1973, **23**(2), 201–204.
148. Dlugach, Zh.M., Yanovitskij, E.G. Light scattering in multi-layer atmospheres. I. The diffuse reflection problem. *Astrofizika*, 1985, **23**(2), 337–348 (Russian).
149. Yanovitskij, E.G. Light scattering in multi-layer atmospheres. II. The Milne problem. Ibid., 1985, **24**(3), 535–548 (Russian).
150. Ivanov, V.V. Radiative transfer in multilayer optically thick atmosphere. II. *Tr. astron. observ. LGU*, 1976, **32**, 23–39 (Russian).
151. Lacis, A.A., Hansen, J.E. A parametrization for the absorption of solar radiation in the Earth's atmosphere. *J. Atmos. Sci.*, 1974, **31**(1), 118–133.
152. Viik, T. Radiation field in multilayer isotropically scattering atmosphere. *Astrofizika*, 1981 **17**(4), 735–748.
153. Viik, T. A generalized principle of invariance and radiation field in multilayer atmospheres. *Astrophys. and Space Sci*, 1982, **86**(2), 169–178.
154. Yanovitskij, E.G. Diffuse reflection and transmission of light in an inhomogeneous atmosphere bounded by reflective surface. *Izv. AN SSSR. Ser. geofiz.*, 1963, N 7, 1140–1146 (Russian).
155. Viik, T. Radiation field in multilayer media. *Tallinn, Valgus*, 1985 (Russian).

156. Davison, B. Angular distribution of neutrons at the interface of two adjoining media. *Can. J. Res.*, 1950, **28A**(3), 303–314.
157. Chandrasekhar S. The angular distribution of the radiation at the interface of two adjoining media. *Can. J. Phys.*, 1951, **29**(1), 14–20.
158. Romanov, Yu.A. Exact solutions of one-speed kinetic equation and their use to calculations of diffusion problems (modified diffusion method). In: Study of critical parameters of reactor system. *Moscow, Atomizdat*, 1960, 3–26 (Russian).
159. Mendelson, M.R., Summerfield, G.C. One-speed neutron transport in two adjacent half-space. *J. Math. Phys.*, 1964, **5**(5), 668–674.
160. Devaux, C., Grandjean, P., Ishiguro, Y., Siewert, C.E. On multiregion problems in radiative transfer. *Astrophys. and Space Sci.*, 1979, **62**(1), 225–233.
161. Saad, E.A., Hendi, A., El Wakil, S.A., Atia M. T. Particle transfer in multiregions. Ibid., 1983, **66**(2), 233–238.
162. Dlugach, Zh.M., Yanovitskij, E.G. Interpretation of Venera-10 optical measurements. *Pisma v astron. zhurn.*, 1978, **4**(10), 471–476 (Russian).
163. Dlugach, Zh.M., Yanovitskij, E.G. Radiation field in the Venusian atmosphere for two-layer atmosphere model. In: Physics of planetary atmospheres. *Kiev, Naukova Dumka*, 1981, 3–27 (Russian).
164. Germogenova, T.A., Konovalov, N.V., Lukashevich, N.L., Feigelson, E.M. On the interpretation of Venera-8 optical measurements. *Kosm. issled.*, 1977, **15**(5), 755–766 (Russian).
165. Konovalov, N.V. Asymptotical regularities of light propagation in optically dense atmospheres. In: Numerical solution of the problems of atmospheric optics. *Moscow*, 1984, 152–164 (Russian).
166. Dlugach, Zh.M. Light reflection by multilayer planetary atmospheres. *Astron. vestnik*, 1983, **17**(1), 20–26 (Russian).
167. Stamnes, K., Conklin, P. A new multi-layer discrete ordinate approach to radiative transfer in vertically inhomogeneous atmospheres. *J. Quant. Spectrosc. and Radiat. Transfer*, 1984, **31**(3), 273–282.
168. Cogley, A.C. Adding and invariant imbedding equations in matrix notation for all the scattering functions. *J. Quant. Spectrosc. and Radiat. Transfer*, 1978, **19**(2), 113–126.
169. Sharma, A., Cogley, A.C., Tonon, S.S. Error analysis of the adding/doubling method for radiative scattering in inhomogeneous media. *J. Quant. Spectrosc. and Radiat. Transfer*, 1981, **26**(1), 39–48.
170. Hulst, H.C. van de, Grossman, K. Multiple light scattering in planetary atmospheres. In: The atmospheres of Venus and Mars, J. Brandt and M. McElroy eds. *New York, Gordon and Breach Sci. Publ.*, 1963, 35–55.
171. Chamberlain, J. The atmosphere of Venus near her cloud tops. *Astrophys. J.*, 1965, **141**(3), 1184–1205.
172. Sobolev, V.V. Theoretical astrophysics. *Moscow, Nauka*, 1985 (Russian).
173. Handbook of mathematical functions. Abramovits, M. and Stegun I., eds., *Natl. Bureau of Stands.*, 1964.
174. Garsia, R.D.M., Siewert, C.E. Radiative transfer in finite inhomogeneous plane-parallel atmospheres. *J. Quant. Spectrosc. and Radiat. Transfer*, 1982, **27**(2), 141–148.
175. King, M.D. Determination of the saled optical thickness of clouds from reflected solar radiation measurements. *J. Atmos. Sci.*, 1987, **44**(13), 1734–1751.
176. Yi, H.C., McCormic, N.J., Sanchez, R. Cloud optical thickness estimation from irradiance measurements. Ibid., 1990, **47**(21), 2567–2579.
177. Gavrilova, L.A. The Ambartsumian functions for cloudless atmosphere with isotropic scattering. *Fiz. atmosf. i okeana*, 1979, **15**(1), 110–114 (Russian).

178. Gavrilova, L.A. On the calculation of the shortwave radiation field at the boundaries of a cloudless atmosphere. Ibid., 1981, **17**(4), 373–383 (Russian).
179. Venus. Hunten, D.M. et al. eds. *Tucson, Arizona, Univ. Arizona Press*, 1983.
180. Ksanfomaliti, L.V. Planet Venus. *Moscow, Nauka*, 1985 (Russian).
181. Irvine, W.M. Monochromatic phase curves and albedos for Venus. *J. Atmos. Sci.*, 1968, **25**(4), 610–616.
182. Morozhenko, A.V., Yanovitskij, E.G. The optical properties of Venus and the Jovian planets. I. The atmosphere of Jupiter according to polarimetric observations. *Icarus*, 1973, **18**(4), 582–592.
183. Hansen, J.E., Hovenier, J.W. Interpretation of the polarization of Venus. *J. Atmos. Sci.*, 1974, **31**(4), 1137–1160.
184. Johnson, H.L. The absolute calibration of the Arizona photometry. *Comm. Lunar and Planet. Lab.*, 1965, N 53, 73–77.
185. Travis, L.D. On the origin of ultraviolet contrasts on Venus. *J. Atmos. Sci.*, 1975, **32**(6), 1190–1200.
186. Tomasko, M.G., Doose, L.R., Smith, P.H., Odell, A.P. Measurements of the flux of sunlight in atmosphere of Venus. *J. Geophys. Res.*, 1980, **85**(A13), 8167–8186.
187. Kuz'min, A.D., Marov, M.Ya. Physics of Venus. *Moscow, Nauka*, 1974 (Russian).
188. Minin, I.N., Tarabukhina, I.M. On the problem of studying optical properties of the Venusian atmosphere. *Fiz. atmosf. i okeana*, 1990, **26**(8), 837–840 (Russian).
189. Moroz, V.I., Golovin, Yu.M., Moshkin, B.E., Ekonomov, A.P. Spectrophotometrical experiment on the probes Venera-11 and Venera-12. *Kosmich. issled.*, 1981, **19**(4), 599–612 (Russian).
190. Avduevskii, V.S., Golovin, Yu.M., Zavelevich, F.S. et al. Preliminary results of direct Venera-9 and Venera-10 measurements of solar radiant fluxes in the atmosphere and on the surface of Venus. *Dokl. AN SSSR*, 1976, **229**(3), 579–532 (Russian).
191. Avduevskii, V.S., Golovin, Yu.M., Zavelevich, F.S. et al. Preliminary results of the study of light regime in the atmosphere and on the surface of Venus. *Kosmich. issled.*, 1976, **14**(5), 735–742 (Russian).
192. Golovin, Yu.M., Optical properties of the Venusian surface: The dependence of albedo on wavelength. Ibid., 1979, **17**(3), 473–476 (Russian).
193. Golovin, Yu.M., Moroz, V.I., Moshkin, V.E. et al. Venusian atmosphere according to Venera-11 and Venera-12 optical measurements. Ibid., 1982, **20**(4), 572–594 (Russian).
194. Marov, M.Ya., Byvshev, B.V., Baranov, B.P. et al. Study of the structure of Venusian clouds using nephelometers mounted on Venera-13 and Venera-14 probes. Ibid., 1983, **21**(2), 269–278 (Russian).
195. Knollenberg, R.J., Hunten, D.M. The microphysics of Venus: results of the Pioneer Venus particle size spectrometer experiment. *J. Geophys. Res.*, 1980, **85**(A13), 8039–8081.
196. Dlugach, Zh.M. The structure of the Venus atmosphere in accordance with Venera-11, -13, -14 optical measurements. *Astron. vestnik*, 1988, **22**(3), 262–271 (Russian).
197. Moroz, V.I., Parfentiev, N.A., San'ko, N.F. Spectrophotometrical experiment on Venera-11 and -12 probes. 2. Analysis of the Venera-11 spectral data by the method of adding layers. *Kosmich. issled.*, 1979, **17**(5), 727–742 (Russian).
198. Ekonomov, A.P., Moshkin, B.E., Golovin, Yu.M. et al. Spectrophotometrical experiment on Venera-11 and -12 probes. 1. Methods, results and preliminary analysis of measurements. Ibid., 1979, **17**(5), 714–726 (Russian).

199. Moshkin, B.E., Ekonomov, A.P., Moroz, V.I. et al. Spectrophotometrical experiment on Venera-13 and -14 probes. 1. Methods, results and preliminary analysis of measurements. Ibid., 1983, **21**(2), 236–245 (Russian).
200. Selivanov, A.S., Gektin, Yu.M., Gerasimov, M.A. et al. Continuation of probe television investigation of Venus surface. Ibid., 1983, **21**(2), 176–182 (Russian).
201. Selivanov, A.S., Gektin, Yu.M., Naraeva, M.K. et al. On dynamical phenomena displayed on panoramas of Venus surface according to data transmitted from Venera-13 and -14 probes. Ibid., 1983, **21**(2), 200–204 (Russian).
202. Dlugach, Zh.M. Estimates of variations of the optical thickness of Venus atmosphere from Venera-13 and Venera-14 data. *Kinemat. i fiz. nebesn. tel*, 1989, **5**(1), 38–44 (Russian).
203. Whitenhill, L.P., Hansen, J.E. On the interpretation of the inverse phase effect for CO_2 equivalent widths on Venus. *Icarus*, 1973, **20**(2), 146–152.
204. Anikonov, A.S. The interpretation of weak absorption lines in the spectrum of Venus. *Astron. zhurn.*, 1977, **54**(1), 169–177 (Russian).
205. Regas, J.L., Boese, R.W., Giver, L.P., Miller, J.H. Does spectroscopic evidence require two scattering layers in the Venus atmosphere? *J. Quant. Spectrosc. and Radiat. Transfer*, 1973, **13**(5), 461–463.
206. Young, L.D.G. High resolution spectra of Venus – a review. *Icarus*, 1972, **17**(3), 632–658.
207. Yanovitskij, E.G., Dlugach, Zh.M. On the effect of width and orientation of spectrograph slit on observed spectrum of Venus. *Kinemat. i fiz. nebesn. tel*, 1995, **11**(3), 20-35 (Russian).
208. Goody, R.M., Young, Y.L. Atmospheric Radiation. Theoretical Basis. Second Edition. *New York, Oxford, Oxford Univ. Press*, 1989.
209. Chamberlain, J.W. Theory of planetary atmospheres. *New York. Academic press*, 1978.
210. Fomin, N.N. Factors of light reflection and transmission for planetary atmosphere with nearly conservative scattering. *Preprint, Kiev, Inst. for theoretical physics, ITF-89-76R*, 1989 (Russian).
211. Michalas, D. Stellar atmospheres. *San Francisco, Freeman and Company, Second edition*, 1978.
212. Kirshner, R.P., Oke, J.B., Penston, M.V., Searle, L. The spectra of supernovae. *Astrophys. J.*, 1973, **185**(1), 303–322.
213. Sobolev, V.V. Multiple light scattering by free electrons in a homogeneous sphere. *Astrofizika*, 1974, **10**(2), 185–191 (Russian).
214. Shakura, N.I. The influence of the electron scattering on the spectrum of emission from semi-opaque plasma. *Astron. zhurn.*, 1972, **49**(3), 652–654 (Russian).
215. Felten, J.E., Rees, M.J. Continuum radiative transfer in a hot plasma, with application to Scorpius X-1. *Astron. and Astrophys*, 1972, **17**(2), 226–242.
216. Slyusarev, S.G. On light scattering by free electrons in stellar atmospheres. *Dokl. AN SSSR*, 1954, **95**(4), 741–744 (Russian).
217. Bochkarev, N., Karitskaya, E.A., Sakhibullin, N.A. Distribution of polarization and intensity of radiation across the stellar disc and numerical values of atmospheric characteristics governing this distribution. *Astrophys. and Space Sci.*, 1985, **108**(1), 15–29.
218. Kurucz, R.L., Peytermann, E., Avrett, E.H. Blanketed model atmospheres for early-type stars. *Washington. Smitsonian Inst. press*, 1974.
219. Kurucz, R.L. Model atmospheres for G, F, A, B and O stars. *Astrophys. J. Suppl. Ser.*, 1979, **40**(1), 1–340.

220. Zeldovich, Ya.B., Shakura, N.I. X-rays emission, accompanying the gas accretion on a neutron star. *Astron. zhurn.*, 1969, **46**(2), 225–236 (Russian).
221. Kondratiev, K.Ya., Bondarenko, V.G., Khvorostianov, V.I. 3-D mesoscopic model of transfer of cloud aerosol with allowance for interaction of radiation, microphysical processes and orography. I. Formulation of the problem. *Opt. atmosf.*, 1989, **2**(2), 115–121 (Russian).
222. Kondratiev, K.Ya., Bondarenko, V.G., Khvorostianov, V.I. 3-D mesoscopic model of transfer of cloud aerosol with allowance for interaction of radiation, microphysical processes and orography. II. Results of calculations. Ibid., 1989, **2**(2), 122–132 (Russian).
223. Stamnes, K. The theory of multiple scattering of radiation in plane parallel atmospheres. *Revs. Geophys.*, 1986, **24**(2), 299–310.
224. Erlov, N.G. Marine optics. *Amsterdam, Elsevier*, 1976.
225. Ivanov, A.P. Physical principles of hydrooptics. *Minsk, Nauka i Tekhnika*, 1975 (Russian).
226. Shifrin, K.S. Introduction in optics of ocean. *Leningrad, Gidrometeoizdat*, 1983 (Russian).
227. Brekhovskih, L.M. Underwater sound waves in the ocean generated by surface waves. *Fiz. atmosf. i okeana*, 1966, **2**(9), 970–980 (Russian).
228. Klyachin, B.I. On the effect of scattering on anisotropy of ocean noise. *Akustich. zhurn.*, 1981, **27**(4), 526–532 (Russian).
229. Klyachin, B.I., Kurianov, B.F. Effect of sound scattering on dependence of low-frequency ocean sounds on depth. *Dokl. AN SSSR*, 1981 **260**(4), 1009–1012 (Russian).
230. Tonoyan, I.P. The effect of sound scattering by volume inhomogeneities of water medium on the angular structure of an ocean noise field. *Fiz. atmosf. i okeana*, 1988, **24**(10), 1117–1119 (Russian).
231. Domke, H., Yanovitskij, E.G. Principles of invariance applied to the computation of internal polarized radiation in multilayered atmospheres *J. Quant. Spectrosc. and Radiat. Transfer*, 1986, **36**(3), 175–186.
232. Loskutov, V.M., Sobolev, V.V. Polarization of radiation scattered by an inhomogeneous atmosphere. *Astrofizika*, 1981, **17**(1), 97–108 (Russian).
233. Sobolev, V.V. Radiative transfer in inhomogeneous medium. *Dokl. AN SSSR*, 1956, **111**(5), 1000–1003 (Russian).
234. Bellman, R., Kalaba, R. On the principle of invariant imbedding and propagation through inhomogeneous media. *Proc. Nat. Acad. Sci. USA*, 1956, **42**, 629–632.
235. Preisendorfer, R.W. Functional relations for the R- and T-operators on plane-parallel media. Ibid., 1958, **44**, 323–327.
236. Ueno, S. The probabilistic method for problems of radiative transfer. I. Diffuse reflection and transmission in a finite inhomogeneous atmosphere. *Astrophys. J.*, 1960, **132**(3), 729–745.
237. Sobolev, V.V. Light scattering in inhomogeneous atmosphere. *Astron. zhurn.*, 1974, **51**(1), 50–55 (Russian).
238. Yanovitskij, E.G. Diffuse reflection and transmission of light by a plane layer of an inhomogeneous medium with anisotropic scattering. Ibid., 1961, **38**(5), 912–919 (Russian).
239. Busbridge, I.W. On inhomogeneous stellar atmospheres. *Astrophys. J.*, 1961, **133**(1), 28–44.
240. Ueno, S. Stochastic equations in radiative transfer by invariant imbedding method. *J. Math. Anal. and Appl.*, 1961, **2**(2), 217–222.
241. Ueno, S. On the S- and T-functions of S. Chandrasekhar in arbitrary stratification. Ibid., 1962, **4**(1), 9–20.

242. Ueno S. The probabilistic method for problems of radiative transfer. XIII. Diffusion matrix. Ibid., 1962, **4**(1), 21–37.
243. Ueno, S., Kalaba, R.E., Kagivada, H.H., Bellman, R.E. Some mathematical aspects multiple scattering in finite inhomogeneous slab with anisotropic scattering. *Publ. Astron. Soc. Jap.*, 1970, **22**, 75–83.
244. Ivanov, V.V. On the standard problem of the theory of radiative transfer. Ibid., 1977, **13**(3), 505–516 (Russian).
245. Viik, T. Source function for inhomogeneous atmosphere. *Publ. Tartu. astrofiz. observ. im V. Struve*, 1978, **46**, 43–54 (Russian).
246. Kawabata, K., Hirata, R. Intensity of radiation reflected by a vertically inhomogeneous medium through multiple scattering. *Astrophys. and Space Sci.*, 1985, **109**(2), 345–356.
247. Wilson, S.J., Wan, F.S. A moment method for radiative transfer in an anisotropically-scattering slab medium with space-dependent albedo $\omega(x)$. *Astrophys. and Space Sci.*, 1987, **137**(1), 107–113.
248. Hunt, G.E., Grant, I.P. Discrete space theory of radiative transfer and its application to problems in planetary atmospheres. *J. Atmos. Sci.*, 1969, **26**(5), 963–972.
249. Hunt, G.E. A review of computational techniques for analyzing the transfer of radiation through a model cloudy atmosphere. *J. Quant. Spectrosc. and Radiat. Transfer*, 1971, **11**(6), 655–690.
250. Dave, J.V. Intensity and polarization of the radiation emerging from a plane parallel atmosphere containing monodispersed aerosols. *Appl. Opt.*, 1970, **9**(12), 2673–2684.
251. Dave, J.V., Canos, J. A direct solution of the radiative transfer equation: application to atmospheric models with arbitrary vertical nonhomogeneities *J. Atmos. Sci.*, 1974, **31**(4), 1089–1101.
252. Stamnes K., Tsay S-Ch., Wiscombe W., Jayaweera K. Numerically stable algorithm for discrete-ordinate-method radiative transfer in multiple scattering and emitting layered media. *Appl. Opt.*, 1988, **27**(12), 2502–2509.
253. Tanaka, M. Radiative transfer in turbid atmospheres. I. Matrix analysis for the problem of diffuse reflection and transmission. *J. Meteorol. Soc. Jap.*, 1971, **49**(4), 296–312.
254. Tanaka, M. Radiative transfer in turbid atmospheres. II. Angular distribution of intensity of the solar radiation diffusely reflected and transmitted by turbid atmospheres. Ibid., 1971, **49**(5), 321–332.
255. Tanaka, M. Radiative transfer in turbid atmospheres. III. Degree of polarization of the solar radiation reflected and transmitted by turbid atmospheres. Ibid., 1971, **49**(5), 333–342.
256. Numerical methods in radiative transfer. Kalkofen, W., ed. *Cambridge Univ. Press*, 1987.
257. Beyond classical models. Clivellari, L., Hubeny, I. and Hummer, D.G., eds. *Dordrecht, Kluwer*, 1991.
258. Ueno S. Diffuse reflection of radiation by a planetary atmosphere. In: Proc. 4-th Internat. Symposium on Space Technology and Science, *Sympos. Committee Office, Japan Rocket Soc., Tokio*, 1963, 451–458.
259. Yanovitskij, E.G. Anisotropic light scattering in an inhomogeneous atmosphere. Asymptotic separation of angular variables in an optically thick layer. *Astron. zhurn.*, 1980, **57**(6), 1277–1286 (Russian).
260. Sobolev, V.V. The Milne problem for inhomogeneous atmosphere. *Dokl. AN SSSR*, 1978 **23**(3), 558–561 (Russian).
261. Stokes, R.A., DeMarcus, W.C. A simple techniques for calculating line profiles of inhomogeneous planetary atmospheres. *Icarus*, 1971, **14**(3), 307–311.

262. Chow, K.S.K., Friedson, A.J., Yung, Y.L. A note on the variational method of Stokes and DeMarcus for radiative transfer in planetary atmospheres. Ibid., 1984, **59**(2), 177–187.
263. Fomin, N.N., Yanovitskij, E.G. Anisotropic light scattering in an inhomogeneous atmosphere. The radiation field for nearly conservative scattering. Ibid., 1987, **64**(5), 992–1003 (Russian).
264. Yanovitskij, E.G. Anisotropic light scattering in an inhomogeneous atmosphere. The case of nearly conservative scattering. Ibid., 1978, **55**(5), 1084–1092 (Russian).
265. Yanovitskij, E.G. Anisotropic light scattering in an inhomogeneous atmosphere. I. Case of pure scattering. *Astron. zhurn.*, 1971, **48**(2), 323–332 (Russian).
266. Yanovitskij, E.G. The albedo and illuminance of the surface of the planet surrounded by purely scattering inhomogeneous atmosphere. *Fiz. atmosf. i okeana*, 1972, **8**(5), 518–525 (Russian).
267. Sobolev, V.V., Loskutov, V.M. Diffuse light reflection from inhomogeneous atmosphere. *Dokl. AN SSSR*, 1975, **224**(6), 47–53 (Russian).
268. Loskutov, V.M., Sobolev, V.V. On the problem of diffuse light reflection by inhomogeneous atmosphere. *Tr. astron. observ. LGU*, 1978, **34**, 3–29 (Russian).
269. Haggag, M.H., Machali, H.M. The Chandrasekhar H-function for exponentially varying atmosphere. *Astron. and Space Sci.*, 1985, **111**(1), 189–195.
270. Sobolev, V.V. The effect of electron scattering on continuous spectrum of a star. *Astrofizika*, 1978, **14**(3), 383–392 (Russian).
271. Sobolev, V.V. On the theory of line formation in stellar spectra. *Astron. zhurn.*, 1976, **53**(4), 681–685 (Russian).
272. Martin, B.J. On a new integral equation arising in the theory of radiative transfer. *SIAM J. Appl. Math.*, 1971, **20**(4), 703–713.
273. Mullikin, T.W., Siewert, C.E. Radiative transfer in inhomogeneous atmospheres. *Ann. Nucl. Energy*, 1980, **7**(2), 205–211.
274. Daniel, R., Badham, V.C., Pomraning, G.C. A review of transport theory methods in nonhomogeneous media. Transp. Theory and Statist. Phys., 1985, **14**(2), 125–166.
275. Ustinov, E.A. Inverse problem of multiple scattering theory and interpretation of measurements of scattered radiation in the Venus cloud layer. *Kosm. issled.*, 1977, **15**(5), 768–775 (Russian).
276. Ustinov, E.A. Inverse problem of thermal probing. 1. Reconstruction of the profile of absorption coefficients of optically active component according to the observations of leaving thermal radiation. Ibid., 1990, **28**(3), 402–412 (Russian).
277. Ustinov, E.A. Inverse problem of thermal probing. 2. Reconstruction of the vertical profile of aerosol extinction coefficient according to the observations of leaving radiation. Ibid., 1990, **28**(5), 725–735 (Russian).
278. Ustinov, E.A. Inverse problem of thermal probing. 3. Reconstruction of the vertical profile of ratio of mixture of small gas component. Ibid., 1991, **29**(2), 289–297 (Russian).
279. Ustinov, E.A. Inverse problem of in situ photometry of scattered solar radiation in planetary atmosphere: reconstruction of characteristics of scattering in atmosphere. Mathematical tools and numerical experiments. Ibid., 1991, **29**(3), 443–457 (Russian).
280. Ustinov, E.A. Inverse problem of the photometry of solar radiation reflected by thick planetary atmosphere. 1. Mathematical tools and weight functions of linearized inverse problem. Ibid., 1991, **29**(4), 604–620 (Russian).

281. Ustinov, E.A. Inverse problem of the photometry of solar radiation reflected by thick planetary atmosphere. 2. Numerical aspects and requirements for geometry of observations. Ibid., 1991, **29**(6), 917–932 (Russian).
282. Ustinov, E.A. Inverse problem of the photometry of solar radiation reflected by thick planetary atmosphere. 3. Distant probing of small gas components and atmospheric aerosol. Ibid., 1992, **30**(2), 212–225 (Russian).
283. Fomin, N.N., Yanovitskij, E.G. Absorption line formation in an optically thick, weakly absorbing planetary atmosphere with nearly conservative scattering. I. Homogeneous slab. *Kinemat. i fiz. nebesn. tel*, 1991, **7**(5), 29–38 (Russian).
284. Fomin, N.N., Yanovitskij, E.G. Absorption line formation in an optically thick, weakly absorbing planetary atmosphere with nearly conservative scattering. II. Vertically inhomogeneous medium. Ibid., 1991, **7**(5), 39–47 (Russian).
285. Fomin, N.N., Yanovitskij, E.G. Absorption line formation in an inhomogeneous atmosphere with nearly conservative scattering. In: Numerical methods for solution of radiative transfer equation. *Tartu, Izd. AN ESSR*, 1988, 201–204 (Russian).
286. Fomin, N.N., Yanovitskij, E.G. Global variations of the optical thickness of the Venus atmosphere. II. Analysis of equivalent widths of CO_2 lines. *Kinemat. i fiz. nebesn. tel*, 1990, **6**(1), 10–18 (Russian).
287. Sobolev, V.V. On the theory of scattering photospheres. *Astrofizika*, 1975, **11**(3), 499–510 (Russian).

Index